FOREWORD

The Urban Water Resources Research Council (UWRRC) of the American Society of Civil Engineers has for 35 years been a leader in the transfer of urban stormwater technology among researchers, practitioners and administrators. One of the principal means of accomplishing this transfer has been through a series of Engineering Foundation Conferences, held in the United States and abroad, as well as International Symposia and technical sessions at professional conferences.

The Council has recognized for some time that there are numerous concerns related to the preservation and enhancement of aquatic ecologic systems in urban areas. This conference builds on an earlier Engineering Foundation conference "Stormwater NPDES Related Monitoring Needs" held in August, 1994 at Crested Butte, Colorado. That conference focused on concerns of the U.S. Environmental Protection Agency's (EPA's) stormwater monitoring requirements contained in its NPDES regulations. An outcome of that conference was that EPA would consider allowing municipalities to substitute biological monitoring for certain of the chemical monitoring requirements of the permit. That raised another question: What ecologic measurements should be taken to assess the impacts of urban runoff on aquatic ecosystems, what are the design parameters to re-establish or improve a degraded urban aquatic ecosystem? The Council organized this conference on watershed development effects on aquatic ecosystems to bring together the experts from the scientific community, plus engineers, municipal officials and state and federal regulators responsible for urban water resource management. What resulted was a mix of perspectives and expertise representing academia, industry, municipalities, and state and federal regulatory agencies.

All of the papers presented in the regular sessions were by invitation, and were reviewed and accepted for publication by both the Conference Chairman and by the Proceedings Editor. All papers are eligible for discussion in the Journal of the Water Resources Planning and Management Division of the American Society of Civil Engineers (ASCE). All papers are eligible for ASCE awards.

The Proceedings are organized by "session", corresponding to the actual conference sessions. Formal papers are presented first, followed by a session discussion which may include both material presented by an author which is not necessarily in his paper, as well as comments/questions from participants, and the answers to those questions furnished by the author (or, in some cases, by other conference participants).

ACKNOWLEDGMENTS

The conference organizers gratefully acknowledge the financial support of the Engineering Foundation and the U.S. Environmental Protection, without which the Conference would not have been possible.

Particular thanks go to the Conference organizing committee:

Roger Bannerman, Wisconsin Department of Natural Resources
Edwin Herricks, University of Illinois
Jonathan Jones, Wright Water Engineers
Eric Livingston, Florida Department of Environmental Resources
John Maxted, Delaware Department of Natural Resources and
    Environmental Control
Robert Pitt, University of Alabama at Birmingham
Tom Schueler, Center for Watershed Protection
Ben Urbonas, Urban Drainage and Flood Control District, Denver
Eric Strecker, Woodward-Clyde Consultants

These individuals gave generously of their time and energy to insure that the Conference would be a success.

Further thanks are due to the Conference co-sponsors, Urban Water Resources Research Council of the American Society of Civil Engineers, American Public Works Association, and the U.S. Environmental Protection Agency.

Thanks also go to Karen R. Jones of Camp Dresser & McKee Inc. for her excellent and timely work in assembling the conference papers and preparing this proceedings.

TABLE OF CONTENTS

FOCUS 1: MONITORING ISSUES

SESSION 1A: APPROACHES TO EXPANDING MONITORING BEYOND WATER
QUALITY
Chair: Eric H. Livingston

SESSION 1B: ADVANCES IN USING TOXICITY BIOINDICATORS IN URBAN
AQUATIC ECOSYSTEM ASSESSMENT
Chair: E.E. Herricks

Assessment of the Response of Aquatic Organisms to Long-term *Insitu* Exposures of

FOCUS 2: WATERSHED DEVELOPMENT EFFECTS IN WATER SURPLUS REGIONS

SESSION 2A: EFFECTS OF WATERSHED DEVELOPMENT ON HYDROLOGY AND AQUATIC HABITAT STRUCTURE
Chair: Eric Strecker

SESSION 2B: IMPACTS OF WATERSHED DEVELOPMENT ON AQUATIC BIOTA, ROUND 1
Chair: Roger Bannerman

FOCUS 4: MANAGEMENT AND INSTITUTIONAL ISSUES

SESSION 4A: WATERSHED MANAGEMENT STRATEGIES TO MITIGATE DEVELOPMENT EFFECTS
Chair: Tom Schueler

SESSION 4B: INSTITUTIONAL ARRANGEMENTS FOR WATERSHED MANAGEMENT
Chair: Jonathan E. Jones

CONFERENCE OVERVIEW AND SUMMARY

Larry A. Roesner, M. ASCE[1]

Introduction

The Urban Water Resources Research Council of the American Society of Civil Engineers sponsored an Engineering Foundation Conference on "Effects of Watershed Development and Management on Aquatic Ecosystems" in Snowbird, Utah, August 4-9, 1996. This conference was the sequel to a previous Engineering Foundation Conference on "Stormwater NPDES Related Monitoring Needs" held in 1994 (Torno, 1995). The 1994 conference was convened "to explore current needs and available technology associated with stormwater monitoring, and to bring together regulators, the regulated community, and their consultants, to determine what we, as professionals and as a nation, need to do to achieve the various goals set forth for stormwater in the Clean Water Act, and the municipal and industrial NPDES regulation that implement that act."

One of the key findings of the 1994 conference was that very little NPDES monitoring of urban runoff was being directed toward measuring the actual effects of stormwater discharges on the short- or long-term health of the environment. These studies are critical if the urban stormwater NPDES program is to be successful. It was decided that a conference dealing with the impacts of urban runoff on aquatic ecosystems would be appropriate. This conference would focus more on the sciences side of ecology and geomorphology, to acquaint the engineering community with studies and practices that the scientific community is undertaking and proposing to integrate the ecology into urban runoff management. These proceedings are the product of that conference.

Thoughts and Observations

Just a glance at the table of contents for this conference reveals how broad the topic of aquatic ecosystems is. First, we see that many different parts of society are involved. There are the *municipal officials* concerned with protecting health and safety of the populace they serve, and providing a healthy urban ecosystem; *regulators* whose charge is to protect and restore urban water environments; *consultants* helping regulators and municipal officials achieve their stormwater management goals; *scientists* studying ways to reestablish aquatic habit in urbanized receiving waters; and lest we forget, there is the *general public* that each of the parties identified above believes they are serving.

---

[1]Chief Technical Officer, Camp Dresser & McKee Inc., 1950 Summit Park Drive, Suite 300, Orlando, Florida 32810

The second thing we see is the diverse subject areas covered. Biology, with *bioassessment* techniques is a major focus area; *toxicity measurement* technologies that focus on finding a protocol for measuring the effects on aquatic organisms of pulsed, random dose exposure to toxic chemicals found in urban runoff. *Morphologic analyses* that apply *engineering methods* to stabilize the receiving water system and reestablish the ecosystem, and other morphologic studies that use a *trial-and-error approach* to receiving water system stabilization using natural and/or man-made materials. After reading the papers, and listening to the speakers and the discussion amongst the participants, one observes that the investigators, while recognizing the diverse spread of subject areas that influence urban water environments, still restrict themselves to studying the issue within their own discipline, without adequately taking into account the effects on their findings of other factors. Discussions between attendees revealed that scientists working with biological indices were hard pressed to explain how they related to toxicity experiments, geomorphologic analyses, and hydrologic studies presented by other presenters. Nor were other disciplines able to quantitatively relate findings in their discipline area to findings in other discipline areas. One of the problems is that the subject area is too broad for one person to understand completely, and there is no continuing forum for persons from these diverse discipline areas to work interactively in their investigations.

What became clear to this engineer during the course of the conference is that engineers view habitat degradation and restoration much differently than does the scientific community. Engineers, being concerned and trained to protect the public health and safety of the community from flood damages, tend to design systems with a high degree of integrity that will withstand the 100-year flood. This has resulted in channel straightening, lined channels, removal of vegetation, and debris removal that are detrimental to the resident aquatic ecology. Scientists on the other hand advocate the use of many techniques for aquatic habitat restoration that cannot withstand the forces of flood flows and thus pose a distinct threat to public health and safety. What is required is the development of *soft engineering* that simultaneously provides for drainage and flood protection, achieves the scientists' criteria for ecosystem protection or restoration, **and** looks and acts like the natural environment. The conference shows that there are a number of issues that need to be addressed to make this happen.

Conclusion

There must be a forum for integrating the diverse disciplines that are involved in developing and managing our urban water resources. The critical needs are to draw together natural sciences of biological assessment (I included toxicology here), geomorphology, and urban hydrology to determine the impacts that each has on the other, then combine this knowledge with institutional and societal considerations to define the design parameters for a sustainable urban water resource.

The first questions to answer are "what constitutes an urban aquatic ecosystem?" and what uses must that ecosystem accommodate? Currently, the development of biological indices is focused on natural waters, but we must take into account the changes in hydrology that urban development causes, and the limits to which we can control or mitigate those changes with BMPs and/or in-stream modifications. We must realize that the waterway must also be used in connection with flood control programs for the urban area; but neither use has exclusive right to the waterway. An *urban aquatic ecosystem* then, might be defined as an ecosystem that is sustainable under a hydrologically modified, but stabilized stream (or other receiving water) system. Perhaps we can study some of the existing urban ecosystems that we judge to be desirable and self-sustaining to determine what defines such a system. Different types of systems may be defined depending upon the desired use of the waterway by the community (e.g., aesthetic enjoyment and/or body contact recreation and/or fishing) and climatology and topography of the area.

The final resolution of these issues will necessitate making difficult decisions relating to:

- our utilization of lands within watersheds,
- the extent of our stormwater management efforts,
- the possible cost of those efforts, and
- what reasonable expectations for resource protection exist or are desired.

These types of decisions can be made only if all potential impacts are considered; and the true potential impacts can be determined only through better integration of diverse disciplines. Forums such as this conference must be continued and expanded if answers to complex questions are to be found.

References

Torno, Harry C. 1995. Stormwater NPDES Related Monitoring Needs, Procceding of an Engineering Foundation Conference, August 7-12, 1994, Crested Butte, Colorado. ASCE, New York, NY.

EFFECTS OF WATERSHED DEVELOPMENT & MANAGEMENT ON AQUATIC ECOSYSTEMS - EPA'S PERSPECTIVE

Alfred Lindsey[1], William F. Swietlik[2] and William E. Hall[3]

ABSTRACT

The United States has made significant progress in improving the quality of our nation's waters during the past twenty years. Most traditional point sources of pollution are in compliance with the Clean Water Act. As we move to address the remaining sources of receiving water use impairment, wastewater management programs face new challenges. One important challenge is to integrate National Pollutant Discharge Elimination System (NPDES) program functions into a broader watershed management approach to successfully manage the program within the context of limited resources and spatially varying environmental impacts and priorities. This is readily apparent in efforts to address one of the primary remaining sources of water quality impairment: urban wet weather discharges, including storm water, combined sewer overflows, and sanitary sewer overflows. To more effectively address these urban wet weather sources of water quality impairment, a watershed management approach is recommended. A number of technical and policy-related issues need to be resolved to ensure successful implementation of a watershed-based approach to managing urban wet weather flows. EPA is supporting a number of initiatives to address these issues.

[1] Deputy Director, Office of Wastewater Management (4201), United States Environmental Protection Agency, 401 M Street, S.W., Washington, DC, 20460.
[2] Phase I Storm Water Program Manager, Office of Wastewater Management (4203), United States Environmental Protection Agency, 401 M Street, S.W., Washington, DC, 20460.
[3] Urban Wet Weather Matrix Manager, Office of Wastewater Management (4203), United States Environmental Protection Agency, 401 M Street, S.W., Washington, DC, 20460.

INTRODUCTION

Since the Federal Water Pollution Control Act (also known as the Clean Water Act) was enacted in 1972, State and EPA efforts in administering the NPDES permit, pretreatment, and sludge programs have resulted in significant water quality improvements. The well-publicized wastewater disasters of the 1960s, for the most part, no longer occur. These accomplishments can be attributed in part to the issuance of approximately 65,000 permits under the NPDES permitting program to control point source discharges of wastewater from industrial factories and sewage treatment plants.

Not all water quality problems, however, have been corrected; nor are all sources adequately controlled. A significant remaining source of impairment is urban wet weather discharges. These discharges result from precipitation events, such as rainfall or snowmelt, and include municipal and industrial storm water runoff, combined sewer overflows (CSOs), and sanitary sewer overflows (SSOs).

Municipal storm water runoff, which is often collected and transported by storm drains to receiving waters, contains pollutants that are accumulated as rainwater or snowmelt flow across the surface of the earth. Discharges of municipal storm water runoff can contain high levels of contaminants, such as sediment, suspended solids, nutrients, heavy metals, bacteria, viruses, oil and grease, toxics, oxygen-demanding substances, and floatables [1].

CSOs occur during wet weather events in approximately 1100 cities that have combined sanitary and storm sewers (these are known as combined sewer systems or CSSs), and contain a mixture of sewage, industrial wastewater, and storm water. CSOs have caused beach closings, shellfish bed closures, and other public health problems.

SSOs are raw sewage overflows from separate sanitary sewer systems that occur when the volume of flows in a sewer system exceeds its capacity due to, among other things, unintentional inflow and infiltration of storm water. Such inflow and infiltration can occur because of inadequate preventative maintenance programs and insufficient sewer rehabilitation. SSOs can also occur during periods of dry weather.

EPA's National Water Quality Inventory: 1994 Report to Congress [2] shows that, for the waters monitored, over a third of the nation's waters are not meeting water quality standards. Thirty-six percent of assessed river miles are impaired, and States attribute 12 percent of this impairment to urban runoff. Similarly, 37 percent of assessed lake acres are impaired with urban runoff causing 18 percent of this impairment. Finally, 37 percent of assessed estuary miles are impaired with 46 percent of this impairment attributable to urban runoff.

Urban runoff is a major pollutant source also adversely affecting shellfish growing waters nationwide [3]. From 1985 to 1990, urban runoff represented the largest increase (from 23 to 38 percent) of adverse effects to harvest-limited shellfish growing waters. Nationwide, urban runoff contributes to the impairment of 38 percent (approximately 3.3 million acres) of shellfish harvest-limited acreage. Urban runoff has the greatest impact on the middle Atlantic area, where harvesting is limited in 58 percent of shellfish acreage because of urban discharges.

Recent NPDES permitting program efforts have focused on characterizing and controlling the pollutant component of urban wet weather discharges. Under the first phase of the NPDES storm water program close to 150,000 industrial facilities and 538 municipal storm sewer systems have been permitted to control pollutant discharges.

In addition to storm water permitting, approximately 1,100 municipalities with overflows from their combined sanitary and storm sewers are being permitted under EPA's 1994 Combined Sewer Overflow Control Policy [4]. This effort should lead to significant reductions in the amount of bacteria, nutrients and oxygen demanding substances that are discharged during wet weather events in cities with combined sewer overflows.

Urban wet weather discharges are different from the process wastewater discharges that have traditionally been addressed through the NPDES program in that they are intermittent, their duration is variable, and they often cause physical impacts, due to their high volumes, even though pollutant concentrations may be minimal in some cases. The physical impacts of urban wet weather discharges, such as scouring, erosion and habitat degradation, are related to the type and amount of human activity that occurs within a given watershed, particularly the amount and type of pollutant sources and activities and the degree of impervious surfaces [5]. The large, variable flows during wet weather events and the very large number of sources make traditional end-of-pipe control strategies and technologies less cost-effective and, in some, cases futile. For example, other than by providing temporary storage, which may or may not be practicable in an urban landscape, effective end-of-pipe controls to reduce peak discharges and prevent stream scouring are difficult and expensive to implement.

Given these challenges, to succeed in managing urban wet weather discharges will require pursuing objectives that go beyond what can be achieved through end-of-pipe pollutant controls. Urban wet weather management should focus on managing the quantitative and qualitative changes that occur in water as it flows over human-made surfaces and interacts with human activities imposed on the watershed landscape. A reoriented management approach should include addressing undesirable chemical changes; changes in flow rates and volumes; changes in other physical characteristics, such as temperature; and maintaining targeted watershed balances for nutrient cycling, groundwater infiltration and other natural watershed processes. EPA believes that a watershed-based approach offers a more effective and efficient opportunity for managing urban wet weather discharges than the traditional focus on individual end-of-pipe pollutant discharges alone.

Overall, the NPDES permitting program is evolving toward a watershed management approach. Initially, permitting authorities are, in many cases, organizing permit issuance so that all permits within a watershed expire and are reissued at roughly the same time. This allows for more consistent and equitable permit requirements within the permits. In addition, by addressing all NPDES facilities at the same time, States or EPA will be able to consider other controls on nonpoint sources as a way to reduce the need for placing more stringent limits on NPDES facilities. Also, the process enables a permitting authority to place increased controls to meet water quality standards on those discharges that have the greatest impact on surface waters, rather than placing these controls on all facilities. A watershed approach also allows States to combine the information collection aspects of permits with ambient monitoring, thus providing a more comprehensive picture of surface water quality.

In subsequent efforts, both point source and nonpoint source discharges of pollutants within a watershed can be evaluated at the same time, usually in the context of determining a total maximum daily load (TMDL) for the pollutants being addressed. Understanding the relationship of point source discharges to other watershed sources offers several opportunities for more efficient and environmentally focused program management.

One way to integrate and coordinate NPDES program functions with other watershed management activities is through a Statewide watershed management approach. Under this approach, a State is divided into geographical management units drawn around river basins. The NPDES Watershed Strategy [6] represents a commitment by EPA to support statewide watershed management. At least twenty States are now developing or implementing a framework to synchronize monitoring, assessment, NPDES permitting, and other activities within geographic management units.

To better manage urban wet weather discharges, EPA is currently developing a watershed framework which can be integrated into overall watershed management at a State level. EPA is also supporting the development of better ways to assess the health of receiving waters and determine the effectiveness of management efforts to control urban wet weather impacts within a watershed. As a result of efforts to date in these areas, EPA has identified a variety of technical and policy issues that still need to be resolved for successful implementation of a watershed-based approach to managing urban wet weather discharges (some of the issues are listed below for consideration). New scientific research and information are critical to provide answers to these complex technical and policy questions.

URBAN WET WEATHER FLOWS FEDERAL ADVISORY COMMITTEE

In 1995, EPA established the Urban Wet Weather Flows Federal Advisory Committee to involve representative stakeholders in addressing the environmental and health problems related to urban wet weather discharges. The Committee is composed of stakeholders representing municipalities, industry, commercial interests, environmental groups, States/Tribes, and technical organizations. The Committee provides a forum to identify and discuss a range of cross-cutting issues associated with urban wet weather discharges including: improvements to the storm water phase I program, water quality standards in a wet weather context, and the coordination of wet weather discharge control efforts on a watershed basis. The Committee is playing an important advisory role in addressing the technical and policy issues inherent in implementing a watershed-based approach for managing urban wet weather discharges.

URBAN WET WEATHER WATERSHED APPROACH

Just as the baseline NPDES program at the Federal and State levels has tended in the past to address individual sources or types of pollutant sources in isolation, local watershed stakeholders, including municipalities, often address storm water management, the operation of sewage collection systems, the operation of wastewater treatment plants, and wet weather overflows separately. They overlook common issues that may cut across these functions and the benefits that may be obtained by addressing them through an integrated approach. In recognition of the need for better coordination, the Advisory Committee is working to determine how best to address the water quality and other environmental impacts associated with urban wet weather discharges on a watershed basis. The Advisory Committee's draft Watershed Framework is serving as a vehicle for the development of recommendations on this subject.

While the framework document is not yet completed, eight important themes have emerged:

- A watershed approach is likely to be the most effective means of addressing the unique characteristics of wet weather discharges, particularly impacts which exist not because of an individual point source, but because of human activity that occurs over a wide geographic area, such as agriculture or construction.

- There is a critical need to begin looking at all water quality and environmental impacts on a watershed basis and to more effectively involve local watershed stakeholders in decisions relative to determining designated uses and implementing controls.

- Since watersheds do not generally follow jurisdictional boundaries, a watershed approach both encourages and necessitates intergovernmental partnerships as well as partnerships among different levels of government and other watershed stakeholders.

- All potential contributors to water quality impairment should be required to undertake certain minimum measures to reduce or eliminate impacts, such as the nine minimum controls under the CSO policy or a storm water management program under the NPDES permitting program.

- A comprehensive watershed assessment reflected in a watershed plan with prescribed minimum components should be used to determine what additional measures are necessary to improve water quality.

- A mechanism for making tradeoffs between controls on different types of discharges, as well as between traditional discharge controls and nontraditional efforts to improve water quality, is necessary. For example, watershed stakeholders may determine that removal of an aging dam might result in much greater water quality improvement than additional controls on CSOs.

- It is important to provide greater flexibility in implementing the NPDES permitting program on a watershed basis, allowing for phased implementation schedules, alternative forms of permits, and the use of pollution prevention and incentive-based mechanisms.

- Finally, there is a recognition that the existing system of designated uses needs to be refined to reflect the reality of urban streams, as well as to protect those areas that have not yet been degraded by physical impacts. For example, some urban streams may not be able to achieve designated uses protecting aquatic habitats, despite effective pollutant reductions, because the preconditions necessary for a thriving aquatic habitat are nonexistent (e.g., river banks are lined in concrete, no riparian foliage exists, habitat is damaged, there is a high degree of impervious surface in the watershed).

URBAN WET WEATHER MONITORING AND ASSESSMENT

To implement urban wet weather flows management on a watershed basis, it is important to assess water quality more holistically. This suggests ambient monitoring as an integral component of an urban wet weather watershed monitoring program. It is also necessary, however, to be able to determine the impact of the various sources (point and nonpoint) on the overall water quality in the watershed for priority-setting and control purposes. Further, pollutant concentration and overall loadings profiles vary markedly because of the highly fluctuating volumes during a wet weather event. This makes end-of-pipe monitoring problematic and limits the value of data obtained. These problems suggest the need to move from an emphasis on chemical-specific monitoring to a focus on biological or other types of monitoring. Clearly, better monitoring and assessment strategies are needed to accomplish this change in emphasis.

The Advisory Committee is wrestling with these problems and is preparing recommendations relative to monitoring. As with the watershed framework, the monitoring recommendations are still a work in progress. Overall, these recommendations will support those of the Interagency Task Force on Monitoring[7] and are expected to help bridge the gap between the State-coordinated monitoring efforts and efforts to develop environmental indicators. Among the draft recommendations are an emphasis on the acquisition of ambient water quality data to determine what kind of problems exist and then conducting additional, more targeted monitoring to reveal the sources of the problems. While there is a recognition that end-of-pipe monitoring may not always yield useful information, compliance monitoring in some form will still be necessary. Information sharing is being promoted as a way to reduce burden and provide greater information to watershed stakeholders, particularly information developed as a result of research projects.

Storm Water and Combined Sewer Overflow Environmental Indicators

In order to implement a watershed approach to managing urban wet weather discharges, managers need better monitoring approaches for assessing receiving water impacts, identifying the sources of those impacts and for assessing and predicting the effectiveness of wet weather control efforts. As previously mentioned, traditional end-of-pipe chemical sampling of storm water discharges or combined sewer system overflows does not, in most cases, provide all the necessary information. Because of this, EPA supported two projects designed to identify and recommend more useful monitoring approaches for storm water and combined sewer system operators. These new approaches, called environmental indicators or performance measures, will be useful in better managing urban wet weather flows.

The Association of Metropolitan Sewerage Agencies (AMSA) undertook a study to identify performance measures for CSO control programs. Working through a stakeholder workgroup, AMSA identified twenty-four performance measures that CSO communities can use to determine the effectiveness of their CSO control programs. They can be used to measure and report progress toward achieving goals and objectives so that the public will be able to assess the investment in CSO control. With these performance measures, CSO communities can better prioritize existing CSO problems and guide future CSO control efforts, especially on a watershed basis. Table 1 summarizes the CSO performance measures.

Table 1: Summary of Recommended CSO Performance Measures

ADMINISTRATIVE	END OF PIPE
- Documented implementation of nine minimum controls - Status of long term control plan(1) - Waste reduction	Flow Measurement - Wet weather flow budget - CSO frequency (1) - CSO frequency in sensitive areas(1) - CSO volume - CSO volume in sensitive areas - Dry weather overflows Pollutant Load Reductions - BOD load - TSS load - Nutrient load - Floatables
RECEIVING WATER	ECOLOGICAL/HUMAN HEALTH/RESOURCE USE
- Dissolved oxygen trend - Fecal coliform trend - Floatables trend - Sediment oxygen demand trend - Trend of metals in bottom sediments	- Shellfish bed closures - Benthic organism diversity - Biological diversity index - Recreational index - Beach closures - Commercial activities

Notes: (1) Appropriate for national tracking.
Source- "Performance Measures for the National CSO Control Program." January 1996. Association of Metropolitan Sewerage Agencies. 1000 Connecticut Ave., NW, Suite 410, Washington DC, 20036.

In the area of urban storm water management, the Center for Watershed Protection, the Rensselaerville Institute and the Water Environment Federation are concluding a project to identify storm water environmental indicators that will help storm water managers better assess existing problems, track the effectiveness of storm water controls and manage storm water at a watershed level. This project identified twenty-six different indicators of particular use. Table 2 describes these indicators.

Storm water environmental indicators can be used to assess the health of a receiving water system, identify stresses on the system, and assess the effectiveness of storm water management activities. These indicators are important in assessing the link between human activities within a watershed, how we manage those activities, and the impacts upon the receiving water environment. As such environmental indicators represent an essential component of a comprehensive watershed management strategy.

Table 2: Storm Water Environmental Indicators

Water Quality Indicators:
- Water quality pollutant monitoring - Toxicity testing - Non-point source loadings - Exceedance frequencies of water quality standards - Sediment contamination - Human health criteria
Physical and Hydrological Indicators:
- Stream widening/down cutting - Physical habitat monitoring - Impacted dry weather flows - Increased flooding frequency - Stream temperature monitoring
Biological Indicators:
- Fish assemblage - Macro-invertebrate assemblage - Single species indicator - Composite indicators - Other biological indicators
Social Indicators:
- Public attitude surveys - Industrial/commercial pollution prevention - Public involvement and monitoring - User perception
Programmatic Indicators:
- No. of illicit connections identified/corrected - No. of BMPs installed, inspected and maintained - Permitting and compliance - Growth and development
Site Indicators:
- BMP performance monitoring - Industrial site compliance monitoring

LINGERING ISSUES TO BE RESOLVED

A number of technical and policy issues will confront resource managers, local governments, industry and Federal and State regulators in both the short-term and long-term as they move to implement a watershed-based approach to addressing urban wet weather discharges. Examples of such issues include the following:

Technical:

• How many and what types of best management practices (BMPs) are necessary to counteract different levels of human activity within different watersheds?

• What is the cumulative effect of different varieties of BMPs, both structural and programmatic, preventative and mitigative, within a watershed when it comes to controlling wet weather discharges?

• What are realistic time frames for planning and implementing wet weather discharge control strategies on a watershed level? How long will it take for improvements to occur and to be measured?

• Do all watersheds, regardless of geographic location, vegetation, geology, soil types, rainfall patterns, etc., react the same way to human activities and wet weather events? If not, how are they different?

• Should management activities be ignored or be limited in highly developed watersheds and should resource managers focus the majority of their effort on managing urban wet weather flows within watersheds not yet exceeding a certain critical impairment threshold? Does such a threshold indicator exist? If yes, would the degree of imperviousness or population density serve as such an indicator?

• How much and what combinations of environmental indicator measurements are necessary to assess the desired management objectives within a given watershed?

• Can all urban wet weather impairment be reversed or prevented through the use of BMPs? Or is there a level beyond which urban wet weather discharge controls will be cost-prohibitive?

• How does a watershed manager identify desirable (baseline or natural) levels of wet weather flow and nutrient cycling to identify a healthy watershed?

• Given that heavily urbanized areas often have high levels of imperviousness and high densities of activities, can fishable and swimmable designated uses ever be achieved? If not, how might appropriate use designations be determined?

• Does the current system of water quality standards need to be re-examined to better address wet weather discharges? If so, what changes would be necessary?

Policy/Programmatic:

• What incentives can serve as the impetus for water resource managers to move to a watershed management approach? How can States and EPA afford to move to watershed management given shrinking resources?

• Watershed management and the use of environmental indicators may be more complex than traditional source by source management. How do managers and their agencies overcome these new complexities?

• Is the current statutory scheme under the Clean Water Act adequate to fully deal with urban wet weather problems? If not, in what respects should it be changed?

• What are effective strategies to address the challenges posed by watersheds that cross multiple jurisdictional boundaries? What roles should different levels of government, industry and the public play?

• How can monitoring data and BMP performance information be comparably generated, usefully accumulated, and fully shared with interested parties?

• How can wet weather data and watershed information be used at a national level for decision making and future program direction?

• How do regulators ensure accountability under a watershed approach that may rely more on performance-based results and less on end-of-pipe compliance monitoring?

IMPLEMENTATION

EPA is confident that a number of these issues will be discussed during this conference and some even resolved. For other issues, we may not have definitive answers for some time. It is thus necessary to move to control impacts in the absence of all of the answers we would ideally like to have. Thus, based on input from our Advisory Committees, and technical forums such as this, EPA is proceeding to develop the necessary policies, regulations and guidance to implement urban wet weather flow controls under a watershed framework.

More specifically, in addition to developing a watershed framework for urban wet weather flows, the Urban Wet Weather Flows Federal Advisory Committee plans to develop recommendations to EPA on how water quality standards can be developed and applied under wet weather conditions to ensure attainment of human health and ecological goals adopted for the water body in the most cost-effective manner possible.

The Advisory Committee also is re-examining the phase I storm water permitting program for ways to make improvements, particularly in streamlining permitting and monitoring requirements, to ensure that the permitting program accomplishes real and measurable environmental results that will contribute to attaining water quality standards. The Committee has identified twenty-seven problems with the phase I program and is now working on several of these including; improving the prioritization scheme for industrial storm water dischargers, fostering more public education and involvement in storm water management, ensuring the effectiveness of storm water BMPs and reducing the cost of the phase I program. The Committee is also examining ways to create a seamless, uniform phase I and phase II storm water program.

The Storm Water Phase II Program Subcommittee is developing recommendations for dealing with phase II storm water sources. This includes identifying important storm water sources not yet controlled, developing mechanisms to deal with these sources, which may or may not include NPDES permitting, and clarifying the roles of dischargers and local, State and Federal governments. Recommendations from the Subcommittee are anticipated this fall. A phase II storm water regulation must be proposed by EPA by September 1997 and finalized March 1999.

The Sanitary Sewer Overflow (SSO) Advisory Subcommittee is exploring different options for defining the role of regulations, program evaluations, technical assistance and outreach materials to enhance the performance of sanitary sewer collection systems. Some of the major issues being considered by the Subcommittee are minimum operational measures for collection systems, reporting, public notification, remediation of collection systems, installation of building laterals and ensuring satellite collection systems are adequately maintained. The SSO Subcommittee is expected to provide recommendations on key issues and the type of products to address them later this fall.

For combined sewer overflows, in 1996 and subsequent years, EPA and States, together with the stakeholders who helped develop the CSO Policy, are focusing on technical guidance for the implementation of CSO controls. This effort is being integrated into the urban wet weather flows initiative.

CONCLUSIONS

The receiving water impacts caused by urban wet weather discharges pose new challenges for water resource managers. Many impacts are the result of growth and development and other human activities within a watershed. The quality and quantity of water as it flows through a watershed can be detrimentally altered by the amount and types of human activity that is encountered. Water resource managers need new management strategies and better assessment tools for dealing with these problems. EPA believes a watershed management approach is the best way to proceed. EPA and its Urban Wet Weather Flows Federal Advisory Committee, and other groups, are developing a watershed strategy and working to resolve technical and policy issues to ensure successful urban wet weather flows management.

REFERENCES CITED

1)   The U.S. Environmental Protection Agency (EPA). 1983. Results of the Nationwide Urban Runoff Program. Vol. I. Final Report. Water Planning Division, Washington, DC 20460.

(2)   National Water Quality Inventory: 1994 Report to Congress. US EPA, Office of Water, Washington, DC 20460. December 1995. EPA 841-R-95-005.

(3)   The Quality of Shellfish Growing Waters on the East Coast, West Coast, and Gulf of Mexico, NOAA, 1988-90.

(4)   Combined Sewer Overflow Control Policy. April 19, 1994. US EPA. Office of Wastewater Management, Washington, DC 20460. Federal Register, Vol. 59, No. 75, pg. 18688.

(5)   "The Importance of Imperviousness", Watershed Protection Techniques, Vol. 1, No. 3, Fall 1994.

(6)   The NPDES Watershed Strategy. March 21, 1994. Office of Water, US EPA, Washington, DC 20460.

(7)   The Strategy for Improving Water-Quality Monitoring in the United States: Final Report of the Intergovernmental Task Force on Monitoring Water Quality. US Geological Survey, Office of Water Data Coordination, Reston, VA, 22092. February 1995.

DISCUSSION

Effects of Watershed Development and Management on Aquatic Ecosystems - EPA Perspective
Alfred Lindsey

In the last 20 years, EPA has come a long way
- 65,000 permits issued, and point sources pretty well under control
- 2/3rds of waterways meet water quality standards

Urban runoff is a significant source of pollution in the impaired waters
- About 15 percent of impaired rivers and lakes are attributed to urban runoff
- Nearly 50 percent of impaired estuary shorelines are due to urban runoff
- Shellfish industry is severely impacted by urban runoff

Stormwater NPDES program has made significant progress, but need to focus more broadly on the NPS problem rather that end of pipe. A holistic approach, or watershed approach is called for. FACA report will address this issue; report is nearly complete (see paper for eight important themes).

EPA has developed a R&D program addressing urban runoff. Half of R&D program in 1997 should go to EPA's Office of R&D in Edison.

Questions/Comments

Comment: Need to balance urban development with protection of the environment. As municipal officials, regulators, academics and consultants, we all come at the problem in different ways. But we must get our head together in a common approach and explain it in lay terms to the community so that they are convinced that we are providing an environment that is robust and healthy. Keep an open mind and please keep an open mind.

Comment: We need to work together to forget the sins of the past and concentrate on obtaining consensus on what should be done. Keep an open mind and look for ways that we can compromise, rather than argue.

Measuring the Health of Aquatic Ecosystems Using Biological
Assessment Techniques: a National Perspective.

Michael T. Barbour[1,2]

Abstract
The detection and assessment of pollution (and other types of
perturbation) in the aquatic environment, and its effects upon the
biological community, is rapidly becoming a central focus of state
agency water resource programs. These bioassessments are intended
to examine the health of the aquatic ecosystem. Bioassessments are
particularly useful because they reflect the condition of the resident
biota from cumulative effects as a result of both nonpoint and point
source impacts. With the advent of the U.S. Environmental Protection
Agency's (EPA) guidance to develop biological criteria, states are
implementing bioassessment approaches that incorporate a regional
reference condition for interpretation of impairment. This approach
lends itself to the watershed management approach because of its
regional orientation. Aquatic systems reflect the condition of their
watersheds, and impacts from differing land use patterns become
cumulative as drainage areas become larger. The use of biological
assessment and monitoring is a crucial component of the monitoring
"toolbox" to effectively measure the health of the overall aquatic
ecosystem.

[1]Director of Ecological Services, Tetra Tech, Inc., 10045 Red Run Boulevard, Suite 110, Owings
Mills, Maryland 21117.

[2]This paper is adapted from Barbour, M.T., J. Diamond, and C. Yoder. 1996. Biological Assessment
Strategies; applications and limitations. In D.R. Grothe, K.L. Dickson, and D.K. Reed (editors). Whole-effluent
toxicity testing: an evaluation of methods and predictability of receiving system responses. SETAC Press,
Pensacola.

Introduction

The central purpose of assessing biological condition of aquatic communities is to determine how well a waterbody supports aquatic life. Biological communities integrate the effects of different pollutant stressors—such as excess nutrients, toxic chemicals, increased temperature, and excessive sediment, and provide an overall measure of the aggregate impact.of the stressors. Biological communities respond to stresses of all degrees over time, providing information on perturbation otherwise not obtained with episodic water chemical measurements or discrete toxicity tests. The assessment of aquatic communities is not a new concept. Use of ambient biological communities, assemblages, and populations to protect, manage, and even exploit water resources have been developing and evolving for the past 150 years (Davis 1995). Despite this history, it has only been in the last decade that a widely accepted technical framework for using biological assemblage data for assessment of the water resource has evolved that is scientifically-sound. Karr (1993) has discussed the major advances made in ambient biomonitoring over the past decade and suggested that biological studies at all levels complement existing approaches in ways that can be cost-effective. Given the relatively recent development of standardized biological assessment methods for small streams and rivers (Plafkin et al. 1989) and the current lack of such methods for other types of aquatic systems, the development and widespread use of formal biological assessment programs has lagged far behind chemical-specific or toxicity-based programs in water resource management (Gibson et al. 1994). We will all be better served by finding ways to integrate WET bioassessment, chemical-specific, and other tools, rather than having them compete.

Biological assessment is defined as an evaluation of the condition of a waterbody using biological surveys and other direct measurements of the resident biota in surface waters (Gibson et al. 1994). A biological survey typically consists of collecting data that can be used to assess the condition of the physical habitat and specific biological assemblages (usually benthic macroinvertebrates and fish), that inhabit the aquatic environment. In situ toxicity testing of resident species is another form of biological assessment used to evaluate exposure to chemical contamination in the receiving stream. *Biological monitoring* is conducted over time and includes toxicity testing, fish tissue analyses, and surveys of single populations for entire assemblages.

Biological monitoring has long been devoted to detecting and assessing risk from chemical discharges into the environment. Regulations fostered by EPA and state water resource agencies have established the impetus and the nature of the monitoring. Biomonitoring approaches that pertain to measuring biological processes of lethality, reproductive viability, growth retardation, bioaccumulation and biomagnification, among others, have been developed and debated among scientists. The complexity of our water resource problems, however, is more than striving for "clean water" (Karr 1993). More sophisticated approaches are needed that address the diversity of human impacts on water resources. There is a growing awareness in the United States that point source water pollution control programs have been very successful, but that nonpoint sources, groundwater contamination (USEPA 1992a, b), and habitat degradation (Judy et al. 1984) continue to degrade the nation's water resources. Government agencies at federal, state, and local levels are broadening their view of water quality protection and are developing and implementing innovative strategies to achieve greater water resource protection. Many of these efforts center on the concept of a "whole basin planning" (WBP) approach, which re-aligns water pollution control programs to operate in a more comprehensive and coordinated fashion (Bowman and Creager 1993). Biological assessment and monitoring is becoming a primary tool for these efforts.

Cumulative Impacts on Watersheds

Karr et al. (1986) outlined five classes of variables influenced by human activities relevant to watershed development (Figure 1). Most water resource agencies are concerned about water quality issues. The Fish and Wildlife Service and sister agencies in the states are concerned about elements of the physical habitat (Karr 1993). This is an especially critical point for National Pollutant Discharge Elimination System (NPDES) discharges in which toxic or physical effects of the discharge must be separated from non-discharge effects in aquatic biota. The whole concept of water resource degradation is dependent on a holistic assessment of the factors influencing that degradation (Karr 1993). Evaluation of the resident aquatic community provides a mechanism for conducting a holistic assessment. Could it be that solutions to our water resource problems in a site-specific situation lie in mitigation activities other than water treatment, or is it possible that water treatment is insufficient for correcting problems?

State water quality standards are composed of three parts: a designated use for each waterbody; a criterion for each physical, chemical, and biological condition based on the designated use; and an antidegradation statement. To effectively attain state water quality standards, all sources of pollution to surface waters must be considered, including nonchemical stresses such as habitat alteration and hydromodification. This requires incorporating evaluations of the physical and biological components of aquatic ecosystems (OST and OWOW 1992). An expansive view of "aquatic ecosystems" should be taken to allow examination of perturbations in areas adjacent to (e.g., riparian forests) and removed from (e.g., upland areas of watersheds) surface waterbodies. These areas can play key roles in determining surface water chemical and habitat quality. For example, removal of riparian vegetation can lead directly to increased bank erosion, stream temperature increases through loss of shading, and increased nutrient input due to loss of vegetative nutrient uptake. Conversion of upland forests and agricultural areas to urban or industrial land uses can cause dramatic changes in receiving waterbody flow patterns and channel morphology, and can cause increased loading of nutrients and heavy metals, depending on the specific land use changes.

Section 303(d) of the Clean Water Act established the total maximum daily load (TMDL) process to provide for more stringent water quality-based controls when technology-based controls are inadequate to achieve State Water Quality Standards. (A TMDL is the estimated assimilative capacity for a waterbody (i.e., the amount of chemical and nonchemical pollution that can enter a waterbody without affecting its designated uses). TMDLs provide for allocation of allowable loads among different pollution sources so that appropriate control actions can be taken, water quality standards achieved, and human health and aquatic resources protected. The watershed approach to managing aquatic ecosystems uses the TMDL process as a framework.)

Habitat quality is an essential measurement the TMDL process and in any biological survey, because aquatic fauna often have very specific habitat requirements that may be independent of water quality. Habitat quality encompasses three of the five classes of anthropogenic stressors: habitat structure, flow regime, and energy source (Figure 1). In this paper, habitat structure is defined as the physical structure of aquatic ecosystem environments. For streams, it includes channel morphology (width, depth, sinuosity); floodplain shape and size; channel gradient; instream cover (boulders, woody debris); substrate types and diversity; riparian vegetation and canopy cover; and bank

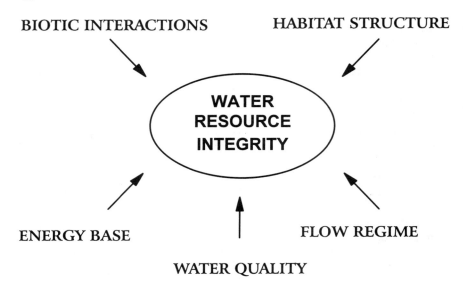

Figure 1. Five classes of variables affecting water resource integrity that are related to human activities (adapted from Karr et al. 1986).

stability. Flow regime is defined by the velocity and volume of water moving through a stream. The energy source for a stream is one of two modes: input of nutrients from either runoff, groundwater, or debris (e.g., leaves) falling into a stream, or photosynthesis by aquatic plants and algae. These three factors are interrelated, and alteration of any or all of these may prohibit the attainment of the aquatic life use. Assuming that water quality remains constant, the predictable relationship between habitat quality (as defined by site-specific factors, riparian quality, and upstream land use) and biological condition can be a sigmoid curve, in theory, as illustrated in Figure 2. Habitat is shown to vary in quality from poor (nonsupporting of an acceptable biological condition) to good (comparable to the reference condition) (x-axis). Biological condition varies from poor (severely impaired) to good (unimpaired) (y-axis). Interpretation of the relationship as depicted can be summarized by three points: (1) the upper right corner of the curve is the ideal situation where optimal habitat quality and biological condition occur. Some variability in habitat quality is possible without affecting the condition of the biological community. (2) However, in the midsectional part of the curve, the decrease in biological condition is proportional to a decrease in habitat quality. (3) In the lower left portion of the curve, habitat quality is poor, and further degradation may result in relatively little difference in biological condition. Communities in this region of the curve are pollution tolerant, opportunistic, thrive in areas of reduced competition, and are able to withstand highly variable conditions (Barbour and Stribling 1991). It will be difficult to differentiate among cumulative impacts in this area of the curve.

Perhaps the two most important areas of the graph are the lower right corner where degraded biological condition can be attributed to something other than habitat quality, and the upper left corner where optimal biological condition is not possible in a severely degraded habitat. The actual determination of these possible outcomes is supported by a reference database adequate to defining the expected relationship between habitat quality and biological condition. The theoretical regression line should be substantiated with a large database sufficient to develop confidence intervals (Figure 3).

In summary, the relationship between habitat quality and biological condition is generally one of three types: (1) the biological condition varies directly with habitat quality—water quality is not the principal factor affecting the biota; (2) the biological community is degraded relative to the potential of its actual habitat—water quality is

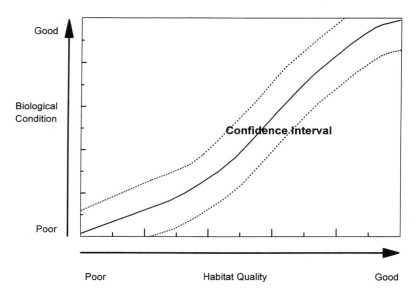

Figure 2. Relationship between habitat quality and biological condition (taken from Barbour and Stribling 1991).

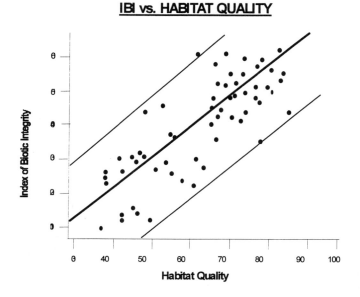

Figure 3. Relationship between habitat quality and the fish IBI (data obtained from Ohio EPA).

implicated as a cause of the degraded biotic condition; or (3) the biological condition is elevated above what actual habitat conditions should support—suspected cause may be organic enrichment or alteration of energy base, both of which will change the structure of the aquatic community without an obvious sign of impairment (Barbour and Stribling 1991). A clear distinction between impacts due to a combination of large-scale habitat alteration and water quality degradation is often not possible. It is difficult to determine with certainty the extent to which biological condition will improve with specific improvements in either water quality or habitat. However, it is likely that biological condition will not improve significantly if water quality improves and habitat degradation remains (Barbour and Stribling 1991).

Implementation of Biological Surveys in State Programs

The use of bioassessments in state water resource programs has been evolving over the past decade, with the most active progress being in streams and having occurred since 1990 with the development of the Rapid Bioassessment Protocols (Plafkin et al. 1989) and the Program Guidance for Biocriteria (USEPA 1990). The current status of bioassessments and biocriteria in state programs is illustrated in a series of maps, which are from a report by EPA (Davis et al. 1996). Numeric biocriteria are in place in the water quality standards for freshwater systems for Ohio and Florida (and recently promulgated in Maine), and under development in 13 other states (Figure 4a). However, the majority of states have yet to move into the biocriteria arena. Nearly three quarters of the states use bioassessment data to measure the attainment of their aquatic life uses (Figure 4b), and all but three states use bioassessments in some manner in their water resource activities (Figure 4c).

The target assemblages used in the state bioassessment programs are primarily benthic macroinvertebrates (all but three states; Figure 5a) and fish (all but 14 states; Figure 5b). Utah is a state that monitors benthic macroinvertebrates in selected stream sites, but does not have a formal bioassessment program. Seven states use algae in their bioassessment programs (Figure 5a). All of these states use another assemblage as well—generally, benthos.

One of the most important contributions of the recent growth in interest in biological assessment has been development of standards

TARGET ASSEMBLAGES USED IN STATE BIOASSESSMENT PROGRAMS

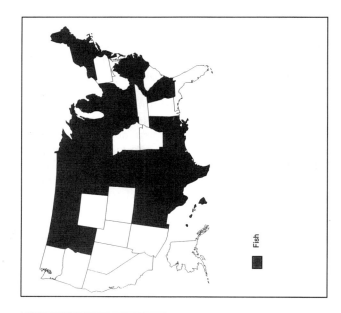

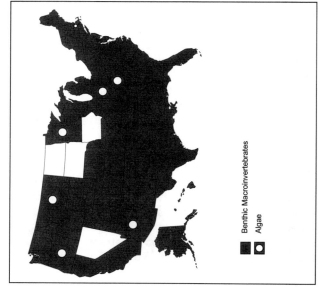

Figure 5. Target assemblages used in state bioassessment programs (taken from Davis et al. 1996): a. benthic macroinvertebrates and algae; b. fish.

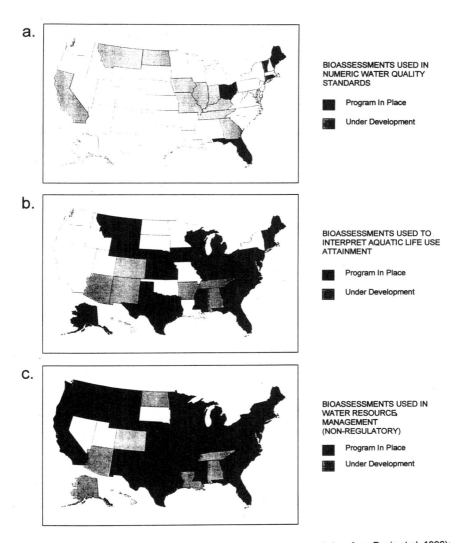

Figure 4. Use of bioassessment in state programs (taken from Davis et al. 1996):
a. numeric water quality standards; b. aquatic life use attainment; c. water resource
management (non-regulatory).

defined by local and regional expectations (Karr 1993). Karr also notes that this approach should have been done for chemical and physical criteria as well. It has been long known that regional differences in background concentrations of various chemicals (e.g., phosphorus, certain metals), provide complications and over-stringent national criteria. Barbour et al. (1995) presented a framework for development of regional reference conditions using the multimetric approach for biological assessments.

The preferred framework for developing reference conditions around the country has been using ecoregions. Nearly 80% of the states have ecoregional reference conditions either in place or under development (Figure 6). Some states, such as those in the Mid-Atlantic Appalachian Mountains (Pennsylvania, Maryland, West Virginia, and Virginia), Mississippi, and Alabama are working together in common ecoregions to develop reference conditions that can be used across state boundaries. Florida DEP has demonstrated how the use of subecoregions has improved their ability to classify streams and conduct bioassessments (Barbour et al. 1996). States are finding that use of a regional framework effectively enhances their watershed approach to managing ecosystems by partitioning natural variability.

Recommendations for Biological Assessment

The choice of a particular biological assessment design to be used in watershed monitoring should be based on the monitoring objectives and ultimate decision process. The recent innovative rapid assessment techniques offer cost efficiencies while maintaining data integrity. However, these approaches are not intended to replace more rigorous approaches. The constraint of resource investment and the requirement of quick "turn-around" of biological data for decisions will dictate the nature of the study design. Regardless of whether a rapid assessment or other approach is chosen, a certain amount of preparation is required. For instance, specific objectives must be defined, sites (or streams) classified into homogeneous classes, reference conditions established, protocols standardized and investigators trained, and a quality assurance plan developed. Recommendations for this basic process are as follows:

Study Objectives   For assessment of anthropogenic perturbation in watersheds, objectives usually pertain to a determination of biological condition or status, monitoring improvements following

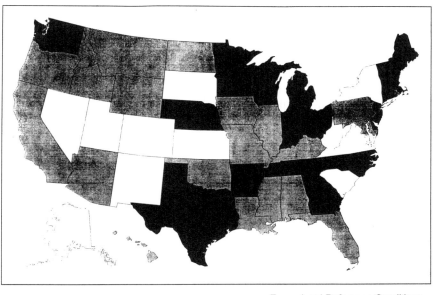

Ecoregional Reference Conditions
In Place

Ecoregional Reference Conditions
Under Development

Figure 6. Use of ecoregional reference conditions in state bioassessment
programs (taken from Davis et al. 1996).

implementation of mitigation measures or restoration practices, and screening for problem or sensitive areas.

Site Classification   A regionalization process is recommended to provide an ecological basis for identifying homogeneous areas from which reference conditions can be established. This process is especially applicable to the watershed approach where portions or reaches of the watershed will have different expectations. Site classes should not be constrained by political boundaries, but relate directly to the protection of the water resource. Both the use of biological metrics and multivariate analyses have been shown to be effective classification tools.

Reference Condition   The establishment of reference condition is based on identification of minimally-disturbed sites that represent the best attainable physical, chemical, and biological condition. The use of biological metrics that represent various elements and processes of the aquatic community have been used successfully to develop reference conditions in the United States.

Standardized Protocols   Rapid assessment approaches used nationally rely on large composite samples (to reduce heterogeneity) that can be collected quickly and efficiently, an organism-based subsampling technique that standardizes sample processing, and a multimetric approach for analyzing data in a consistent format. Thorough training of investigators enhances the ability to provide a consistent unit of effort and area sampled, thus reducing the qualitative nature of the methods.

Data Analysis   For the multimetric approach adapted by the majority of state water resources agencies, metrics chosen for relevancy and sensitivity are calibrated for the site class and reference condition. This means that regional validation efforts are required. A transformation from values of different scales to score categories is done from calibrated thresholds based on a selected percentile of the site class. Information from the individual metrics as well as an aggregated index allows judgment of the biological status of the assemblage.

Habitat Assessment   The assessment of the quality of the physical habitat structure is critical to the interpretation of biological condition. A visual-based assessment technique is being used successfully in the United States to obtain a holistic evaluation of the habitat. Although rather subjective, investigator variability is controlled through the selection and definition of straightforward parameters and the development of a systematic process for judging the quality of the individual parameters.

Quality Assurance Plan   The development of an effective quality assurance (QA) plan is paramount to the success of the program. This plan should assign responsibility, establish protocol, define preventive and corrective action, and ensure that study objectives are met. QA plans are especially important in rapid assessment techniques to regulate the performance and maintain data integrity.

Literature Cited

Barbour, M. T., J. Gerritsen, G. E. Griffith, R. Frydenborg, E. McCarron, J. S. White, and M. L. Bastian. 1996. A framework for biological criteria for Florida streams using benthic macroinvertebrates. Journal of the North American Benthological Society 15:185-211.

Barbour, M. T., and J. B. Stribling. 1991. Use of habitat assessment in evaluating the biological integrity of stream communities. Pages 25-38 *in* biological criteria: research and regulation, proceedings of a symposium, December 12-13, 1990, Arlington, Virginia. EPA-440-5-91-005. Office of Water, U. S. Environmental Protection Agency, Washington, DC.

Barbour, M. T., J. B. Stribling, and J. R. Karr. 1995. Multimetric approach for establishing biocriteria and measuring biological condition. Pages 63-77 *in* W. S. Davis, and T. P. Simon (editors). Biological assessment and criteria: tools for water resource planning and decision making. CRC Press, Inc., Boca Raton.

Bowman, M. L., and C. S. Creager. 1993. Whole basin planning: practical lessons learned from North Carolina, Delaware, and Washington state. To be published in proceedings of the national conference on urban runoff management, March 30 - April 2, 1993. Chicago, Illinois.

Davis, W. S. 1995. Biological assessment and criteria: building on the past. Pages 15-29 *in* W. S. Davis, and T. P. Simon (editors). Biological assessment and criteria: tools for water resource planning and decision making. CRC Press, Inc., Boca Raton.

Davis, W. S., B. D. Snyder, J. B. Stribling, and C. Stoughton. 1996. Summary of state biological assessment programs for streams and wadeable rivers. EPA 230-R-96-007. U. S. Environmental Protection Agency, Office of Policy, Planning, and Evaluation, Washington, D. C.

Gibson, G. R., M. T. Barbour, J. B. Stribling, and J. R. Karr. 1994. Biological criteria: technical guidance for streams and small rivers. EPA-822-B94-001. Office of Science and Technology, U. S. Environmental Protection Agency, Washington, DC.

Judy, R. D., P. N. Seeley, T. M. Murray, S. C. Svirsky, M. R. Whitworth, and L. S. Ischinger. 1984. 1982 National fisheries survey. Volume 1. Technical report: initial findings. FWS/OBS-84-06. U. S. Fish and Wildlife Service.

Karr, J. R. 1993. Defining and assessing ecological integrity: beyond water quality. Pages 1521-1531 in C. H. Ward (editor). Environmental toxicology and chemistry. Pergamon Press, Tarrytown, New York.

Karr, J. R., K. D. Fausch, P. L. Angermeier, P. R. Yant, and I. J. Schlosser. 1986. Assessing biological integrity in running waters: a method and its rationale. Illinois Natural History Survey Special Publication 5, Champaign, Illinois.

OST and OWOW. 1992. TMDL framework for action. Draft final. Office of Science and Technology and Office of Wetlands, Oceans, and Watersheds. Washington, D. C. May 18, 1992.

Plafkin, J. L., M. T. Barbour, K. D. Porter, S. K. Gross, and R. M. Hughes. 1989. Rapid bioassessment protocols for use in streams and rivers: benthic macroinvertebrates and fish. EPA-440-4-89-001. Office of Water Regulations and Standards, U. S. EPA, Washington, DC.

U. S. Environmental Protection Agency. 1990. Biological criteria: national program guidance for surface waters. EPA-440-5-90-004. Office of Water Regulations and Standards, U. S. EPA, Washington, DC.

U. S. Environmental Protection Agency. 1992a. The quality of our nation's water: 1990. EPA 841-K-92-001. Office of Water, U. S. EPA, Washington, DC.

U. S. Environmental Protection Agency. 1992b. National water quality inventory: 1990 report to Congress. EPA 503-9-92/006. Office of Water, U. S. EPA, Washington, DC.

Using Bioassessments to Evaluate Cumulative Effects

Ellen McCarron, Eric H.Livingston, and Russel Frydenborg[1]

ABSTRACT

Assessing the cumulative environmental effects of multiple pollution sources within a watershed presents many new and complex challenges. This is especially true for nonpoint sources which, unlike traditional point sources of pollution, are intermittent, creating temporally and spatially variable shock loadings to receiving waters. Consequently, traditional assessment techniques which rely solely upon sampling and characterization of the water column are ineffective in determining the environmental effects of stormwater and other nonpoint discharges. This paper will discuss the need and rationale for alternative sampling and assessment procedures that provide a more ecologically-based manner of determining the cumulative environmental effects of nonpoint sources of pollution. Activities undertaken by the Florida Department of Environmental Protection in the past six years to develop biological community assessment tools to evaluate nonpoint discharges will be summarized. The development, use, and implementation of riverine biological community assessment tools, based on comparisons between impacted sites and ecoregion reference sites, will be reviewed.

INTRODUCTION

During the late 1970s, stormwater and other nonpoint sources (NPS) of pollution were identified as major contributors to the degradation of Florida's surface and ground water resources. To minimize stormwater pollutant loadings discharged from new land use activities, the Florida Environmental Regulation Commission adopted a statewide stormwater treatment regulation in February 1982. This rule, implemented cooperatively by the state's Department of Environmental Protection and five regional water management districts, establishes permitting procedures and, for various types of stormwater management practices, design criteria presumed to achieve a specified

---

[1]Florida Department of Environmental Protection, 2600 Blair Stone Road, Tallahassee, Florida 32399-2400

treatment level. This rule is one of numerous statutes and regulations that have been implemented during the past 20 years to minimize the detrimental environmental effects associated with the state's extremely rapid growth. Collectively, the individual laws and programs enacted during this period can be considered "Florida's Watershed Management Program" (Livingston, 1993, 1995).

An essential component of this watershed management program is monitoring, to evaluate environmental conditions and the program's environmental benefits. In the past, water quality management actions focused on traditional point sources of pollution, such as domestic or industrial wastewater discharges. These point sources typically discharge effluents of uniform, known quality at continuous design flows, making them relatively easy to monitor, assess and control. Point source assessments generally have relied almost solely upon water column chemistry monitoring. On the other hand, stormwater and other nonpoint sources of pollution, because of their intermittent, diffuse, land use specific nature, are highly variable in effluent quality and environmental effects. Of particular environmental concern is the cumulative impact on a water body from the numerous stormwater/nonpoint sources within a watershed.

The traditional water quality monitoring and management efforts used for point discharges generally suffer from several deficiencies when trying to understand and manage stormwater/NPS pollution. These deficiencies include difficulty in:

1. Assessing intermittent, shock loadings of pollutants.
2. Assessing cumulative impacts of multiple sources.
3. Comparing water bodies and establishing priorities for management actions.
4. Distinguishing actual or potential problems from perceived problems.
5. Discriminating anthropogenic loadings from natural watershed loadings of metals and nutrients.
6. Establishing cost-effective ways to assess pollution sources and trends on a watershed basis.

An answer to these problems has come from the area of biological monitoring. This paper will review the development and implementation of biological monitoring protocols in Florida which are being used to improve evaluation and management of stormwater and other intermittent pollution sources, along with traditional point sources. The paper will focus on efforts in assessing river systems although a similar effort recently has been initiated for lakes.

BIOLOGICAL COMMUNITY MONITORING

It has long been known that resident biota in a waterbody function as continual natural monitors capable of detecting the effects of both episodic as well as cumulative pollution and habitat alteration. Some advantages of using the biota to assess environmental quality include the following.

1. Biological communities reflect overall ecological integrity (chemical, physical and biological).
2. Over time, biological communities integrate the effects of different stressors, providing a measure of fluctuating environmental conditions.
3. By integrating responses to highly variable pollutant inputs, biological communities provide a practical approach for monitoring stormwater/nonpoint source impacts and the effectiveness of best management practices.
4. Routine monitoring of biological communities can be relatively inexpensive, particularly when compared to the cost of assessing toxic substances.
5. The public is very interested in the status of biological communities as a measure of environmental health.
6. Biological communities offer a practical way to evaluate the habitat degradation typically associated with stormwater discharges.

Determining Biological Integrity - The Rapid Bioassessment Concept

Karr and Dudley (1981) define biotic integrity as "the ability of an aquatic ecosystem to support and maintain a balanced, integrated, adaptive community of organisms having a species composition, diversity and functional organization comparable to that of the natural habitats within a region". This practical definition is based on measurable characteristics of aquatic communities and comparisons to a regional reference site thus providing a framework for bioassessments.

In 1985, EPA conducted a survey to identify states that routinely performed biological assessments and to evaluate the field methods being used. A workgroup of state and EPA biologists was formed to review the existing methods and to refine protocols for monitoring benthic macroinvertebrates and fish. In May 1989, "Rapid Bioassessment Protocols for Use in Streams and Rivers" (EPA, 1989a) was published to which the reader is referred for a more comprehensive discussion of this topic.

The rapid bioassessment protocols (also known as community bioassessment protocols) advocate an integrated assessment, comparing habitat (physical structure and flow regime) and biological measures with empirically defined reference conditions. Reference conditions are established through systematic monitoring of actual sites (ecoregion reference sites) that represent the natural range of variation in "least disturbed" water chemistry, habitat, and biological condition. The concept of ecoregions and ecoregional reference sites is discussed in "Regionalization as a Tool for Managing Environmental Resources" (EPA, 1989b).

With the publication of these two landmark documents, several states began intensive work to refine the ecoregions and protocols to suit local conditions and needs. The state-of-the-art in this rapidly developing field continues to evolve into many more variations. The original authors of the documents strongly advocate customizing both ecoregions and the bioassessment protocols. The most important common element required in these efforts is the use of a scientifically defensible approach.

Consequently, before this protocol can be used and biological community assessment programs implemented, state specific analyses must be undertaken. This presents many unique challenges, requiring special expertise, adequate funding and several years. For example, state specific subecoregions must be delineated, appropriate ecoregion reference sites selected and sampled, community bioassessment sampling and evaluation methods modified, appropriate biological community metrics selected, and each of these must be verified.

These techniques offer the best means of accurately assessing cumulative impacts, especially of stormwater and other nonpoint sources of pollution. The Florida Bioassessment Program was designed to address many of the above issues by developing a refined bioassessment protocol for use in Florida streams. This tool can be used with traditional water column monitoring, toxicity testing, and sediment monitoring to document impairment from nonpoint sources of pollution and to determine the effectiveness of management programs (Livingston et. al., 1995).

For the following reasons, the stream benthic macroinvertebrate community was selected as the target indicator group for the Florida bioassessment initiative:

* they have limited migration patterns, thus well-suited for site-specific impacts,
* they integrate the effects of short-term environmental variations,
* degraded conditions can often be detected by an experienced biologist with only a cursory examination of the macroinvertebrate community,
* sampling is relatively easy and inexpensive,
* they are abundant in most streams, and,
* most states that routinely collect biosurvey data focus on macroinvertebrates.

Biological assessment involves an integrated analysis of functional and structural components of aquatic communities. Bioassessments are best used to detect aquatic life impairments and assess their relative severity. Once an impairment is detected, additional chemical and biological toxicity testing can identify the causative agent and its source. Both biological and chemical methods play critical roles in successful pollution control and environmental management programs. They are complementary, not mutually exclusive, approaches that enhance overall program effectiveness.

THE FLORIDA STORMWATER/NPS BIOASSESSMENT PROJECTS

In 1990, the Stormwater/Nonpoint Source Management Section of the Florida Department of Environmental Protection began a multi-year effort to refine and enhance current biological community assessment methods. This work consists of six primary phases: I. Define Management Objectives, II. Develop Assessment Techniques, III. Standardize Methods and Begin Sampling, IV. Analyze Data, and V. Develop Standard Reporting Format for Management. A sixth phase will be the development of revised state biocriteria which reflect the use of the new methodologies. This paper discusses program phases I through V.

Phase I - Management Objectives

Most water resource managers today realize the need to improve their water quality information used for decision making at all levels of management. The nonpoint source bioassessment program began at this most basic level - to clearly articulate how the development of improved biological assessment methods can meet the needs of agency management.

As a result of this coordination effort with management, two broad areas of need emerged. The first is to develop a bioassessment tool that provides a screening mechanism for identifying potential biological impairment. This particular approach is not intended to quantify the degree of impairment nor provide definitive data that would be used to establish cause-and-effect. This is called a *Tier 1* assessment. Tier 1 assessments include the use of multiple metrics but are more cursory evaluations incorporating the cost and time efficiencies needed to cover a large geographic area.

A second management objective is to develop an approach that will provide a consistent, well-documented quantitative evaluation of stream condition. The goal is to produce repeatable results which can serve as a basis for comparing sites over time (trend monitoring). The ability to discriminate the level of impairment in this approach is enhanced by performing taxonomic identifications to the lowest practical level. This provides information on population as well as community level effects. The methodology required for this more rigorous investigation are termed *Tier 2* methods.

In the following section, the two approaches, Tiers 1 and 2, are described in detail.

Phase II - Develop Assessment Techniques

Recognizing the need to clearly define the spatial framework for the development of both Tier 1 and Tier 2 bioassessment methods, the first task was to regionalize Florida's diverse ecological communities to the lowest level possible.

Component II-1: Delineate Ecoregions and Select Reference Sites

Spatial frameworks can profoundly influence the effectiveness of research, assessment, and management of many water resource problems, especially those caused by stormwater and nonpoint sources. Traditionally, we have relied on spatial frameworks based on political boundaries, watersheds, hydrologic units, or physiographic regions. However, these units do not correspond to patterns in vegetation, soils, land surface form, land use, climate, rainfall or other characteristics that control or reflect spatial variations in surface water quality or aquatic organisms.

Effective water quality management programs must recognize the significance of land/water interactions, nonpoint sources, and regional variations in attainable water quality. Water quality assessments need a regional framework to:

1. compare regional land and water patterns;
2. compare ecological and habitat similarities and differences;
3. establish realistic, achievable chemical and biological standards;
4. assess the effects of all pollution sources within a watershed, especially intermittent discharges;
5. predict the effectiveness of management practices;
6. prioritize assessment and management efforts;
7. locate monitoring and special study sites; and
8. extrapolate site-specific information to larger areas.

Omernik (1987) proposed using spatial frameworks based on ecological regions (ecoregions) to assess the health of aquatic systems. Ecoregions are areas of relative homogeneity in ecological systems and relationships between organisms and their environments. Ecoregions usually are defined by patterns of homogeneity in a combination of factors such as climate, physiography, geology, soils, vegetation and dominant land uses. These regions also define areas within which there are different patterns in human stresses on the environment and different patterns in the existing and attainable quality of environmental resources. Ecoregions reflect similarities in the type, quality and quantity of water resources and the factors affecting them. Therefore, regional patterns of environmental factors reflect regional patterns in surface water quality.

Omernik (1987) originally identified 76 ecoregions in the conterminous United States including three in Florida. These ecoregions were useful for stratifying streams in Arkansas, Nebraska, Ohio, Oregon, Washington, and Wisconsin. They were used to set water quality standards in Arkansas, lake management goals in Minnesota, and to develop biocriteria in Ohio. However, in many states, the resolution of the ecoregions was of insufficient detail leading to collaborative projects involving states, EPA regions and the EPA Environmental Research Lab-Corvallis to refine ecoregions and delineate subregions.

Delineating regions or subregions typically involves compiling and reviewing relevant materials, maps and data; outlining the regional characteristics; drafting the regional and subregional boundaries; digitizing the boundary lines, creating digital coverages, and producing maps; and revising after review by state managers and scientists. To delineate subregions in Florida, aerial and satellite images, maps, and other documents were obtained describing environmental characteristics including physiography, geology, soils, climate, land use, vegetation, wetlands, and various biological communities. Analysis of this information led to the definition of the following ecoregions and subregions in Florida (Griffin et. al., 1994):

1. The Southeastern Plains Ecoregion, with three subregions - Southern Pine Plains and Hills, Dougherty/Marianna Plains, and Tifton Upland/Tallahassee Hills
2. The Southern Coastal Plain Ecoregion, with six subregions - Gulf Coast Flatwoods, Southwestern Florida Flatwoods, Central Florida Ridges and Uplands, Eastern Florida Flatwoods, Okefenokee Swamps and Plains, and Sea Island Flatwoods.
3. The Southern Florida Coastal Plain Ecoregion, with four subregions - Everglades, Big Cypress, Miami Ridge and Atlantic Coastal Strip, and Southern Coast and Islands.

Once ecoregions and subregions are delineated and field verified, ecoregion reference sites must be selected. An essential component of the management framework, these sites allow us to evaluate the environmental health of a locale by comparing it to a known reference site - a key concept in Karr and Dudley's definition of biotic integrity, which compares site evaluations to the aquatic community of "natural habitats within a region". Ecoregion reference sites used in water resources management must have two essential components: they must represent the ecoregion, and have ecological conditions that can be reasonably attained given current background conditions.

Reference sites must be carefully selected because they will be used for two purposes: (1) Benchmark for establishing regional biocriteria; and, (2) Control sites to which test sites will be compared. The two main criteria for selecting reference sites are that they be minimally impaired and that they represent the region's natural biological community. The ideal reference site will have extensive, natural, riparian vegetation; a diversity of substrate materials; natural physical structures; a natural hydrograph; a representative and diverse abundance of naturally-occurring biological communities; and a minimum of known, human induced disturbances or discharges. General guidelines for selecting reference sites are given in EPA (1989b).

To select stream subecoregion reference sites in Florida, the following steps were taken:

1. Using GIS techniques, information about the general characteristics of each ecoregion and subregion was analyzed to better understand representative conditions. Information reviewed included topographic maps, land use and soil maps, county highway maps, vegetational coverage maps, Landsat imagery, and the 1988 and 1990 Florida Water Quality Assessment 305(b) reports.

2. A set of stream sites with surface watersheds that appear relatively undisturbed and entirely within a subecoregion was chosen in which candidate reference sites were located. The actual number of sites per watershed is a function of the apparent homogeneity or heterogeneity of the region, the size of the region, hydrologic characteristics, and the number of candidate sites available. Access is a major factor in selection of the final reference sites. The number of candidate sites per subregion varied, ranging from only eight in subregion 75C, the Central Florida

Ridges and Uplands where relatively few streams are found, to twenty sites in subregion 75D, the Eastern Florida Flatwoods. A list of the candidate sites was developed that included the subregion, site number, stream name and location, major basin, county, GIS map name, watershed area, and other information.

3. Department and water management district biologists reviewed the information for each candidate site and then conducted site visits. This ground reconnaissance allowed staff to get a sense of the usefulness of the subecoregions, the characteristics that comprise reference sites in each region, the range of characteristics and types of disturbances in each region, and how site characteristics and stream types vary between regions. Using this process, sites were dropped that were found unsuitable because of disturbances not apparent on aerials or maps or because of anomalous situations while additional sites were identified.

4. Aerial reconnaissance was conducted to identify disturbances not observable from the ground, to get a better sense for spatial patterns of disturbances and geographic characteristics in each region, and to photograph typical characteristics, site locations, or disturbances.

5. Over 100 subecoregional candidate reference sites originally were selected by EPA and FDEP biologists. A thorough review process for each site to determine its representativeness and an analysis of available staff hours to conduct bioassessments was performed, reducing the final number of reference sites to 83.

6. To characterize and validate the reference sites they were sampled both in summer and winter for several years. This also helped test for a possible seasonal effect.

The distribution of reference sites varies among DEP districts as well as among subecoregions. The number of sites in the districts range from three to 30 while the number of subecoregion sites varies from six to 13. This information is summarized below:

Subecoregion	Reference Sites
65F Southern Pine Plains and Hills	8
65G Dougherty/Marianna Plains	6
65H Tifton Upland/Tallahassee Hills	8
75A Gulf Coast Flatwoods	3
75B Southwestern Florida Flatwoods	12
75C Central Fla Ridges & Uplands	8
75D Eastern Fla Flatwoods	10
75E Okefenokee Swamp & Plains	8
75F Sea Island Flatwoods	9

It is important to remember that reference sites represent the least or minimally disturbed ecosystem conditions. All of them have some level of disturbance which is a moving target because of ongoing human activity and natural processes. Since levels of impact are relative on a regional basis, the characteristics of appropriate reference sites will be different in different ecoregions and subregions and for different waterbody and habitat types. It is desirable, therefore, to have a large number of reference sites for each region to help define the different types of streams, to characterize the natural variability within similar stream types, and to clarify the factors that characterize the best sites from factors present in the lower quality sites.

The second component in Phase II is the development of the Tier 1 and Tier 2 bioassessment protocols.

Component II-2:   Develop Multi-Metric Approaches for Tier 1 and Tier 2 Bioassessments

The biological attributes that are measured represent elements of the structure and function of the macroinvertebrate assemblage and are called *metrics*. A metric is defined as a characteristic of the biota that changes in come predictable way with increased human influence (Barbour et al. 1995). These metrics are specific measures of diversity, composition, and functional feeding group representation and include ecological information on pollution tolerance.

The purpose of using multiple metrics in assessing biological condition is to maximize the information available regarding the elements and processes of aquatic communities. Metrics allow the ecologist to use meaningful indicator attributes in assessing the status of communities in response to perturbation. The validity of an integrated assessment using multiple metrics is supported by the use of measurements of biological attributes firmly rooted in sound ecological principles (Karr et al. 1986; Fausch et al. 1990; Ferraro, et. al. 1989, Lyons 1992).

Tier II Bioassessment Protocol - The Stream Condition Index (SCI) for Florida: The information derived from selected biological metrics are aggregated into an index of biological condition called the *Stream Condition Index (SCI)* for Florida. The SCI is used primarily as an indicator of ecosystem health and to identify impairment with respect to the reference condition.

The development of appropriate metrics follows a determination of (1) taxa to be sampled, (2) the biological characteristics of reference conditions, and to a certain extent, (3) the anthropogenic influences being assessed. In many situations, multiple stressors impact ecological resources, and specific "cause-and-effect" assessments may be difficult. However, changes in individual metrics or suites of metrics in response to perturbation by certain stressors (or sets thereof) are important diagnostic assessment indicators. For this reason, use of a multimetric approach for evaluating nonpoint source effects upon the biota is a more powerful tool than traditional approaches to bioassessment.

The basic approach to developing metrics is modeled after EPA's technical guidance for biocriteria (Barbour et al., in review). Candidate metrics are selected based on knowledge of aquatic systems, flora and fauna, literature reviews, and historical data. Candidate metrics are evaluated for efficacy and validity for implementation into the bioassessment program. Less robust metrics, or those not well-founded in ecological principles, are excluded as a result of this research process. Metrics with little or no relationship to stressors are rejected.

Core metrics are those that provide useful information in discriminating between good or poor quality ecological conditions. It is important to understand the effects of various stressors on the behavior of specific metrics. Metrics that use the relative sensitivity of the monitored populations to specific pollutants, where these relationships are well-characterized, can be useful as a diagnostic tool.

Core metrics should be selected to represent diverse aspects of structure, composition, individual health, or processes of the aquatic biota. Together they form the foundation for a sound, integrated analysis of the biotic condition to judge the attainment of biological criteria. For a metric to be useful, it must have the following attributes: (1) relevant to the biological community under study and to the specified program objectives; (2) sensitive to stressors and provides a response that can be discriminated from natural variation; (3) environmentally-benign to measure in the aquatic environment; and (4) cost-effective to sample and to implement into water resource programs.

To select metrics for Florida streams, a two phase process was used. In the optimization phase the metrics are evaluated for their relevance and natural variability, while the calibration phase is used to determine their discriminatory power and sensitivity to perturbation. In the first phase, all potential metrics having relevance to Florida stream macroinvertebrate communities were identified, a total of 47. These metrics were classified into categories roughly corresponding to various elements and processes of the macroinvertebrate assemblage (Table 1). Categories used for this metric classification corresponded to the following:

A. Richness measures, which signify the relative variety or diversity of the aquatic assemblage.
B. Composition measures, which provide information on the make-up of the assemblage and the relative abundance of particular taxa to the total community.
C. Tolerance measures, which relate to the relative sensitivity or tolerance of the assemblage and component populations to various types of perturbation.
D. Trophic measures, which are surrogates of more complicated processes, such as biotic trophic interaction, production and food source availability. Trophic metrics primarily are related to functional feed group designation and density, both

difficult to evaluate. Therefore, these metrics are in an evolutionary status around the country and will continue undergoing refinement.

The SCI for Florida was developed by aggregating the metrics that proved responsive to independent measures of impacts. Aggregation simplifies management and decision-making so that a single index value is used to determine whether action is needed. The exact nature of the action needed (e.g., restoration, mitigation, enforcement) is not determined by the index value, but by analysis of the component metrics.

The approach used to define the index for Florida was to develop expectations for the values of each of the metrics from the reference data set, and to score metrics according to whether they are within the range of reference expectations. Metrics within the range receive a high score, those outside receive a low score. The index value is then the sum of the metric scores. The index is further normalized to reference condition, such that the distribution of index values in the reference sites forms the expectations for the region.

In an assessment, streams can be judged for impairment based on the summed index value. If the index value is below a criterion, then the stream is judged impaired. The index value criterion is based on the index value distribution in reference streams. For example, the 25th percentile (lower quartile) of reference expectations is commonly used. Reference sites had been carefully selected to be representative of least impacted conditions in each ecoregion, and investigators involved in site selection and sampling were confident that the reference sites represented best available conditions in Florida streams. Therefore, the lower quartile of each metric distribution in reference sites was selected as the criterion for the minimum value of the metric representative of reference conditions. Thus, any metric value above the lower quartile of the reference distribution received the highest possible score. Using this rationale, scoring criteria were developed which are a modification of the methodology of Karr et al.(1986; Karr 1991).

Tier I Bioassessment Protocol - Bioreconnaissance (BioRecon): The purpose of the BioRecon protocol is to provide a rapid bioassessment technique to maximize the use of limited technical resources, minimize costs of assessing the condition of streams, increase the geographic coverage of stream assessments, and provide for a fast response time to management. This approach is not intended to replace the more rigorous Stream Condition Index (SCI). The two techniques simply correspond to different levels of resolution and management concerns. BioRecon can be used independently of the SCI as a screening tool, or as a precursor to it to identify sites that may need more intensive evaluation.

The two major unique components of BioRecon are the Water Survey and the Biological Survey, as depicted in Figure 1.

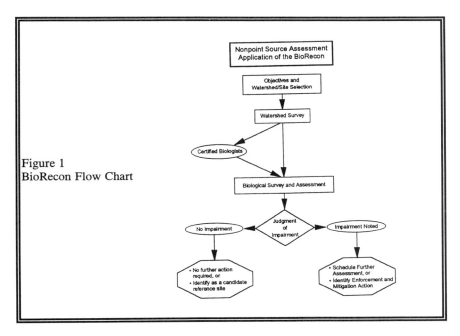

Figure 1
BioRecon Flow Chart

A BioRecon is initiated with a planning phase called the Objectives and Watershed/Site Selection. In this stage, specific objectives are established for the BioRecon and the target watershed and actual sampling site(s) are determined. In this phase, activities may be conducted entirely within the office setting if resources allow.

Establishing objectives for a BioRecon requires close coordination between technical and management perspectives. The information which will come from a BioRecon must immediately be transferred to the program manager or technical manager for further action. The following is a hypothetical list of BioRecon objectives:

* (Technical) - To close a data gap identified in the 305(b) Report.
* (Management/Technical) - To investigate the effects of a target land use activity.
* (Technical) - To verify qualitative assessments of past nonpoint source problems.
* (Management) - To survey a broad geographic area for indications of problems.
* (Technical) - To investigate the overall effectiveness of management practices.

As can be seen from this partial listing, objectives may easily fall into both management and technical categories. This maximizes the information from a BioRecon and enhances the cost-effectiveness of this assessment strategy even further.

Depending on the specific objective(s) that have already been established for the BioRecon, a watershed and/or number of sites may already be known. If they are known, the Watershed Survey can be implemented.

The Watershed Survey provides: (1) an initial evaluation of the conditions and suspected problems, if any, (2) a screening of sites which may require additional investigation, (3) logistical planning of the extent and nature of resource investment in an area, and, (4) development of a monitoring and assessment plan for management. It is likely that the data comprising a Watershed Survey are available for most stream sites from routine monitoring and other programs. A site visit may not be necessary to complete the Watershed Survey.

The Biological Survey involves the actual biological collection and evaluation. The staff time needed to conduct the Biological Survey and analyze the data is substantially less than for the more rigorous SCI. It typically requires 17 hours to collect, sort, identify Tier II samples while only 3.5 hours are required for a BioRecon. The field identifications done for BioRecon take only 1.5 hours compared to the 15 hours of lab work for SCI samples. Because BioRecon is done quickly and without comprehensive vouchering and documentation, it is extremely important that an experienced biologist perform the Biological Survey portion.

The following general strategy is used for the Biological Survey:

* identify major habitats for sampling.
* collect macroinvertebrates using a dip net (4 sweeps composited).
* identify organisms and calculate three metrics: Florida Index, EPT, and total taxa.
* determine final site rating.

If two or three of the metrics pass predetermined threshold values, the site is considered *healthy*. If less than two pass threshold, the site is listed as *potentially impaired*. Pilot studies are underway at present to evaluate the BioRecon metric thresholds and whether or not laboratory identification will be required for the organisms collected.

Component II-3: Habitat Evaluation

One of the most important aspects of the community bioassessment procedure for either Tier I or Tier II assessment types is the evaluation of stream habitat, which includes physical characteristics and water quality. Since conditions in the watershed determine conditions in the stream, habitat quality is dependent on land use, channel and riparian features as well as instream factors such as substrate types and velocity.

In 1991, the FDEP developed its first habitat evaluation methods (Frydenborg, 1991) based on EPA's Rapid Bioassessment Protocols document (EPA, 1989b). As part of the NPS bioassesment program these initial habitat evaluation methods have been refined and they continue to undergoe assessment and evolution. A standardized habitat assessment form has been developed to expedite the process and to increase consistency (Figure 2). The components of habitat evaluation are: (1) physical/ chemical characterization, and, (2) habitat assessment.

Physical/chemical characterization includes determining predominant surrounding land uses; identifying local watershed erosion, nonpoint, and point source pollution sources; estimating stream depth, width, high water mark, temperature, and velocity; noting stream alterations such as impoundment or channelization; estimating canopy cover; and evaluating sediment/substrate. Water quality parameters measured include pH, dissolved oxygen, conductivity, and secchi disk depth; and, noting water clarity, color, odors and surface oils, the measurement and observation of land use, riparian zone conditions, channel and substrate features and water quality.

Habitat assessment includes evaluating water velocity, substrate and cover, channel conditions, bank stability and riparian zone vegetation based on their capacity to support a stable, well balanced benthic community.

The overall asssessment of ecosystem condition will first focus on the habitat evaluation and then the biological components. Habitat, then, is the principal determinant of biological potential, sets the context for interpreting the survey and cna be used as a general predictor of biological condition. Therefore, the role of the Habitat Evaluation cannot be over-emphasized.

PHASE III. DEVELOP STANDARDIZED BIOLOGICAL ASSESSMENT METHODS

Standarized, cost-effective methods were developed to collect and process benthic macroinvertebrates from Florida Streams for Tier I (BioRecon) and Tier II (SCI) assessment approaches and for habitat and physical/chemical evaluations which ccompany the SCI and BioRecon. Standard operating procedures for all of these program components are contained in FDEP's Standard Operating Procedures for Biological Assessment, April 1996.

Using DEP training funds, DEP's biologists attend quarterly "Biocriteria Committee Meetings" where they participate in workshops, discussions, and field exercises to learn about the bioassessment protocols. Beginning in the summer of 1992, DEP biologists from the Tallahassee and district offices conducted bioassessments, following the procedures set forth in the SOP, at all of the candidate reference sites. Sampling at these sites, and at additional reference sites, has continued on a summer-winter sampling schedule since then.

STATE OF FLORIDA
DEPARTMENT OF ENVIRONMENTAL PROTECTION
FRESHWATER BENTHIC HABITAT ASSESSMENT FIELD DATA SHEET   (4-22-96)

SUBMITTING AGENCY CODE:_____   STORET STATION NUMBER: DATE (M/D/Y):   RECEIVING BODY OF WATER:
SUBMITTING AGENCY NAME:_____

REMARKS: COUNTY: LOCATION: FIELD ID/NAME:

Habitat Parameter	Optimal	Suboptimal	Marginal	Poor
Substrate Types & Availability	Greater than 30% snags, logs, tree roots, aquatic vegetation, leaf packs (partially decayed), undercut banks, rock, or other stable habitat.	16% to 30% snags, logs, tree roots, aquatic vegetation, leaf packs, etc. Adequate habitat. Some substrates may be new fall (fresh leaves or snags).	5% to 15% snags, logs, tree roots, aquatic vegetation, leaf packs, etc. Less than desirable habitat, frequently disturbed or removed.	Less than 5% snags, logs, tree roots, leaf packs, etc. Lack of habitat is obvious, substrates unstable or smothered.
	20 19 18 17 16	15 14 13 12 11	10 9 8 7 6	5 4 3 2 1
Water Velocity	Max. observed at typical transect: >0.25 m/sec. but < 1 m/sec	Max. observed at typical transect: 0.1 to 0.25 m/sec	Max. observed at typical transect: 0.05 to 0.1 m/sec	Max. observed at typical transect <0.05 m/sec, or spate occurring; > 1 m/sec
	20 19 18 17 16	15 14 13 12 11	10 9 8 7 6	5 4 3 2 1
Artificial Channelization	No artificial channelization or dredging. Stream with normal, sinuous pattern	May have been channelized in the past (>20 yrs), but mostly recovered, fairly good sinuous pattern	Channelized, somewhat recovered, but > 80% of area affected	Artificially channelized, box-cut banks, straight, instream habitat highly altered
	20 19 18 17 16	15 14 13 12 11	10 9 8 7 6	5 4 3 2 1
Habitat Smothering	Less than 20% of habitats affected by sand or silt accumulation	20%-50% of habitats affected by sand or silt accumulation	Smothering of 50%-80% of habitats with sand or silt, pools shallow, frequent sediment movement	Smothering of >80% of habitats with sand and silt, a severe problem, pools absent
	20 19 18 17 16	15 14 13 12 11	10 9 8 7 6	5 4 3 2 1
Bank Stability	Stable. No evidence of erosion or bank failure. Little potential for future problems.	Moderately stable. Infrequent or small areas of erosion, mostly healed over.	Moderately unstable. Moderate areas of erosion, high erosion potential during floods.	Unstable. Many (60%-80%) raw, eroded areas. Obvious bank sloughing.
	20 19 18 17 16	15 14 13 12 11	10 9 8 7 6	5 4 3 2 1
Riparian Buffer Zone Width	Width of native vegetation (least buffered side) greater than 18 m	Width of native vegetation (least buffered side) 12 m to 18 m	Width of native vegetation 6 to 12 m, human activities still close to system	Less than 6 m of native buffer zone due to intensive human activities
	20 19 18 17 16	15 14 13 12 11	10 9 8 7 6	5 4 3 2 1
Riparian Zone Vegetation Quality	Over 80% of riparian surfaces consist of native plants, including trees, understory shrubs, or non-woody macrophytes. Normal, expected plant community for given sunlight & habitat conditions.	50% to 80% of riparian zone is vegetated, and/or one class of plants normally expected for the sunlight & habitat conditions is not represented. Some disruption in community evident.	25% to 50% of riparian zone is vegetated, and/or one or two expected classes of plants are not represented. Patches of bare soil or closely cropped vegetation, disruption obvious.	Less than 25% of streambank surfaces are vegetated and/or poor plant community (e.g. grass monoculture or exotics present. Vegetation removed to stubble height of 2 inches or less.
	20 19 18 17 16	15 14 13 12 11	10 9 8 7 6	5 4 3 2 1

Add 5 points if cross-sectional area of flow is estimated   Comments
to be > one square meter during periods of normal flow.

TOTAL SCORE

ANALYSIS DATE: ANALYST: SIGNATURE:

PHASE IV - ANALYZE DATA

Evaluation of Metrics: From the biological data collected at the candidate reference sites in Summer 1992, a total of 47 metrics initially were calculated and entered into the data base. These parameters were analyzed using a number of statistical methods including covariate and autocorrelation analysis, analysis of variance, and cluster analysis. Two key graphical displays, scatter plots of physicochemical variables versus biological metrics and box-and-whisker plots, were relied on for evaluating site classifications and discriminatory power. All of the 47 metrics fit the condition of biological relevance since they represent elements and processes of the macroinvertebrate assemblage and are thought to change in a predictable fashion in response to perturbation.

By evaluating the inherent variability within the reference site database, 35 candidate metrics were identified. A highly variable metric would not be useful because the discriminatory power would be diminished. Conversely, a metric that has a narrow variance within a maximum or optimal range for reference conditions would be a useful metric. Box-and-whisker plots of the sites classified by subecoregions were used to depict the natural variability of each metric.

With the inclusion of additional data in subsequent years, seven of the 35 candidate metrics appear to be appropriate as final metrics (Table 1). These are the metrics that illustrate a relatively tight range of values among the various subecoregions and are at the high end of their range of values for the reference sites. Three of the metrics are from the richness measures - the Number of Total Taxa, EPT Index, and the Number of Chironomidae Taxa. The Shannon-Wiener Index, Percent Diptera, and Percent Dominant Taxon are from the composition measures. The only metric from the tolerance measures is the Florida Index. Likewise, Percent Filterers is the only metric from the trophic measures. The seven core metrics are then summed to determine the Stream Condition Index allowing evalution of the biological condition of the sampling sites.

Metric Classification: Once *candidate* metrics were identified, analyses was performed to develop a classification system for Florida streams that would aggregate the streams into a small number of classes that could be managed and monitored with similar expectations. The classes account for significant variation in the biological metric data with classes that do not contribute to the variation explained separated from those that do.

The consistent pattern that emerged from this analysis was that biological metrics of Florida streams tend to aggregate in three *bioregions*: the subecoregions of the Florida panhandle (65f, g, h, and most of 75a), the subecoregions of peninsular Florida (75b, c, and d and part of a), and the two subecoregions in the northeast of Florida (75e and f). A *bioregion* is an aggregation of multiple subecoregions where macroinvertebrates

Table 1 Candidate and Selected Core Metrics

Metric Category	Rejected Candidate Metrics	Final Core Metrics	Definition Summary	Response to Increasing Biological Condition
Richness Measures	% Ephemeroptera Taxa # Coleoptera Taxa # Orthocladiinae Taxa # Tanytarsini Taxa # Crustacean/Mollusc Taxa	% Total Taxa	Measures overall variety of macroinvertebrate assemblage	↑
		EPT Index	Sum of no. taxa in 3 insect orders: Ephemeroptera, Plecoptera, Trichoptera	↑
		# Chironomidae Taxa	Sum of no. larval midge taxa	↑
Compositio n Measures	% Oligochaeta, % Odonata, % Ephemeroptera, %Isopoda, % Trichoptera, % Plecoptera, % Coleoptera, % Gastropoda, % Pelecypods, % Amphipoda, % Orthoclads to Chironomids, % Tanytarsini to Chironomids, %Crustacean/Mollusc, Shannon-Wiener Index	% Dominant Taxon	Measures dominance of single most abundant taxon	↑
		% Diptera	Relative abundance of dipterans	↓
				↓
Tolerance Measures	# Class 1 and Class 2 Taxa % Class 1 and Class 2 Hilsenhoff Biotic Index	Florida Index	Uses abundance and pollution tolerance values for some invertebrates, heavily weighted to arthropods	↑
Trophic Measures	# Scraper/Piercer Taxa % Scrapers, %Predators, Density % Collector-Gatherers % Shredders	% Collector-Filterers	Relative abundance of this functional feeding group	↓

show distinct patterns of physical, chemical and biological characteristics. Figure 3
depicts the three bioregions in Florida.

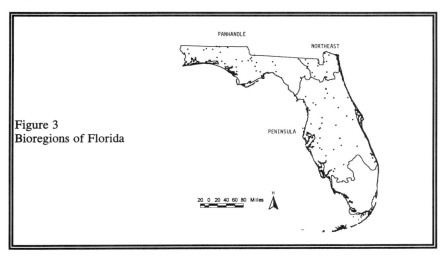

Figure 3
Bioregions of Florida

Each of the three bioregions is also a contiguous geographic region (panhandle,
peninsula, and northeast), with observable physical and chemical differences between
the three regions. The observed regional biological differences are partly related to
acid-base chemistry of the regions. Peninsula Florida is dominated by limestone
bedrock, and surface and ground waters are typically well-buffered. Water in the
panhandle and northeastern Florida are more often poorly-buffered or acidic.
Relationships between ambient pH and several metric values are seen and can be
explained by strong correlations between certain faunal groups and water with a certain
pH.

Metric Calibration: The candidate metrics were then calibrated through an evaluation
of both reference and impaired sites. The ability of a metric to discriminate between
a reference and a known impaired site is necessary for the calibration of the metric for
monitoring and assessment purposes.

Data from Florida DEP's Point Source Program were used to evaluate the ability to
discriminate between "good" and "bad" biological condition. Sites in this point source
program were either upstream or downstream of known point source dischargers.
Many of the upstream control sites in the point source program were not necessarily
good reference sites, because of habitat degradation or some other reason. Two
considerations in using the point source data are that although the season of collection
was the same as that of the reference sites, some point source data were collected the
previous year; and the methods used in the point source program were similar, but not

identical in all respects, to those employed in the nonpoint source program. However, these considerations did not prevent integration of the data from the two programs to evaluate discriminatory ability.

A corollary study was performed to compare biological sampling methods to determine whether (1) Hester-Dendy artificial substrate samplers provide data representative of stream biological status, and (2) whether dip nets or Hester-Dendy substrates are more powerful for detecting biological impairment. Results from statistical analysis of metrics suggest that *dip net data are as good or somewhat better than Hester-Dendy data in distinguishing biological impairment in Florida streams.* These results, together with the fact that use of a Hester-Dendy is more labor intensive and costly, support the use of dip net collections as the primary macroinvertebrate collecting gear for the NPS monitoring program in Florida.

PHASE V - DEVELOP STANDARDIZED REPORTING FORMAT

FDEP has begun the development of a standardized, statewide reporting format called "EcoSummary" to present the findings from an SCI or BioRecon bioassessment. The EcoSummary report is designed specifically for a management-level audience to convey information quickly, accurately and to the point.

FDEP's new Geographic Information System (GIS) will play a major role in producing the reports. Biologists will receive training on GIS software to compile a variety of report graphics ranging from site location to field and lab analytical results. The graphics together with explanatory text will be assembled in an automated manner for each report to provide quick turn-around time for management. In addition, the EcoSummary will include site photographs and will be written with a simple, easy to understand style.

Although the EcoSummary is a brief report which comes at the end of an assessment event, it is seen as an extremely important element of the bioassessment program. Efficient, responsive information management which is linked to program goals and objectives is the key to any successful monitoring effort.

DISCUSSION AND CONCLUSIONS

Adding biological community assessments to traditional water quality monitoring and evaluation approaches greatly enhance the ability of these tools to ascertain the cumulative environmental effects of point and nonpoint source pollution discharges. The importance of this biological assessment framework in accurately assessing the environmental health of surface waters is seen by comparing Ohio's use attainment conclusions. The Ohio Environmental Protection Agency incorporated biocriteria into its water quality standards regulations in recent years (Yoder, 1989). These biocriteria are based on a system of tiered aquatic life uses representing five classes. These include coldwater habitat, warmwater habitat, exceptional warmwater habitat, modified warmwater habitat and limited resource waters. These designations have been qualitatively defined in ecological terms, and chemical criteria, either quantitative or narrative, have been established for each. Using both the water chemistry and bioassessment data, conclusions about the attainment of beneficial uses in Ohio water bodies include (EPA, 1989a):

- Based on chemical data, 52% of the segments fully attained aquatic life uses;
- Based on biosurvey data, only 23% achieved full attainment;
- The two types of assessment agreed on full attainment in 17% of the cases with overall agreement on 46% of the cases;
- In 35% of the cases, chemical data indicated full attainment but biosurvey data indicated partial or non-attainment. In nearly half of these cases, impairments were due to habitat or flow modifications, or siltation.

Unfortunately, as can be seen from the activities described in this paper, conducting the preliminary technical analyses that are essential to establish the scientific rationale for these assessment tools is not easy, quick, nor inexpensive. Ultimately, the goal of these efforts is to develop quantitative biocriteria which can be used to more effectively assess, manage and evaluate stormwater/NPS pollution sources and management efforts. Metrics reflecting community characteristics may be considered appropriate in biocriteria programs if their relevance can be demonstrated, response range is verified and documented, and the potential for application in water resource programs exists.

However, before the FDEP can adopt biocriteria, lots more work must be done to further refine, calibrate and evaluate the biometrics. Frequent evaluation of metrics and indices is an essential feature of the use of biocriteria. However, once established, the multimetric approach for assessing biological condition offers the following attributes: (1) relies on information about several populations or species assemblages, rather than just target species; (2) relates to a community-level potential or expectation based on a reference condition; (3) uses multiple metrics to function as surrogate measures of more complicated elements and processes; and (4) incorporates ecological principles that enable an interpretation of exposure/response relations.

Sediment assessment (Livingston et. al., 1995), together with watershed characterization and mapping of pollution sources, can be used to screen watersheds and sub-basins to determine potential "hot spots". Bioassessment and water chemistry sampling can then be done to assess the actual health of the aquatic system in these locations. The initial focus of a bioassessment should be on habitat quality. Based on a regional reference, the habitat at an impacted site may be equal to or less than the desired quality for that particular system. If the habitat at the impact site and reference are equal, then a direct comparison of biological condition can be made. If the habitat at the impact site is lower in quality than the reference, the habitat potential should be evaluated as a first step.

A site-specific control may be more appropriate than a regional reference for an assessment of an impact site. If so, then care must be taken in selecting an appropriate site-specific reference site to assure that its habitat and sediment characteristics are representative for the area. Once a determination of the appropriate reference site type is made, possible outcomes of the bioassessment are: (1) no biological effects; (2) effects due to habitat degradation; (3) effects due to sediment or water quality; or (4) effects due to a combination of sediment, water quality and habitat degradation.

The projects described in this paper greatly contributed to the development, refinement, calibration, and testing of several essential biological community assessment tools in Florida. The Department is anxious to begin using these tools to better assess the effects of intermittent pollutant sources, evaluate the effectiveness of BMPs and management programs, prioritize watersheds and subbasins for management activities, and, in conjunction with the water management districts, to develop and implement the stormwater pollutant load reduction goals (PLRGs) required by State Water Policy and being established through the state's Surface Water Improvement and Management (SWIM) Program. The biological community assessment methods, in conjunction with the recently started effort by the Department to receive delegation of EPA's NPDES permitting program and a future initiative for basin wide monitoring, permitting, and compliance, provide the technical and institutional tools needed to cost-effectively reduce stormwater/NPS pollutant loadings on a watershed basis.

However, before these initiatives can be fully implemented, additional analysis and evaluation needs to be done to refine the assessment methods. For the NPS bioassessment program, issues still to be resolved include (FDEP, 1994):

1. Validate the division of subecoregion 75a into both panhandle and peninsular bioregions. This subecoregion is under-represented by reference sites. An evaluation of this area could be improved with the sampling of additional sites.

2. Evaluate the Hilsenhoff Biotic Index (HBI) as a sensitivity measure. The HBI was evaluated as part of the present data analyses. However, too many assumptions had to be made regarding the tolerance assignment to the various taxa. Logically, the HBI should be more meaningful than similar measures in assessing biological condition, because the index incorporates information from the whole assemblage. The proper assignment of tolerance scores should be addressed and the efficacy of the HBI metric re-evaluated. Similarly, tolerance assignments should address a broader range of pollutants.

3. Conduct habitat evaluations at reference and habitat-limited sites to determine the resolving power of the habitat parameters along a gradient of impact. This kind of analysis has been done for the biometrics using macroinvertebrate data from point source impact studies. Sites with nonpoint source habitat limitations e.g., erosion, deforestation. etc. should be included and the evaluation will require site-

specific chemical analyses to separate chemical and physical limitations to the biota.

4. Develop a software program to handle the storage, sorting, and analysis of the biological and habitat data. For the department to successfully implement the bioassessment program, a "user-friendly" program should be developed. A contractor has been hired to review the current Florida DEP computer programs for handling specific biological data sets and to modify them to include the present suite of metrics and assessment approach.

ACKNOWLEDGEMENTS

The authors gratefully acknowledge the creativity, dedication and hard work on these projects by many individuals at the FDEP, especially the biologists and by our contractors, especially Mike Barbour, James Omernik, Glenn Griffith and Mike Bastian. The projects described in this paper were funded in whole or in part by grants from the U.S. EPA under Section 104(b)(3) and Section 319 of the Federal Clean Water Act.

INFORMATION SOURCES

Barbour, M.T., J.B. Stribling, and J.R. Karr, (in press). The Multimetric Approach for Establishing Biocriteria and Measuring Biological Condition. In W. Davis, T. Simon (eds.), Biological Assessment and Criteria: Tools for Water Resource Planning and Decision Making. Lewis Publishers.

Barbour, M.T., J.B. Stribling, and J.R. Karr, (in review). Biological Criteria: Technical Guidance for Streams. U.S. Environmental Protection Agency, Office of Science and Technology, Health and Ecological Criteria Division, Washington, DC.

EPA. 1989a. Rapid Bioassessment Protocols for Use in Streams and Rivers: Benthic Macroinvertebrates and Fish. EPA/444/4-89-001. Washington D.C.

EPA. 1989b. Regionalization as a Tool for Managing Environmental Resources. EPA/600/3-89-060. Corvallis, OR.

Fausch, K.D., J. Lyons, J.R. Karr, P.L. Angermeier. 1990. Fish Communities as Indicators of Environmental Degradation. Am. Soc. Symp. 8:123-44.

FDEP. 1994. Bioassessment for the Nonpoint Source Program. Final report submitted by EA Engineering, Science, and Technology, and Tetra Tech, Inc. Tallahassee, Fl.

FDEP, 1996. Standard Operating Procedures for Biological Assessment.

Ferraro, S.P., F.A. Cole, W.A. DeBen, and R.C. Swartz. 1989. Power-cost Efficiency of Eight Macrobenthic Sampling Schemes in Puget Sound, Washington. Can. J. Fish. Aq. Sci. 46: 2157-2165.

Frydenborg, R. 1991. Impact Bioassessment Investigations Draft Document. Fla. Dept. Environ. Reg. Tech. Ser. Tallahassee, Fl.

Griffith, G, J. M. Omernik, C. Rohmand, S. Pierson. 1994. Florida Regionalization Project. USEPA Environmental Research Laboratory. Corvallis, Oregon. Final Report Prepared for Fla. Dept. Environ. Protection. Tallahassee, Fl.

Karr, J. and D. Dudley. 1981. Ecological perspective on water quality goals. Environ. Mgmt. 5(1): 55-68.

Karr, J.R., K.D. Fausch, P.L. Angermeier, P.R. Yant, and I.J. Schlosser. 1986. Assessing Biological Integrity in Running Waters: A Method and Its Rationale. Special Publication 5. Illinois Natural History Survey, Urbana, Illinois.

Karr, J.R. 1991. Biological integrity: A Long-neglected Aspect of Water Resource Management. Ecological Applications 1:66-84.

Livingston, E.H. 1993. Local Government Model Stormwater Management Program. Stormwater/NPS Management Section, FDEP. Tallahassee, Fl.

Livingston, E.H.   1995. The Evolution of Florida's Stormwater/Watershed Management Program. In Proceedings of the National Conference on Urban Runoff Management: Enhancing Urban Watershed Management at the Local, County, and State Levels. EPA 625/R-95/003. Cincinatti, Ohio.

Livingston, E.H., E. McCarron, T. Seal, and G. Sloane. 1995. Use of Sediment and Biological Monitoring.   In Stormwater NPDES Related Monitoring Needs, Proceedings of an Engineering Foundation Conference. ASCE, New York, NY.

Lyons, J. 1992. Using the Index of Biotic Integrity (IBI) to Measure Environmental Quality in Warmwater Streams of Wisconsin. General Technical Report, NC-149. U.S. Department of Agriculture, Forest Service. St. Paul, Mn.

Omernik, J. 1987. Ecoregions of the Conterminous United States. Annals of the Assoc. Amer. Geogr. 77: 118-125.

Yoder, C. 1989. The Development and Use of Biological Criteria for Ohio Surface Waters. Pp 139-146. In: Water Quality Standards for the 21st Century.

Bioassessment for Intermittent Central Texas Streams

Robert Hansen[1]

Abstract

Since June 1993 the City of Austin, TX, Drainage Utility Department has developed and implemented bioassessment techniques on two local intermittent streams with varying degrees of urbanization and development. The two streams, Onion Creek and Barton Creek, provide more than sixty percent of the total recharge to the Barton Springs segment of the Edwards Aquifer. These creeks become losing streams over the Edwards Aquifer recharge area and cease to flow during prolonged periods of decreased rainfall. The City is currently evaluating the effectiveness of biological monitoring techniques in relation to water chemistry, land use, temperature, and flow velocities. Evaluation of local creeks using the bioassessment techniques developed for perennial streams indicates that these methods fail to accurately assess the degree of impact on the local lotic community structure due to biological seasonality and intermittent flow. Current assessment techniques need to be modified or new techniques developed for intermittent streams that will enable investigators to distinguish between aquatic biota impacts due to natural, environmental variations and impairments related to anthropogenic activities in the local watershed.

Introduction

The three year study of two nonpoint source polluted creeks in the Central Texas area near Austin, TX is funded by a grant from the Texas Clean Rivers Act

[1] Environmental Quality Specialist, City of Austin, TX, Drainage Utility Department, 206 East 9th Street, Austin, TX, 78701

(Senate Bill 818). The grant is provided by the Texas Natural Resource Conservation Commission (TNRCC) and administered through the Lower Colorado River Authority (LCRA). The study will be completed in August 1996. Onion and Barton creeks are the two largest contributors to the Barton Springs segment of the Edwards Aquifer which discharges an average of 32 million gallons a day into the Colorado River near downtown Austin, TX (USGS, 1986). Barton Springs contributes between one and eighty percent of the flow in the Colorado River through downtown Austin depending on rainfall and LCRA release rates from upstream dams (USGS, 1986). This segment of the aquifer is the sole source of drinking water for approximately 35,000 Central Texas residents. Currently these two contributing creeks have varying degrees of development and impairment within their catchments. This potential for variation in the level of impact due to nonpoint source pollution provides for an excellent in situ study and the evaluation of various bioassessment techniques for the assessment of impairment to water quality and biological integrity.

During the study design phase of the project, City researchers and staff reviewed existing methodologies for the analysis and assessment of water quality, habitat, and biological community structure. Bioassessment techniques for benthic macroinvertebrates, diatoms, and fishes were reviewed. Even though many of these indices were developed for perennial streams, their wide usage and acceptance make it important to evaluate their usefulness in assessing the impacts of nonpoint source pollution in intermittent streams. Due to the intermittent nature of Barton and Onion creeks and the lack of perennial fish communities except in the deepest pool areas of the stream channel, methods to assess the integrity of the fish communities (e.g. the Index of Biotic Integrity, Karr, et.al. 1986) were not included in the study. On a quarterly basis, water quality samples, benthic macroinvertebrates, diatoms, and chlorophyll a samples are collected at seventeen selected sites along the two creeks. Additionally, transects to quantify percent algae cover and physical habitat assessments are performed during each sampling event. Following laboratory analysis of the samples, the resulting data are evaluated using various multi-parameter indices, and univariate and multivariate statistical analysis. Analytical results indicate that the major factor influencing the benthic macroinvertebrate community is physical habitat disturbances such as flow and temperature. Diatom communities respond to fluctuations in physical habitat disturbances (flow and temperature) and variations in nutrient levels while chlorophyll a responses vary with nutrient levels.

Central Texas Plateau Ecoregion Streams

The Central Texas Plateau ecoregion lies west of a line along Interstate Highway 35 approximately from McLennan to Bexar counties (Waco to San

Antonio). This region is underlain by the Edwards Plateau and the dominant land surface forms are tableland with reliefs, plains, and open high hills. Natural vegetation includes juniper and oak savanna, mesquite and oak savanna, and bluestem and non-native grass plains. Most of the natural vegetation is supported by a thin layer of reddish-brown, gravelly sandy loam prairie soils. Much of the region is open woodland, forest, or native grassland that has been grazed by cattle or sheep. Creeks and streams in this ecoregion tend to exhibit clear upland flow with alternating runs and riffles and intermittent glides and pools. Riffle and run areas with clear flow and appropriate cobble substrate provide excellent habitat for algae, periphyton and benthic macroinvertebrate communities.

Base flow in many of these Central Texas Hill Country streams is sustained by spring and artesian discharges where the Edwards Limestone is overlain by the low permeability Del Rio clay. In areas of Edwards Limestone outcrops (e.g. the Barton Springs Recharge Zone) these streams can become losing streams where they intersect the high permeability Edwards karst limestone. During periods of extended low rainfall, many of these streams can become intermittent resulting in the dewatering of riffle and run areas. Under moderate to extreme drought conditions such as during the first and third years of the study, flow in the main stream channel is reduced to intermittent, isolated standing pools. Subsequently, many of the study sites along with a significant length of the stream mainstem have experienced periods of dewatering during the study period. Only the largest, deepest isolated pools are able to sustain fish populations during these extended periods of low rainfall.

In contrast, during storm events, the stream channels can be subjected to high velocity flow rates with intense scouring of the channel substrate and erosional banks. These high flow regimes can persist for extended periods of time during extremely wet years (e.g. 1991 - 1992) resulting in a widening of the main stream channel and downstream deposition of gravel and sediment. Additionally, these storm events can introduce significantly increased loading of nutrients into these streams which naturally exhibit low levels of nutrients during base flow conditions in rural and undeveloped areas. Studies conducted by the Texas Water Commission on Barton Creek in 1985 and the James River (perennial reference stream for the ecoregion) in 1987 reported nitrate-nitrite levels of less than 0.01 mg/L to 1.21 mg/L. Orthophosphorous levels varied from less than 0.01 mg/L to 0.02 mg/L (TWC, 1989). During storm events, the maximum values for these nutrients reported by the City of Austin's Environmental and Conservation Services Department are 2.48 mg/L for nitrate-nitrite and 0.90 mg/L for orthophosphorous (COA, 1995).

As noted above, low order, unimpacted streams in the Central Texas Hill Country exhibit low level concentrations of nutrients during base flow conditions.

These reduced nutrient levels are an important factor in the level of primary productivity and the abundance and diversity of benthic macroinvertebrates in these streams. However, when limiting nutrient concentrations increase and optimal growth conditions exist, naturally low levels of algae growth can increase dramatically and produce nuisance levels of algae cover, primarily composed of filamentous green algae of the orders Cladophorales and Zygnematales. These algae blooms, dominated by <u>Cladophora</u>, <u>Rhizoclonium</u>, and <u>Spirogyra</u> tend to be episodic and can result in deleterious impacts on the physical, chemical and biological integrity of the stream. Such unpredictable cycles of spates and dewatering in these Central Texas streams appear to be the driving dynamic force in shaping the structure of the aquatic biological communities.

<u>Study Streams</u>

Both Barton and Onion creeks have their headwaters in the Central Texas Plateau ecoregion west of Austin, TX. The waters of Barton Creek begin in Hays County and traverse a total channel length of 83 kilometers. The creek drains a watershed area of approximately 310 square kilometers and contributes approximately 28 percent of the total recharge for Barton Springs (USGS, 1986). Land use in the upstream segments of the stream remain rural with low intensity livestock grazing of improved and unimproved pastureland constituting the dominant land use activity. Land use in the middle and downstream segments of the stream varies from rural to high density commercial and residential development. New commercial and residential developments and major highway construction projects are planned or are currently under construction over the environmentally sensitive recharge zone of the aquifer. In the Barton Creek waterhsed eight study sites were selected. All of the study sites lie upstream of the aquifer recharge zone and land uses in the watersheds include rural livestock grazing, low to medium density residential and commercial development, and golf courses with treated wastewater effluent irrigation.

In contrast, the Onion Creek watershed contains approximately 890 square kilometers and the main stream channel measures approximately 130 kilometers in length. Onion Creek is the largest contributing stream for Barton Springs with approximately 34 percent of the total recharge to the springs (USGS, 1986). Land uses are similar to those found in the Barton Creek watershed with the exception of numerous residential developments of medium to high density that rely on septic systems for wastewater disposal. Unlike Barton Creek, Onion Creek extends into the Texas Blackland Prairie ecoregion before its confluence with the Colorado River east of Austin, TX. The nine study sites selected on Onion Creek lie both upstream and downstream of the aquifer recharge zone and the most downstream sites lie in the boundary area of the Central Texas Plateau and the Blackland Prairie ecoregions. The variation in the degree of development in the subwatersheds, along with ecoregion location allows for the comparison of study results between sites with varying degrees of anthropogenic activity in the subwatersheds.

Study Watershed Boundaries and Sites

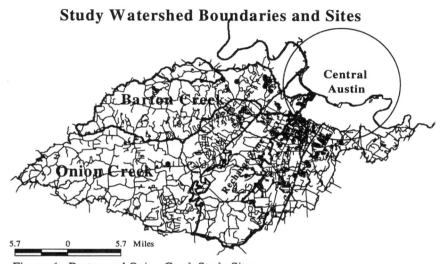

Figure 1. Barton and Onion Creek Study Sites

Monitoring Methods

        Sampling on a quarterly basis at all 17 study sites includes benthic macroinvertebrate collections, diatom collections, chlorophyll a samples, water quality samples, percent algae cover transects, and physical habitat assessments. The following discussion of sampling techniques will focus on the macroinvertebrate and diatom sampling protocols. Sampling methodologies are modified versions of existing protocols (e.g. EPA Rapid Bioassessment Protocols for Use in Streams and Rivers, Plafkin, et.al.,1989) or techniques developed by City of Austin researchers.

        The benthic macroinvertebrate bioassessment protocol is a modified version of the USEPA Rapid Bioassessment Protocol III. The protocol is designed to accomplish three major goals: establish baseline data of the aquatic fauna for Barton and Onion creeks, determine the degree of impairment at various study sites along the creeks, and provide data for the refinement of sampling and analysis techniques for increased sensitivity in the detection of degradation due to nonpoint source pollution in local watersheds. A minimum of 200 organisms are collected from the riffle area at each site using a 0.09 m² Surber sampler with a mesh size of

700 microns. Due to the depauperate state of these low nutrient streams, the number of Surber samples required to collect 200 organisms varies from one to ten. As a result, subsampling is not required. Substrate immediately upstream of the net opening is vigorously agitated for a minimum of 60 seconds. At each site, two 100 organism samples are collected and stored individually in 95 percent ethanol in 30 ml polypropylene vials. The number of organisms in the two vials is totaled for final data analysis for each site. In this manner, a comparison can be made between the 100 organism EPA protocol and the City of Austin 200 organism collection method. The coarse particulate organic material (CPOM) sample was eliminated from the sampling protocol due to the paucity of areas with accumulations of large particulate matter and the general lack of shredder organisms in this region. Likewise, due to the low number of chironomids collected, the EPT/Chironomidae metric was modified to EPT/EPT+Chironomidae. Benthic macroinvertebrate data for each study site are analyzed using the USEPA RBP III multi-metric criteria (Plafkin, et.al., 1989).

Additionally, at each monitoring site, 3 diatom samples are collected from substrate randomly selected from the study riffle. A petri dish is used to outline the area on the rock to be sampled. The demarcated area is scraped with a wire brush and the particulate matter is deposited into a shallow collecting pan. A sufficient amount of ambient creek water is used to flush the finer plant material from the substrate and the scraping tools. Extreme care is taken to ensure that the total volume of water required to flush the sample rock, tools, and pan does not exceed the volume of the sample bottle (approximately 30 ml). The diatom sample material is cleaned in concentrated nitric acid and one slide is prepared for each sample. Diatom taxa are identified and counted along y-axis transects under oil immersion at 1000x magnification. 500 individuals are counted per sample and taxa identification is made to species level using Patrick and Reimer (1966 and 1975) and Krammer and Lange-Bertalot (1986, 1988, 1991a and 1991b). Species counts are analyzed using multivariate statistical analyses such as correspondence analysis (CA) and principal components analysis (PCA). In addition, data results are analyzed for taxa richness, evenness, and Simpson diversity.

The biomass and chlorophyll a samples are collected from three rocks selected in the riffle areas at each monitoring site. Each rock is collected as a unique, discrete sample. A 100 cm^2 area is demarcated and scraped in the same manner as the diatom collection. The resulting sample is laboratory analyzed for chlorophyll a, total suspended solids, and total volatile solids. The amount of combustible organic matter in a sample provides an important estimate of standing crop or periphyton production. Since heterotrophic organisms may comprise a significant portion of the mass of combustible organic matter in a sample, chlorophyll a measurements are included with the biomass data.

In addition to the quarterly sampling schedule for benthic macroinvertebrates, a quantitative, sequential sampling event was performed in March - May 1994. At two upstream and one downstream site on Barton Creek, 10 - 12 Surber samples were collected throughout the site riffle area. All collections were processed and identified using the same methods as the benthic macroinvertebrate protocols. This intensive sampling effort was performed to evaluate the efficiency of the 200 organism macroinvertebrate sampling technique in its representation of the total macroinvertebrate community at three specific riffle areas.

Study Results

The original study design involved an upstream - downstream approach with site selection based on comparability and appropriate riffle area for benthic macroinvertebrate and periphyton habitat. Reference sites were selected upstream of areas with major development or new highway construction. As mentioned above, the main anthropogenic activity in these upstream areas is low intensity cattle grazing and clearing of native vegetation for pasture land. Many of the study sites on both creeks lie downstream of confluences with tributaries that drain subwatersheds with varying levels of residential or commercial development, major highways, or golf courses that utilize treated wastewater effluent for irrigation. Eight sites along Barton Creek are monitored during the three year study. Along Onion Creek, nine sites are monitored the final two years of the project.

As of June 1996, approximately 30,000 benthic macroinvertebrates were collected at the 17 study sites and approximately 26,000 identified to genus level when possible. Oligachaetes, amphipods, ostracods, hydracarinas, hirudineas, chironomids, and ceratopogonids were identified to order or family level. Scores based on the multi-metric USEPA RBP III were calculated for all study sites. Score results indicate the lack of an upstream - downstream trend with respect to degree of biological impairment in Barton Creek (Figure 2). Moreover, scores for the most downstream site are as high or higher than the reference site for 3 of the quarterly monitoring periods.

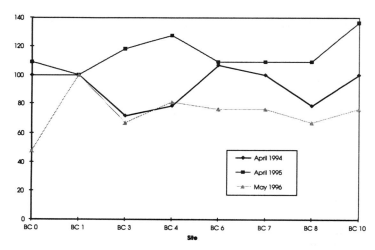

Figure 2. RBP III scores for Barton Creek sites (April-May, 1994 - 1996)

Analysis at the individual univariate metric level fails to describe an upstream - downstream trend with respect to biological integrity. In fact, no relationship is discernible between site location and taxa richness, while evenness and Simpson diversity fail to show any trend with regard to invertebrate populations. During the three year study period, Austin, TX and the surrounding area experienced two of the longest drought periods in recorded history. The summer of 1993 included 55 days without measurable precipitation. During this drought period there was no appreciable flow on Barton Creek. Flow returned to the creek in October 1993 and sampling events were scheduled for November and December.

Previous to this study, researchers from the University of Texas - Austin and the City of Austin predicted that many of the benthic macroinvertebrate groups migrated into the alluvium as water levels declined. Study data for October 1993 - January 1994 indicate that the macroinvertebrate communities failed to survive such an extended hot, dry period. Field surveys in October and November 1993 failed to detect a sufficient number of organisms to warrant sampling. Even in the first half of December, eight Surber samples were required to collect 50 organisms. Many of the sites with a sufficient number of organisms to sample were dominated by water mites or black fly larvae while mayflies and caddisflies were scarce. These data indicate that benthic communities are lost during extended, unpredictable drought periods and that sampling during the wet periods of the year document the recolonization and recovery of the benthic community. As the period of continuous flow in the stream lengthens, the taxa richness tends to increase and the benthic community structure appears similar to assemblages in Central Texas perennial streams.

Benthic macroinvertebrate data indicate also that benthic community structure in Barton and Onion creeks does not follow the ecological model in which upstream unimpacted sites exhibit higher biological diversity and abundance than downstream, impacted sites. In fact, the reverse has been true during some of our quarterly sampling events, as evidenced in Figure 3. Barton Creek Site #10, the most downstream site, exhibited higher taxa richness scores than the most upstream reference site, #0. This trend was also evident in the sequential sampling data from March - May, 1994, where 342 organisms and 27 taxa were collected at the most upstream site and 463 organisms and 38 taxa were collected at Site #10.

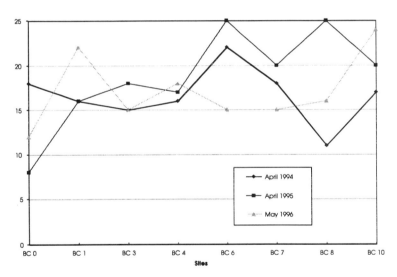

Figure 3. Taxa richness for Barton Creek sites (April-May, 1994 - 1996)

Detrended correspondence analysis (DCA) and principal component analysis (PCA) were used to condense the multi-dimensional biological information into three or four dimensions or axes. Multiple regressions of the DCA and PCA species axes onto the environmental parameters were used to develop environmental models of taxa variation. This empirical environmental modeling technique is useful in the evaluation of the influence of specific environmental parameters on benthic community structure. Each of the species axes were compared through multiple regression with the water quality parameters. The statistical environmental model with the best fit accounted for 15 percent of the species condensation as variation in water quality values. Of this, almost 13 percent was accounted for by flow and temperature. Likewise, regression of PCA analysis onto water quality values accounted for 38 percent of the species variation. Once again, temperature and flow accounted for the largest portion of this variation. The responses of the benthic invertebrate assemblages to these physical perturbations make it difficult to evaluate the impact of anthropogenic activities on the biological communities.

In addition to the benthic macroinvertebrate data analysis and evaluation, diatom data were analyzed using univariate and multivariate techniques. In general, diatom data analysis indicates that diatom communities and benthic macroinvertebrate communities respond similarly to flow and temporal variation. However, it should be noted that since the aquatic biological communities in intermittent streams are in a constant state of flux and recovery due to the physical perturbations of dewatering and spates, organisms with short life histories and rapid colonization rates may present more stable community assemblages in a shorter study period than longer lived organisms.

Even though Barton Creek is characterized by a relatively low species richness with respect to diatoms, analysis of diatom data indicates a slight increase in species richness along an upstream - downstream gradient. In a multiple regression of taxa richness on water quality parameters, 53 percent of the variation (p=0.008) was explained by flow and pH values. In an analysis of diatom data using the Simpson diversity index, no significant upstream - downstream gradient was found. Additionally, taxa richness increases with taxa diversity and it appears that taxa richness is not correlated with taxa evenness. The low taxa richness may be due to the low nutrient levels and the frequency of physical disturbances in these intermittent streams.

After condensing the diatom species information using DCA and PCA, the resulting axes were regressed on the water quality parameters in the same manner as the invertebrate data analysis in order to create empirical models of environmental variables that best fit the biological information. This analysis indicates that 38 percent of the DCA species variation is explained by water quality parameters. Site temperature, pH, and nitrogen levels account for 37 percent of this variation. Likewise, water quality information accounts for 41 percent of the PCA variation. Temperature by itself describes 36 percent of the total variation in this regression. Additionally, the correlation coefficient between temperature and flow is nearly one (0.96) indicating that these two variables are nearly equivalent with respect to diatom regressions. These analytical results indicate that seasonal variation in the physical characteristics of the aquatic environment accounts for a significant percentage of the change in the biological community structure. As with the invertebrate data analysis, the diatom community response to flow and temperature increases the difficulty of evaluating the impacts of anthropogenic activities on the integrity of the biological communities.

Discussion

Central Texas Plateau ecoregion intermittent streams with low nutrient concentrations exhibit low levels of abundance and taxa richness with regard to diatom and invertebrate communities. This condition is the inverse of the traditional model of ecological integrity. High abundance and high diversity are assumed to be necessary components of a healthy, aquatic ecosystem in the traditional model. However, along the study reach of Barton Creek, sampling data indicate a trend of increasing taxa richness from upstream to downstream. During the study period, Barton Creek Site #0, the most upstream site, had a mean value of 10 taxa per sampling event. In contrast, Barton Creek Site #10, the most downstream site, had a mean value of 19 taxa per sampling event. Future studies during extended periods (e.g. two to three years) of continual flow in the creeks will be necessary to ascertain if these trends will continue under conditions that simulate perennial stream flow.

Sampling methods also indicate a trend of decreasing sampling effort required to collect the minimum number of organisms at the more downstream sites. The mean values for the number of Surber samples required to collect 200 organisms at Barton Creek Sites #0 and #1, the most upstream sites, were 14 and 13, respectively. In contrast, Barton Creek Sites #8 and #10, the most downstream sites, had mean values of 3 and 4, respectively. The decrease in sampling effort at the downstream sites is indicative of the increased abundance of benthic organisms at these sites, perhaps due to increased flow and nutrient levels.

Additionally, data from the quantitative, sequential sampling event in March - May 1994 support the trends of increasing abundance and diversity at the downstream sites on Barton Creek. Thirteen Surber samples were collected at a site on the Shield Ranch, 11.5 km upstream of Barton Creek Site #0. Land use around the Shield Ranch site is restricted to low intensity livestock grazing. This sampling event resulted in 342 total organisms and 27 total taxa. However, eleven Surber samples collected at Barton Creek Site #10, the most downstream study site, resulted in a total of 463 organisms and 38 total taxa. Again, observed increases in abundance and diversity may be the result of various environmental factors including increased flow volume at downstream sites and slight increases in nutrient levels from natural and anthropogenic sources.

Recommendations

Future bioassessment strategies for intermittent Central Texas streams need to be based on an ecological model of community structure that accounts for lower abundance and diversity at unimpacted sites. These strategies must take into account also that benthic and diatom communities may be in a perpetual state of disturbance, recolonization and recovery. Metrics need to be developed that distinguish between community structure response to temporal variation and physical perturbations and impacts from anthropogenic activities. If multi-parameter indices based on characteristics of benthic aquatic communities are to be used in evaluating impairments to water quality and ecological integrity in intermittent stream systems, such as Barton and Onion creeks, upstream and downstream reference sites need to be evaluated along with long term monitoring efforts that document the seasonality and dynamics of the biological commentates during both wet and dry conditions. In addition, various indices such as the Hilsenhoff Biotic Index (HBI) and the ratio of Shredders/Total Organisms need to be modified for the appropriate ecoregion.

Acknowledgements

This project was funded through a grant provided by the Texas Clean Rivers Act. The grant was provided to the City of Austin, TX by the Texas Natural Resource Conservation Commission and admisitered by the Lower Colorado River Authority. Mateo Scoggins, Elizabeth Borer, Ellen Geismar, and Osvaldo Hernandez dedicated countless hours to the success of this project. Ed Peacock provided valuable editorial comments for the report.

References

City of Austin, 1995. Bioassessment Strategies for Nonpoint Source Polluted Streams. Second Annual Report, unpublished. City of Ausitn Environmental and Conservation Services Department.

Karr, J. R., 1981. Assessment of biotic integrity using fish communities. Fisheries 6(6): 21-27.

Krammer, K. and H. Lange-Bertalot, 1986,1988, 1991a, 1991b. Band 2/1 - 2/4 Bacillariophyceae. Gustav Fischer Verlag, Stuttgart.

Merritt, R. W. and K. W. Cummins, 1984. An Introduction to the Aquatic Insects of North America, 2nd edition. Kendall/Hunt, Dubuque, Iowa.

Patrick, R. and C. W. Reimer, 1966 and 1975. Diatoms of the United States exclusive of Alaska and Hawaii, Volumes 1 and 2. The Academy of Natural Sciences of Philadelphia.

Plafkin, J.L., M.T. Barbour, K.D. Porter, S.K. Gross and R.M. Hughes, 1989. Rapid Bioassessment Protocols for Use in Streams and Rivers: Benthic Macroinvertebrates and Fish. USEPA Office of Water. EPA/44/4-89-001.

Texas Water Commission, 1989. An Assessment of Six Least Disturbed Unclassified Texas Streams, LP 89-04.

The Control of Toxicants at Critical Source Areas

Robert Pitt[1], M. ASCE

Summary

The Department of Civil and Environmental Engineering at the University of Alabama at Birmingham is engaged in a multi-year cooperative agreement with the Storm and Combined Sewer Program of the U.S. EPA. Part of this cooperative agreement includes the testing of a special treatment device (a multi-chambered treatment train, or MCTT) to treat runoff from critical source areas. Additional funding was also provided by the U.S. Army-Construction Engineering Research Laboratory in Champaign, Illinois. This paper reviews the design of the MCTT and presents monitored performance information for a broad list of stormwater constituents. A previous description of the MCTT was given by Robertson, et al. (1995) and more detail was given in his MSCE thesis (Robertson 1995). We expect to complete a detailed design manual for the MCTT (along with different filter media, which tests were summarized previously by Clark, et al. (1995) in 1996 and 1997.

Runoff from paved parking and storage areas, and especially gas station areas, has been observed to be contaminated with concentrations of many critical pollutants. These paved areas are usually found to contribute most of the toxicant pollutant loadings to stormwater outfalls. Polycyclic aromatic hydrocarbons (PAHs), the most commonly detected toxic organic compounds found in urban runoff, along with heavy metals are mostly associated with automobile use, especially during starting vehicles. The major goal of this research was to test a method of how these toxicants can be controlled at small critical source areas.

Earlier bench scale treatability studies sponsored by the U. S. Environmental Protection Agency (EPA) found that the most beneficial treatment for the removal of stormwater toxicants (as measured using the Microtox™ test) included quiescent settling for at least 24 hours (generally 40% to 90% reductions), screening through at least 40 μm screens (20% to 70% reductions), and aeration and/or photo-degradation for at least 24 hours (up to 80% reductions) (Pitt, et al. 1995). These processes were combined in

---

[1] Associate Professor, Department of Civil and Environmental Engineering, The University of Alabama at Birmingham. Birmingham, AL 35294.

70

the MCTT. The MCTT contains aeration, sedimentation, sorption, and sand/peat filtration and has been shown to provide excellent toxicant removals.

The MCTT is most suitable for use at relatively small and isolated paved critical source areas, from about 0.1 to 1 ha (0.25 to 2.5 acre) in area. These areas would include vehicle service facilities (gas stations, car washes, oil change stores, etc.), convenience store parking areas and areas used for equipment storage, along with salvage yards. The MCTT is an underground device that has three main chambers: an initial grit chamber for removal of the largest sediment and most volatile materials; a main settling chamber (containing initial aeration and sorbent pillows) for the removal of fine sediment and associated toxicants and floating hydrocarbons; and a sand and peat mixed media filter/ion exchange unit for the removal of filterable toxicants. A typical MCTT requires between 0.5 and 1.5 percent of the paved drainage area, which is about 1/3 of the area required for a well designed wet detention pond.

A pilot-scale MCTT was constructed in Birmingham, AL, and tested over a six month monitoring period, from May to October, 1994. Two additional full-scale MCTT units have recently been constructed and are currently being monitored as part of Wisconsin's 319 grant from the U.S. EPA.

Complete organic and metallic toxicant analyses, in addition to conventional pollutants, are included in the analysis program. During monitoring of 13 storms at a parking facility, the Birmingham pilot-scale MCTT was found to have the following overall median removal rates: 96% for total toxicity (as measured using the Microtox™ screening test), 98% for filtered toxicity, 83% for suspended solids, 60% for COD, 40% for turbidity, 100% for lead, 91% for zinc, 100% for n-Nitro-di-n-proplamine, 100% for pyrene, and 99% for bis (2-ethyl hexyl) phthalate. The color was increased by about 50% due to staining from the peat and the pH decreased by about one-half pH unit, also from the peat media. Ammonia nitrogen was increased by several times, and nitrate nitrogen had very low removals (about 14%). The MCTT therefore operated as intended: it had very effective removal rates for both filtered and particulate stormwater toxicants and suspended solids. Increased filterable toxicant removals were obtained in the peat/sand mixed media filter/ion exchange chamber, at the expense of increased color, lowered pH, and depressed COD and nitrate removal rates. The preliminary full-scale test results substantiate the excellent removals found during the pilot-scale tests, while showing better control of COD and nutrients and less detrimental effects on pH and color.

Introduction

The historical use of oil and grease traps to treat stormwater at gas stations and other areas has been shown to be ineffective for various reasons, especially lack of maintenance and poor design for the relatively low levels of free floating oils present in most stormwaters (Schueler 1994).

Small Prefabricated Oil Separators. Numerous manufacturers have developed small prefabricated separators to remove oils and solids from runoff. These separators are rarely specifically designed and sized for stormwater discharges, but usually consist of modified grease and oil separators. The solids are intended to settle within these separators, either by free fall or by counter-current or cross-current lamellar separation. Many of these separators have been sold and installed in France, especially along highways (Rupperd 1993). Despite the number of installations, few studies have been carried out in order to assess their efficiency (Aires and Tabuchi 1995).

Available results from Fourage (1992), Rupperd (1993) and Legrand, *et al.* (1994) for stormwater treatment show that :

• these devices are usually greatly undersized. They should work reasonable well at flow rates between 20 and 30% of their design hydraulic capacity. For higher flow rates, the flow is very turbulent (Reynolds numbers can be higher than 6000) and the removal efficiency is very poor.

• these devices need to be cleaned very frequently. If they are not cleaned, the deposits are scoured during storm events, with negative efficiencies. But the cleaning work is usually manual, and expensive, and not very easy because the separators are very compact. Some new devices are equipped with an automatic pump which can be a significant improvement for maintenance. Currently, the cleaning frequencies are very insufficient and the stormwater pollutant control efficiencies are very limited.

Prefabricated separators could be used for stormwater treatment if the following conditions are respected:

• realistic design hydraulic capacity in terms of maximum flow rates, flow distribution and flow regime;

• realistic solids removal efficiency: the finest and polluted solids will usually not be trapped with a high efficiency because of too high hydraulic velocities;

• frequent cleaning and/or an automatic extraction to assure good overall efficiency;

• specific conception for stormwater which takes into account the solids characteristics, the rapid flow variations, and the maintenance requirements.

The Multi Chambered Treatment Tank. The MCTT was developed to specifically address many of the above concerns. It was developed and tested with specific stormwater conditions in mind, plus it has been tested at several sizes for the removal of stormwater pollutants of concern. Figure 1 shows a general cross-sectional view of a MCTT. It includes a special catchbasin followed by a two chambered tank that is intended to reduce a broad range of toxicants (volatile, particulate, and dissolved). The runoff enters the catchbasin chamber by passing over a flash aerator (small column packing balls with counter-current air flow) to remove highly volatile components. This catchbasin also serves as a grit chamber to remove the largest (fastest settling) particles. The second chamber serves as an enhanced settling chamber to remove smaller particles and has inclined tube or plate settlers to enhance sedimentation. This chamber also contains fine bubble diffusers and sorbent pads to further enhance the removal of floatable hydrocarbons and additional volatile compounds. The water is then pumped to the final chamber at a slow rate to maximize pollutant reductions. The final chamber contains a mixed media (sand and peat) slow filter/ion exchange device, with a filter fabric top layer. The MCTT is typically sized to totally contain all of the runoff from a 6 to 20 mm (0.25 to 0.8 in) rain, depending on interevent time, typical rain size, and rain intensity.

Catchbasin | Main Settling Chamber | Filtering Chamber
— Packed Column | — sorbent pillows | — sorbent filter fabric,
aerators | — fine bubble aerators | — mixed media filter layer
| — tube settlers | (sand and peat)
| | — filter fabric
| | — gravel packed
| | underdrain

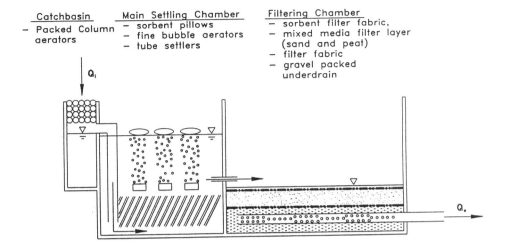

Figure 1. General Schematic of MCTT

Methodology

A pilot-scale MCTT was constructed and tested in Birmingham, Alabama. Two full-scale MCTT units have also been recently constructed in Wisconsin (in Milwaukee at a city public works yard, and in Minocqua at a municipal parking lot) as part of their EPA 319 grant. These full-scale MCTT units are currently being monitored. Samples were collected before and after each chamber of the device. To better estimate fate and treatability of toxicants, samples were partitioned into filterable ("dissolved") and non-filterable ("particulate") components before being analyzed for a wide range of toxicants (using detection limits from about 1 to 10 µg/L) and conventional pollutants, as listed on Table 1.

The pilot-scale MCTT was set up to capture runoff from a parking and vehicle service area on the campus of the University of Alabama at Birmingham. The catchbasin/grit chamber is a 25 cm vertical PVC pipe containing about 6 L of 3 cm diameter packing column spheres. The main settling chamber is about 1.3 m² in area and 1 m deep which with a 72-hour settling time was expected to result in a median toxicity reduction of about 90%. The filter chamber is about 1.5 m² in area and contains 0.5 m of sand and peat directly on 0.15 m of sand over a fine plastic screen and coarse gravel that covers the underdrain. A Gunderboom™ filter fabric also covers the top of the filter media to distribute the water over the filter surface by reducing the water infiltration rate through the filter and to provide additional pollutant capture.

During a storm event, runoff from the parking lot is pumped into the catchbasin/grit chamber automatically. During filling, an air pump supplies air to

TABLE 1. COMPOUNDS ANALYZED DURING MCTT TESTS

Organic Toxicants by GC/MSD - filtered and unfiltered (1 to 10 μg/L MDL)
    Polycyclic aromatic hydrocarbons
    Phthalate esters
    Phenols

Organic Toxicants by GC/ECD - filtered and unfiltered (0.01 to 0.1 μg/L MDL)
    Pesticides

Heavy Metals by GFAA - filtered and unfiltered (1 to 5 μg/L MDL)
    Cadmium
    Copper
    Lead
    Zinc

Toxicity Screening by Microtox™ - filtered and unfiltered

Nutrients by Ion Chromatography - filtered (1 mg/L MDL)
    Nitrate
    Nitrite
    Ammonia
    Phosphate

Major Ions by Ion Chromatography - filtered (0.1 to 1 mg/L MDL)
    Cations (calcium, magnesium, potassium, sodium, and lithium)
    Anions (chloride, sulfate, and fluoride)

Conventional Analyses
    COD
    Color
    Specific Conductance
    Hardness
    Alkalinity
    pH
    Turbidity
    Solids (total, suspended, dissolved, and volatile forms)
    Particle size (Coulter Multisizer IIe)

aeration stones located in the main settling chamber. When the settling chamber is full, all pumps and samplers cease. After a quiescent settling period of up to 72 hours, water is pumped through the filter media and discharged.

Design of the MCTT

Catchbasin. Catchbasins have been found to be effective in removing pollutants associated with coarser runoff solids. Moderate reductions in total and suspended solids (up to about 45%, depending on the inflowing water rate) have been indicated by prior studies (Lager and Smith 1976, Pitt and Bissonnette 1985). While relatively few pollutants are associated with these coarser solids, their removal decreases maintenance problems of the other MCTT chambers. The size of the MCTT catchbasin sump is controlled by three factors: the runoff flow rate, the suspended solids (SS) concentration in the runoff, and the desired frequency at which the catchbasin will be cleaned so as not to sacrifice efficiency.

Main Settling Chamber. The main settling chamber mimics completely mixed settling column bench-scale tests and uses a treatment ratio of depth to time for removal estimates. In addition to housing plate or tube settlers, the main settling chamber also contains floating sorbent "pillows" to trap floating grease and oil and contains a fine bubble diffuser. The settling time in the main settling chamber typically ranges from 1 to 3 days.

Peat/Sand Ion Exchange Chamber. Based on literature descriptions of stormwater filtration, especially by the City of Austin (1988), Galli (1990) and Shaver (undated and 1991), earlier UAB bench-scale treatability tests (Pitt, et al. 1995), and the preliminary UAB filter media column tests (Clark, et al. 1995), it was determined that a mixed media sand and peat "filter" should be used as a polishing unit after the main settling chamber. This unit was expected to provide additional toxicant reductions, especially for filtered forms of the organics and metals. The surface hydraulic loading rate of this filter/ion exchange chamber was between 1.5 and 6 m per day (5 and 20 ft per day). The 50%/50% mixture of the sand and peat had a depth of 0.5 m (18 in), resting on 0.15 m of sand. The sand had the following size: 71% finer than #30 sieve (0.6 mm), 65% finer than #40 sieve (0.425 mm), and 0.5% finer than #50 sieve (0.18 mm). The effective size (D_{10}) of the sand was 0.31 mm and the uniformity coefficient (D_{60}/D_{10}) was 1.45. A filter fabric was used to separate these layers from the gravel and perforated pipe underdrain. In order to better facilitate surface spreading of water on top of the media and to prevent channelization, another filter fabric (Gunderboom™) was placed on top of the media.

Example Design. The design of the MCTT is very site specific, being highly dependent on local rains (rain depths, rain intensities, and interevent times). A computer model was therefore developed to determine the amount of annual rainfall treated, the toxicity reduction rate for each individual storm, and the overall toxicity reduction associated with a long series of rains for different locations in the U.S. Table 2 shows the resultant required main settling chamber sizes for 21 cities having rain depths ranging from 180 mm (7.1 in) (Phoenix) to 1500 mm (60 in) (New Orleans) per year. Example

Table 2. MCTT Main Settling Chamber Required Sizes
(all 48 hours holding times, except as noted, with 5 foot settling depths).

City	Annual Rain Depth (in)	Runoff Capacity (in) for 70% Toxicant Control	Runoff Capacity (in) for 90% Toxicant Control
Phoenix, AZ	7.1	0.25 (24 hours)	0.35
Reno, NV	7.5	0.20 (18 hours)	0.20
Bozeman, MT	12.8	0.25	0.40
Los Angeles, CA	14.9	0.30	0.45
Rapid City, SD	16.3	0.20 (18 hours)	0.22
Minneapolis, MN	26.4	0.32	0.50
Dallas, TX	29.5	0.50	0.96
Madison, WI	30.8	0.32	0.52
Milwaukee, WI	30.9	0.36	0.65
Detroit, MI	31.0	0.24	0.50
Austin, TX	31.5	0.22 (18 hours)	0.32
St. Louis, MO	33.9	0.30	0.49
Buffalo, NY	37.5	0.35	0.50
Seattle, WA	38.8	0.25	0.40
Newark, NJ	42.3	0.48	0.96
Portland, ME	43.5	0.42	0.72
Atlanta, GA	48.6	0.55	0.95
Little Rock, AR	49.2	0.52	0.85
Birmingham, AL	54.5	0.37	0.53
Miami, FL	57.6	0.40	0.73
New Orleans, LA	59.7	0.80	0.92

detailed design plots are shown in Figures 2 through 4 for Phoenix, Milwaukee, and New Orleans.

The overall range in MCTT size varies by more than three times for the same level of treatment for the different cities. The required size of the main settling chamber generally increases as the annual rain depth increases. However, the interevent period and the rain depth for individual rains determines the specific runoff treatment volume requirement. As an example, Seattle requires a much smaller MCTT than other cities having similar annual total rains because of its small rain depths for each rain. Rapid City requires a smaller MCTT, compared to Los Angeles, because Los Angeles has much larger rains when it does rain. Similarly, Dallas requires an unusually large MCTT because of its high rain intensities and large individual rains, compared to upper Midwest cities that have similar annual rain depths.

In all cases, the most effective holding time is 2 days for 90% toxicant control (for the 1.5 m, 5 ft, settling chamber depth). In most cases, a toxicity removal goal of

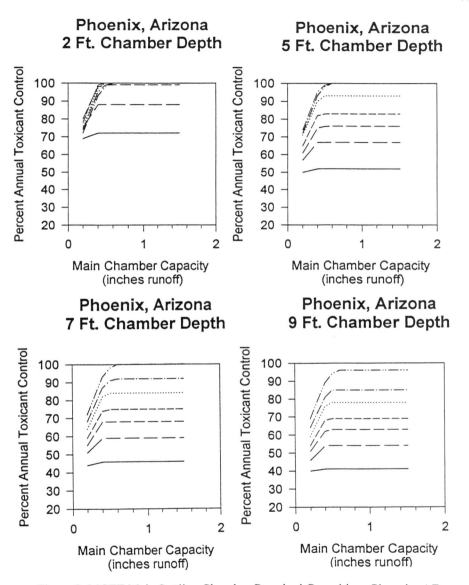

Figure 2. MCTT Main Settling Chamber Required Capacities - Phoenix, AZ

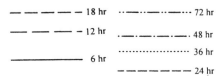

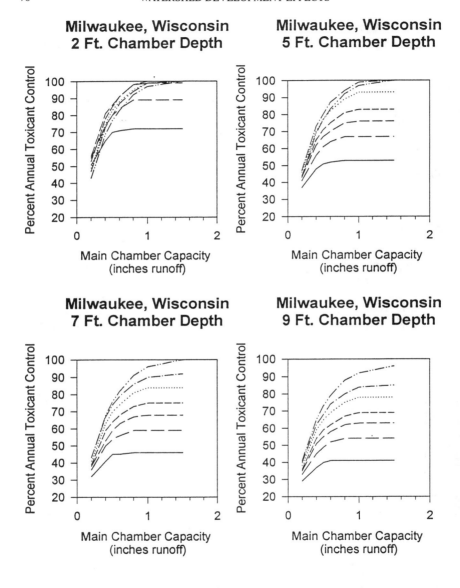

Figure 3. MCTT Main Settling Chamber Required Capacities - Milwaukee, WI

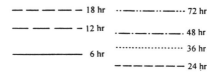

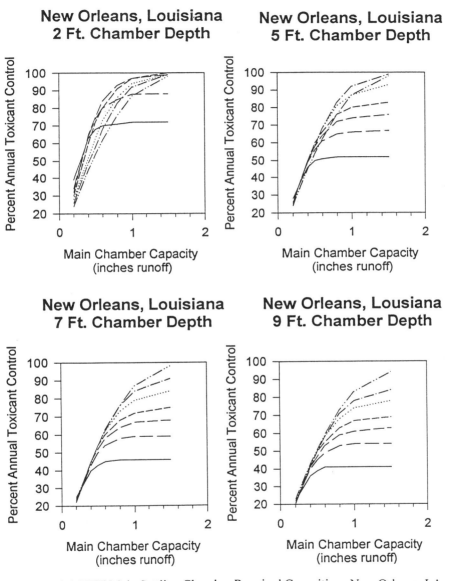

Figure 4. MCTT Main Settling Chamber Required Capacities - New Orleans, LA

about 70% in the main settling chamber is probably the most cost effective choice, considering the additional treatment that will be provided in the sand/peat chamber.

Figures 2 through 4 show the runoff volume requirements for an MCTT having a 0.6, 1.5, 2.1, or 2.7 m (2, 5, 7, or 9 foot) settling depths in the main settling chamber for three cities. These extreme and typical examples show that the required runoff depth storage capacity increases as the depth of the main settling chamber increases. As an example, for 90% toxicant control at Milwaukee, the storage requirement for a 1.5 m (5 ft) settling depth was shown to be 16.5 mm (0.65 in) on Table 2. Figure 3 indicates that the required storage volume for a 0.6 m (2 ft) settling chamber would only be 14 mm (0.55 in) of runoff, while it would increase to 19 mm (0.75 in) of runoff for a 2.1 m (7 ft) settling depth and to 23 mm (0.9 in) for a 2.7 m (9 ft) settling depth. The greater depths require more time for the stormwater particulates to settle and be trapped in the chamber, while the shallower tanks require a greater surface area. The best tank design for a specific location is based on site specific conditions, especially the presence of subsurface utilities or groundwater and hydraulic grade line requirements. A large surface tank is usually much more expensive, even though the required volume is less, especially if heavy traffic will be traveling over the tank.

A combination of a 48 hour holding time and 11 mm (0.45 in) runoff storage volume would satisfy a 75% treatment goal for Milwaukee conditions, as shown on Figure 3. This 11 mm runoff volume corresponds to a rain depth of about 13 mm (0.51 in) for pavement (Pitt 1987). The 11 mm runoff storage volume corresponds to a live chamber volume of 22 m^3 (770 ft^3) and a surface area of 10 m^2 (110 ft^2) for a 0.2 ha (0.5 acre) paved drainage area. The surface area of the MCTT would therefore be about 0.5 percent of the drainage area. This device would capture and treat about 80% of the annual runoff at a 95% level, resulting in an annual toxicity reduction of about 75% (0.8 X 0.95). The size of the main settling chamber would need to be greater than this because about 0.7 m (two feet) of "dead" storage must be added to provided for standing water below the outlet orifice (or pump) which would keep the inclined tubes submerged. About a 0.2 m (6 inch) height is also needed below the inclined tubes for the flow distribution system and for long-term storage of fine material that will accumulate.

Additional treatment beyond the 75% level would result from the filter/ion exchange device. The pumped effluent from the main settling chamber would be directed towards a mixed peat/sand filter/ion exchange chamber, which must provide a surface hydraulic loading rate of between 1.5 and 6 m per day (5 and 20 ft per day), and have a depth of at least 0.5 m (18 in). In addition to the pumped effluent, any excess runoff after the main settling chamber is full would also be directed towards the filter.

Each of the treatment chambers need to be vented, mosquito proofed, and be easily accessible for maintenance. The device needs to be inspected, the initial catchbasin should be cleaned, and the sorbent pillows should be exchanged, at least every six months. It is expected that the ion exchange media should last from 3 to 5 years before requiring replacement (as determined during our filtration experiments).

Observed Performance

Table 3 summarizes the statistical test results (Wilcoxon signed rank test, using StatExact™) showing the significance of the differences in sample concentrations between the influent and effluent samples for each major process in the MCTT (the catchbasin/grit chamber, the main settling chamber, and the sand/peat filter). Table 4 summarizes some of the significant percentage changes in concentrations of the constituents as they passed through the MCTT. No data is shown for the catchbasin/grit chamber because of the lack of significant concentration changes observed.

Figures 5 through 8 are plots of the concentrations of suspended solids, unfiltered toxicity, unfiltered zinc, and unfiltered bis(2-ethylhexyl) phthalate as the stormwater passed through the MCTT. The four data locations on these plots correspond to the four sampling locations on the MCTT. The sample location labeled "inlet" is the overall inlet to the MCTT (and the inlet to the catchbasin/grit chamber). The location labeled "catch basin" is the effluent from the catchbasin (and inlet to the main settling chamber). Similarly, the location labeled "settling chamber" is the outlet from the settling chamber (and the inlet to the sand/peat chamber). Finally, the location labeled "peat-sand" is the outlet from the sand/peat chamber (and the outlet from the MCTT). Individual samples are traced through the MCTT on separate lines. Therefore, the slopes of the lines indicate the relative removal rates (mg/L reduction) for each sample and for each individual major unit process in the MCTT. If the lines are all parallel between two sampling locations, then the removal rates are similar. If a line has a positive slope, then a concentration increased occurred. If the lines have close to zero slope, then little removal has occurred (as for the catchbasin/grit chamber for most constituents and samples).

The suspended solids trends shown on Figure 5 show the significant reductions in suspended solids concentrations through the main settling chamber, with little benefit from the catchbasin/grit chamber and the sand/peat chamber. However, the first storm had a significant increase in suspended solids concentration as it passed through the sand and peat due to flushing of fines from the incompletely washed media.

The relative toxicity changes (as measured using the Microtox™ unit) are shown on Figure 6 and indicate significant reductions in toxicity, especially for the moderate and highly toxic samples. No effluent samples were considered toxic (all effluent samples were "non toxic", or causing less than a 20% light reduction after 25 minutes of exposure). Figures 7 and 8 are for zinc and bis(2-ethylhexl) phthalate, a metallic and an organic toxicant, and show significant and large reductions in concentrations, mostly through the main settling chamber (corresponding to the large fraction of stormwater toxicants found in the particulate sample fraction). Zinc also had further important decreases in concentrations in the peat/sand chamber.

Numerous other organic compounds were also analyzed, but only about 15 of the 70 target compounds were detected in sufficient frequency, or significant concentrations to be reported. The organic analyte described above was representative of the 15 compounds that were detected in sufficient quantities. In all cases, the concentrations observed were representative of stormwater concentrations expected to be found in similar parking areas. However, the frequency of the organic compounds detected were

Table 3. One-Sided Probabilities that Inlet Equals Effluent Concentrations
(Wilcoxon Signed Rank Test)

Constituent	Catch-basin	Main Settling Chamber	Peat/Sand Chamber	Overall Device
Common Constituents				
total solids	0.24	**0.0017***	0.18	**0.0005**
total solids (volatile)	0.24	**0.0049**	**0.015**	**0.013**
suspended solids	0.15	**0.0010**	0.12	**0.0002**
suspended solids (volatile)	0.23	**0.0024**	0.16	**0.0027**
dissolved solids	0.39	0.23	0.082	0.078
dissolved solids (volatile)	0.23	**0.038**	**0.031**	0.46
turbidity	**0.02**	**0.0005**	**0.0005**	0.13
conductivity	0.35	0.066	**0.0005**	**0.028**
color (uf)**	0.52	**0.0044**	**0.0010***	**0.0007***
color (f)**	0.31	**0.0015**	**0.0005***	**0.0032***
pH	0.45	0.30	**0.0010**	**0.0046**
COD (uf)	0.40	**0.0093**	0.34	**0.031**
COD (f)	0.19	**0.0017**	0.44	0.17
Nutrients (f)				
nitrite	0.32	**0.0093**	**0.024**	0.13
nitrate	0.19	**0.0046**	0.16	**0.011**
ammonia	0.23	**0.018***	0.12	**0.0034***
phosphate	0.31	0.31	0.31	0.19
Major Ions (f)				
hardness	0.13	0.20	**0.0078**	**0.011**
calcium	0.34	0.17	**0.0005**	**0.0017**
potassium	0.25	0.18	0.074	**0.046**
magnesium	0.50	**0.0081**	0.16	**0.017**
sodium	0.11	0.19	0.10	0.065
bicarbonate	0.27	**0.0024**	**0.0005**	**0.0007**
chloride	0.47	0.26	0.25	**0.039**
carbonate	0.15	**0.016**	**0.0005**	**0.0049**
fluoride	0.25	0.50	**0.039**	0.15
lithium	0.16	0.25	0.50	0.33
sulfate	0.15	0.50	0.32	**0.011**
Toxicants				
Microtox (uf)	0.45	**0.050**	**0.0078**	**0.0022**
Microtox (f)	0.24	**0.0049**	**0.050**	**0.0015**

* bold values are significant at the 95% level, or greater.

** uf = unfiltered sample
   f = filtered sample

*** ammonia and color values increased, as expected

Table 4. Median Observed Percentage Changes in Constituent Concentrations

Constituent	Main Settling Chamber	Sand/Peat Chamber	Overall Device
Common Constituents			
total solids	31%	2.6%	32%
suspended solids	91	-44	83
turbidity	50	-150	40
conductivity	-15	21	11
color	19	-61	-46
pH	-0.3	6.7	7.9
COD	56	-24	60
Nutrients			
nitrate	27	-5	14
ammonia	-155	-7	-400
Toxicants			
Microtox™ (uf)	18	70	96
Microtox™ (f)	64	43	98
lead	89	38	100
zinc	39	62	91
n-Nitro-di-n-propylamine	82	100	100
hexachlorobutadiene	72	83	34
pyrene	100	n/a	100
bis (2-ethylhexyl) phthalate	99	-190	99

substantially greater (being from 30 to 80% for the 15 primary compounds, compared to 10 to 30% for most past stormwater studies). As expected, few samples had detectable filterable organic toxicant concentrations. The use of the Microtox™ toxicity screening procedure (for both filterable and total sample fractions) was therefore important as an indicator of the "treatability" of the toxic components of the samples.

Preliminary Full-Scale MCTT Test Results

Preliminary results from the full-scale Wisconsin tests have been encouraging and collaborate the high levels of treatment observed during the Birmingham pilot-scale tests. Table 5 shows the treatment levels that have been observed, based on seven tests in Minocqua (during one year of operation) and three tests in Milwaukee (during the first several months of operation). This initial data indicates very high removals (generally >90%) for suspended solids, COD, turbidity, phosphorus, lead, zinc, and many organic toxicants. None of the organic toxicants were ever observed in effluent water from either full-scale MCTT, even considering the excellent detection limits available in the

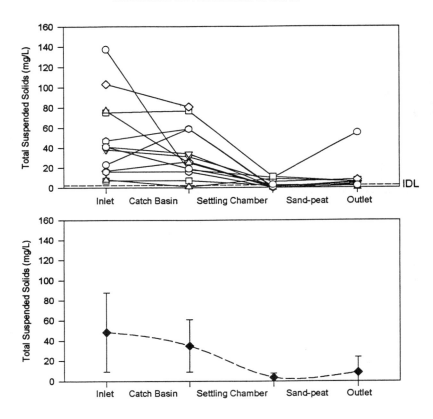

	Catch Basin Chamber	Settling Chamber	Sand-peat Chamber	MCTT Overall
Concentration Difference				
1-sided P Value	0.1543	0.0010	-0.1191	0.0002
Min. Percent Reduction	-157	-800	-500	25
Max. Percent Reduction	88	100	45	100
Median Percent Reduction	17	91	-400	83
Std. Dev. of Percent Reduction	65	257	240	22
COV of Percent Reduction	7.4	19	-1.5	0.28

Figure 5. MCTT Performance for Suspended Solids

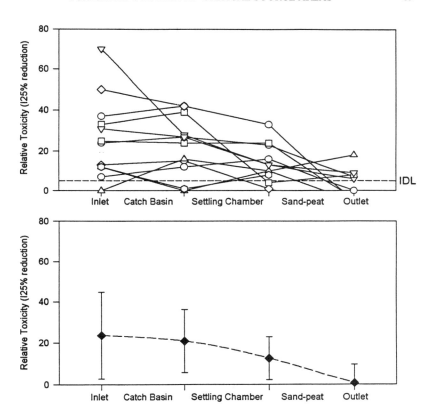

	Catch Basin Chamber	Settling Chamber	Sand-peat Chamber	MCTT Overall
Concentration Difference				
1-sided P Value	0.4464	0.0537	0.0078	0.0022
Min. Percent Reduction	-71	-700	-175	-83
Max. Percent Reduction	100	93	1200	185
Median Percent Reduction	4	18	70	96
Std. Dev. of Percent Reduction	53	238	368	66
COV of Percent Reduction	2.7	-3.9	2.7	0.74

Figure 6. MCTT Performance for Relative Toxicity (by Microtox™) - Unfiltered Sample

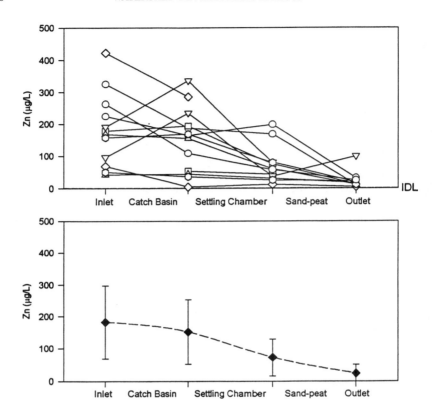

	Catch Basin Chamber	Settling Chamber	Sand-peat Chamber	MCTT Overall
Concentration Difference				
1-sided P Value	0.1219	0.0046	0.0874	0.0005
Min. Percent Reduction	-144	-171	-5908	-3
Max. Percent Reduction	99	84	94	97
Median Percent Reduction	27	39	62	91
Std. Dev. of Percent Reduction	65	68	1796	31
COV of Percent Reduction	5.7	2.9	-3.6	0.42

Figure 7. MCTT Performance for Zinc - Unfiltered Sample

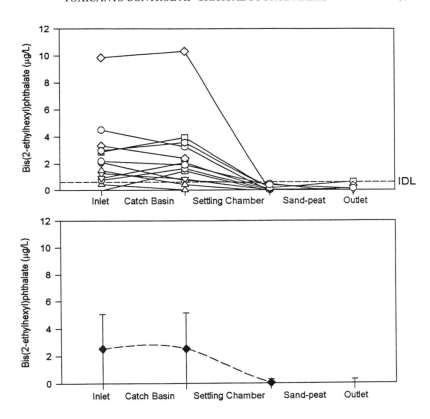

	Catch Basin Chamber	Settling Chamber	Sand-peat Chamber	MCTT Overall
Concentration Difference				
1-sided P Value	0.5000	0.0020	0.1563	0.0020
Min. Percent Reduction	-121	34	-650	-667
Max. Percent Reduction	5033	1667	167	193
Median Percent Reduction	28	99	-188	99
Std. Dev. of Percent Reduction	1397	454	300	226
COV of Percent Reduction	3.6	2.0	-1.7	5.2

Figure 8. MCTT Performance for Bis(2-ethylhexyl)phthalate - Unfiltered Sample

Table 5. Preliminary Performance Information for Full-Scale MCTT Tests
(median removals and median effluent quality)

	Milwaukee MCTT	Minocqua MCTT
suspended solids	>95 (<5 mg/L)	85 (10 mg/L)
COD	90 (10 mg/L)	na
turbidity	90 (5 NTU)	na
pH	-7 (8 pH)	na
ammonia	50 (<0.03 mg/L)	na
nitrates	0 (0.3 mg/L)	na
phosphorus	90 (0.03 mg/L)	80 (0.1 mg/L)
cadmium	90 (0.1 μg/L)	na
copper	90 (3 μg/L)	65 (15 μg/L)
lead	95 (2 μg/L)	nd (<3 μg/L)
zinc	>85 (<20 μg/L)	90 (15 μg/L)
benzo(a)anthracene	>45 (<0.05 μg/L)	>65 (<0.2 μg/L)
benzo(b)fluoranthene	>95 (<0.1 μg/L)	>75 (<0.1 μg/L)
dibenzo(a,h)anthracene	>80 (<0.02 μg/L)	>90 (<0.1 μg/L)
fluoranthene	>95 (<0.1 μg/L)	>90 (<0.1 μg/L)
indeno(1,2,3-cd)pyrene	>90 (<0.1 μg/L)	>95 (<0.1 μg/L)
phenanthrene	>70 (<0.05 μg/L)	>65 (<0.2 μg/L)
pyrene	>80 (<0.05 μg/L)	>75 (<0.2 μg/L)

Wisconsin laboratories. The influent organic toxicant concentrations were all less than 5 μg/L and were only found in the unfiltered sample fractions. The MCTT effluent concentrations were also very low for all of the other constituents monitored: <10 mg/L for suspended solids, <0.1 mg/L for phosphorus, <5 μg/L for cadmium and lead, and <20 μg/L for copper and zinc. The pH changes in the Milwaukee MCTT were much less than observed during the Birmingham pilot-scale tests, possibly because of added activated carbon in the final chamber in Milwaukee. Color was also much better controlled in the full-scale Milwaukee MCTT.

The Milwaukee installation is at a public works garage and serves about 0.1 ha (0.25 acre) of pavement. This MCTT was designed to withstand very heavy vehicles driving over the unit. The estimated cost was $54,000 (including a $16,000 engineering cost), but the actual cost was $72,000. The high cost was likely due to uncertainties associated with construction of an unknown device by the contractors and because it was a retro-fit installation. It therefore had to fit within very tight site layout constraints. As an example, installation problems occurred due to sanitary sewerage not being accurately located as mapped.

The Minocqua site was a 1 ha (2.5 acre) newly paved parking area serving a state park and commercial area. It was located in a grassed area and was also a retro-fit installation, designed to fit within an existing storm drainage system. The installed cost of this MCTT was about $95,000.

It is anticipated that MCTT costs could be substantially reduced if designed to better integrate with a new drainage system and not installed as a retro-fitted stormwater control practice. Plastic tank manufactures have also expressed an interest in preparing pre-fabricated MCTT units that could be sized in a few standard sizes for small critical source areas. It is expected that these pre-fabricated units would be much less expensive and easier to install than the custom built units tested to date.

Conclusions

The development and testing of the MCTT showed that the treatment unit provided substantial reductions in stormwater toxicants (both in particulate and filtered phases), and suspended solids. Increases in color and a slight decrease in pH also occurred during the filtration step at the pilot-scale unit. The main settling chamber resulted in substantial reductions in total and dissolved toxicity, lead, zinc, certain organic toxicants, suspended solids, COD, turbidity, and color. The filter/ion exchange unit is also responsible for additional filterable toxicant reductions. However, the catchbasin/grit chamber did not indicate any significant improvements in water quality, although it is an important element in reducing maintenance problems by trapping bulk material. The use of the MCTT is seen to be capable of reducing a broad range of stormwater pollutants that have been shown to cause substantial receiving water problems (Pitt 1995).

Acknowledgments

This research was mostly funded by the Urban Watershed Management Branch of the U.S. EPA, Edison, New Jersey. Rich Field provided much guidance and assistance during the research. Additional funding was also provided by the U.S. Army-Construction Engineering Research Laboratory in Champaign, Illinois. Rick Scholtz's efforts are greatly appreciated. Much of the work reported in this paper was carried out by Brian Robertson as part of his MSCE thesis at the University of Alabama at Birmingham. Additional contributions were also made by the following UAB personnel: Dr. Keith Parmer, Shirley Clark, Olga Mirov, Jay Day, and Holly Ray. Special thanks is also extended to the cities of Minocqua and Milwaukee, the state of Wisconsin, and Region V of the EPA for constructing and monitoring the full-scale MCTT installations. Roger Bannerman and Tom Blake of the Wisconsin Department of Natural Resources along with Steve Corsi of the USGS in Madison were especially instrumental in carrying out these full-scale tests.

References

Aires, N, and J.P. Tabuchi. "Hydrocarbons separators and stormwater treatment." (in French) *TSM, Special Issue «Stormwater»*, no 11, pp 862-864. 1995.

Austin, Texas (City of). *Design Guidelines for Water Quality Control Basins*. Environmental DCM. City of Austin Transportation and Public Services Department. 1988.

Clark, S., R. Pitt, and R. Field. "Stormwater Treatment: Inlet Devices and Filtration." Presented at the *Stormwater NPDES Related Monitoring Needs* ASCE conference, Mt. Crested Butte, CO. Proceedings edited by H. Torno. Published by ASCE, New York. pp. 641-650. 1995.

Fourage, M. "Assessment of the efficiency of a prefabricated separator for stormwater treatment. Thoughts to and tests of materials to trap hydrocarbons." (in French) *Unpublished DESS student report, Universities of Nancy and Metz*, September 1992.

Galli, John. *Peat-Sand Filters: A Proposed Stormwater Management Practice for Urbanized Areas*. Prepared for the Coordinated Anacostia Retrofit Program and Office Of Policy and Planning, D.C. Department of Public Works. 1990

Lager, J. and W. Smith. *Catchbasin Technology Overview and Assessment*. U.S. Environmental Protection Agency. 1976.

Legrand, J., H. Maillot, F. Nougarède, and S. Defontaine. "A device for stormwater treatment in the urban development zone of Annœullin." (in French) *TSM*, no 11, pp 639-643. 1994.

Pitt, R. and P. Bissonnette. *Characterizing and Controlling Urban Runoff through Street and Sewerage Cleaning*. U.S. Environmental Protection Agency. Storm and Combined Sewer Program, Risk Reduction Engineering Laboratory. EPA/600/S2-85/038. PB 85-186500. Cincinnati, Ohio, June 1985.

Pitt, R. *Small Storm Urban Flow and Particulate Washoff Contributions to Outfall Discharges*. Ph.D. dissertation. Dept. of Civil and Environmental Engineering, the University of Wisconsin, Madison, Wisconsin. Listed in Dissertation Abstracts International, University Microfilms International, Vol. 49, No. 1, 1988. November 1987.

Pitt, R., R. Field, M. Lalor, and M. Brown. "Urban Stormwater Toxic Pollutants: Assessment, Sources, and Treatability." *Water Environment Research*. May/June 1995.

Pitt, R. "Biological Effects of Urban runoff Discharges." In: *Urban runoff and Receiving Water Systems: An Interdisciplinary Analysis of Impact, Monitoring, and Management*.

Engineering Foundation and ASCE. Lewis Publishers, Chelsea, Michigan. to be published in 1995.

Robertson, B., R. Pitt, A. Ayyoubi, and R. Field. "A Multi-Chambered Stormwater Treatment Train." In: *Stormwater NPDES Related Monitoring Needs* (Edited by H.C. Torno). Engineering Foundation and ASCE. pp. 631-640. 1995.

Robertson, B. *Evaluation of a Multi-Chambered Treatment Train (MCTT) for Treatment of Stormwater Runoff from Critical Pollutant Source Areas.* MSCE Thesis, Department of Civil and Environmental Engineering. University of Alabama at Birmingham. 268 pages. 1995.

Rupperd, Y. "A lamellar separator for urban street runoff treatment." (in French) *Bulletin de Liaison des Laboratoires des Ponts et Chaussées*, no 183, pp 85-90.1993.

Schueler, T. (editor). "Hydrocarbon Hotspots in the Urban Landscape: Can they be Controlled?" *Watershed Protection Techniques*. Vol. 1, No. 1, pp. 3 - 5. Feb. 1994.

Shaver, Earl. "Sand Filter Design for Water Quality Treatment." *Proceedings from an Engineering Foundation Specialty Conference*. Crested Butte, Colorado, 1991.

DISCUSSION

Measuring the Health of Aquatic Ecosystems Using Biological Assessment Techniques
Mike Barbour

The required approach to ecologic assessments
- Cost-effective
- Multiple site investigations
- Rapid turn-around for data
- Readily obtained decisions or judgements
- Easily translated to management and public
- Use environmentally benign procedures

Only three states have biological guidelines in place: Florida, Delaware, Ohio

Using Bioassessments to Evaluate Cumulative Effects
Ellen McCarron

There is a need and rationale for alternative sampling and assessment procedures that provide a more ecologically-based manner of determining the cumulative environmental effects of nonpoint sources of pollution. Florida Department of Environmental Protection has been working on a methodology for six years.

Developing bioassessment tools is not quick nor inexpensive. Tools developed to date need refinement, calibration and evaluation of the biometrics

Issues still to be resolved: 1)Validate the division of the panhandle and peninsula into sub-ecoregions; 2) Evaluate the Hisenhoff Biotic Index, as a sensitivity measure; 3) Conduct habitat evaluations at reference and habitat-limited sites; 4) Develop a "user-friendly" data management program.

Questions/Comments:

Question: How often do you get false positives?
Answer: The protocol is still under development, should know better in about
             a year

Question: How do you incorporate flow, and wet weather events into the
             technology?
Answer: Just try not to go out after large rains that might disrupt the
             ecosystem?

Question: Emphasis is on biology, but how do you determine by this method
               the cause of the impairment?
Answer: We don't know this at this time, but will be working on this in the
               next few years. Using GIS to record land use and biological
               assessment; this helps know where to target additional efforts.

Question: How will this be used for regulation and determining how to get
               there?
Answer: Don't know yet, but are working toward this, Eric will address this
               more in his closing talk.

Bioassessment for Intermittent Central Texas Streams
   Robert Hanson

City of Austin has developed and implemented bioassessment techniques on two
local intermittent streams with various degrees of urbanization and development.
Current bioassessment techniques do not work well, and future strategies must be
based on a ecological model of community structure that accounts for low
abundance and diversity at unimpacted sites.

Questions/Comments

Question: What is the stormwater sampling budget of for City?
Answer: The budget is about $1M dollars per year, but does not include any
               biological sampling.

Question: How did diatoms correlate to nutrient levels?
Answer: Diatoms correlate more closely to phosphorus than nitrogen

The Control of Toxicants at Critical Source Areas
   Robert Pitt

This paper reviews the design of the MCTT (a multi-chambered treatment train)
and presents monitored performance information for a broad list of stormwater
constituents.

A pilot-scale MCTT was constructed in Birmingham, AL, and tested over a six
month monitoring period, from May to October, 1994. Two additional full-scale
MCTT units have recently been constructed and are currently being monitored.

The MCTT had very effective removal rates for both filtered and particulate
stormwater toxicants and suspended solids.

Questions/Comments

Question:	Do you have any cost data on cost of implementation and O&M?
Answer:	No

Question: Any look at controls to bypass or automatic controls to better control flows through these devices?

Answer: No, communities installing the two pilot facilities wanted completely passive facilities, but design is such that they do by-pass when they are full.

Comment: Getting about 50% removal of phosphorus, but export of nitrogen by denitrification to ammonia occurs

Comment: The two devices installed to date were bid high. Approx. $90,000 for each of the two pilot installations, but city engineers estimate the cost can be reduced by a factor of 2.

Assessment of the Response of Aquatic Organisms to Long-term
Insitu Exposures of Urban Runoff

Ronald Crunkilton[1], Jon Kleist , Joseph Ramcheck,
William DeVita and Daniel Villeneueve

Abstract

   Lincoln Creek, an urban stream, located in Milwaukee Co., contains just two
fish species and 12 families of benthic macroinvertebrates. It is classified as severely
impaired relative to a similar-sized, reference stream with 20 fish species and 21
families of benthic macroinvertebrates. A 1993 study of toxicological impacts of
stormwater runoff in Lincoln Creek failed to detect toxicity in over 300, conventional
48-96 hr acute and 7 d chronic tests with *Ceriodaphnia dubia* and *Pimephales
promelas* performed on 24 consecutive runoff events during a five month period
beginning with the first snowmelt runoff. Substantial mortality was observed,
however, for juvenile and adult *P. promelas* exposed for greater than 14 d,
suggesting that effects of long-term exposure could be responsible for the extensive
biological degradation observed in Lincoln Creek and that conventional wastewater
toxicity testing procedures could not be used to predict this effect. In 1994, toxicity
tests were modified to include longer-term *insitu* exposures. Tests were also
designed to separate toxicity associated with base flow from that associated with high
flow. Reproduction and survival in *Daphnia magna*, and growth and survival in *P.
promelas* were monitored over seven consecutive 14 d periods to assess acute and
chronic effects. Tests were performed between June and September in flow-through
aquaria housed within a U.S. Geological Survey gauging station located adjacent to
Lincoln Creek. Additional tests assessed mortality of juvenile and adult *P. promelas*
during long-term exposures that lasted up to 61 d.
   Mortality of *D. magna* exposed to stream water measured after the typical
interval of 48 hr was never significantly ($p < 0.05$) greater than controls. At seven
days, 36 % of the tests showed significant mortality and at the end of 14 d, 93% of
the tests detected significant increases in mortality. Production of neonates

[1] Associate Professor of Water Resources, College of Natural Resources, University of Wisconsin
Stevens Point, WI 54481

beginning with seven day old *D. magna* following seven days of exposure to river water was also significantly reduced in 36% of the tests and in 93% of the tests by day 14.

Mortality in *P. promelas* larva exposed to stream water measured after the typical 96 hour interval was never significantly ($p<0.05$) greater than controls and was greater than the controls in only 14% of the tests at the end of 14 d. Growth as measured by a change in biomass was a more sensitive measure of toxic effects than mortality in this species. By day seven, 60 % of the tests showed significant biomass reductions and by day 14, 75% compared to controls. Longer exposures of 17-61d, however, with juvenile and adult *P. promelas* were effective in detecting increased mortality. Organisms in all four long-term tests exhibited significantly ($p<0.05$) increased mortality ranging from 30% to 95% in 17-61 d exposures.

The above toxicity tests were also assessed under high flow and base flow conditions in Lincoln Creek. There were few differences (less than 5% of tests) in toxic effects identified under the two different flow regimes. Although Lincoln Creek water at base flow demonstrated the same magnitude of toxic effects in test organisms as high flow, stormwater runoff was clearly responsible for the many potentially toxic contaminants detected in stream water during the study. This study suggests that typical wastewater effluent toxicity tests were ineffective at detecting the type of degradation seen in this urban stream. Longer-term tests, greater than seven days and preferably more than 14 days, and more sensitive endpoints such as measures of biomass were needed to identify adverse biological impacts.

Introduction

Stormwater runoff is a major source of contaminants in urban waterways. Levels of suspended solids, heavy metals, polycyclic aromatic hydrocarbons (PAHs), pesticides, and bacteria often exceed federal water quality standards (Bannerman et al., 1993). Sources of contaminants include automobile emissions, streets, parking lots, rooftops, lawns, as well as construction sites, stream bank erosion, and wet and dry atmospheric deposition (Bomboi and Hernandez, 1991; Cheong, 1991; Zanoni, 1986; Novotny et al., 1985; Cole et al., 1984). These contaminants are known to have detrimental impacts on the biota of urban waters (Pitt, 1995; House et al., 1993; Clements et al., 1988; Medeiros and Coler, 1983).

In addition to chemical inputs, many urban stream channels suffer from physical degradation. Urbanization leads to fundamental changes in the hydrologic, hydraulic, erosional and depositional characteristics of fluvial systems causing increased stream channel instability (Rhoads, 1995). Stream channels are often modified to prevent flooding and expedite storm flow passage (House et al., 1993). Channels are often lined with concrete contributing to extreme fluctuation in river discharges. Unstable or uniform channels lack structure that allows colonization of aquatic organisms. Water control structures such as culverts and low head dams regulate flow, but can prevent upstream movements by fish. Low discharge volumes at base flow in urban streams can lead to high water temperatures with associated biological effects (Herricks, 1995). The combined effect of habitat instability, habitat alterations, temperature changes and chemical toxicity can drastically alter natural aquatic communities in urban waterways.

The purpose of this study was to assess the biological effects of urban runoff in Lincoln Creek located in a metropolitan drainage in Milwaukee, WI. Substantial degradation to physical habitat and the biotic community of the stream had been documented in a previous study (Masterson and Bannerman,1994). The role of toxicants in limiting development of the biotic community was unclear. The goal of this study was to determine if toxicants alone could play a role in the observed degradation of the stream community. This was a practical question because most stormwater remediation efforts implemented to protect streams have focused on mitigating physical degradation of habitat. Little is known about how stormwater remediation structures might ultimately influence toxicity to aquatic organisms. Results from studies performed in 1993 and 1994 are reported here.

Study Area

Lincoln Creek is an urban stream located in Milwaukee, Wisconsin. It is approximately 14.5 km long and drains a 50 square km watershed. In 1993, land use in the drainage basin was about 50% residential, 12% industrial, and 7% commercial. The remaining area was open space including U.S. Army property, Havenwoods Forest Preserve, Milwaukee County Lincoln Creek Parkway, municipal parks and golf courses (WDNR, 1991). Impervious cover in the watershed is approximately 30%.

Lincoln Creek originates in a series of parking lots which drain into a concrete channel. Alternating concrete-lined and earthen sections are found along the stream course from the headwaters to the mouth at the Milwaukee river. Concrete-lined sections provide little habitat for fish or macroinvertebrates and exacerbate discharge fluctuations and scour in downstream earthen sections of the channel. Drop structures and culverts have been installed to regulate water flows.

Much of the original fauna indigenous to Lincoln Creek has disappeared with urbanization. The existing fauna is dominated by pollution tolerant taxa and the community is rated as "severely impaired" (Masterson and Bannerman, 1994). In 1993, the stream supported two species of fish and 12 taxa of macroinvertebrates compared to 20 species of fish and 21 taxa of macroinvertebrates in a nearby similar-sized unimpacted reference stream.

Methods 1993

Two automated samplers were maintained in a U.S. Geological Survey gauging station located about midway between the origin and mouth of Lincoln Creek. One sampler was programmed to collect a flow-proportional composite sample representative of an entire storm, and a second sampler was programmed to collect up to eight discrete samples sequentially during a storm. Both samplers were activated by a rise in stream level 15 cm above base flow. Sampling terminated when the stream level fell below the 15 cm rise. All samples were collected from the stream. Virtually all flow in the stream during runoff events originated from runoff.

Water was also obtained biweekly at base flow with a grab device that sampled the entire water column. If a storm was in progress during the scheduled sampling time, collection was delayed until the stream returned to base flow. Upon collection, the high flow composite sample and the base flow sample were split for chemical analysis and toxicity testing. Discrete samples collected during high flow were used only for toxicity testing. All samples were held on ice or preserved with approved procedures before testing or chemical analysis.

Water Quality:

Twenty high flow composite and nine base flow grab samples from Lincoln Creek were analyzed for a variety of inorganic and organic contaminants between March and September of 1993 (Ramcheck, 1994). Total recoverable metals were measured on unfiltered samples and dissolved metals were determined by passing samples through a 0.45 micron filter. Total polycyclic aromatic hydrocarbons were measured on unfiltered samples and dissolved PAHs were determined from the equilibrium partitioning model described by DeVita (1994). For metals, water hardness from composite and base flow samples was used to calculate both the Acute Toxicity Criteria (ATC) and Chronic Toxicity Criteria (CTC) for cadmium, copper, zinc, and lead (WDNR 1989).

Laboratory Toxicity Tests:

Laboratory acute toxicity tests were performed on high flow discrete, high flow composite and base flow grab samples collected from Lincoln Creek. Tests and analysis followed standard procedures commonly known as acute and chronic whole effluent toxicity tests (WET) tests (U.S. EPA, 1993; 1991). Tests were conducted with five serial dilutions (100%, 50%, 25%, 12.5%, 6.25%) and control water. Test organisms were *Ceriodaphnia dubia, Daphnia magna* (cladocerans) and *Pimephales promelas* (fathead minnows). Control and dilution water consisted of laboratory reconstituted water of hardness equivalent to sample hardness.

On-site Toxicity Testing:

Indigenous fish, marcoinvertebrates, and long-term exposures with *P. promelas* were conducted on-site in flow-through tanks. Eighteen 37-L aquaria located inside the gauging station were supplied with a continuous metered flow of water from Lincoln Creek. Filtration was limited to 4 mm perforations in the stream intake. Turnover rate was about two tank volumes/hour. Control exposure tanks were similar except clean (dechlorinated aged tap water) was recirculated through a sand filter and another set of 18 tanks. Temperature in all tests was maintained at ambient stream temperature by external baths supplied with creek water.

Native fish believed to have been present in Lincoln Creek before urbanization were selected for use in 7 d flow-through toxicity tests. The species were selected from historical records of Lincoln Creek and from records of similar streams in the same region. They were: common shiner (*Notropis cornutus*), golden shiner (*Notemigonus crysoleucas*), blacknose dace (*Rhinichthyes atratulus*), rainbow darter (*Etheostoma caeruleum*), fantail darter (*Etheostoma flabellare*), white sucker (*Catostomus commersoni*), yellow perch (*Perca flavescens*), largemouth bass, (*Micropterus salmoides*), bluegill (*Lepomis macrochirus*) and fathead minnow (*Pimephales promelas*). Fish were collected from wild stock in nearby streams, transported to the study site and exposed for 7 d. Temperature was maintained at ambient stream level with an external water bath. Fish were not fed. Test design required two replicates of 10 fish each in exposure and control tanks. Data were analyzed for statistical significance with Fishers Exact Test (p < 0.05) which compares the proportion of animals surviving in the control to those surviving in the exposure.

Macroinvertebrates were collected with artificial leaf retaining devices (ALRDs). Baskets (15 X 21 cm.) were assembled from 1 cm square mesh, polyethylene net material. Colonization substrate was 10 grams of air-dried leaves of three tree species (*Acer saccharium, Quercus alba* and *Populus deltoides*). Leaves were weighed and then layered into the mesh baskets. The ALRDs were closed and transported to a reference stream for a two week colonization. They were mounted on a 90 cm section of 1 cm steel bar which was driven into the bottom of the stream with the leaf baskets positioned about 8 cm above the bottom. The colonized ALRDs were then placed in 300 micron mesh bags for transport to Lincoln Creek. The colonized ALRDs were temperature-acclimated before being placed in exposure aquaria. Experimental design entailed four replicate ALRDs in exposure and four replicates in control tanks. Turnover of water in all tanks was about two times per hour and exposure duration was 48 hr. After exposure, contents of the ALRDs were placed in a pan, and the live macroinvertebrates were sorted from the dead and preserved in 80% isopropyl alcohol for later identification to Family level. Mortality data were analyzed with non-parametric one-way analysis of variance (SAS Institute, Cary, NC).

Juvenile *P. promelas* were exposed under conditions identical to indigenous fish species tests except for a longer 30 d exposure and daily feeding with granular trout chow.

Methods 1994

Based on the limitations of standard WET testing procedures and other short-term tests in assessing biological effects in the 1993 field season, toxicity assessments in 1994 were modified to include long-term exposures. An additional objective in 1994 was to determine if there was a difference in effects to organisms exposed at high flow and those exposed to base flow in the stream. All exposures were performed between June 8 and September 16, 1994 in on-site flow-through tanks located inside the U.S. Geological Survey gauging station previously described. Toxicity tests performed included *D. magna* survival and reproduction over 14 d, *P.*

promelas fry survival and growth over 14 d, and juvenile *P. promelas* survival for up to 60 d. All tests were performed at ambient stream temperature.

Flow-Through Exposure System:

The river flow-through system was designed to compare exposures of organisms to base flow with high flow water for extended periods of time. Test organisms placed in six 37 L aquaria designated as base flow received stream water only when the stream was at a normal low flow of about 0.08 cubic meters/s. Delivery of water to base flow aquaria was automatically stopped whenever the stream level rose more than 15 cm above base flow which corresponded to 0.25 cubic meters/s or greater and resumed whenever the stream dropped below that level. To maintain dissolved oxygen levels in base flow tanks when water flow was stopped, tanks were automatically aerated.

Another set of six 37 L tanks were used for the high flow tests. These tanks received water (base flow and high flow) continuously from the river regardless of water level. Because of the sporadic nature and short duration of high flow events, test organisms in high flow aquaria received a continuous supply of water from Lincoln Creek. Thus, organisms exposed at high flow were exposed to normal base flow as well as any high flow events. This system allowed the added effects caused by stormwater runoff to be measured. Turnover rate of water for all tanks was about two times per hour.

Toxicity Tests:

Mortality and reproduction were measured in *D. magna* over a 14 d period. Seven consecutive tests were performed between June 9 and September 15, 1994. Tests were performed for each water type, control, base flow, and high flow. Each test consisted of 10 replicates with one 7 d old *D. magna*. Survival and number of young produced by each organism were recorded at 24 hour intervals. Organisms were fed twice daily. In addition, 39 consecutive *D. magna* 48 hr. tests with *D. magna* neonates were performed every seven days over the study duration. The tests consisted of two replicates with 10 organisms per replicate. Mortality was recorded daily.

Mortality was recorded daily in *P. promelas* over seven consecutive 14 d test periods between June 9 and September 15, 1994. Growth (biomass) was measured at day 7 and day 14. The test consisted of four replicates with 10 organisms per replicate. Fish were fed daily. Biomass was calculated by dividing the fry weight by 10 (the number of *P. promelas* fry that were present at the start of each test). Survival of juvenile *P. promelas* was also assessed in longer exposures that lasted up to 60 days. Thirty juvenile *P. promelas* per water type were placed in flow-through aquaria. Mortality was recorded daily. Fish were fed daily.

A recirculating water system as previously described provided clean control water for all bioassay tests. Ambient stream temperatures were maintained in all exposure and test aquaria with an external water bath supplied with stream water. Statistical calculations were performed with Toxstat (Gulley and West, 1994).

Water Quality:
Dissolved oxygen and temperature readings were recorded hourly throughout all tests. Thirteen flow-proportional water samples (high flow) and nine composite (base flow) samples were collected for chemical analysis as in 1993.

Results

Base flow in Lincoln Creek was 0.08 cubic meters/s. The average high flow, based on 47 measured runoff events greater than 0.25 cubic meters/s (a 15 cm rise in the water level) during the 1993-94 runoff season, was 12.6 cubic meters/s. The average interval between runoff events during this time was 3.5 d and the average duration of a high flow event was 11 hr. During 1994 the number of runoff events ranged from two to six per each 14 d test period. The stream was at high flow(> 0.25 cubic meters/s for 13% of the runoff season between March and November.

Stream temperatures ranged from 14C to 27C during the study period with a daily fluctuation of 2C to 5C. Dissolved oxygen ranged from 1-15 mg/L and dropped below 5 mg/L during four of the seven 14 d test periods during 1994. Generally, reductions were small and lasted for only one or two hours between 12 am. and 9 am. However, one test period in September showed reductions to less than 5 mg/L for 18 % of the total 14 d test period.

Total recoverable metal concentrations in high flow samples averaged 7X greater than base flow samples and dissolved metals about 2X greater (Table 1). About half of the metals were in the dissolved form at base flow and about 10 % at high flow. Total recoverable metal concentrations were strongly correlated with suspended solids content (Table 2). Acute Toxicity Criteria based on total recoverable Pb, Zn, Cu, and Cd were exceeded from 0% (Cd) to 39% (Zn & Cu) of the time for these metals in high flow samples, but never in base flow samples (Table 3). Chronic Toxicity Criteria were exceeded from 33% (Cd) to 88% (Pb & Zn) of the time in high flow samples and between 0% (Pb) to 6% (Zn, Cu, Cd) in base flow samples (Table 4). Total and dissolved PAHs were about 20X greater at high flow than base flow (Table 5). About half of PAHs at high flow and half at base flow were classified as dissolved. Total PAHs and suspended solids content were strongly correlated (Table 2).

Toxicity tests 1993:
Acute mortality during high flow or base flow was rarely identified beginning with the first snowmelt runoff in March and continuing through September in 1993. Out of 95 separate tests based on WET procedures with *C. dubia, D. magna* and *P. promelas,* only seven showed greater than 10% mortality (Table 6). Included in this summary were first flush samples, subsequent discrete samples, whole-storm runoff composite samples, as well as base flow samples. There was no discernible pattern in observed toxicity with respect to sample type nor was there any correlation of

Metal (µg/L)	Total Recoverable		Dissolved	
	High Flow	Base Flow	High Flow	Base Flow
Lead	35	3	1.7	1.2
Zinc	133	22	13	8
Copper	23	7	5	4
Cadmium	0.6	0.1	0.1	0.1

Table 1. Average total recoverable and dissolved metals for 13 stormwater runoff and 9 base flow samples collected from Lincoln Creek in 1994.

	r-square
Lead	0.97
Zinc	0.90
Copper	0.92
Cadmium	0.86
PAHs	0.95

Table 2. Coefficient of determination (r-square) values for total suspended residue solids vs total recoverable metal and polycyclic aromatic hydrocarbon concentrations for 21 samples collected in 1994.

Metal (total recoverable)	Percent of samples Acute Toxicity Criteria (ATC) exceeded	
	High flow (n=33)	Base flow (n=17)
Lead	3%	0%
Zinc	39%	0%
Copper	39%	0%
Cadmium	0%	0%

Table 3. Acute toxicity criteria (ATC) exceedences of 33 unfiltered high flow and 17 base flow samples collected from Lincoln Cr. 1993-94. Acute toxicity criteria were based on water hardness.

Metal (total recoverable)	Percent of samples Chronic Toxicity Criteria (ATC) exceeded	
	High Flow (n=33)	Base Flow (n=17)
Lead	88%	0%
Zinc	88%	6%
Copper	76%	6%
Cadmium	33%	0%

Table 4. Chronic toxicity criteria (CTC) exceedences of 33 unfiltered high flow and 17 base flow samples collected from Lincoln Cr. in 1993-94. Chronic toxicity criteria were based on water hardness.

Polycyclic Aromatic Hydrocarbon (PAH) concentration (µg/L)			
High flow		Base flow	
Total	Dissolved	Total	Dissolved
13.4	6.5	0.65	0.36

Table 5. Average polycyclic aromatic hydrocarbon (PAH) concentrations of nine unfiltered stormwater runoff and eight base flow samples collected from Lincoln Creek in 1994.

		n	> 10 % mortality
High flow	C. dubia	83	2
	D. magna	83	0
	P. promelas	83	4
Base flow	C. dubia	12	0
	D. magna	12	0
	P. promelas	12	1

Table 6. Summary of 48-96 hour acute laboratory toxicity tests with *Ceriodaphnia dubia*, *Daphnia magna* and *Pimephales promelas* performed with Lincoln Creek water in 1993.

toxic effects among species for any particular sample. Mortalities for individual tests ranged from 15% to 80%. The greatest observed mortality was *80 % for C. dubia*, but this coincided with a high flow event that was attributed to a break in a municipal water line that spilled chlorinated water into Lincoln Creek. The second greatest observed mortality was 53 % for *P. promelas* and occurred during a high flow event. The remaining mortalities were all less than 50%. None of the 31 seven day reproductive tests with *C. dubia* showed significant decreases in production of young in either water type (Table 7).

Flow-through tests with indigenous species of fish (7 d exposures) and macroinvertebrates (2 d exposures) were also ineffective in detecting toxicity (Table 8). Only one fish species, blacknose dace, showed significantly greater mortality than the control. On another test date with the same species, no mortality occurred. Both 7 d exposures with this species included a high flow event and there were no discernible differences identified in dissolved oxygen, temperature, or other water quality measurements. Only one macroinvertebrate family, Hydropsychidae, (caddisflies) showed a statistical increase in mortality (Table 8). This increased toxicity occurred during a base flow exposure, but was not repeated during subsequent base flow and high flow exposures with the same taxa. Again, there were no discernible patterns identified in water quality measurements that could explain this mortality. Other pollution intolerant Families of insects such as Pteronarcidae (stonefly), Perlidae and Ephemerillidae (mayflies) were also used in multiple tests with no increased mortalities.

Although mortality was rarely identified in previously described tests performed in 1993, it was greater than 50% in two consecutive flow-through

		n	Statistical difference from control (p<0.05)
High flow	C. dubia	21	none
Base flow	C. dubia	10	none

Table 7. Summary of 7 day chronic laboratory tests of *Ceriodaphnia dubia* performed with Lincoln Creek water in 1993.

	Number of tests	Duration (days)	n	Statistically significant mortality (p<0.05)
Fish (11 species)	10	7	26	1
Invertebrates (23 Families)	9	2	41	1

Table 8. Summary of indigenous species flow-through tests performed at the creek-side flow-through exposure station on Lincoln Creek in 1993.

exposures with *P. promelas* that lasted 30 d. Mortality in both tests was not encountered until after about 14 d of exposure, but was consistent and repeatable in both tests. Because of these findings, testing procedures in 1994 were modified to include longer exposures and all tests were performed on-site in flow-through aquaria.

Toxicity tests 1994:

Mortality for *D. magna* exposed for the longer duration of 14 d at high flow and base flow was substantial (Fig. 1). Cumulative mortality ranged from 30% to 100% in the seven consecutive tests. There were no apparent differences in toxicity of organisms exposed to base flow and those that were exposed to high flow. Both were equally toxic. If these tests had been terminated at 48 hr, the typical endpoint for an invertebrate lethality test, no statistical differences between exposed and control organisms would have been detected, at the end of 7 d effects would have been identified in 36% of the tests and at the end of 14 d effects would have been detected in 93% of the tests (Table 9). No further increases in the ability to detect statistical differences were realized by using production of young as an endpoint, although total production of young in control exposures was approximately twice that in base flow or high flow exposures at the end of the test.

Mortality for *P. promelas* exposed under similar base flow and high flow conditions at the end of 14 d ranged from 5% to 100% in seven tests (Fig. 2). Overall, mortality was less than for *D. magna*. If these tests were terminated at 4 d or 7 d, the typical endpoints of an acute toxicity test with fish, no statistical

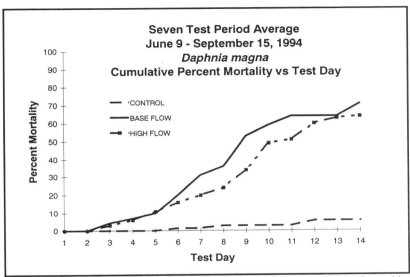

Figure 1. Average mortality in *Daphnia magna* during seven 14 day tests performed in 1994.

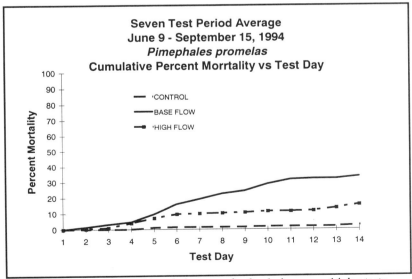

Figure 2. Average mortality in *Pimephales promelas* fry during seven 14 day tests performed in 1994.

	Percent of tests in which significant (p<0.05) toxic effects compared to control were detected				
Exposure	48 hours	96 hours	7 days	14 days	17-61 days
D. magna- mortality	0	-	36	93	-
D. magna - reproduction	0	-	36	93	-
P. promelas - mortality	-	0	0	14	100
P. promelas - biomass	-	-	60	75	-

Table 9. Time effects relating exposure time and the ability to statistically detect toxicity in *Daphnia magna* and *Pimephales promelas* exposed to Lincoln Creek water in flow-through toxicity tests performed in 1994.

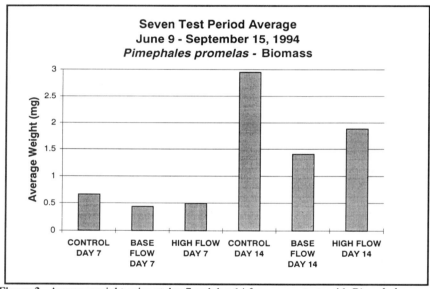

Figure 3. Average weight gains at day 7 and day 14 from seven tests with *Pimephales promelas* fry follwing exposure to Lincoln Creek water in 1994 .

differences between exposed and control organisms would have been detected, at the end 14 d effects would have been identified in 14% of the tests and at the end of 17 d to 61 d effects would have been detected in all of the tests (Table 9). The ability to detect adverse effects increased when growth (biomass), measured as an increase in weight over the 14 d period, was considered. At day seven 60% of the tests showed a statistical decrease in growth compared to the control and at 14 d, 75%. Average weights at the end of 14 d in control exposures were about twice that of base flow or high flow exposures (Fig. 3).

Mortality could not be attributed to any particular pattern of runoff or water quality measurement. Patterns of toxic responses during 14 d tests that had up to six runoff events were no different than those that had two runoff events. Although dissolved oxygen occasionally dropped below 5.0 mg/L, the reduced levels did not appear to be related to any immediate mortality.

Discussion

Short-term tests based on acute and chronic whole effluent toxicity (WET) protocol appeared to underestimate the toxic effects of urban runoff in Lincoln Creek. WET tests are of limited duration, up to 7 days in length, and employ a few standard test species that can be easily handled under laboratory conditions. These species may not be as sensitive as indigenous fauna to potential toxic effects, therefore may underestimate the potential for stream degradation. However, similar results were obtained when we employed tests that used pollution sensitive fish and macroinvertebrates from nearby reference streams. Only when test duration was extended beyond seven days in flow-through tests was substantial mortality observed. Long-term exposures were needed to detect potential toxicity from stormwater runoff. The ability to detect statistical differences when they exist (statistical power) is a function of the ratio of the variability between different treatments and variability within treatments which in turn is a function of sample size or replication. A complete evaluation of statistical power has not been completed for the statistical analysis described here, but it appears that increased replication would not have increased the ability to detect adverse effects if the tests were terminated at the typical endpoints of 48 hours for invertebrates and 96 hours for fish. Mortality simply did not occur until after these endpoints had passed.

Pitt (1995) has urged caution in the use of standard laboratory bioassays to detect toxicity in urban runoff because long-term exposure may be more important than short-term effects associated with specific runoff events. He recommended use of field studies in preference to bioassays to study effects of urban runoff. In contrast, Dickson et al. (1989) in a study of the Trinity River in Texas reported that ambient toxicity was a good predictor of in-stream effects. Our studies indicated that the typical WET approach could not detect in-stream effects. However, by lengthening exposure time and performing these tests *insitu*, useful information about the presence of toxic conditions were detected in 7-14 days.

Although not demonstrated here, routine WET tests may also overestimate potential toxicity. The typical runoff event lasts only a fraction of the time required to complete a whole-organism effluent toxicity test with *C. dubia* or *P. promelas*. The average high flow event in Lincoln Creek lasted 11 hours. Organisms used in WET tests must be exposed 2-4 days for lethality endpoints and 7 days for chronic endpoints. Because concentrations of many potentially toxicants at high flow are greater than base flow, exposing test organisms to samples collected at high flow may result in an overestimation potential harmful effects. Toxicant concentrations during high flow may also change dramatically within minutes, thus grab or even composite flow-weighted samples do not provide realistic exposure scenarios for test organisms. This potential problem is part of what Herricks and Clarke (1996) have coined "timescale toxicity effects" which asserts sampling and analysis procedures and toxicity assessments must be fully integrated because of issues related to rapidly changing toxicant concentrations and exposure times. Timescale concerns beg adoption of surrogate tests such as bacterial assays, biochemical assays or other measures of organism well-being that can be performed in minutes rather than days. Alternatively, *insitu* tests are warranted when long-term effects are of interest. Burton and Stemmer (1988) demonstrated the usefulness of caging surrogate organisms in polluted streams to assess biological effects. We chose to use a modified *insitu* approach because it provided the most biologically relevant information about in-stream effects. By exposing organisms to flow-through conditions off-stream in an enclosed building we were able to carefully control the experimental environment. This allowed us to isolate "toxicity" and remove the physical factors associated with channel degradation associated with urban runoff that is often not possible with interpretation of biological data from field surveys. The off-stream exposure facility permitted us to replicate experimental treatments and to expose appropriate control organisms in clean water. Organisms could also be fed daily, a necessity for tests lasting for more than a few days, and for tests which assess growth or reproduction.

The strong correlation between total recoverable metals or total PAHs and suspended solids suggest that most contaminants within the stream channel are particulate-bound. A portion of these particulates may be resuspended sediments deposited from previous runoff events. Resuspended sediments may be more important than short-term discharges from first flush runoff and contribute to delayed toxicological effects. Heany and Huber (1978) reported scouring and resuspension of unconsolidated bottom sediments as the dominant source of pollutants in a stream in Springfield, Missouri. Ellis (1986) showed accumulated nutrients, metals, hydrocarbons, and benthal deposits caused delayed toxicological effects in an urban stream. Our studies have consistently shown significant toxicity to organisms exposed under base flow conditions. This suggests a mechanism where urban runoff delivers contaminants to the stream channel with the toxic effects being manifested after base flow has resumed (Mederios et al., 1983).

Because of the complex toxicological interactions that may occur between potential toxicants and changing water chemistry, and physiological interactions within an organism that result from exposure to complex mixtures, it is difficult to determine with certainty what substances are contributing to toxicity. Concentrations of total recoverable lead, copper, cadmium, and zinc often exceeded the acute and chronic toxicity criteria for metals established for the protection of aquatic life. Suitable criteria for PAHs do not exist for the protection of aquatic life, but water concentrations were within the range known to cause adverse effects in some aquatic organisms. In other related studies, we have determined with *invitro* and *invivo* tests that organisms are metabolically challenged upon exposure to PAHs found in Lincoln Creek (Villeneueve et al., in press). Biotransformation and excretion of sublethal concentrations of PAHs (or metals) exerts a metabolic cost that may only manifest itself in longer exposures by compromising the organism's immune system, perhaps leading to death from secondary causes.

Applications:

This was an intensive study focused on one stream. The experimental design does not allow us to make broad generalizations about the potential role of toxicity in other urban streams. Although chemical characteristics of urban runoff from widely differing geographic areas may be similar, the physical factors which control the transport, fate and toxicity of these substances once in the stream channel may differ widely. Effects of urban runoff on stream biota are likely site-specific as has been stated by Pitt (1995) and others, however, our findings may have some application to other urban drainage assessments. We demonstrated short-comings of WET tests or other short-term tests in detecting whole-animal toxicity in Lincoln Creek; their use in detecting toxicity on streams more highly influenced by urban runoff may be more successful. By simply lengthening exposure time and using flow-through tests we detected substantial mortality related to toxicant exposure. The extent to which this observed mortality can explain the paucity of aquatic organisms in Lincoln Creek is not entirely clear, but does indicate that toxicants in addition to physical degradation play an important role in shaping communities of aquatic organisms in urban streams and that toxicity reduction must be an integral part of stormwater remediation efforts.

References

Bannerman, R.T., D.W. Owens, R.B. Dodds and N.J. Hornewer. 1993. Sources of pollutants in Wisconsin stormwater. Wat. Sci. Tech. 28:241-259.

Bomboi, M.T. and A. Hernandez. 1991. Hydrocarbons in urban runoff: their contributions to the wastewaters. Wat. Res. 25:557-565.

Burton, G.A. Ju. and B.L. Stemmer. 1988. Evaluation of surrogate tests in toxicant impact assessments. Toxicity Assessments: An International Journal 3:255-69.

Cheong, C.P. 1991. Quality of stormwater runoff from an urbanized watershed. Environ. Monitor. and Assess. 19:449-456.

Clements, W.H., Cairns, J. Jr. and D.S. Cherry. 1988. Impacts of heavy metals on insect communities in streams: a comparison of observational and experimental results. Can. J. Fish Aquat. Sci. 45:2017-2025.

Cole, R.H., Fredrick, R.E., Healy, R.P. and R.G. Rolan. 1984. Preliminary findings of the priority pollutants monitoring project of the nationwide urban runoff program. Wat. Pollut. Control Fed. 546:898-908.

DeVita, W.M. 1994. Use of semipermeable polymeric membrane devices to monitor organic contaminants in Lincoln Creek, Milwaukee, Wisconsin. Masters thesis, University of Wisconsin-Stevens Point. 73pp.

Dickson, K.L., Waller, W.T., Kennedy, J.H., Arnold, W.R. Desmond, W.P. et al. 1989. A water quality and ecological survey of the Trinity River. Vol.1. Results of the Trinity River Studies. University of North Texas, Denton, TX

Ellis, J.B. 1986. Pollutional aspects of urban runoff. In: Urban Runoff Pollution. eds., H.C. Torno, J. Marsalek and M. Desbordes. Springer-Verlag, Berlin.

Gulley, D. and West, Inc. 1994. Toxsat 3.4. Western Ecosystems Technology, Inc. Cheyenne, WY.

Herricks, E.E. and S. Clarke. 1995. A contest for understanding stormwater effects in receiving streams. pp.3-9 in Stormwater Runoff and Receiving Systems, Impact Monitoring and Assessment ed. by E.E. Herricks, Lewis Pub. 458 p.

Heany, J. P. and W.C. Huber. 1978. Nationwide assessment of receiving water 8impacts from urban storm water pollution. U.S. Environmental Protection Agency, Cincinnati, OH.

Herricks, E.E. and S. Clarke. 1996. Protocol for wet weather effect assessment and technology performance. Draft document for Water Environment Research Foundation.

House, M.A., Ellis, J.B., Herricks, E.E., Hvitved-Jacobson, T., Seager, J., Lijklema, L., Aalderink, H. and I.T. Clifforde. 1993. Urban drainage - impacts on receiving water quality. Wat. Sci. Tech. 12:117-158.

Masterson, J. P. and R.T. Bannerman. 1994. Impacts of storm water runoff on urban Streams in Milwaukee Co. Wisconsin. pp. 123-133, In Proceedings of the American Water Resources Association, National Symposium on Water Quality ed. by G.L. Pederson. Chicago, IL., American Water Resources Association, Herndon, VA.

Mederios, C.R., Leblanc, R. and R.A. Coler. 1983. An insitu assessment of the acute toxicity of urban runoff to benthic macroinvertebrates. Env. Toxicol. and Chem. 2:119-126.

Novotny, V., Sung, H,A., Bannerman, R.T. and K. Baum. 1985. Estimating nonpoint pollution from small urban watersheds. J. Wat. Pollut. Control Fed. 57:339-348.

Pitt, R.E. 1995. Biological effects of urban runoff discharges. pp. 127-162 in Stormwater Runoff and Receiving Systems, Impact Monitoring and Assessment, ed. by E.E. Herricks, Lewis Pub. 458 p.

Ramcheck, J.M. 1995. Toxicity evaluation of urban stormwater runoff in Lincoln Creek, Milwaukee, Wisconsin. Masters Thesis, University of Wisconsin, Stevens Point,WI. 87p.

Rhoads, B.L. 1995. Stream power: a unifying theme for urban fluvial geomorphology. pp. 65-76 in Stormwater Runoff and Receiving Systems, Impact Monitoring and Assessment ed. by E.E. Herricks, Lewis Pub. 458 p.

U.S. Environmental Protection Agency. 1993. Methods for measuring the acute toxicity of effluents to freshwater and marine organisms. EPA-600/4-90/027F. Cincinnati, OH.

U. S. Environmental Protection Agency. 1991. Short-term methods for estimating the chronic toxicity of effluents and receiving waters to freshwater organisms. EPA-600/4-91/002, Cincinnati, OH.

Wisconsin Department of Natural Resources (WDNR). 1991. A non-point source control plan for the Milwaukee River South Branch Priority Watershed Project. PUBL-WR-245-91. Madison, WI.

Wisconsin Department of Natural Resources (WDNR). 1989. Surface water quality criteria for toxic substances. Wisconsin Administrative Code Chapter NR 105. Registrar No. 398, Madison, WI.

Villeneueve, D.V. Crunkilton, R.C. and W. DeVita. In press. Ah receptor mediated potency of dissolved lipophyllic organics collected from Lincoln Cr. Milwaukee, WI to PLHC-1 hepotama cells. Environ Tox. Chem.

Zanoni, A.E. 1986. Characteristics and treatability of urban runoff residuals. Wat. Res. 20:651-659.

Assessing the response of aquatic organisms to short-term exposures to urban runoff.

E. E. Herricks[1], R. Brent[2], I. Milne[3], and I. Johnson[3]

<u>Abstract</u>

Urban runoff, particularly discharges during wet weather, present a contaminant source that has specific analysis needs. During a storm event, concentrations of contaminants may vary over very short time intervals requiring modification of standardized test procedures. To adequately assess the potential toxicity of short-term exposure to storm-related contaminants in urban runoff a test system must meet several criteria: 1) fast response time with a time scale for the response matching the time scale of the exposure, 2) capacity to respond to changing conditions (e.g. track changing toxicity), and 3) respond to the effects of a single stressor, as well as integrate the effect of multiple physical, chemical, and biological stressors (Herricks et al. 1994). Although these criteria may be contradictory for a single test system, all criteria can be met in a test battery involving multiple test systems (Milne et al. 1996). Where urban runoff produces short exposure times with large concentration transients, rapid test procedures such as Microtox and an enzyme inhibition assay (IQ testing) are appropriate. Lethality tests with commonly used species such as *Ceriodaphnia dubia* and *Hyalella azteca* have also been developed to screen toxicity and provide a time scale of exposure that is matched to that of the runoff event. For discharges modified by best management practices, or where sampling provided time-based composites, more standard whole effluent-based test designs with C. *dubia* and H. *azteca* are appropriate. Toxicity testing conducted in the absence of field validation may lack a connection to observed receiving system effects. To establish a foundation

[1] Professor of Environmental Biology, Department of Civil Engineering, University of Illinois at Urbana Champaign.
[2] Graduate Research Assistant, Department of Civil Engineering, University of Illinois at Urbana Champaign.
[3] Water Research Centre, Henley Road, Medmenham, Marlow, Bucks, SL7 2HD, UK

for interpretation of laboratory-based toxicity testing with field-based studies, the laboratory-based testing is compared with in situ testing using caged organisms, or field bioassessments. Test application results indicate different degrees of concordance between tests in measured degree of toxicity; some tests give consistently similar results whereas others show more complex and variable patterns. Toxicity testing in single events may not be predictive of long-term effects in receiving waters, but multiple event analysis provides information on sources and variability of toxicity that is useful in watershed management.

Introduction

Toxicity is produced both by the concentration of a contaminant and the duration of exposure. Typically, higher concentrations are required to produce effects when the duration of exposure is short. This fact is critical when considering the effect of urban runoff, particularly runoff associated with wet weather events. Toxicity testing for impact analysis and regulation development has progressed over the past fifty or so years from single chemical testing to complex effluent analysis. In single chemical testing, it is possible to carefully define contaminant concentrations and develop dose/concentration-response relationships for test organisms. Acute testing typically requires high contaminant concentrations and short exposure periods, such as 96 hours for many fish species, and 24 to 48 hour exposure periods for many invertebrates. Although chemical specific testing provides accurate and precise results for a single contaminant, testing of contaminants in natural waters, or evaluation of actual contaminant effects in receiving systems usually finds that the contaminant concentrations tolerated in natural waters or receiving systems is much greater than expected from laboratory-based toxicity analysis. As testing progressed to add more and more complexity, reflecting the presence of multiple contaminants and variable receiving systems, typical of most discharge situations, it was recognized that chemical specific testing should be supplemented, or replaced, by testing that accounted for complex mixtures of contaminants in receiving waters.

The existing, standard methods for evaluating the effects of discharges on receiving waters emphasize what is called whole effluent toxicity (WET) testing. In WET testing a sample of the effluent, or discharge, containing complex mixtures of contaminants is used as the toxicant. In these tests, a percentage dilution, rather than a defined change in chemical concentration, is used to define a dose-response relationship for an entire effluent. Thus a WET test provides a more realistic appraisal of general toxicity and is often used to define general toxicity characteristics of a discharge.

Unfortunately, a difficulty still exists when examining the toxic effect of wet weather related urban runoff. The discharge is complex and can be tested as an effluent in WET procedures, but the expected change in concentration of contaminants and contaminant composition during a wet weather event requires a change in thinking about exposure conditions in the typical WET test. For example, storm events typically produce a hydrograph that may have a duration from minutes to hours, but seldom 24 to 96 hours. Within the hydrograph, there are also expected changes in contaminant concentration. At the initiation of overland flow, contaminants accumulated on the land surface since the last storm are carried in what is termed a "first flush" of contaminants. This first flush may produce high concentrations of contaminants, but these high concentrations are quickly diluted by additional runoff.

An investigation of the response of toxicity testing systems to wet weather discharges thus requires a change in thinking about test design. In a "time scale toxicity testing," it is important to consider both the test system, and the duration of exposure. A test system is defined as a measurement or analysis unit that integrates a complex response and is used to examine or assess the effect of toxicants (Herricks, et al. 1994). Test systems can range from simple biochemical assays to experimental manipulation of ecological systems. As might be expected, this duration of exposure can be quite variable and involve event specific responses, responses to multiple events, and responses to residuals of contaminants produced by storm events. It is the examination of these time-related exposure issues that has been the focus of the wet weather toxicity research sponsored by the Water Environment Research Foundation "Region Specific Time Scale Toxicity" that has provided the basis for the following discussions. The research has involved regional analysis of test system performance and has involved analysis of over 100 storm events. Some events were sampled sequentially with sampling intervals ranging from 60 seconds to 60 minutes. These samples were then subjected to laboratory-based toxicity analysis. Other analyses included in situ testing and receiving system bioassessments that included continuous monitoring (at one minute intervals) of organism response to storm events and the response of caged organisms to single and multiple events. This laboratory and field, storm event analysis has included investigations of time scale effects on a range of test systems, including species (such as Ceriodaphnia dubia) that are commonly used in WET analysis. The following discussion reviews general findings of event, multiple event, and receiving water analyses of urban runoff effects.

Event-Based Analysis

Storm event analysis completed in this study has included sampling at a number of locations in a single watershed, and at locations that provide a regional perspective on urban, wet weather discharges. The sampling at multiple locations within a watershed has provided a time scale context associated with location. This time scale context recognizes that different locations may be subject to both different concentration times in a storm event, and different event durations. For example, the concentration time and duration of elevated discharge will be short in headwaters areas and longer at downstream locations. The differences in watershed location also reflect the influence of hydrologic change on channels and channel stability, and the level of chemical complexity that may be expected in runoff samples. For example, samples from headwaters locations or from the outfall of a sewer network may be less complex because these discharges contain runoff from one or few sources. Samples from downstream locations in receiving systems may be more complex, because of influence from multiple sources. Further, the sampling location within a watershed will assist in defining the biological or ecological testing needed to address differences in flora and fauna along ecosystem gradients where the species composition and function of upstream flora and fauna differs from downstream locations.

To illustrate these differences, the results from sampling several storm events are provided in Figures 1-5. Figures 1 and 2 provide plots of stage height and conductivity plotted with percent mortality of *Ceriodaphnia dubia* observed in the toxicity testing of discrete samples collected sequentially during an event. The toxicity analysis utilized a screening procedure similar to a WET testing procedures. The sampling location is near the headwaters of the watershed from a 60 inches (150 cm) storm sewer. The duration of the event can be estimated from both stage height and conductivity as approximately 2 hours. Figure 3 provides a similar plot for a storm event monitored at a location approximately 2.5 miles (4 km) downstream. At this sampling site, similar rainfall events produce a hydrograph of much longer duration, with an expected reduction in contaminant concentration. The differences between upstream and downstream locations are also found in the time interval between onset of rainfall and the increase in discharge in the stream. A relatively short lag period of 10 to 20 minutes has been observed upstream with a lag of 30 to 60 minutes downstream.

Event related toxicity has produced some general patterns. This pattern typically finds an increase in toxicity early in the hydrograph and a later peak, often associated with the hydrograph peak, or the descending

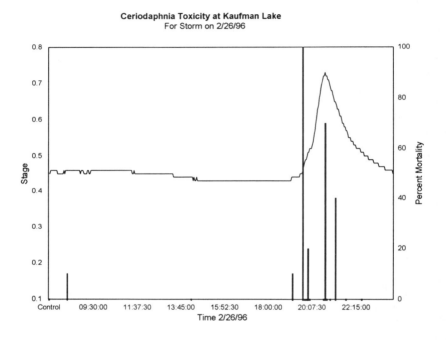

Figure 1. Plot of toxicity and stage during an event at an upstream location.

limb of the hydrograph. The initial toxicity has been associated with the first flush of contaminants. This is illustrated in Figure 2, where the specific conductivity initially drops, then peaks. This is associated with the runoff of road deicing salts washed off during the first flush and then diluted by later runoff. The second peak in toxicity is problematic. From observation, the later samples tend to have higher concentrations of suspended materials suggesting that this later peak may be due to the resuspension of previously deposited contaminants. An alternative explanation is that different sources in the system deliver contaminants to the sampling location at different times. This delivery sequence may be more variable at downstream locations where the timing of contaminant delivery from different watershed locations will affect both concentration (with relation to changing dilution capacity associated with higher discharge volumes downstream) and duration. This may be the case in Figure 3 where the downstream location also shows some pre-storm toxicity. Note too that Figure 3 provides data from a different test system, the Microtox assay. This test system has consistently produced both inhibition and stimulation during storm events.

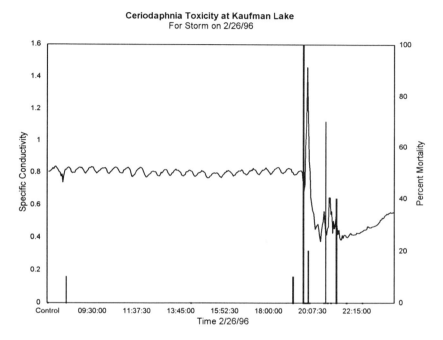

Ceriodaphnia Toxicity at Kaufman Lake
For Storm on 2/26/96

Figure 2. Plot of toxicity and specific conductivity during an event at an upstream location

Other research indicates that the stimulation is due to the high degree of sensitivity of this test system to certain contaminants (Biggs, 1996).

A question that may be addressed in this data is "What is the significance of the observed mortality?" It is difficult to answer that question in the absence of field data, but it is possible to develop a sense of responses to categorize the percent mortality observed in toxicity testing using the screening methods employed.   For the test systems employed in this research, a 10% mortality in controls is viewed as acceptable. Based on that standard, a 20 - 30% mortality is considered a low mortality; 40 - 60% is considered a moderate mortality; and 70 - 100% high mortality.

In summary, an event analysis using changes in physical conditions (stage/discharge), chemical conditions (monitored both with continuous logging and grab sampling), and toxicity has been successful. Physical and chemical measurements have been useful in defining event characteristics, that can then be used to develop a sampling program that will provide the optimum design for toxicity assessment. This research has demonstrated that test systems can be selected that provide response times consistent with exposure times. For example, Microtox response is determined for 5, 10, and 20 minute exposures. The *Ceriodaphia* IQ response is measured after a one hour exposure.   While the 24 to 48 hour exposures of *Ceriodaphnia dubia* and *Hyalella azteca* exceed a realistic exposure period, test procedures are being developed for samples that produce high toxicity that will assess a time scale toxicity (Brent and Herricks, 1996). The use of a test battery assists in meeting the other criteria of a time scale toxicity analysis. Multiple test systems allow an evaluation of response to changing conditions, and assist in interpreting results to sort out effects produced by a single stressor or multiple stressors.

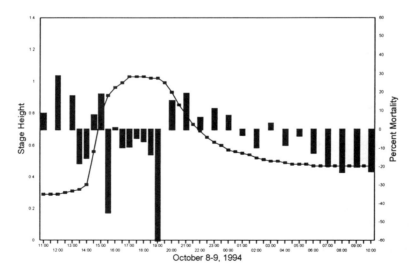

Microtox Results at Downstream Site
for Storm on October 8, 1994

Figure 3. Plot of toxicity and stage at a downstream location.

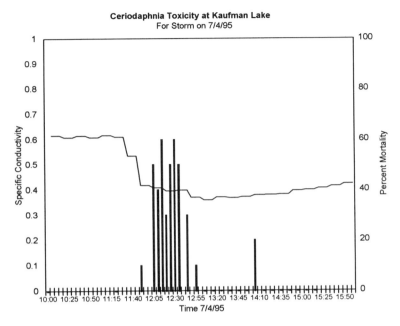

Figure 4.   Plot of *Ceriodaphnia dubia* mortality and specific conductivity for the event on 7/4/95.

Multiple Event Analysis

Multiple events are common in urban runoff. Depending on regional climate, it is not unusual for a period of storms to occur where rainfall may be produced by passing storms over a given time period. General guidance given for stormwater sampling is that an interval between storms of 72 hours is needed. A reasonable assumption supporting this specification is that a minimum period is required to allow for the accumulation of materials on surfaces that may be "cleaned" by runoff. Figures 6 - 9 illustrate issues associated with multiple event analysis. A storm sequence was sampled in late May, 1996. A dry period of more than 72 hours was followed by a series of rainfall events in which passing storms, with rainfall intensity of 0.25 in to 1.5 in occurred. Figure 6 illustrates toxicity of the initial event as measured by the Microtox assay. These results illustrate a moderate to high level of toxicity in the first flush, followed by a second peak in toxicity later in the storm. A second storm was monitored three days later, Figure 7. The plot of specific conductivity over a two day period in Figure 7 illustrates the effects of repeated rainfall, and the toxic response observed   indicated   a   low   level   of

**Comparison of Toxicity Test Data for
Storm Event at KLSS on 7/4/95**

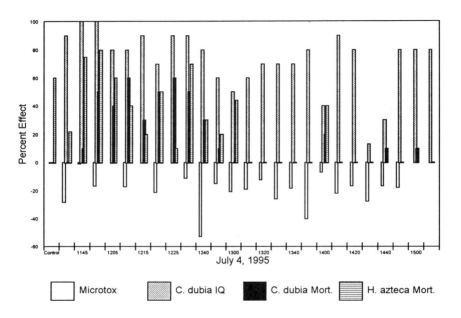

Figure 5. Comparison of different test system results for storm samples collected 7/4/95.

toxicity, both during the first flush period and later in the event monitored. Figures 8 and 9 provide details of the change in toxicity observed during these two events. Note should be taken in Figures 8 and 9 of the changing response observed over the 20 minute exposure period for Microtox. This increase in percent effect with time of exposure is typical of metal toxicity and can be a useful tool in source identification.

Multiple event analysis can clearly be affected by the interval between storm events. Observations at the same site for different intervals between storms suggests there is a correlation between observed toxicity and dry period. Repeated events may serve to "flush" the system. Although a 72 hour dry period is a useful criterion, our research indicates that longer time intervals between storms may produce higher levels of toxicity. Depending on the objectives of the program to assess short term toxicity, storm interval will be a critical issue in program design.

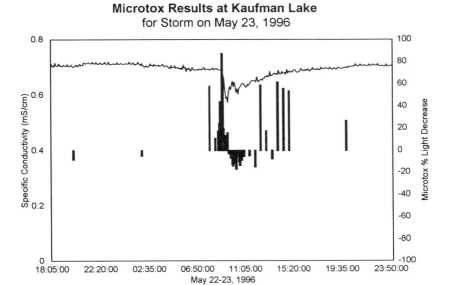

Figure 6. Toxicity assessed in the initial storm of a multiple storm sequence.

Receiving Water Analysis

The connection between observed, event toxicity and a receiving system effect is at times problematic, but critical to management. The issue of laboratory to field extrapolation has been a focus of environmental toxicology and ecotoxicology for some time (National Research Council, 1981; Levin, et al. 1988; USEPA, 1991). The primary difficulty is connecting a field bioassessment, which has long time scale characteristics involving influences from multiple events as well as baseflow conditions, with event-based measurements. One approach is to design receiving system analyses or assessment activities to allow an event-based interpretation. This requires the replication of field sampling before and after events, or the use of in situ testing, such as the deployment of caged organisms, to observe event specific responses under natural receiving system conditions.

As part of the time scale research, studies were conducted to connect observed toxicity with receiving system condition. The field site selected (the same site that provided event toxicity information) provided a stormwater source that consistently produced moderate to high levels of toxicity. A reference site adjacent to the stormwater source, and a

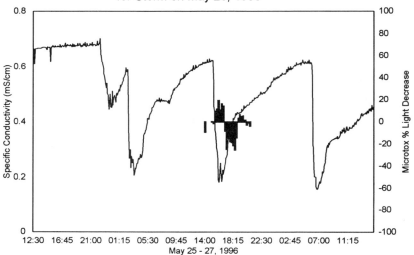

Figure 7. Toxicity assessed after an interval of three days with multiple rainfall events in the interim.

downstream location that can be used to assess receiving system effect. The results from field bioassessments and in situ studies are provided in Figures 10 -12. Figure 10 provides information on the seasonal change in insect abundance at the three study sites. This data shows a clear effect of the storm sewer on downstream fauna.   It also indicates a seasonal difference in abundance. Although there is a clear effect shown in this data, questions of event effects are unanswered.

To address event related questions in a field setting, short term exposures of caged organisms at the same locations used for event toxicity analysis were sampled.   Figure 11 provides the result of several in situ deployments with similar deployment times (3, 7, and 14 days) using *Hyalella azteca,* a species commonly used in laboratory testing, and a local isopod, *Asellus sp.,* indicative of resident species effects. Short duration exposures typically produced low mortality, while longer exposures, which may have included a storm event, produced moderate toxicity. In general, no event related toxicity was observed even though deployment was made through events that produced moderate to high levels of toxicity in laboratory analyses.

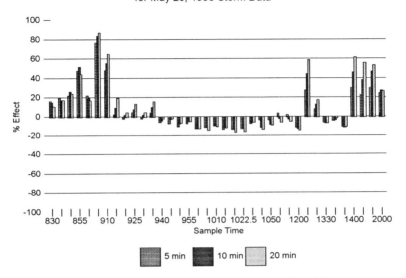

Figure 8. Microtox assay results for the storm sampled on May 23,

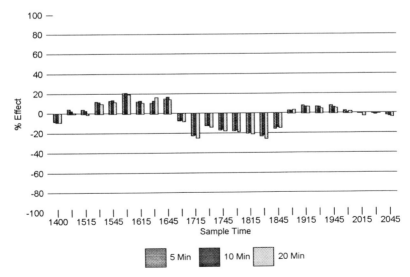

996.
Figure 9. Microtox results for a storm sampled May 26, 1996.

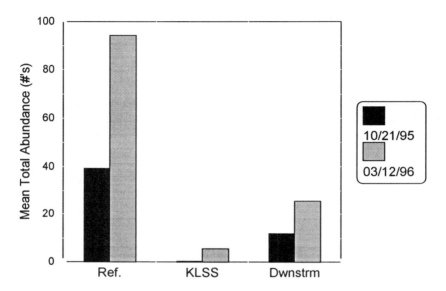

Figure 10. Abundance of insect taxa at reference, storm sewer, and downstream sampling locations.

The results of field assessments and in situ deployments of caged organisms illustrate the difficulty of connecting event toxicity to receiving system quality or condition. In these analyses, both short term exposures and exposure to actual receiving system conditions present during a storm event produce little toxicity. It is only in longer term exposures that toxicity is observed. In these experiments, toxicity was observed at both the reference and downstream locations, not in the storm sewer dominated stream reach, which would be expected to have the greatest effect. Unfortunately, this contradictory evidence is typical of field-based studies where it is not possible to control all exposure conditions. As in most urban runoff analysis programs, there is an evident degradation of receiving system quality and there are adjacent outfalls or urban runoff discharge locations. Also true of most urban runoff analysis programs, toxicity in wet weather events is observed, but it is variable, and not profound (typically low to moderate) while in situ testing fails to provide consistent indication of event-related toxicity.

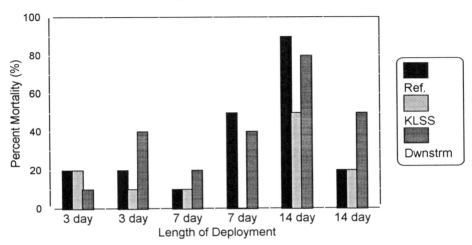

Figure 11. Percent mortality of a commonly used laboratory species in field, in situ exposures.

Conclusions

It is well recognized that urban runoff produces short term changes in contaminant concentration in wet weather discharges and the systems that receive these discharges. Our existing water quality management programs are built on a scientific foundation provided by toxicity testing that determines   the acceptable concentration of contaminants in receiving waters. With a focus on stormwater, it is also critical to develop a similar scientific foundation for the duration of exposure to contaminants.   This is particularly true for storm-related discharges in urban settings.   This research has shown that toxic effects are observed in water samples collected during storm events. This toxicity is related to both the first flush, and post hydrograph peak periods.   Correlative studies involving field assessments and in situ exposures of common laboratory and indigenous species do not exhibit an event related toxicity, although there is a clear effect/impact indicated by longer term studies. Although there is growing evidence that short term exposure can produce longer term effects (Brent and Herricks, 1996, Krunkelton, et al. in press) there is a clear need for additional time scale toxicity research supplemented by evaluation of the effect of multiple stressors both during and after event time scale exposures.

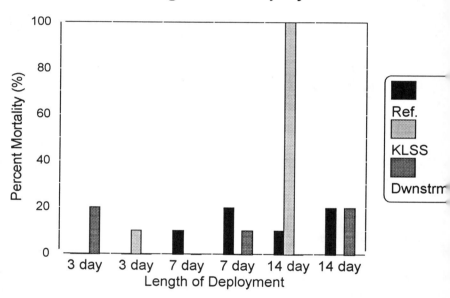

Figure 12. Percent mortality of indigenous organisms in field, in situ, exposures at the reference, storm sewer reach and downstream study sites.

The measurement of short-term toxicity does provide a definitive assessment of toxic potential. To provide the strongest case for the definition of toxic characteristics, a test battery involving multiple test systems will meet criteria for time scale toxicity determination. The measurement of multiple events at the same location provides a basis for variability in toxicity as well as a foundation for analyses that assist in identifying sources of toxicity in a watershed.

Acknowledgments

Research leading to this publication was supported, in part, by the Water Environment Research Foundation as part of a four year research project "Region Specific Time-Scale Toxicity in Aquatic Ecosystems" Project 92-BAR-2, which is a recipient of federal funding under WERF's EPA Cooperative Agreement No. CR 818249, and U. S. EPA Fellowship funding awarded to Mr. Brent.

References

Biggs, Lara. (1996) An Investigation of Hormesis Using the Microtox Assay. MS. Thesis, Department of Civil Engineering, University of Illinois at Urbana-Champaign.

Brent, R. N. and E. E. Herricks. (1996) Time of exposure toxicity as an element of criteria for wet weather discharge events. Proceedings of WEFTEC '96, October 5-9, 1996, Dallas, TX.

Crunkilton, R., J. Kleist, J. Ramcheck, W. DeVita, and D. Villeneueve. (1996) Assessment of the response of aquatic organisms to long-term Insitu esposures of urban runoff. Proceedings of an Engineering Foundation Conference "Effects of Watershed Development and Management of Aquatic Systems:" August 4-9, 1996.

Herricks, E. E., Milne, I. and Johnson, I. (1994) Selecting biological test systems to assess timescale toxicity. Water Environment Research Foundation, Project 92-BAR-1, Interim Report.

Levin, S. A, M. A. Harwell, J. R. Kelly, and K. D. Kimball (eds) (1988) Ecotoxicology: Problems and Approaches. Springer-Verlag, New York. 547 pp.

Milne, I, E. E. Herricks, and I. Johnson. (1996) Protocols for Wet-Weather Toxicity Testing and Assessment. Water Environment Research Foundation, Interim Report WERF Project 92-BAR-1.

National Research Council. (1981) Testing for Effects of Chemicals on Ecosystems. National Academy Press, Washington, D. C. 103 pp.

U. S. Environmental Protection Agency. (1991)   Technical Support Document for Water Quality-based Toxics Control. EPA/505/2-90-001.

DISCUSSION

Assessment of the Response of Aquatic Organisms to Long-term Insitu Exposures of Urban Runoff
Ronald Crunkilton

A 1993 study of toxicological impacts of stormwater runoff in Lincoln Creek failed to detect toxicity in over 300, conventional 48-96 hr acute and 7 day chronic tests. Substantial mortality was observed, however, for juvenile and adult organisms exposed for greater than 14 days.

Although not demonstrated here, routine WET tests may also overestimate potential toxicity. We demonstrated short-comings of WET tests or other short-term tests in detecting whole-animal toxicity. By simply lengthening exposure time and using flow-through tests we detected substantial mortality related to toxicant exposure.

Questions/Comments

Question: What was the "control" for the long term tests in terms of environment?

Answer: Control was carbon filtered tap water.

Comment: The control is carbon filtered, and the tap water was chlorinated, so there should be no bacteria in the control. The organisms in the stream tanks are subject to various types of bacteria and fungus plus variations in water quality that might occur in the stream between storms. What kinds of chemical analysis were run on the water during the long term tests? This data is necessary before any conclusion can be drawn regarding the cause of mortality. If the stream water was disinfected before it was introduced to the tank, would the results be different?

Assessing the Response of Aquatic Organisms to Short-Term Exposures to Urban Runoff
Ed Herricks

The measurement of short-term toxicity does provide a definitive assessment of toxic potential. To provide the strongest case for the definition of toxic characteristics, a test battery involving multiple test systems will meet criteria for time scale toxicity determination. The measurement of multiple events at the same location provides a basis for variability in toxicity as well as a foundation for analyses that assist in identifying sources of toxicity in a watershed.

Panel Discussion – Barbour, Bannerman, Crunkilton, Morrisey

Crunkilton:	Just hypothesizing that heavy metals are the cause of long term mortality. But believe that there is some desoption there. Some tests have been run in which a storm flow was run through the test facility and a fine sediment remained in the bottom. Subsequently death was experience whether river water was circulated through or clean water.
Comment:	Have you started to muse about water quality criteria, and toxicity? There seems to be a lot of question as to what the real causative agent is.
Comment:	What does toxicity mean? What does bioassessment mean? Which is the better indicator. Bioassessment has been used as a alternative for determining if designated uses have been attained.
Comment:	The type of work that Bannerman and Crunkilton are doing is very expensive. The work on one site was about $40,000 for sampling and another $40,000 for laboratory analyses.
Comment:	We need to look carefully at what the potential causes of these long term deaths and toxic results are. Do not be too hasty to point to a specific culprit. Shock introduced to the test species when it is subjected to the test tank, may make it more susceptible to attack by bacteria or fungus.
Comment:	What is an acceptable increase in mortality as a result of urbanization? This is not just a scientific question but a social question, i.e. what percent mortality is acceptable to the community.
Question:	Is toxicity the proper measure of health in an urban stream system? How do the toxicity studies support our efforts to improve a urban stream system. Bioassessment done by Florida is an attempt to provide simple tests of impairment.
Bannerman:	You cannot ignore habitat, and flow conditions, but you need to integrate toxicity considerations into the question.
Comment:	The debate on whether the problem is sediment, water column, flow, etc. will continue for many years. Another question is whether the "control" truly represents the "clean stream" environment, since we

base all our conclusions on the comparison of the test tanks with the control.

The Alluvial Progress of Piedmont Streams

Bruce K. Ferguson[1]

In the Piedmont region of the southeastern United States, cycles of development have impacted streams cumulatively, and are continuing to demand an informed response. This paper reviews development impacts and their connections over time as seen in two rehabilitation and protection projects, and suggests objectives for healing streams that continue to be impacted these ways.

Before European settlement, Piedmont streams were in equilibrium with their watersheds. In the late eighteenth century, just prior to major settlement, William Bartram (1988) recorded the region's healthy vegetative cover, the clear stream water, and the wetland vegetation growing on the stable banks. Beaver damming and harvesting added storage, diversity and resilience throughout the drainage system. But in the nineteenth century trapping and habitat loss exterminated the beaver from most of the Piedmont (Parrish 1960), eliminating one ecological counterbalance from the impacts of the development that would follow.

The first cycle: Clearing and cultivation

The first major wave of development started about the beginning of the nineteenth century, with clearing of the woods and cultivation, primarily for cotton. Between the woody stems of cotton plants, cotton is essentially a bare-soil agriculture. For a century the bare soil was exposed to the region's abundant rainfall. Essentially the entire topsoil layer, encompassing the region's entire seedbed and rooting zone, was peeled away, eliminating all the propagules of all the native species that could have reclaimed the land promptly, while leaving the exposed subsoil surface shifting with further erosion. Where runoff concentrated, it dug gullies. At its most catastrophic, runoff dug what is now called Providence Canyon near Columbus, Georgia: with the absorptive layer of soil and vegetation no longer counterbalancing against entropy, the entire drainage system collapsed more than 100 feet down to a new base level. Streams flowed red with suspended sediment.

Trimble's (1974, p. 5) research showed that the Piedmont's erosive land use expanded to its greatest intensity between the Civil War and the early years of this century, and then abruptly declined at the time of the Great Depression. The full cycle of settlement added sediment to the stream system, and then stopped adding it—both changes are important, as this paper will show.

The region's average depth of total erosion (Trimble, 1974, p. 3) varied from 7 to more than 12 inches. The average depth seems like a small quantity, but when it is

---

[1]Professor and MLA Coordinator, School of Environmental Design, University of Georgia, Caldwell Hall, Athens, GA 30602.

multiplied by the entire area of a stream's watershed, it is a catastrophically large addition to the sediment regime of the streams and riparian corridors. As the lowland corridors occupy only about 10 percent of the area of a watershed, the sediment volume concentrated as it fell downslope, converging into masses yards deep.

Scull Shoals

The magnitude and relevance of the anthropogenic sediment load is exemplified at a water mill built in the early nineteenth century at Scull Shoals, Georgia. Historic preservationists initiated a study of this site by inquiring about the feasibility of protecting the mill's remains from overbank flooding of the Oconee River.

A shoal is a bedrock fall in a stream; it is a logical place to locate a water mill. In the middle of the nineteenth century the Scull Shoals mill was the biggest mill on the Oconee River, supporting 500 laborers (Coulter, 1964). It was the first brick cotton mill in Georgia. It was also the first paper mill in Georgia; Georgia is now the principal paper-producing state.

But today the town of Scull Shoals is empty, and overgrown with trees and brush. All the structures are in ruins; some are marked off as archaeological sites. At the river, no shoal or mill dam is visible; the river grades past the town site at a single uninterrupted slope.

The shoal that gave the mill its power and the town its name today is buried under 14 feet of sediment, poured down upon the river from the watershed's cotton fields during the last century and a half. The appearance of the channel today is consistent with the rapid aggradation that occurred during the cotton period (Happ, Rittenhouse and Dobson, 1940): the channel is bounded by large sand levees, while meander scrolls are absent from the floodplain.

The great mill is visible today only in a backswamp of the floodplain, in the form of a few stone piers and part of the headwater intake channel. Much of the mill, like the shoal, is buried. The visible remains represent the mill's second story. The mill's entire power system is now below ground; this conclusion has been confirmed by subsurface radar and magnetometer surveys. As one walks around on the floodplain surface, the bulk of the historic remains are under one's feet.

The question of the feasibility of flood protection had to be answered in the negative, because the mill remains are located today in a sump in today's backswamp. Holding back surface floods would require a dike extensive enough to conflict with the preservation objectives of the site. Even after perimeter diking, standing water in the mill would continue to arise from ground water that is recharged through the entire floodplain surface during overbank flows. The bulk of the mill remains—the buried portion—is saturated with ground water.

A simple model of how the river profile at Scull Shoals changed, diagrammed in Figure 1, explains the failure of the mill and the features that are visible on the site today. Before the cultivation period, the channel flowed at a relatively low gradient to the crest of the bedrock shoal, fell sharply down the shoal, and then continued downstream at a low gradient. Farming-generated sediment raised the channel to a relatively steep gradient in accord with the heavy sediment load. It quickly filled the mill pond. After overflowing the dam, it filled the valley below, particularly deeply at the base of the shoal where the mill's tailwater must have been located. Written records from the late nineteenth century describe the rising flooding of the mill: it had to be shut down a few days per year, then a few weeks per year, and then it was abandoned entirely about 1890, barely 40 years after it was built. The rising sediment eliminated the fall that had given the mill its power, and brought the life of the town to an end.

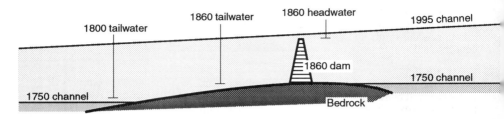

Figure 1. Profile of the Oconee River at Scull Shoals, Georgia, showing the change since 1750 (schematic only, not to scale).

The evolution of the river is now directed away from flooding of the mill. Since the middle of this century, cotton fields have been abandoned to vegetative succession, depriving the streams of their sediment load. The streams are tending to incise into their sediment deposits. At Scull Shoals, both the occurrence of overbank flows and the height of ground water will tend gradually to decline as the Oconee channel incises and widens. The incision process here is lagging behind that of typical headwaters, because the secondary sediment from the eroding headwaters continues to pour down upon the main stream, slowing net sediment removal. Nevertheless the day will come, many human generations in the future, when the mill will be left again free-standing except for an apron of sediment with the water of the river lapping at its base.

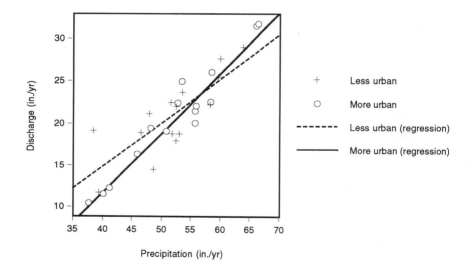

Figure 2. Changes in annual runoff during urbanization of Peachtree Creek in Atlanta, Georgia (data from Ferguson and Suckling, 1980).

The second cycle: Reclearing and urbanization

In the last few decades, the Piedmont has been one of the fastest-urbanizing regions of the country. Urban construction clears the second-growth woods and momentarily bares the soil to erosion, giving already-disrupted downstream channels more sediment to deal with (Wolman, 1967). It continues by paving the soil and heating the air, depriving the streams of base flow via infiltration and ground water replenishment, and diverting rainwater during storms into faster and higher flows that accelerate erosion of stream banks and aquatic habitats.

Ferguson and Suckling (1990) directly observed these effects on Peachtree Creek in Atlanta. Figure 2 summarizes the change in flow regime as the creek's watershed urbanized over a 30 year period. In times of low rainfall, urbanization lowered the flow, making the period of base-flow stress in the stream ecosystem worse. In times of abundant rain, the flow is higher with increased frequency and intensity of peak flows. This is the worst of all possible hydrologic regimes: the stabilizing aquatic resource is withdrawn in times of shortage while its destructive power grows in times of excess.

The alluvial sequence now in progress on the Piedmont is summarized in Figure 3. Farm-generated sediment buried presettlement equilibrium channels and riparian corridors. When field abandonment withdrew the sediment source, streams incised their previously deposited alluvium. Some still have headcuts that are migrating headward into the sediment masses, digging out new channels. When they incise down to their original bedrock-controlled elevations, they meander laterally, undercutting their banks. In the process they produce secondary sediment out of their channels: the streams are their own polluters as they continue to adjust to the alluvial relics of the cotton period. With urbanization, their erosive power is increasing, collapsing their beds and banks faster, carrying the material downstream and tending eventually to excavate and remold the entire riparian corridor in a brown flurry of erosion, sediment, and habitat loss. We can look ahead in the sequence (Harvey and Watson, 1986) to a future in which the entire floodplain has been remolded into equilibrium with the new regimes, with water flowing in a large channel to accommodate urban peak flows and only a few abandoned terraces at the sides of the valleys remaining from the first cycle's impact.

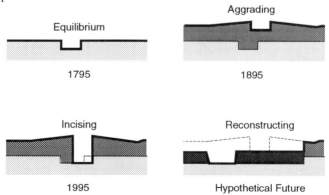

Figure 3. Alluvial history of typical Piedmont headwater streams (schematic cross-sections only, not to scale; hypothetical future is based on Harvey and Watson, 1986).

Coggins Park

The interacting impacts of successive watershed land uses are exemplified at Coggins Park, an industrial park in the 140 acre watershed of one of the Oconee's unnamed tributaries in Athens, Georgia (Figure 4). A project to stabilize the watershed (Ferguson and Gonnsen, 1993; Jesitus, 1995) was initiated by a lawsuit complaining of excessive sediment and stormwater discharging through residential properties downstream.

This watershed, like the rest of the Piedmont, had been cultivated through the early years of this century. Elevated alluvial terraces occupied the two stream valleys. When field inspection began in 1988, the larger of the two branch streams had incised into its alluvium to a bedrock-controlled base level and was meandering laterally, caving in its 5 ft high banks. The smaller stream corridor was marked only by a level swampy lowland of alluvial valley fill, terminated at the downstream end by a 6 ft deep headcut actively eroding the cotton sediment and bringing the drainage down to the base level established by the larger, faster eroding branch.

At this time the industrial park was half developed, with all streets and utilities in place and half the industrial lots occupied by buildings and parking lots. Erosion during construction had been poorly controlled, adding sediment to the already unstable streams. Locally developed parts of the park were 90% or more impervious; the overall watershed was 30% impervious. Urban impervious surfaces alone increased volumes and rates of storm runoff, increased channel erosion, and diminished base flow in dry summers. Urban encroachments directly on the stream corridors had further unsettled them. Streets and structural fills encroached on floodplains. Some stream segments were confined in culverts; one was replaced by a deep, narrow, newly excavated channel that was rapidly widening, destroying the remaining riparian bench. A channel had been dug through an old farm dam, apparently to drain the pond; the pond's bottom sediment was exposed to erosion while the stream's fall through the eroding breach was now the biggest single sediment source in the watershed. A detention reservoir with a perforated standpipe outlet had proven ineffective at either capturing sediment or suppressing downstream peak flows.

This watershed's need for rehabilitation were comprehensive. Mobile sediment had to be prevented from moving downstream, and the sources of further sediment in the eroding stream channels had to be stabilized—in these respects this project was like stream rehabilitation in an abused rural watershed (Heede, 1977). In addition, this project had to suppress urban peak storm flows, augment base flow, establish wetlands for water quality improvement, and fit within established urban land use and infrastructure.

In the smaller tributary, two small reservoirs and a check dam were fit into the only spaces available in the stream corridor. The reservoirs suppressed peak flows in the tributary, and permitted some infiltration from small permanent pools. The check dam supplemented sediment capture and flow attenuation to the degree that space permitted. These features involved excavation and disturbance, but within a year or two after installation they were mantled in wetland vegetation, transforming water quality, stabilizing accumulating sediment and adapting to changing hydrologic regimes.

In the large tributary, the dam breach was stabilized with a gabion weir. Below the weir crest, permanently retained sediment forms a moist substrate where native wetland vegetation has been establishing; the wetland area expands as the accumulation grows. Also in the large tributary, two large check dams were installed. Sediment has been visibly accumulating behind the dams, raising the base level in incised and dug channels, reducing the height of channel sides, and providing a stable, moist substrate for wetland vegetation.

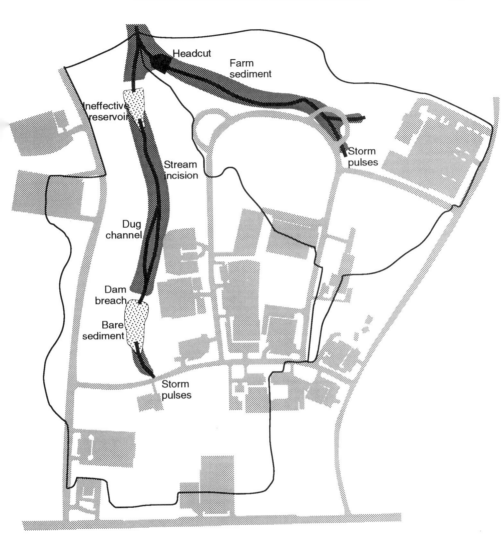

Figure 4. Cumulative disturbances to the streams of Coggins Park, Athens, Georgia, as of 1988.

The lowest structure in the large tributary is the old detention basin, where the perforated standpipe was wrapped in filter fabric and the reservoir was enlarged by excavation, with the lower elevations reserved for permanent sediment storage. Captured sediment is visibly forming a sandy delta at the head of the reservoir and a deep, fine-textured layer over the bottom and around the outlet filter. Frequent and prolonged ponding indicates that peak flows are being effectively suppressed. Cost concerns have prevented vegetation of the basin's margins with as much diversity as planned (Dapkus, 1993). Preexisting trees in the frequently inundated zone have died, as they would in a new beaver pond. However, opening of the tree canopy admits light to the substrate floor, encouraging herbaceous and shrubby plants to stabilize the substrate.

Stormflow modeling indicated that the stream rehabilitation measures adequately reduced 10-year peak flows from the partly developed watershed due to the storage volumes in basins, the limited capacities of basin outlets, and attenuation in low-velocity flow. However, although the measures capitalized on the capacity of the stream corridors to the degree that land use allowed, they were not, by themselves, sufficient to suppress the runoff from both completed and future industrial development. Future development must include on-lot provisions to control hydrologic and sediment discharges. Design criteria for on-lot infiltration were worked out that would be effective in suppressing future peak flow.

That the stream system is becoming more stable is indicated by the rate of sediment arrival at the big basin at the bottom of the large tributary. Sediment accumulation has been calculated from successive topographic surveys of the basin floor. Sediment capture was rapid (0.92 af/yr) in the first year after construction, declined by four fifths (to 0.19 af/yr) for the next two years, and has visibly continued declining since then. The only explanation for the declining rate of capture that is consistent with everything that has been observed on the site is that the production of sediment has been reduced at its sources in the formerly eroding channels. Stabilization can be attributed to the permanent sediment storage in basins and reservoirs, the stabilized and raised base levels, reduced stream gradient, and reduced peak flows and velocities. Unfortunately, one headcut has not been stabilized, because it is outside the boundary of the industrial park; it remains today the single major source of downstream sediment.

Wetland and riparian vegetation has been colonizing all sediment surfaces and stream banks, enhancing and prolonging stability. Selectively located plantings of grass and willows have been very successful. Exposed sediment in the drained farm pond has vegetated very thoroughly despite only incidental seeding, possibly due to the presence of buried weed seeds in the sediment and the release, by aeration, of formerly trapped nutrients to them. Beaver tracks have been seen on the sediment substrate. Most of the Coggins site could be modified by beaver without conflict with people, because the wooded stream corridor is separated from buildings and yards by steep bluffs. Beaver may be taking over the maintenance of the check dams and other structures throughout the project, eliminating the need for perpetual human vigilance. The riparian system that had been successively devastated by cultivation and urbanization is now relatively capable of managing itself with its own self-regenerating biota and natural equilibrating processes.

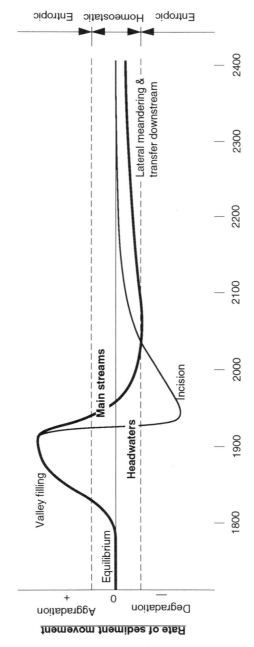

Figure 5. The adjustment of Piedmont streams over time.

An equilibrating approach

In Bartram's day, when Piedmont streams were in dynamic equilibrium, discharges of water and sediment equaled inflows to the system. Mild disturbances were compensated for with continuous development of interacting soil, water, microclimate and vegetation. Over geologic time, the streams were in Mackin's (1948) condition of "shifting equilibrium", maintaining approximate equilibrium during adjustment to long-term change. Increasing degrees of disturbance led to increasingly vigorous responses.

Figure 5 shows the changes since then. The dashed horizontal lines outline a zone near equilibrium in which stress is mild enough that the system can respond homeostatically, imposing negative feedback on perturbations and thus maintaining health (Ferguson, 1994). Outside these bounds, disturbances exceed the capability to respond, and the system collapses passively under the control of entropy. The figure shows that valley filling in the cotton period devastated streams with its great external load of sediment. It was so rapid that it was essentially simultaneous in both the headwaters and the main streams.

When the external supply of sediment was withdrawn, headwater and tributary streams quickly began incising, washing their sediment downstream. The adjustment is a natural process of equilibration to contemporary regimes of sediment and water, and we need not, on principle, stand in its way. However, the process that we are witnessing today is as destructive as the anthropogenic stresses that set it in motion and today's flow regimes. On certain headwater streams, the process is in its highly destructive initial stages, as headcuts excavate into artificially elevated farm sediment, violently reforming their profiles and their floodplains. The process is relatively advanced in some other small streams, where incision was completed years ago and some constructional forms and stabilizing vegetation have begun to take hold. Essentially all Piedmont headwater streams lack counterbalancing influences proportional to the disturbance from which they must adjust.

The main streams are reacting relatively slowly, because they continue to receive secondary sediment eroded from their tributaries. Figure 6 illustrates the gradual shift of sediment from headwaters to main streams. Mainstream degradation, although inevitable, has barely begun, and its progress will be slow. Much later, when the upstream sediment supply is exhausted, some main streams may go through a period of visible degradation. Nevertheless the great volumes of sediment now filling the stream and river corridors will buffer and attenuate change for a long time.

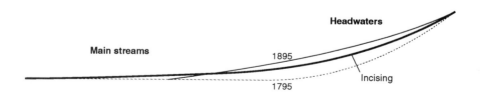

Figure 6. Change in Piedmont stream profiles over time.

The overall rate of adjustment in the region will decline, as shown in the right part of Figure 5. As sediment shifts downstream, vegetation will take over relatively stable banks, beaver will occupy stabilizing reaches, turbidity will decline, and the system will approach equilibrium.

For the moment, where headwater streams are seeking equilibrium, we can aid their effort by adding counterbalancing mechanisms, while not denying the inevitability of long-term change. An approach to accomplishing this is highly selective grade control. To be successful, such grade control must be selectively located, properly implemented, and given only realistic expectations.

Check dams are grade control structures with intervening pools. Properly located check dams terrace the stream profile from dam to dam. They stabilize the base level, slowing downcutting. In the pools between dams, flow gradient and erosive energy decline and sediment settles out. The pools capture and accrete everything that comes down the channel—water, sediment, organic debris, dissolved constituents, organic propagules. The stable banks and moist sediment provide substrate for riparian vegetation. Stream water infiltrates the stable sediment surface, replenishing the alluvial aquifer and enhancing base flow. Thus check dams can interact with the natural processes of streams to counterbalance external forces. While allowing inevitable transition, they induce maintenance of health through absorption, storage, diversity and self-regulation (Ferguson, in press). With them, the curves in Figure 5 need not dip so far into entropic incision; they can stay in the homeostatic zone of shifting equilibrium. The system can be resilient and self-regulated, while continuing to change in the long run.

An equilibrating approach is different from that of sediment basins that are dredged at intervals of time, and from detention basins that drain after storms are past. Captured sediment, water, and organic matter are intended to remain largely in place. Dredging would resuspend sediment and rejuvenate the upstream base level. Drainage would discard the water resource.

Grade controls should not be added to streams that are already stable and healthy, those where advancing headcuts would undermine the structures (Heede, 1960, 1986), or those that are unstable in ways other than incision (Rosgen and Fittante, 1986). Stream segments that are incising can be recognized from site-specific rates and directions of change. Any proposal for stream alteration or management should be investigated for its potential effects on flow, stability and habitat in upstream, downstream and laterally adjacent areas.

In the Piedmont's anthropogenic alluvium, the purpose of grade control is to enable the taking over of the stream corridors by natural equilibrating mechanisms. In the long run, those mechanisms must operate while the streams continue to adjust. The purpose is not to halt inevitable change. That all grade control structures are temporary to some degree must be accepted. Selection of construction materials specifically for flexibility, biodegradation and organic adaptation is appropriate where long-term change is expected.

All such grade control must include or accommodate revegetation of the stream corridor. Unlike fixed structures, communities of self-regenerating biota have unlimited life spans. Grade control structures turn out to be of only accessory and momentary effect. The long-term work of health is carried out by self-regenerating biota that take over the corridor when the opportunity is given, and adapt resiliently to continuing changes in sediment and flood regimes. A keystone species in riparian landscapes is beaver. Like properly placed check dams, beaver damming and harvesting add diversity, resilience, and dynamic equilibrium to the riparian corridor. Appropriate placement of check dams simulates beaver dam siting and pond formation (Furqueron,

1993). Volunteer seedings and appropriate plantings in an environment momentarily stabilized by check dams introduce communities that seize the habitat and regenerate themselves while adapting dynamically to change.

References

Bartram, William, 1988, *Travels through North and South Carolina, Georgia, East and West Florida*, New York: Penguin Books.

Coulter, E. Merlon, 1964, Scull Shoals: An Extinct Georgia Manufacturing and Farming Community, *Georgia Historical Quarterly* vol. 48, p. 33-63.

Dapkus, Kathleen Marie, 1993, *Vegetational Alternatives for a Detention Basin with Severe Conditions*, MLA thesis, Athens: University of Georgia.

Ferguson, Bruce K., (in press), Maintaining Landscape Health in the Midst of Land Use Change, accepted for publication in *Journal of Environmental Management*.

Ferguson, Bruce K., 1994, The Concept of Landscape Health, *Journal of Environmental Management* vol. 40, p. 129-137.

Ferguson, Bruce K., and P. Rexford Gonnsen, 1993, Stream Rehabilitation in a Disturbed Industrial Watershed, pages 146-149 of *Proceedings of the 1993 Georgia Water Resources Conference*, Athens: University of Georgia Institute of Natural Resources.

Ferguson, Bruce K., and Philip W. Suckling, 1990, Changing Rainfall-Runoff Relationships in the Urbanizing Peachtree Creek Watershed, Atlanta, Georgia, *Water Resources Bulletin* vol. 26, no. 2, p. 313-322.

Furqueron, Thomas Christopher, 1993, *A Wetland Enhancement Design, Modeled on Beaver Ponds, for a Channelized Piedmont Stream*, MLA thesis, Athens: University of Georgia.

Happ, S.C., Gordon Rittenhouse, and G.C. Dobson, 1940, *Some Principles of Accelerated Stream and Valley Sedimentation*, Technical Bulletin 695, Washington: U.S. Department of Agriculture.

Harvey, Michael D., and Chester C. Watson, 1986, Fluvial Processes and Morphological Thresholds in Incised Channel Restoration, *Water Resources Bulletin* vol. 22, no. 3, p. 359-368.

Heede, Burchard H., 1960, *A Study of Early Gully-Control Structures in the Colorado Front Range*, Station Paper No. 55, Fort Collins: U.S. Forest Service Rocky Mountain Forest and Range Experiment Station.

Heede, Burchard H., 1977, *Case Study of a Watershed Rehabilitation Project: Alkali Creek, Colorado*, Research Paper RM-189, Fort Collins: U.S. Forest Service Rocky Mountain Forest and Range Experiment Station.

Heede, Burchard H., 1986, Designing for Dynamic Equilibrium in Streams, *Water Resources Bulletin* vol. 22, no. 3, p. 351-357.

Jesitus, John, 1995, Geosynthetics Anchor Projects from the Ground Up, *Erosion Control* vol. 2, no. 6, p. 28-30.

Mackin, J.H., 1948, Concept of the Graded River, *Bulletin of the Geological Society of America* vol. 59, p. 463-512.

Parrish, William F., 1960, *Status of the Beaver (Castor canadensis carolinensis) in Georgia, 1959*, MS thesis, Athens, University of Georgia.

Rosgen, Dave, and Brenda L. Fittante, 1986, Fish Habitat Structures—A Selection Guide Using Stream Classification, p. 163-179 of *Fifth Trout Stream Habitat Improvement Workshop, Proceedings of a Conference*, Jack G. Miller, John A. Arway and Robert F. Carline, editors, Harrisburg: Pennsylvania Fish Commission.

Trimble, Stanley Wayne, 1974, *Man-Induced Soil Erosion on the Southern Piedmont 1700-1970*, Ankeny, Iowa: Soil Conservation Society of America.
Wolman, M. Gordon, 1967, A Cycle of Sedimentation and Erosion in Urban River Channels, *Geografiska Annaler* vol. 49A, p. 385-395.

Experience From Morphological Research on Canadian Streams: Is Control of the Two-Year Frequency Runoff Event the Best Basis for Stream Channel Protection?

C.R. MacRae[1]

Abstract

The increase in runoff rate and volume associated with urbanization has been related to accelerated rates of geomorphic activity often resulting in the destabilization and adjustment of the stream channel. Stormwater Management ponds have been constructed to control instream erosion potential based on the 1:2 year frequency post- to pre-development peak flow shaving concept. An alluvial channel downstream of such a facility was monitored for flow and assessed using a geomorphic survey. The hours of exceedance of geomorphically significant mid-bankfull flows increased by 4.2 times after 34% of the basin had been urbanized. The channel is in adjustment and it has enlarged by as much as 3 times pre-development bankfull cross-sectional area. Theoretical analyses of the magnitude and transverse distribution of excess boundary shear stress indicated that despite operation of the facility, erosion potential in the mid-bed increased by 2.4 pre-development values under existing land use conditions. These data are supported by observations from 7 streams downstream of similar facilities in Surrey, British Columbia. The need for a multi-criterion concept based on discharge and boundary material characteristics and a 2-dimensional approach is evident. An alternate design criteria based on a zero net change in the transverse distribution of shear stress about a channel perimeter using an erosion index method is proposed.

[1]Senior Technical Specialist, Aquafor Beech Limited, 132 Ontario Street, Kingston, Ontario, K7L 2Y4

Introduction

Urbanization can lead to an increase in both runoff volume and rate which can be translated into an increase in instream erosion potential and a concomitant adjustment in channel form (Leopold, 1968; Hammer, 1972; Graf, 1975; Allen and Narramore, 1985; Neller, 1988; Booth, 1990). To mitigate against this impact, Stormwater Management (SWM) policies have been developed based on control of the post-development peak flow rate of the 1:2 year frequency design storm event to pre-development conditions. This is achieved through the design of facilities with sufficient storage volume and control of the outflow rate to provide the required degree of attenuation. This approach is referred to as the Zero Runoff Increase (ZRI) criteria (control does not apply to runoff volume).

In Canada, millions of dollars have been spent and thousands of SWM ponds have been built to control instream erosion potential based on the ZRI concept. While the incorporation of erosion control measures into SWM policy represents a significant step forward in the management of urban water resources, the ZRI criteria has meet with limited success and may actually aggravate erosion hazard (McCuen, 1979, McCuen and Moglen, 1988; Lorant, 1983 and 1988; MacRae, 1993). The lack of success of this approach has been attributed to the use of peak flow rate in a single criterion design approach (Whipple et al., 1981; McCuen and Moglen, 1988) and the lack of consideration of the frequency and duration of flows and boundary material sensitivity to scour (McCuen and Moglen, 1988; MacRae and Rowney, 1992).

In this study, the effectiveness of an existing ZRI facility located in an urbanizing area on Morningside Tributary in Markham, Ontario is assessed in terms of the change in instream erosion potential and the stability of the receiving channel. Alteration in channel morphology is explained using a theoretical analysis of instream erosion potential based on a 2-dimensional erosion index approach (MacRae, 1991). Results from this case study are compared to survey results for 7 stream channels located downstream of similar peak flow shaving facilities in Surrey, British Columbia.

Urbanization And Channel Morphology

The effect of land use alteration, such as the urbanization of a basin, on the flow regime may be generalized as a change in the natural storage of a catchment (McCuen, 1979). This alteration may lead to an increase in both runoff volume and rate with an associated increase in erosion potential. Although Hollis and Luckett (1976) and Leopold (1973) reported studies to the contrary, most case studies have shown that channels tend to enlarge following urbanization (Savini and Krammerer, 1961; Hammer, 1972; Allen and Narramore, 1985; Lee and Ham, 1988; MacRae et al., 1994). Morisawa and LaFlure (1979) observed that channel cross-sectional form begins to enlarge after 20 to 25% of the basin achieve more then 5% impervious cover with channel enlargement increasing rapidly to 5 to 7 times pre-development channel cross-sectional area after complete development of the basin. Similarly, Booth and Reinelt (1993) reported that bank de-stabilization began after about a 10% increase in average basin imperviousness. Schueler (1995) noted that these findings were supported by fisheries habitat studies. The demarcation between stable and unstable channels after an increase in basin imperviousness of approximately 10%, indicates that a threshold may exist after which morphological change in cross-section and planimetric form will occur in the receiver. This threshold also appears to be relatively low, indicating that most streams channel systems are highly sensitive to a perturbation in the prevailing sediment-flow regime.

The Importance of Bank Materials and Mid-Bankfull Flows

Sediment transport and erosion potential increase in a non-uniform, but directly proportional manner with flow depth and hence flow rate (Leopold et al., 1964; Vanoni, 1975). Using the product of the flow frequency and sediment transport rate functions, Leopold et al., (1964) generated an effective work curve with a maximum value corresponding to a bankfull stage which recurs on average once every year to once every two years. They noted, however, that the range of geomorphically significant flows lie between the lower limit of competence and an upper limit established as those flows which are no longer contained within the active channel. A compilation of studies by Hollis (1975) showed that the increase in runoff rate due to urbanization is non-uniform such that the increase in flood frequency diminishes with return period. MacRae and Rowney (1992) demonstrated that this non-uniformity causes the maximum point on the effective work curve to shift toward the mid-bankfull events - those events that occur less then once in every year, but are above the level of competence. Consequently, mid-bankfull events are the events which do the most work in urban streams and an erosion control philosophy which does not adequately address this shift in the effective work curve may not satisfactorily meet the intended purpose.

In addition, channel banks are often composed of stratified, heterogeneous materials with the basal stratigraphic unit typically being the least resistant to scour (Klimek, 1974; Andrews, 1982; and, Thorne and Lewin, 1982). Maintenance of a stable channel form, consequently, is strongly related to the sensitivity of the basal layer to a disturbance in the flow regime (Thorne and Tovey, 1981; Harvey et al., 1979; Knighton, 1987; MacRae and Rowney, 1992). Consequently, a 1-dimensional analyses based on average flow rate may be less indicative of the streams ability to do work in terms of the erosion of intact boundary materials than a 2-dimensional approach (MacRae, 1991), and stream power (τv: the product of shear stress (τ) and mean flow velocity (v)), may be more significant than discharge alone (Knighton, 1987).

Basis For The 1:2 Year Erosion Control Approach

Urban Stormwater Management (SWM) consists of a set of Best Management Practices (BMPs) which are applied, primarily in new urban developments, as a means to 'protect' the environment from detrimental impacts which have been associated with traditional urban forms and drainage practices. Leopold (1968) noted that the increase in the magnitude and frequency of flows due to urbanization could be compensated through storage controls such that the flows experienced by the channel downstream of the facility could be maintained within the range of flows experienced prior to development of the tributary area. It was also observed that a stream channel will develop a form sufficient to convey, without overflow, a discharge with a recurrence interval of 1.5 to 2 years and that this flow is in accord with the bankfull stage. This work was influential in developing the ZRI concept which was adopted as the basis for instream erosion control in Ontario and many other jurisdictions across Canada.

The ZRI concept, however, is a generic rather then case specific design procedure (the same criteria apply to all watersheds and stream channels regardless of size and boundary material composition), with flow as the single criterion for design based on a single flow rate (1:2 year event). Further, it does not address the increase in the frequency of mid-bankfull to bankfull flows, the increase in duration of high flow rates caused by attenuation within the storage facility or the sensitivity of the boundary materials in the receiving channel to erosion (McCuen, 1979; Lorant, 1983; McCuen and Moglen 1988; MacRae, 1994). The increase in flow frequency and duration was shown to result in an increase in sediment transport potential by McCuen and Moglen (1988) and Lorant (1988), who noted that this was sufficient to cause enlargement of the channel. Despite these concerns, numerous erosion control facilities continue to be constructed based on this hydrologic criteria.

Surrey, British Columbia Data

Lee and Ham, (1988) surveyed 30 reaches totalling 6.6 km of channel through both urban and rural streams (with and without ZRI control) in the District of Surrey, British Columbia. Using the rural streams to represent a stable, pre-development baseline condition, they concluded that stream channel width increased on average by 1.7 times the pre-development value for urbanized watersheds without erosion control facilities. They also concluded that channel widening, although reduced, still occurred in those streams urbanized with erosion control facilities in place.

Seven of the subwatersheds from the 1985 study were re-surveyed in 1988 along reaches immediately downstream of ZRI facilities to determine boundary material characteristics and sensitivity to scour. These observations indicate that these facilities maintained pre-development channel width in one case (Enver Creek), with various degrees of success for the remain 6 channels. The ratio of post- to pre-development channel width was found to range from $1.63 \leq (W_R) \leq 3.8$ (the average value exceeds the value reported above for urban streams without erosion controls). The parameter, W_R was found to be correlated with the strength of the least resistant, bank toe stratigraphic unit (MacRae and Rowney, 1992). These results imply that additional storage control is required as the sensitivity of the basal unit to scour increases and that erosion control criteria should include a measure of sensitivity of channel boundary materials to an increase in erosion potential. The District is currently undertaking a review of their drainage policy with respect to their erosion control criteria (Paul Ham, 1996 pres. comm.).

Morningside Creek Case Study

Morningside Tributary, drainage area 21.4 km^2, is a subwatershed of the Rouge River, a valuable cold water fishery within the Greater Toronto Area. The lower portion of the Tributary basin is fully developed and the headwater portion, drainage area 10 km^2, is currently 34% developed. The remaining 66% of the headwater area is planned for development under a mix of residential, light industrial and commercial land uses. Stormwater Management for the existing development within the headwater area, which is primarily medium to high density residential, consists of a 2 centralized, detention facilities (the North and South Ponds: Figure 1). These facilities were designed for flood and erosion control with the later based on the ZRI approach.

Erosion problems within the headwater area of the Tributary, which sustained a resident population of brook trout prior to urbanization, lead to concern with regard to the stability of the channel and its impact on the Rouge River. Two of the primary objectives of this study were to: gain insight into the sensitivity of the system to an alteration in the mechanisms controlling channel form; and, evaluate the success of the ZRI facility in terms of controlling instream erosion.

The Study Area

The watershed occupies areas of the Peel Plain and South Slopes physiographic regions characterized by the silty sand deposits of the Leaside Till overlaying inter-bedded lacustrine sands and silty clays (Chapman and Putnam, 1984). Approximately 80% of the soils consist of the Millikan and Woburn loams (Hoffman and Richards, 1955), which have imperfect to good drainage characteristics, respectively. The balance of the area is covered with Peel Clay soils. The topography is characterized as gently rolling upland plain.

Upper Morningside Tributary is a misfit stream which is moderately entrenched in the Leaside Till. Prior to development, tributaries upstream of the North Pond consisted of rills and gullies leading to an intermittent channel which ended at the farmer's ford approximately 0.6 km downstream of the North Pond (Figure 1). Upstream of the North Pond these tributaries are now sewered or

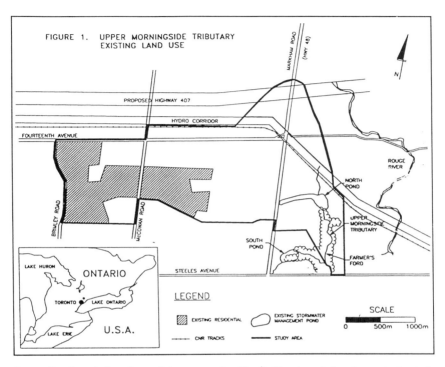

FIGURE 1. UPPER MORNINGSIDE TRIBUTARY EXISTING LAND USE

channelized and the channel between the North Pond and the farmer's ford now flows perennially. Downstream of the ford, baseflow is maintained throughout the year by groundwater inputs with two springs located at the west side of the ford. The channel from the North Pond to Steels Avenue, approximately 1.3 km in length, is referred to as the upper Morningside Tributary and this length of channel defines the limits of the study area (Figure 1).

A silty sand alluvial soil covers the valley bottom which is underlain by layers of silt, cross-bedded sand and fine gravel. A deposit of sand (massive bedding structure) exists beneath the recent alluvium and a layer of grey stoney clay was observed at the farmer's ford. Vegetation within the valley consists of slope and bottomland forest which generally increase in density and decrease in degree of disturbance downvalley. Currently sediment supply to the study reach from the majority of the tributary area is limited due to the settling of coarse particulates in the pond. Silt, clay and detritus is prevalent for a 100 m stretch downstream of the control facility indicating that a supply of fine sediment is passing through the pond and into the study reach. Sediments, primarily in the sand size fraction, are supplied within the valley through erosion of the active channel and the floodplain banks where the active channel is in contact with the valley wall.

Study Methodology

Geomorphic surveys were conducted in the summers of 1993 and 1994 and a flow monitoring station was established at the pond outlet and operated continuously over the above 2 year period. The study reach was divided into 5 Response Segments (RSs) based on their sediment transport potential, the nature of local sediment sources, and 'like' morphological characteristics and processes (Figure 2).

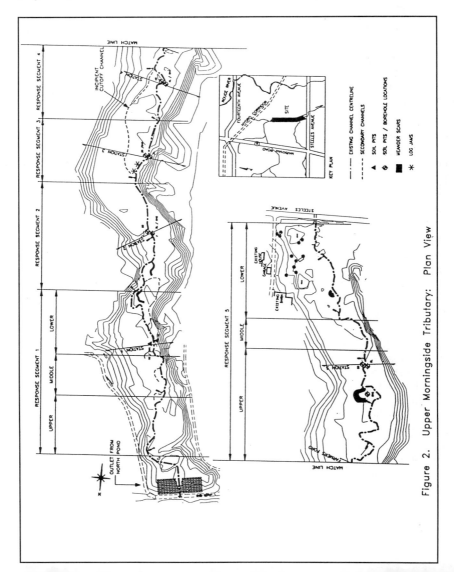

Figure 2. Upper Morningside Tributary: Plan View

Within each RS, channel form variables were collected along a representative reach to characterize: planimetric form (meander wavelength and amplitude, pool-riffle sequence, and bar forms), longitudinal profile; hydraulic geometry (channel and valley cross-sections along straight channel segments between meander bends); hydrologic characteristics (discharge rate, Manning's roughness coefficient, primary flow velocity); boundary material composition (bank stratification and material composition, critical shear stress (by stratigraphic unit), substrate composition (particle size analyses of bar forms and non-descriptive, loose alluvium) and substratum composition (particle size characteristics of the intact bed materials or bed armour beneath the loose alluvial deposits); and, descriptive accounts of riparian vegetation. Rates of geomorphic activity were inferred from fluvial forms, such as transient bars, paleofluvial investigations of abandoned meander features, comparison of historic longitudinal profiles and oblique photographs, erosion pin data, and empirical relations published in the literature. Collective consideration of these data provided insight into the evolutionary path of the system given a change in the driving mechanisms.

<u>Baseline Geomorphic State</u>

The interpretation of observed morphological features, which indicate an unstable channel environment, differ between RSs due to changes in valley gradient, the hydraulic affect of Large Organic Debris (LOD) jams, significant local sources of coarse sediment, niche points, pre-development hydraulic geometry, and the relative erodability of the bed and bank materials. A summary of the dominant processes and morphological response are presented in Table I for each RS. Taken collectively, however, morphological evidence suggest that the entire length of the upper Morningside Tributary is currently unstable and either undergoing a change in form or adjustment is pending.

Erosion pins (1.0 m in length) were installed in the mid-bed, bank toe and mid-bank locations at two sites, E2 (RS 1) and E3 (RS 5) in the spring of 1993 (Figure 3). Results from the monitoring of these sites over a 1 year period are reported in Table II. High rates of bed and bank erosion were observed at site E2 where a niche point, which is migrating upstream has come to within 5 m of the pin locations. A moderate rate of downcutting and a high rate of basal scour were reported at site E3. The lower rate of downcutting is attributed to: the increase in flow resistance due to LODs within the channel and an increase in the influx of coarse sediment originating in RS 3. The high rate of basal scour is attributed to an increase in shear stress at the bank toe due to greater flow depths and deflection of flows against the banks by LODs accumulating within the channel. An inspection of the channel in April, 1995 noted that enlargement of the channel had resulted in the complete removal of all six pins.

Table I Summary of Morphological Erosion Index Analyses

REACH/ STATION	COMMENT	EROSION PIN EXPOSURE (cm)
RS 1 (E2)	Entrenched due to the upstream migration of niche point - rate of downcutting decreasing - high influx of coarse sediment - banks have widened due to basal scour - geomorphic activity declining	BED = 13 TOE = 14 BNK = 29
RS 2	Aggrading due to backwater from LOD jam in RS 3 - banks saturated and failing through mass wasting and basal scour - cutoff channel forming at LOD jam - moderate rate of geomorphic activity	
RS 3	Cutoff at u/s end has isolated RS 3 from flow increase - this is balanced by a low sediment load immediately d/s of the LOD - channel temporarily stable - influx of sand in middle segment from erosion of valley wall causing aggradation.	
RS 4	Flow energy reduced by cutoff at LOD jam in RS 3 - sediment supply from RS 3 has created aggrading condition - low rate of geomorphic activity but high rates pending failure of LOD	
RS 5 (E3)	Entrenched - rate of downcutting diminishing due to a high sediment load and the hydraulic effect of local LODs - high rate of basal scour causing bank failure through oversteepened -aggravated by deflection of flows against banks by LOD jams	BED = 3 TOE = 16 BNK = -1

RS = Response Segment; E2 & E3 are erosion pin stations; BED, TOE and BNK refer to the mid-bed region and the basal and mid-bank stratigraphic units respectively.

Changes in longitudinal profile between 1993 and 1994 also indicate that the channel is in a period of adjustment (Figure 3). Due to differences in survey protocol between the two years, only elevation differences over successive riffle sections can be interpreted as morphologic change. These data indicate that degradation is actively occurring in RS 1 between Stations 200 and 240 due to the upstream migration of a niche point and valley formation. Aggradation is occurring in RS 2 between Stations 480 and 590 due to the backwater affect of the LOD jam. Aggradation is also occurring in RS 3 (Stations 640 to 700) due to the bypass of high flows through the cutoff at the LOD jam and a significant influx of coarse sediment from basal erosion of the valley wall. Aggradation observed in RS 4

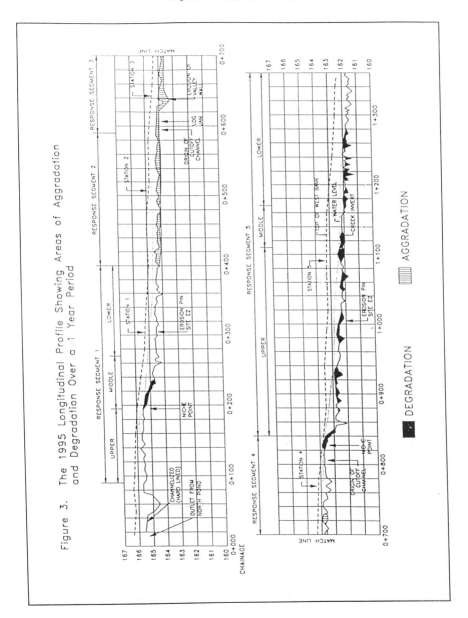

Figure 3. The 1995 Longitudinal Profile Showing Areas of Aggradation and Degradation Over a 1 Year Period

Table II   Summary of Morphological Erosion Index Analyses

REACH/ STATION	COMMENT	EROSION PIN EXPOSURE (cm)
RS 1 (E2)	Entrenched due to the upstream migration of niche point - rate of downcutting decreasing - high influx of coarse sediment - banks have widened due to basal scour - geomorphic activity declining	BED = 13 TOE = 14 BNK = 29
RS 2	Aggrading due to backwater from LOD jam in RS 3 - banks saturated and failing through mass wasting and basal scour - cutoff channel forming at LOD jam - moderate rate of geomorphic activity	
RS 3	Cutoff at u/s end has isolated RS 3 from flow increase - this is balanced by a low sediment load immediately d/s of the LOD - channel temporarily stable - influx of sand in middle segment from erosion of valley wall causing aggradation.	
RS 4	Flow energy reduced by cutoff at LOD jam in RS 3 - sediment supply from RS 3 has created aggrading condition - low rate of geomorphic activity but high rates pending failure of LOD	
RS 5 (E3)	Entrenched - rate of downcutting diminishing due to a high sediment load and the hydraulic effect of local LODs - high rate of basal scour causing bank failure through oversteepened -aggravated by deflection of flows against banks by LOD jams	BED = 3 TOE = 16 BNK = -1

RS = Response Segment; E2 & E3 are erosion pin stations; BED, TOE and BNK refer to the mid-bed region and the basal and mid-bank stratigraphic units respectively.

(Stations 700 to 790) is associated with the lower energy environment created by the diversion of flows, as noted above, and the movement of the sediment wave from RS 3 through this reach. The behaviour of RS 5 is as noted above for site E3. While the behaviour of the five RSs appears to be highly variable, aggradation in RS 2, 3, and 4 is a temporary phenomena linked to the life expectancy of the LOD jam in RS 3. Failure of the LOD, headcutting of the cutoff channel or upstream migration of the niche point in RS 5 will eventually result in the degradation of these reaches. Consequently, the overriding process within all segments is characteristic of valley formation. The complex response of the channel system also serves to illustrate the need for a geomorphic interpretation of the fluvial forms and processes operating within the channel system.

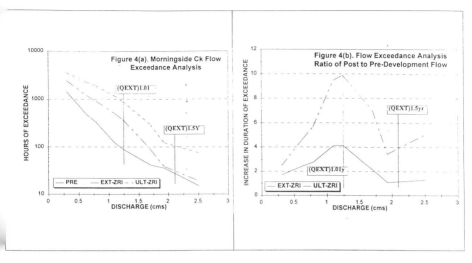

Figure 4(a). Morningside Ck Flow Exceedance Analysis

Figure 4(b). Flow Exceedance Analysis Ratio of Post to Pre-Development Flow

Exceedance Analyses

QUALHYMO (Rowney and MacRae, 1991), a continuous hydrologic simulation model developed for the planning level design and evaluation of BMPs in urbanizing watersheds, was setup and calibrated to observed flows (1993 and 1994) at the pond outlet. The model was used to generate a 5 year time series at a one hour time step based on Atmospheric Environment Service precipitation and temperature records reported for L.B. Pearson International Airport. These data and the time step of one hour were considered appropriate given the size of the sub-watersheds being modelled and the amount of flow routing created by the Pond. The fit between predicted and observed flow rates was considered to be good. Flows under ultimate development conditions were obtained by adjusting the fraction of impervious area in the model. All other parameter values remained the same.

An exceedance analyses of the 5 year period was undertaken to assess the change in the duration of time in which the flow series exceeded a set of specified flow rates. This exercise was applied to pre-development conditions (PRE), existing land use characteristics with the existing ZRI facility in place (EXT-ZRI), and the ultimate land use scenario with the ZRI pond (ULT-ZRI; Figure 4(A)). The duration of events in the mid-bankfull ((Q_{EXT})$_{1.01YR}$=1.24 m^{3}s^{-1}) flow range increased by 4 and 10 times pre-development levels for existing and ultimate land use conditions respectively (Table III; Figure 4(B)). A smaller increase in the hours of exceedance for bankfull flow ((Q_{EXT})$_{1.5YR}$=2.1 m^{3}s^{-1}) was also noted. These data indicate that a significant alteration in the flow regime has occurred, particularly with respect to flows in the mid-bankfull flow range. This increase in flows is consistent with the current classification of the channel system as unstable and this state will be aggravated under future land use conditions.

Table III   Hours Of Exceedance For The $Q_{1.01YR}$ and $Q_{1.5YR}$ Flow Rates

RI (yrs)	Q_{EXT} (m^3s^{-1})	EXCEEDANCE (hours)		
		PRE	EXT-ZRI	ULT-ZRI
1.01	1.24	90	380	900
1.5	2.1	30	34	120

RI = Recurrence Interval; Q_{EXT} = Existing Flow Rate; PRE, EXT and ULT refer to pre-development, existing and ultimate land use; ZRI = 1:2 year peak flow shaving facility

Instream Erosion Potential

To assess instream erosion potential (E_S), a comparative approach was adopted wherein indices of E_S for the pre-development condition were adopted as a baseline to determine the relative impact of a land use alteration and the effectiveness of the ZRI facility. Paleofluvial techniques were used to reconstruct an historic meander scar which was used to approximate pre-development E_S channel form. This procedure assumes that the cross-section and planimetric geometry of the meander scar is in equilibrium with land use conditions and the associated flow-sediment regimes circa 1950.

An index of E_S is defined in the SHEAR1 command in QUALHYMO as:

$$(\Delta E_S)_P = \left[\int \left(f(q_S)_o - (q_S)_I \right)_{POST} dt - \int \left(f(q_S)_o - (q_S)_I \right)_{PRE} dt \right]_P$$

in which $(\Delta E_S)_P$ is the change in instream erosion potential at point P about the channel perimeter, q_S is the sediment transport potential, the subscripts I and O stand for Input and Output from the control reach and POST and PRE represent post- and pre-development land use conditions respectively (MacRae, 1994).

The SHEAR 1 command requires: data describing cross-section and planimetric geometry; and, critical shear stress (τ_{crt}) and shear stress as a function of flow depth for each boundary station. Critical shear stress was determined based on the physical properties of the material within each bank stratigraphic unit and the intact bed materials. Shear stress versus flow depth relations were based on minimizing the error between predicted and observed flow and the average boundary shear stress computed by the model for 40 stations about the channel perimeter and average boundary shear stress ($\tau = \rho g R S$). Using this approach, E_S was then determined for the meander scar and each RS at the representative cross-sections for the PRE, EXT-ZRI, and ULT-ZRI conditions for the 5 year simulation period. This technique allows for the assessment of E_S about both the bed and banks for channels formed in stratified, heterogeneous boundary materials in a pseudo 3-dimensional manner.

It has been established that sediment transport potential serves as a measure of erosion potential (Vanoni, 1975; Simons and Li, 1982) and that the shearing action of the sediment-water mixture is the dominant erosion process (Leopold et al., 1964; Simons and Li, 1982; Harvey and Watson, 1986). Various explicit forms of the relation for q_S were applied including the Meyer-Peter Muller, stream power, and DuBoys bed load transport functions (MacRae, 1991). In this application these relations were found to produce similar results, consequently, stream power as defined below was adopted for this analysis,

$$\left(q_s\right)_P = \left[a(\tau_o - \tau_{crt})\bar{v}\right]_P$$

in which a is a coefficient (given the comparative approach a was set to unity), τ_o is the instantaneous shear stress, and v is the average primary flow velocity.

The results of the analysis are presented in Table III for the BED and TOE stations (results are not presented for the BNK station for which $(E_S)_{BNK} \ll (E_S)_{TOE}$). Data in Table III report the *erosion index ratio* ($R_{ES(BED)}$), expressed as $(E_S)_{BED}$ for each representative cross-section over $(E_S)_{BED}$ at the meander scar for EXT-ZRI and ULT-ZRI conditions. A value of $R_{ES(BED)}=1$ indicates that (E_S) has been reduced back to pre-development levels. Assuming the pre-development channel form to be stable, this would indicate that further enlargement of the channel would be minimal. The ratio, however, was found to be $1.25 \leq R_{ES(BED)} \leq 2.76$ for EXT-ZRI implying that enlargement is still occurring. This prediction was confirmed during the April, 1995 inspection.

The complexity of dealing with a channel system that is in adjustment can be simplified by using the pre-development channel as a basis for comparison of the change in erosion potential. Flows for PRE, EXT-ZRI and ULT-ZRI conditions were used to determine E_S at the meander scar. Results were expressed in terms of the *erosion index ratio* wherein $R_{ES(BED)}=2.4$ for EXT-ZRI and $R_{ES(BED)}=4.6$ for ULT-ZRI conditions. These data are consistent with the magnitude of the enlargement ratio ($1.29 \leq R_E \leq 2.94$), which is defined as post- divided by pre-development channel cross-section area at bankfull stage (Table II). Flood frequency analysis of pre-development flows also indicate that enlargement of the channel has occurred.

Table IV. Summary of E_S And Hydraulic Geometry Data

PARAMETER		RESPONSE SEGMENT CROSS-SECTION NUMBER					
		1	2	3	4	5	SCAR
W_{BFL} (m)		6.36	5.31	4.70	3.50	6.40	4.70
D_{BFK} (m)		0.53	0.40	0.65	0.44	0.89	0.58
A_{BFL} (m^2)		3.50	5.15	2.83	2.25	5.00	1.75
Q_{BFL} (m^3/s)		2.9	3.2	2.34	1.25	6.05	1.36
R_E^{1}		2.0	2.94	1.62	1.29	2.86	1.00
$R_{ES(BED)}^{2}$	EXT^3	2.25	1.57	1.25	1.57	2.76	2.41
	ULT^3	4.38	4.10	1.67	2.41	6.10	4.58
$R_{ES(TOE)}$	EXT	0.98	1.00	2.74	1.28	10.11	1.00
	ULT	12.17	1.00	7.55	5.82	45.71	1.00

1. RE is the enlargement ratio expressed as the post- to pre-development bankfull channel area; 2. R_{ES} is the ratio of E_S at the RS cross-section to E_S at the meander scar under pre-development conditions and TOE and BED refer to the basal stratigraphic and mid-bed regions of the channel respectively; 3. EXT and ULT are the existing and ultimate land use scenarios with the 1:2 year peak flow shaving facility in place.

Bankfull flow, which ranges from $3 \leq RI \leq 10$ years, exceeds that reported by Leopold et al., (1964) and Leopold (1968) for natural, stable channels. The exception was RS 4 (RI=1.6 years), where isolation from the increase in high flows due to the cutoff channel in RS 3, small cross-section profile and broad, flat floodplain along with high boundary material resistance to scour have combined to minimize the degree of impact on this RS. Under ULT-ZRI conditions the enlargement ratio, $1.67 \leq R_{ES(BED)} \leq 6.10$, indicates that further enlargement will occur despite operation of the ZRI facility.

Alternate Erosion Control Design Criteria

The change in E_S, the comparison of longitudinal profiles, an exceedance analysis of frequent flow events, a geomorphic survey and erosion pin data were used to demonstrate that progressive enlargement of Upper Morningside Tributary is occurring despite the presence of a ZRI facility. Secondly, the degree of enlargement is consistent with streams which have no stormwater controls. These data support similar observations from Surrey B.C., qualitative observations reported by many municipalities and other studies reported in the literature which conclude that the ZRI concept has not satisfied the intent of erosion control and that alternate methods must be considered. Whipple et al., (1981) concluded that provision must be made so that the aggregate erosional effect of a series of storms will not be greater than that prescribed to a known stable state. They rationalized that this may be achieved through provision of additional detention storage over and above that required for flood control. This concept, referred to as 'Extended Detention Control', was demonstrated using a 1-dimensional discrete event analysis of sediment transport capacity by McCuen and Moglen (1988). Peak outflow rate from a control pond was reduced until the computed post-development sediment transport rate, approached the pre-development level. In their particular case study, this represented an 85% reduction in post-development flows below the pre-development 1:2 year peak flow rate. Balancing pre- and post-development erosion potential based on average hydraulic parameters may not, however, be adequate by itself because of the heterogeneous and stratified nature of channel boundary materials. It is important to understand, consequently, how the potential for erosion and the resistance to disruption are distributed about the channel perimeter (MacRae and Rowney, 1992).

Based on the importance of flow rate and bed load movement, McCuen and Moglen (1988) proposed a policy which states the post-development bed load cannot exceed the pre-development amount for the 1:2 year recurrence interval (RI) flow event (assuming the pre-development channel to represent a stable state). Erosion, however, is an on-going although highly discontinuous process and bank structure is an important consideration in the determination of channel sensitivity to a disturbance. MacRae (1991) proposed an expansion of McCuen and Moglen's (1988) design criteria wherein channel erosion is minimized if the alteration in the transverse distribution of erosion potential about a channel perimeter is maintained constant with pre-development values, over the range of available flows, such that the channel is just able to move the dominant particle size of its bed load. This criteria resulted in the development of an approach referred to as Distributed Runoff Control (DRC; MacRae, 1994). A form of this approach was recently adopted as the preferred methodology for erosion control by the Ontario Ministry of the Environment and Energy (P'ng, 1994).

Conclusions

High rates of geomorphic activity, elevated sediment yields from bed and bank erosion, and channel enlargement of up to 3 times pre-development conditions were documented in a channel after a 34% urbanization of the basin. This activity was correlated with an increase in erosion potential in the mid-bed and bank toe regions of up to 6 and 10 times amounts determined under pre-development conditions respectively. An exceedance analysis of a 5 year flow series also indicated that geomorphically significant mid-bankfull flows increased in duration by 4.6 times. These impacts, which are similar to streams which have no Stormwater Management program, were documented for a channel downstream of a ZRI facility. Channel instability is predicted to increase further under the ultimate development scenario resulting in additional adjustment of the channel system. These results, which are supported by findings from 7 stream systems in Surrey, British Columbia, indicate that the ZRI approach fails to meet the intent of Stormwater Management. It also points to the need for a multi-criterion concept which considers both discharge and boundary material characteristics. An alternate multi-criterion design approach is proposed based on a zero net change in the transverse distribution of shear stress about a channel perimeter using an erosion index method.

References

Allen, P.M. and Narramore, R. (1985). "Bedrock Controls on Stream Channel Enlargement With Urbanization, North Central Texas," Water Resources Bulletin, 21:6, pp. 1037-1048.

Andrews, E.D. (1982). "Bank Stability and Channel Width Adjustment, East Fork River, Wyoming" Water Resources Research, 18, pp. 1184-1192.

Booth, D.B. (1990). "Stream Channel Incision Following Drainage Basin Urbanization," Water Resources Bulletin, 26:3, pp. 407-417.

Booth, D.B. and Reinelt, L. (1993). "Consequences of Urbanization on Aquatic Systems. - Measured Effects, Degradation Thresholds, and Corrective Strategies," in Proceedings Watershed '93 A National Conference on Watershed Management, (Mar. 21-24), Alexandria, Virginia.

Chapman and Putnam (1984). "The Physiography of Southern Ontario, 3rd Edition," Ontario Geol. Survey, Special Vol. 2, Ontario Ministry of Natural Res., 270 pp.

Graf, W.L. (1975). "The Impact of Suburbanization on Fluvial Geomorphology," Water Resources Research, 11(14), pp. 690-692.

Hammer, T.R. (1972). "Stream Channel Enlargement Due to Urbanization," Water Resources Research, 8, pp. 1530-1537.

Harvey, A.M., Hitchock, D.H. and Hughes, D.J. (1979), "Event Frequency and Morphological Adjustment of Fluvial Systems In Upland Britain," In: Adjustments of the Fluvial System, D.D. Rhodes and G. Williams ed., Kendall Hunt Dubuque, Iowa, pp. 139-167.

Harvey, D.M. and Watson, C.C. (1986). "Fluvial Processes and Morphological Thresholds In Incised Channel Restoration," Water Resources Bulletin, 22(3), pp. 359-368.

Hoffman, D.W. and Richards. N.R. (1990). "Soil Survey of Peel County," Report No. 18, Ont. Soils Survey, Ministry of Agric. and Food, 85 pp.

Hollis, G.E. (1975). "The Effect of Urbanization on Floods of Different Recurrence Interval," Water Resources Research, 11(3), pp. 431-435.

Hollis, G.E. and Luckett, J.K. (1976). "The Response of Natural Stream Channels to Urbanization," Two Case Studies From Southwest England," Jour. of Hydrology, 30, pp 351-363.

Klimek, K. (1974). "The Retreat of Alluvial River Banks in the Wisloka Valley (South Poland)," Geographia Polonica, 28, pp. 59-75.

Knighton, A.D. (1987). "Downstream River Channel Adjustments," In *River Channels Environment and Process*, K. Richards (ed.), Basil Blackwell, pp. 95-128.

Lee, K. and Ham, P.J. (1988). "Effects of Surrey's Storm Water Management Policy on Channel Erosion," Proc., International Symp. on Urban Hydrology and Municipal Eng., Town of Markham, Ont.

Leopold, L.B., Wolman M.G. and Miller, J.P. (1964). "Fluvial Processes in Geomorphology," W.H. Freeman and Co., San Francisco. 522 pp.

Leopold,L.B. (1968), "Hydrology for Urban Planning - A Guidebook on the Hydrologic Effects of Urban Land Use," U.S. Geol. Sur. Circ. 554, 18 pp.

Leopold, L.B. (1973). "River Channel Change Through Time: An Example," Bulletin 84, Geological Soc. of America, pp. 1845-1860.

Lorant, F.I. (1983). "Erosion Control by Storm Water Management is it Worth a Dam," Proc., Canadian Society of Civil Eng., Annual Conf., June 2-3, Ottawa., Ont., pp. 723-727.

Lorant, F.I. (1988). "Erosion Process In Urban Areas," Proc., Internat. Symp. On Urban Hydro, and Municipal Eng., Sec. C, Town of Markham, Markham, Ont.

MacRae, C.R. (1991). "A Procedure For The Design Of Storage Facilities For Instream Erosion Control In Urban Streams," Unpublished Ph.D. Thesis, Dept. of Civil Eng., Univ. of Ottawa.

MacRae, C.R. and Rowney, A.C. (1992). "The Role of Moderate Flow Events and Bank Structure in the Determination of Channel Response to Urbanization," 45th Annual Conf., Resolving Uncertainty in Water Management, D. Shrubsole (ed.), Proc., Canadian Water Resources Assoc., Kingston, Ont.

MacRae, C.R. (1993), "An Alternate Design Approach for the Control of stream Erosion Potential in Urbanizing Watersheds," Sixth International Conf. on Urban Storm Drainage, Sept. 12-17, 1993, Niagara Falls, Ont.

MacRae, C.R., Smylie, J.L., and Levesque, R. (1994). "Sawmill Creek Natural Channel Design Case Study: Stream Survey Techniques and Observations," First International Conf. fro Natural Channel Systems, Niagara Peninsula Conser. Auth., American Fisheries Soc., Canadian Water Res. Assoc., Soil & Water Conser. Soc., and the Ontario Ministry of Natural Res., Niagara Falls, Ontario (Mar. 2-4).

McCuen, R.H. (1979), "Downstream Effects of Storm Water Management Basins," Journal of the Hydr. Div., ASCE, 1,1 pp. 21-42.

McCuen, R.H. and Moglen, G.E. (1988). "Multicriterion Storm-water Management Methods," Journal of Water Resources Planning & Management, ASCE, 114. 4., pp. 414-431.

Morisawa, M. and LaFlure, (1979). "Hydraulic Geometry, Stream Equilibrium and Urbanization," In Adjustments of the Fluvial System, D.D. Rhodes and G.P. Williams (eds.), Proc. 10th Annual Geomorphology Symp. Series, Binghamton, N.Y., (Sept. 21-22), pp. 333-350.

Neller, R.J. (1988). "A Comparison of Channel Erosion in Small Urban and Rural Catchments, Armidale, New South Wales," Earth Surface Processes and Landforms, 13, pp. 1-7.

P'ng, J. (1994). "Stormwater Management Practices, Planning and Design Manual," Environmental Science and Standards Division, Program Development Branch, Ontario Ministry of Environment and Energy (June).

Rowney, A.C. and MacRae, C.R. (1991). "QUALHYMO User's Manual Version 2.1: A Language For Continuous Hydrologic Simulation," Dept. of Civil Eng., Royal Military College of Canada.

Savini, J. and Krammerer, J.C. (1961). "Urban Growth and the Water Regime," USGS Water Supply Pap. 1591-A (Part 2), 42 pp.

Schueler, T. (1995). "Site Planning for Urban Stream Protection. Centre for Watershed Protection, Metropolitan Washington Council of Governments, Silver Spring, MD, 222 pp.

Simons, D.B. and Li, R.M. (1982). "Engineering Analysis of Fluvial Systems," Simons, Li & Associates, Inc., Colorado.

Stevens, M.A., Simons, D.B. and Richardson, E.V. (1975). "Nonequilibrium River Form," Journal of Hyd., Div., ASCE, 101 (HY5), pp. 557-567.

Thorne, C.E. and Tovey, M.K. (1981). "Stability of Composite River Banks," Earth Surface Processes and Landforms, 6, pp. 469-484.

Thorne, C.R. and Lewin, J. (1982). "Bank Processes, Bed Material Movement, and Planform Development," In: Adjustments of the Fluvial System, D.D. Rhodes and G.Williams ed., Kendall Hunt Dubuque, Iowa, pp. 117-138.

Vanoni, V.A. (1975). "Sedimentation Engineering," ASCE, 54, NY, NY.

Whipple, W. Jr., DiLouie, J.M. and Pytler, T. Jr. (1981). "Erosion Potential of Streams in Urbanizing Areas," Water Res. Bull., AWRA, 17, 1 pp. 36-45.

Effects of Urban Growth on Stream Habitat

Douglas T. Sovern, P.E.[1]
Percy M. Washington, Ph.D.[2]

Abstract

The general impacts of urbanization on stream habitat are well-known among professionals responsible for stream habitat and fish populations. Most attempts to reconstruct damaged streams focus on replacing damaged habitat to pre-urban conditions or on stream modifications that reflect the physical characteristics of the stream as it currently presents itself. In general, both approaches are subject to higher failure rates and higher maintenance costs. Because the changes brought by urban growth are dynamic, the fundamental concepts of how the stream is adjusting to new hydrologic conditions must be understood before effective stream remediation can occur. "Living systems" do not adapt to constantly changing environmental conditions, and streambed instability is often the dominant limiting factor for fish populations in urban streams.

This paper presents a logical picture for stream response to urban growth, including the hydraulic relationships to habitat conditions. Once the stream response is understood, a "new urban stream" can be developed that manages the effects of increased runoff from urban watersheds. The "new urban

[1]Director of Engineering Services, Gaia Northwest, Inc., 1420 S.W. 306th Street, Federal Way, Washington 98023-3420

[2]President, Gaia Northwest, Inc., 10522 Lake City Way N.E., C-201, Seattle, Washington 98125

stream" concept is presented through experience with completed projects on
two Pacific Northwest streams.

Introduction

Salmon populations in the Pacific Northwest are dwindling. The causes of
population reductions are many. The demise of small streams is one
significant cause of the problem. Small streams provide spawning for
several species. For species such as coho and steelhead, small streams
provide a home for up to a year prior to the smolts leaving the stream for
saltwater. Understanding of the impacts of urbanization and general
deforestation on the health of small Western Washington streams is an
important step to reestablishing healthy fish populations.

Larger rates and volumes of stormwater runoff result from urbanization,
increasing both the rate of erosion and volume of sediments. Sedimentation
and erosion are natural processes that can be dominant limiting factors for
fish populations. Erosion is the physical process by which sediments are
produced. Excessive erosion results from soil instability, and excess stream
erosion destroys redds, rearing habitat, and food production areas. The
presence of erosion, rather than sediment deposition, can be a much larger
factor causing stream habitat deterioration.

Sediments are derived from terrestrial and riparian sources. Terrestrial
sources are influenced by the condition of the whole watershed. Factors
include the type and extent of vegetative cover, percent imperviousness, soil
types, and watershed shape, slope, and gradient. Often, the two most
dominant terrestrial sediment sources are gully and surface erosion. Erosion
from other incidental sources, such as stormwater flows, trails and mass
wasting (land slides), can also be factors.

Riparian sediments are generated from the banks and beds of streams and
rivers. Riparian sediments combine with terrestrial sediments to affect the
resulting stable channel shape, large organic debris (LOD) present in the
stream, bed and bank stability, and the amount and type of bank vegetation
that provide shading and cover.

Pacific Northwest fish derive most of their food from living bottom
organisms (or benthos) that live in, or on, the substrate of the stream.
Stream benthos are most affected by erosion and sedimentation. Suspended
or dissolved pollutants can cause water quality to be toxic to fish and the
benthic populations through erosional processes in the watershed. The

benthic system can be limited both by erosion (which provides conditions of constant change) and by sedimentation (which smothers redds and food production areas, and fills rearing habitat). Control of erosion is the dominant method of sediment management, but passage of sediments through the system is also an important management feature.

Stream Classification. A classification system for natural rivers was developed where the morphological arrangement of stream characteristics are organized into relatively homogeneous stream types (Rosgen, 1994). Morphologically similar stream reaches are divided into seven major stream type categories for various land forms that differ in entrenchment (ratio of width of floodplain to width of dominant discharge channel), streambed gradient, width/depth ratio (W/D) of the dominant discharge channel, and sinuosity. Within each major category, six stream types are defined that are delineated by dominant discharge channel materials ranging from bedrock to silt/clay along a continuum of gradient ranges.

Rosgen's classification concept is that a stream type can be generally categorized according to a set of defined parameters, and that changes in physical conditions (e.g. sediment load or channel width) will result in a corresponding predictable change in channel classification. Detailed discussion of Rosgen's classification concept is not included in this paper, but the concept has significant utility.

Sediment Transport. Sediment transportation concepts are useful to help understand changes that occur and to predict hydraulic equilibrium conditions. Any stream will respond to imposed changes. Six basic hydraulic relationships describe stream responses to hydrologic changes, regardless of stream size (Simons and Sentürk, 1991):

- Depth of flow in the dominant discharge channel is directly proportional to discharge

- Width of the dominant discharge channel is directly proportional to water discharge and sediment discharge.

- Dominant discharge channel shape (width divided by depth) is directly related to sediment discharge.

- Channel gradient is inversely proportional to water discharge and directly proportional to sediment discharge and grain size.

- Sinuosity is directly proportional to valley gradient and inversely proportional to sediment discharge.

- Transport of bed material is directly related to stream power and concentration of fine material, and inversely proportional to the fall diameter of the bed material.

Stream Habitat Characteristics. The bank-full discharge is known as the dominant discharge, which has a recurrence period of approximately once each 1.5 years. The dominant discharge channel is the channel formed by the dominant discharge, or the bank-full discharge. The aquatic habitat channel is the permanently wetted portion of the dominant discharge channel and is the area that contains the food web, fish, and associated fish habitat.

Within the dominant discharge channel, environmental conditions must promote habitat suitable for fish populations, which are strongly affected by a stream's hydraulic characteristics. Hydraulic/habitat criteria in five general categories must be satisfied:

- Estuaries/Deltas
- Passage
- Refuge
- Rearing
- Spawning and Incubation

Each species and age of fish have different limiting criteria. In urban streams, these conditions may be dominated by the stream's hydraulic characteristics. At issue are the effects of urbanization on the stream's hydraulic characteristics and what, if anything, can be done to mitigate the adverse effects of urbanization.

Pre-Development Conditions

In Western Washington small streams are often in steep-sided canyons with fairly straight valley bottoms. In their natural (forested) state, small streams have a slightly meandering, small low-flow channel in a narrow valley bottom. Growth along banks is often dense, providing shade, channel stability and cover. Debris jams are common, slowing flows during higher runoff events. Soils in Western Washington are often products of glacial activity comprising both smooth cobbles and stones, as well as fine materials. Clayey bank materials, heavy root structure, and a dominant steady base flow cause dominant discharge channels to exhibit nearly vertical sides and small widths (3 to 6 feet wide, sometimes as small as one foot) relative to the width of the valley bottom. Typical streams in old-growth, coniferous forests have low nutrient levels and low annual sediment yields.

As many as 250 plant and animal species may comprise the aquatic biological community. In its original condition, the aquatic biological community evolved to use stable aquatic habitat associated with old growth, coniferous watersheds. Much of this aquatic community functions as the food web, with fish populations representing the mega fauna. Some salmon and trout species are the top (or apical) predators. Biological diversity occurs because natural watersheds have three qualities:

- Forested watershed
- Habitat diversity
- Habitat stability

Stream Characteristics. Streambed gradients gradually increase from the outlet of a stream to the upstream reaches of the watershed (Leopold, Wolman, and Miller, 1964). Because flow rates are generally greater, streambed gradients in the lower parts of the watershed are flatter than the streambed gradients in the upper reaches of a watershed.

Bed features vary according to streambed gradient (Rosgen, 1994). Figure 1 is illustrative of the variable streambed features in a typical Western Washington watershed. Pool/riffle habitat dominates streambeds with gradients less than 2 percent. Pool/drop (step/pool) habitat dominates streambeds with gradients of 4 to 10 percent.

Exclusive of estuaries, habitat can be characterized as either pool/drop or pool/riffle. Pool/drop habitat is dominant in reaches with smaller flows and steeper valley gradients. The pools are formed by large organic debris (LOD) or rock formations. As the stream flow rates increase, valley gradients tend to decrease and the streams tend to be dominated by pool/riffle conditions. Pools form on the outside of bends and riffles form between pools. Alternating pools and riffles are present in practically all perennial channels. In straight or meandering streams, pools form at 5 to 7 channel widths (As width increases, the number of pools decrease).

Fish Species. Salmon and trout in Western Washington evolved to take advantage of these conditions. Figure 1 lists different species that typically utilize the various regions of the watershed. Less athletic species, such as chum and pink salmon, cannot utilize the steeper-gradient, upper areas of the watershed to spawn. The fry of these species migrate to Puget Sound or the

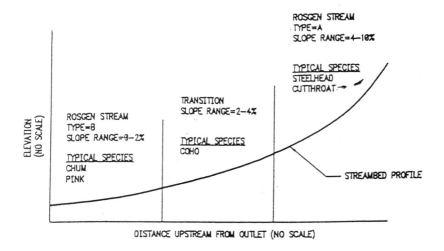

FIGURE 1: RELATIONSHIP OF STREAM HABITAT TO STREAMBED
SLOPE IN WESTERN WASHINGTON STREAMS

Pacific Ocean shortly after emergence. More athletic species (e.g., coho, steelhead, and cutthroat) occupy the upper (steeper gradient) reaches of the watershed. The young coho and steelhead reside in the stream for a year before migrating to saltwater. Cutthroat may reside their entire lives in fresh water, but many migrate to saltwater after two years in the stream. While in the stream, the young salmon, as well as young and adult trout, will utilize any part of the watershed in order to meet habitat requirements. For example, if unsuitable habitat is available in the upper reaches of a watershed, young coho may seek winter refuge in lower segments.

Effects of Urbanization

For Western Washington urban basins, the increase in the frequently occurring discharges can range from 10 to 100 times pre-development flow rates, but urbanized (and deforested) basin low flow rates are equal to, or often less than, the low flow rates for pristine conditions.

Watershed Conditions. The effects of urbanization are dramatic, in most cases permanently altering the pre-development hydrologic balance in significant ways:

- Total water passing through urban streams increases

- Runoff rates and volumes increase

- During late summer and fall, groundwater available in the system to generate base flows are often less

- Stormwater runoff increases all forms of erosion and carries significant amounts of sediments and pollutants, including oil, grease, and polluted fine sediments from streets and parking lots

Urban storm flows have higher pollutant levels, including nutrients. Vegetation loss and exposure of native soils to water can reduce pH and cause the loss of buffering for stormwater pollution. In addition, streams in urban or deforested watersheds are often significantly short in representation of the food web species and lack biological diversity. Fewer than 50 plant and animal species may comprise the aquatic habitat community in urban watersheds.

In most cases, terrestrial sediment volumes are dramatically increased by urbanization. Excess sediment increases the width of the dominant discharge channel. While the focus of this paper is on hydraulic responses of the streambed, terrestrial sediments and pollutants should be included in any analysis of stream conditions.

Stream Conditions. Most natural Pacific Northwest streams can be described as sediment starved. These streams have small width to depth (W/D) ratios, with most of the dominant discharge channel acting as an aquatic habitat channel. Most natural small Western Washington streams were generally dominated by pools and drops, formed by large organic debris (LOD). The width of the dominant discharge channel is coincident with the width of the aquatic habitat channel. As flow rates and volumes increase, streambeds become unstable. When streambeds become unstable, the aquatic habitat channel has a small W/D ratio but is substantially less than the width of the dominant discharge channel.

"Living systems" do not adapt to constantly changing environmental conditions. Constant changes of the aquatic habitat channel stifle food production and destroy spawning and incubation habitat. Compared to the previously described pristine conditions, streambed instability can be observed through several visual indicators:

- Wider and flatter gradient dominant discharge channel

- Increased braiding and head-cutting
- Reduced pool frequency and less diverse habitat
- Increased sediment from terrestrial sources
- Increased bank undercutting and shallow landsliding
- Reduced large organic debris (LOD)
- Banks adjacent to the aquatic habitat channel are bare (no protection from predation)
- Drastically reduced aquatic community diversity
- Minimal periphyton (biological growth) on streambed cobbles

A braided stream has a divided flow at normal stage with small, mid-channel bars or small islands and has the aspect of a single large channel within which are subordinate channels. Head-cutting describes channel bottom erosion moving upstream through the basin, indicating that a readjustment of the basin's gradient and its stream discharge and sediment load characteristics is taking place (Simons and Sentürk, 1991).

Stable urban streambed gradients are often one or two percent compared to two to ten percent gradients in natural watersheds. As the streambed seeks a stable gradient, the streambed can lower several feet, inducing significant volumes of shallow landsliding. Measurements taken on several Western Washington streams show streambed gradient to flatten from four percent or more to one percent. In 1000 feet of a stream whose gradient changes from one percent to four percent, the induced streambed erosion could equal 30 feet. During the transition from steeper to flatter gradients, the habitat is in a perpetual state of change.

When portions of a stream approach stability, the resulting aquatic habitat channel will be too wide, too shallow, and too homogeneous to support fish populations. In some streams, flow depths are insufficient to submerge returning adults.

Effect on Fish Species. An important factor resulting from streambed degradation is the loss of pool/drop habitat. The Rosgen classification system illustrates that an urban stream will not evolve to support the same types of habitat that existed prior to urbanization. As shown in Figure 1, the percentage of pool/drop habitat is partly a function of streambed gradient, with flatter gradients located lower in the watershed where flows are larger. Larger flows and flatter gradients favor pool/riffle habitat. The altered conditions favor less athletic fish species.

Many of the sediments from gullies and surface erosion sources are entering the stream as the storm hydrograph recedes. Sediment particles move at a slower average velocity than water and the wider channels move sediment less efficiently. The slower moving sediments tend to settle in critical habitat areas, particularly rearing, spawning, and incubation. The sediment smothers the substrate, depleting water in the interstices of dissolved oxygen needed to support eggs and propagate a diverse food supply. When food supplies are limited, fish populations are similarly limited.

For all practical purposes deforestation of watersheds permanently alter watershed hydrologic and biological conditions. Once the urbanization process starts, effects on stream shape are not completely reversible. Developing and maintaining quality fish habitat is possible under urbanized or deforested conditions, but without human intervention the potential for quality aquatic habitat in urban and deforested streams is very low.

The New Urban Stream

The divergence between urban watershed conditions and ideal watershed conditions can be narrowed through addressing three conditions:

- Hydrologic management
- Terrestrial erosion management (e.g., reforestation, eliminate gully erosion)
- Stream rehabilitation

Stream rehabilitation is the focus of this paper. To maintain the same pre-development species in urban and deforested streams, a "new urban stream" is needed to counter the tendency of urban streams to change toward lower percentages of pools. Without intervention, the habitat would shift from pool/drop to pool/riffle. Pools should occupy 30 to 60 percent of the stream, tending toward 60 percent for species like coho and steelhead that prefer steeper stream habitat. A basic requirement of the "new urban stream" is to provide pool/drop habitat equal to that which supported species prior to watershed changes.

The primary impacts to habitat occur within the dominant discharge channel. The stream flows that affect the size and shape of the dominant discharge channel are relatively low, less than the 2-year runoff rate. While larger storms are very damaging and often produce overwhelming amounts of sediment, relatively small flow rates govern the hydraulic habitat characteristics. Therefore, the quality and number of the fish are mostly controlled by lower flow rates.

Restoration is defined as the reestablishment of the whole ecosystem (National Research Council, 1992). In highly urbanized watersheds, ecological restoration is not achievable and is an unlikely achievement even when relatively small amounts of urbanization and other deforestation are present. The concepts presented in this paper are better described as stream rehabilitation, or "reclamation" where fish populations have been totally decimated.

Restoration of the exact conditions that used to exist in an urban stream is not possible. It is the authors' opinion that too many past stream rehabilitation projects emphasized streambank rehabilitation and ignored the root cause of the problems with deteriorating stream habitat. Often, that root cause is streambed instability.

Urban stream rehabilitation must be accomplished within the context of the historic conditions that can be re-created versus the conditions that cannot be meaningfully restored. With three notable exceptions, most historic environmental conditions can be re-created in urban streams:

- The width of the dominant discharge channel will be greater

- The banks of the aquatic habitat channel cannot be coincident with the dominant discharge channel (in pool/riffle habitat)

- Higher flood flows are more frequent, depths of flow are greater, and velocities are higher

Urban changes to the watershed are often a moving target; as more urbanization occurs, hydrologic and biological changes accumulate. The extent that the dominant discharge channel spreads is a direct function of the amount of urbanization in the watershed.

The objective of urban stream rehabilitation is to stabilize the streambed with appurtenances that also create a hospitable environment for fish populations. Accompanying the prime objective of stream rehabilitation is the need to develop sufficient food mass and diversity to support desired fish populations. The hydraulic/habitat criteria noted in the introduction are an idealized description of conditions that can readily support wild salmon and trout populations. Quality salmon and trout habitat exists when the hydraulic/habitat criteria are met, the streambed is stable, and the aquatic habitat channel is confined.

The "new urban stream" creates an aquatic habitat channel with characteristics similar to the old dominant discharge channel. The aquatic habitat channel forms in a permanent location within the bed of the new dominant discharge channel. The dominant discharge channel has a flatter streambed gradient and a wider dominant discharge channel than the same stream in pristine conditions.

Five basic elements describe the principles of stream rehabilitation:

1. Stabilize streambed
2. Confine aquatic habitat channel (not a required element for pool/drop habitat)
3. Revegetate dominant discharge channel banks
4. Reduce watershed generated sediment (reduce dominant discharge channel width)
5. Vegetate dominant discharge channel bed not occupied by the aquatic habitat channel (not a required element for pool/drop habitat)

The first element stabilizes the dominant discharge channel by increasing the roughness and/or flattening the streambed gradient. Increasing roughness will only be practical for very small streams. For nearly all unstable stream reaches, flatter streambed gradients are required.

Increasing roughness is accomplished by placing obstructions in the stream and is an essential part of channel confinement (Element 2). Flattening the streambed gradient is accomplished by placing structures that span the stream with elevations higher than the current streambed. For habitat welfare and aesthetics, these "spanning structures" should be fabricated from timber and boulders. Concrete, metal, and wire baskets may have application in a few locations.

Ideally, long reaches of unstable streams will be stabilized. Near the spanning structures, aggradation will replace degradation (as long as sediment is available). If only short reaches are stabilized, large storms can be expected to deposit substantial sediments through the stabilized reach. Although temporarily unsightly, the materials can be incorporated into the final stream geometry (small storms may not reveal any aggradation tendencies). If long reaches of stream are stabilized, the amount of sediment that will move is substantially reduced and stream aggradation is minimized.

For most streams, the sediments most likely to accumulate are the larger sizes that are moved as bed load. Once stable streambed gradients are established, the amount of sediment that can move and cause aggradation is finite. After a few larger storms, bed load movement will be minimal.

The object is to sculpt the incoming sediment deposition to form bars and banks and confine the aquatic habitat channel to maintain the same location. The second element confines the aquatic habitat channel. Concentrating flows in the aquatic habitat channel helps keep the substrate size optimal and clear of fines. Optimum-sized substrate will supply appropriate habitat for a diverse stream aquatic community that will ultimately support salmon and trout populations. The gradient of the aquatic habitat channel will vary with the flow rate, but the range is limited by the gradient established by spanning structures.

Deflectors, bank logs, and boulders can be used to confine the aquatic habitat channel and force it to maintain the same location within the dominant discharge channel after each storm. These structures deflect flows, protect banks, and create additional roughness. The initial presence of bed load can be a positive factor in channel confinement. Deposited sediments will form point bars on the inside of bends. Judicious siting of deflectors and bank logs will cause the sediment deposition to confine and define the aquatic habitat channel.

Elements 3 and 5 address stream revegetation and will only be effective after Elements 1 and 2 are in place. Bank revegetation (Element 3) can be undertaken after completion of streambed control (Element 1), but will be most effective after channel confinement features are in place.

Revegetation of the bed of the dominant discharge channel (Element 5) should wait until aggradation has occurred on the stream banks. This step may have to wait several years until enough large storms have occurred to move available bedload materials into position.

Regardless of the habitat type, plant species used in urban conditions may require different vegetation than natural conditions. The frequency of flooding and the added exposure to wet root conditions limit the diversity of the riparian vegetation that can survive in urban watersheds.

Element 4 should be an ongoing activity for any stream rehabilitation program and will produce positive results regardless of how much in-stream rehabilitation has been completed.

Observations of Typical Pacific Northwest Streams

The concept of a "new urban stream" has been applied to three Western Washington watersheds:

- Pipers Creek (and tributaries) in Seattle
- Mill Creek in Kent
- Thornton Creek in Seattle

Some improvements have been completed on Pipers Creek and Mill Creek. The results of Pipers Creek are described in this paper and are representative of both projects.

The Pipers Creek watershed has an area of approximately three square miles. The upper plateau contains community shopping and commercial development. Although significant areas are drained by roadside ditches and culverts across roads, extensive storm drainage pipe systems exist in much of the upper plateau. The upper watershed is 100 percent developed and contains a high percentage of paved and impermeable surfaces.

The lower section of the watershed is surrounded by steep hillsides inside the boundaries of Carkeek Park, which contains three named creeks and several ravines. Low flows are from a large storm drain and groundwater. Venema Creek and Mohlendorph Creek are significant tributaries of Pipers Creek that also have fish populations.

Flow rates for Pipers Creek and tributaries have been estimated by the City of Seattle (Seattle), At the outlet to Puget Sound, the 1-year peak flow is 330 cubic feet per second (cfs) and the 100-year event is 1,000 cfs. No flow rates for pre-developed conditions have been made, but it is unlikely that the historic 1-year event exceeded 20 cfs. Base flows in late summer are estimated at 2.5 cfs at the outlet to Puget Sound.

In the natural watershed condition, Pipers Creek was probably dominated by pools and drops formed by large organic debris (LOD), and the width of the dominant discharge channel was coincident with the width of the streambed that supported aquatic habitat (aquatic habitat channel).

In 1971, 14 boulder bed control structures were installed in the upper reaches of Pipers Creek to stabilize the streambed against the addition of stormwater from storm drain projects. Although most structures deteriorated, stream degradation was less than would have occurred in their absence. The existing structures provided a wealth of boulders for stream rehabilitation work completed since 1991 by the Drainage and Wastewater Utility of Seattle and other groups.

Streambed gradients have been measured for various reaches of Pipers, Venema, and Mohlendorph Creeks, noting locations of head-cutting. The streambed of lower Pipers Creek has a nearly stable dominant discharge channel with an unconfined aquatic habitat channel. The streambed gradient varies from 1 to 1.3 percent. In the middle and upper reaches of Pipers Creek, overall gradients between the 14 structures vary from 4 to 8 percent. Head-cutting is occurring between structures, with the average gradient between head-cutting sites averaging approximately 1.4 percent.

The estimated storm flow rates in the middle reaches of Pipers Creek are only 10 percent lower than the flow rates at the outlet to Puget Sound. Streambed gradients between the head-cutting sites are approximately equal to the stable streambed gradient for the reach. The rehabilitation strategy locates new spanning structures at head-cutting locations. As long as passage criteria for the appropriate fish species are met, few additional bed control facilities are required. If the height of the head cut exceeds the passage criteria, additional spanning structures are installed.

In recent years, the chum population had a significant resurgence. Due to an aggressive fish imprinting program, record returns of chum salmon returned to Pipers Creek, Venema Creek, and Mohlendorph Creek since 1993. In addition, in 1995 coho and steelhead observations were made for the first time in decades. The return of the coho and steelhead coincide with the return of fish that would have migrated to Puget Sound after completion of the initial improvements to Pipers Creek.

The presence of thick algal growth (periphyton) on the stones in much of the streambed in several reaches of Pipers Creek indicates that substantial substrate area is currently stable. The algal growth demonstrates recent colonization and sufficient production to support much larger and more diverse aquatic invertebrate populations than was present earlier.

Conclusions

Urbanization irretrievably alters the hydrologic and hydraulic relationships of all watersheds. Because small Western Washington streams are not influenced by snowmelt runoff, the hydrologic influence of urbanization is exaggerated. The urbanized one- and two-year stormwater flow rates can range up to 100 times the historic flow rates. The effects of urbanized flow rates on small Western Washington streams demonstrate the extreme results of large scale urbanization (or deforestation).

Stream rehabilitation that will sustain salmon populations is possible as long as the unavoidable changes in the hydraulic character of a stream are recognized. The changes include a significantly wider dominant discharge channel, flatter streambed gradients, and the need to establish a permanent location for the aquatic habitat channel within the bottom of the dominant discharge channel. Stream rehabilitation can take place through stabilization of the streambed and confining the aquatic habitat channel. These steps are necessary to reestablish and preserve steeper gradient channel reaches required by some species.

Bibliography

Bell, Milo C., **Fisheries Handbook of Engineering Requirements and Biological Criteria**, US Army Corps of Engineers, Portland, Oregon, January, 1984

Kerr Wood Leidal Associates LTD., D.B. Lister & Associates LTD, **Stream Enhancement Guide**, Province of British Columbia, Vancouver, British Columbia, 1980

Leopold, Luna B., Wolman, Gordon M., Miller, John P., **Fluvial Processes in Geomorphology**, W. H. Freeman and Company, San Francisco and London, 1964

National Research Council, **Restoration of Aquatic Ecosystems**, Washington, D.C., 1992

Rosgen, David L., **A Classification of Natural Rivers**, Catena, Elsiver Science, B.V., Amsterdam, The Netherlands, 1994

Simons, Daryl B. and Senturk, Fuat, **Sediment Transport Technology, Water and Sediment Dynamics**, Water Resources Publications, Littleton, Colorado, 1991

Large Woody Debris in Urban Streams of the Pacific Northwest

Derek B. Booth[1,2], David R. Montgomery[2], and John Bethel[3]

ABSTRACT

Large woody debris (LWD) performs key functions in undisturbed streams that drain lowland forested watersheds, including dissipation of flow energy, stabilization of bedforms and channel banks, entrapment of sediment, and formation of pools. These functions vary between individual channels, however, depending on the size and morphology of the stream, which in turn depend on climate, watershed size, valley slope, geologic substrate, and relative inputs of water and sediment. Loss of LWD will alter channel form and processes, yielding greater sediment fluxes, more rapid bank erosion and incision, and loss of heterogeneity in bed morphology. Just as LWD is ubiquitous in undisturbed lowland streams of the Pacific Northwest, it is significantly depleted in urbanized systems where it is lost through washout, downcutting, and direct removal. Given the dramatic changes in runoff processes and sediment delivery that typify urban watersheds, we doubt that simple reintroduction of LWD will fully restore the lost functions of urban streams. Instead, projects that replace LWD may be best suited to recover a more limited set of rehabilitation goals; they are also necessary components of more comprehensive restoration efforts in once-forested lowland landscapes. Project designs range from the visually pleasing to the hydraulically engineered, but most approaches nonetheless fail rapidly in a dynamic stream environment. Stable configurations of LWD are best recognized through the careful observation of form and riverine context in natural systems, where years of varying water and sediment discharges have obliterated all but the most stable arrangements.

ROLE OF LWD

Logs, branches, and stumps are ubiquitous in undisturbed lowland humid-region stream and river channels. These pieces of large woody debris (LWD) can trap

---

[1] Director, Center for Urban Water Resources Management, Department of Civil Engineering, University of Washington, Box 352700, Seattle, Washington 98195 Phone (206) 543-7923; Fax (206) 685-3836; Email dbooth@u.washington.edu

[2] Department of Geological Sciences, University of Washington, Seattle, WA 98195

[3] King County Surface Water Management Division, Seattle, WA 98104

sediment, divert low and high flows, and provide cover and shading for aquatic organisms. Although the influence of any particular piece of LWD is difficult to predict and likely to change over time, both functionally and spatially, the collective effects of LWD can be substantial and very persistent. This discussion focuses on the *physical* effects of wood in channels, not because we judge the biological effects to be of lesser importance but because many of those biological effects result from the physical modifications to stream habitat that LWD imparts.

In the urban environment LWD is rare, and for many years the beneficial role of such material was ignored. Logs and branches were recognized only for their ability to block culverts or to lodge under bridges. Their presence in the active flow complicated any estimates of roughness or channel capacity, and the scour imparted by associated flow diversions was seen as a threat to bank stability and to an orderly channel geometry.

Those perceptions of LWD in urban channels are now undergoing significant revision, for several reasons:

- Recognition of adverse changes in stream-channel morphology and stability following the removal of LWD that normally accompanies land-use changes;

- Realization of the inadequacy of "traditional" mitigation for urban effects on stream channels (typically upland flow control via detention or retention ponds); and

- Increasing public interest in restored biological productivity, particularly fish, in urban stream channels.

Although the influence of LWD can be substantial, that influence is not uniform along the channel network from headwaters to river mouth. The same log that fully bridges a small mountain stream channel, suspended many feet above the water surface, could easily float down the center of the river in the broad alluvial valley below without ever touching the bed or banks. In between, that same log could wedge into both streambanks at the level of the bankfull flow, forming a step in the channel that controls bed elevations for many decades. Thus the position in the channel network is a critical determinant on the function(s) that LWD may perform: observations, analyses, management proscriptions, and rehabilitation strategies are *not* necessarily transportable from one location to another. The methodology for determining whether different sites on different streams are "equivalent" is the subject of *channel classification*, which we recognize as a necessary framework for any detailed discussion of the role and manipulation of LWD in urban stream channels.

Channel Types and Classification

Principles and Limitations. Geomorphologists and biologists have been organizing and categorizing the myriad array of stream channels for about a century. The purpose of such an organization is fundamental: if a channel of interest can be placed in a group, and the properties of that group are already known, then the properties of

our new channel will also be known (Kondolf, 1995). Those "properties" depend on
the organizational scheme, but they include such attributes as the channels' response
to environmental change (increased sediment load, placement of an artificial habitat-
enhancement structure, or removal of LWD) or its importance in supporting stream
biota (Mosley, 1987).

Classification also can affect the communication between different workers, and
between different disciplines, about the nature and condition of stream channels. The
influence can be positive, by providing a common basis for discussion and a general
agreement about the important attributes of a channel that must be recognized and
evaluated in order to understand the channel's function (Platts, 1980). Yet the
influence of classification also can be detrimental, by suggesting an overly simplistic
range of channel conditions that obscures critical differences between channels that
are ostensibly "the same." It may also impart a false understanding if the
classification method is taken outside of where it was developed to where the
dominant landscape processes are significantly different: channels may be "classified"
but the predictive power of that classification will be low or misleading.

Two examples, both relevant to urban stream channels of the Pacific Northwest,
illustrate this problem. The classification method of Rosgen (*e.g.*, 1994 and prior
informal publications) has been applied widely throughout the United States by land
managers because of its broad range of physical channel conditions that are included
in the framework, its relatively straightforward application, and the hypothesized yet
detailed channel-response matrix that accompanies the classification method. Yet
nowhere in this method is the role of LWD recognized, reflecting the non-forested
environment in which this method was first developed. This does not negate its
utility in many settings but should remind us of its potential limitations in some.

The classification of Montgomery and Buffington (1993) was established
explicitly to address the channels found in forested watersheds of the Pacific
Northwest, where LWD is ubiquitous and one of the most important management
needs is to evaluate the addition or removal of logs in the stream channel. Yet the
same channel types in different lithologic units may appear very similar but *respond*
very differently to changes in flow regime or sediment inputs. Some of the sediment-
delivery processes and sources of channel roughness are very different in lowland
urban channels than in the headwater channels of the Cascade Range, and so the
utility of this (or any) classification may be limited for predicting specific channel
response in different watershed settings.

Criteria. Despite these caveats, the related issues of channel types and channel
classification are inescapable in addressing the role of LWD in urban channels. We
will use the classification of Montgomery and Buffington (1993) because of its
explicit recognition of the influence of wood and other such obstructions on channel
morphology. We recognize, however, that the urban lowlands of the Pacific
Northwest present a range of vegetation and geologic conditions that are necessary
to consider in addition to the particular channel "type."

In an idealized watershed, the position of a channel in the drainage network is
well correlated with (a) the sources of sediment, (b) the channel slope, (c) the role of

LWD, and (d) the visible channel morphology (Figure 1). Thus a recognition of any of these characteristics, most commonly channel morphology, immediately yields information on the other characteristics as well. Conversely, in a disturbed watershed where channel morphology is no longer a good indication of the "appropriate" channel type, a less disturbance-sensitive parameter (normally channel slope) can suggest the type of channel morphology for which future restoration efforts should aim.

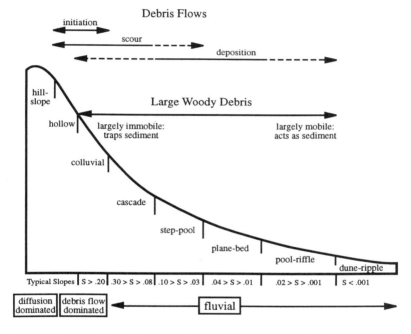

Figure 1. Characteristics of the channel types of Montgomery and Buffington (1993).

The different channel types are:

- **Colluvial Channels:** The small channels that are wholly surrounded by colluvium (*i.e.*, sediment transported by hillslope processes such as creep or landsliding and not by stream transport) that generally lie at the tips of the channel network.

- **Cascade Channels:** The steepest of the alluvial channels, characterized by large clasts that form the primary roughness elements and impose a strongly three-dimensional structure to the flow. Tumbling flow around individual boulders dissipates most of the energy of the flow; bed morphology is disorganized with at most small pools that span a fraction of the total channel width.

- **Step-Pool Channels:** Channels displaying full-width-spanning accumulations of coarse sediment that form a sequence of steps, typically one to four channel-widths apart, that separate low-gradient pools filled with finer sediment. The step-forming sediment is mobile but only at very high discharges; in contrast, sediment in the pools can be rapidly flushed downstream over the intervening steps. The spacing of the steps appears to maximize the flow resistance (Whittaker and Jaeggi, 1982) suggesting that this morphology is essential for maintaining a stable low-flow bed under slope and discharge conditions that would otherwise readily transport sediment downstream. Both "free" and "forced" step-pool channels can be identified, depending on whether alluvial (*i.e.*, episodically transported) sediment or immovable obstructions (*e.g.*, bedrock or large logs) form the majority of the steps.

- **Plane-Bed Channels:** Channels lacking well-defined bedforms and instead displaying long, and commonly channel-wide, reaches of uniform "riffles" or "glides." In contrast to the steeper channels any flow oscillation is generally horizontal, not vertical, but the lateral variations are insufficient to produce pronounced meanders and associated pools.

- **Pool-Riffle Channels:** The most common of the lowland stream channels, with laterally oscillating flow producing a sequence of pools at the outside of bends with corresponding bars on the inside of bends. In the relatively straight reach between each bend a more laterally uniform riffle forms. Analogous to step-pool channels, the classification recognizes "free" pool-riffle channels, where this distinctive morphology forms simply by virtue of the inertial characteristics of the water moving in a sinuous or meandering channel; and "forced" pool-riffle channels where the presence of pools is closely tied to obstructions, such as LWD, but where the removal of such obstructions could yield a morphology more closely akin to plane-bed channels.

- **Dune-Ripple Channels:** The classic lowland sand-bedded channels typical of large rivers, where the character of the predominant bedform will change in response to increasing discharge from plane bed at low flows to ripples, sand waves, dunes, high-energy plane bed, and antidunes at highest flows.

In addition to channel type, the specific response of a channel to watershed changes or restoration efforts, will also depend on the geological context and the nature of bank-forming materials. For example, the effect of removal of LWD from a forced step-pool channel differs dramatically between channels in the Cascade Range, where there is a large supply of boulders capable of reforming steps, and in the Puget Lowland, where the available glacial sand and gravel are too fine to stabilize steps. Dramatic downcutting consequently follows log removal from steep, forced-alluvial channels underlain by glacial deposits, in contrast to the response of the "same" type of channel where a supply of large clasts is available. Hence, channel classifications based on channel morphology (Montgomery and Buffington, 1993) or on grain size and slope (Rosgen, 1994) do not always provide sufficient information to predict specific channel response.

The Role of LWD in Different Sizes and Types of Channels

When a tree falls into a stream channel, a variety of hydraulic changes occur. Typically, current is diverted over or around the obstruction, scouring the bed and banks. Other floating debris, too short to lodge against opposing banks of the channel, may be trapped against the log. Disruption of the flow will cause a local reduction in the transporting power of the flow, and so a sediment wedge may build on the upstream side of the log. The overall down-channel flux of sediment will reduce, because some of the elevation drop of the water surface in the channel will occur by abrupt wood-armored plunges which will reduce the capacity to transport sediment.

Biological changes follow hydraulic changes (see also Harmon and others, 1986). Fish will seek out the relatively deeper water in the scoured pools adjacent to the log, for refuge from both high flows and terrestrial predators. In addition, the decaying wood introduces nutrients into the aquatic food chain, and traps additional organic material that otherwise would be flushed downstream and out of the system altogether. Fish populations decline rapidly and precipitously following removal of LWD (*e.g.*, Bryant, 1983; Elliot, 1986). These changes, however, are not uniform across the range of channel types and sizes (Figure 2).

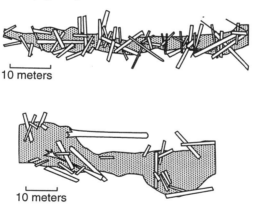

10 meters

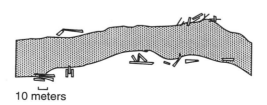

10 meters

Figure 2. Position of large woody debris in channels of different sizes (redrawn from Salo and Cundy, 1987).

High Gradient Channels. In cascade and step-pool channels the role of LWD is highly variable. Where valleys are tightly confining, fallen trees are often suspended well above the active channel, bridging between the valley walls. Where they do contact the flow they become another obstruction, whose importance depends on their relative abundance: in a boulder-dominated channel, for example, the presence or absence of what minimal LWD falls into the stream may be of little consequence. Where large bedrock boulders are sparse or absent, however, logs may provide the primary control on bed morphology, forming the steps that account for most of the vertical drop, and dissipate most of the flow energy, in a step-pool channel (Heede, 1972 a, b; Swanson and others, 1976; Keller and Swanson, 1979; Keller and Tally, 1979; Bilby, 1981). In some channels, the distribution of bedrock channel reaches is controlled by log jams that force local deposition (Montgomery and others, 1996). If the steps are removed from the flow, either by direct human action or by channel incision that leaves the logs suspended above the now-confined channel, the consequences on channel morphology can be very dramatic (see below).

Dune-Ripple Channels. Where channels are large and individual pieces of LWD span but a fraction of the channel width, logs tend to move as sediment particles in most situations. Unlike most sediment particles, however, individual pieces have a somewhat greater opportunity to interact and interlock, which under favorable conditions can build a debris jam of massive, channel-spanning proportions (Russel, 1909; Lobeck, 1939; Keller and Swanson, 1979; Abbe and Montgomery, 1996). These conditions lie outside of the range of our concern in this discussion, but the urban environment is not immune to the consequences of debris jams on large rivers. Traditionally those jams have been viewed as threats to navigation and to efficient passage of flood flows, and so they have been removed whenever possible. They also can stabilize channel banks, however, and so their absence may also have negative consequences for land uses adjacent to the river.

Lowland Streams: Plane-Bed and Pool-Riffle Channels. It is within these channel types, typically spanning a range of gradients between about 0.1 and 3 percent, that LWD has the greatest range of functions (*e.g.*, Lisle and Kelsey, 1982; Keller and others, 1985; Montgomery and others, 1995). By observation of relatively undisturbed lowland channels, these functions include:

- *Hydraulic roughness*, which may increase by 50 percent or more through the disruption of flow imposed by a high concentration of logs, stumps, or debris jams;

- *Sediment storage* behind channel-spanning logs that create a stepped bed profile with a wedge of trapped sediment just upstream, or individual obstructions that result in a (normally smaller) zone of deposition in the eddy just downstream (Keller and Swanson, 1979);

- *Bank protection* through a combination of flow deflection and general hydraulic roughness which reduces the rate of sediment transport (note also, however, that flow deflection may locally *increase* bank erosion); and

- *Creation of habitat diversity* by the variety of bedforms, sediment-transport zones, and sheltered areas that typically accompany LWD in the active flow. Pool formation, in particular, can be almost completely determined by the presence and location of LWD in forested channels of this type, with pool frequencies two or three times greater than in LWD-free streams (*e.g.*, Andrus and others, 1988; Robison and Beschta, 1990; Smith and others, 1993; Montgomery and others, 1995).

Conversely, the removal of LWD that commonly accompanies urbanization of a watershed and its associated riparian areas produces the opposite tendencies. Channels experience greater erosion and lateral instability, the flux of sediment is greater and more closely tied to individual high-flow events, and a diversity of physical habitat is replaced by uniformity in channel profile and channel cross-section. Many of these effects are also a product of the increased discharges that normally accompany urbanization, and we cannot always distinguish unequivocally the relative significance of increased flows and decreased LWD abundance in these settings. We will return to the implications of this uncertainty in our discussion of stream-channel restoration.

Quantifying the Role of LWD in Channel Hydraulics and Sediment Transport

Traditionally, hydraulic analyses of open-channel flow and sediment transport have assumed such "normal" simplifications as uniform channel cross sections and steady flow. In part these assumptions have been made to mimic original experimental conditions in laboratory flumes, on which these analyses are based. They also are made in recognition of the simplifications needed to find analytical solutions to the equations of motion in a three-dimensional fluid. Even the best field-based experimental data have generally been conducted on the most uniform, obstruction-free channels (*e.g.*, Milhous, 1973; Dietrich and Smith, 1983; Kinerson, 1990) to avoid the "complications" that would be inescapably imposed by irregularities such as LWD. Only recently has the role of log roughness been incorporated into some fluvial field studies (*e.g.*, Buffington, 1995).

In an effort to achieve analytical simplicity, however, we have sometimes lost site of the underlying goals: to understand the functions of real stream channels, to evaluate the ability of channels to achieve those functions, and to construct or reconstruct channels that achieve those functions to the best of their ability. It is precisely the *least* easily quantified aspects of open-channel flow and sediment transport that produce the heterogeneity necessary for habitat diversity and the attenuation of flow erosivity. So for example the channel types that must dissipate the greatest amount of flow energy per unit bed area, steeply dropping cascade or step-pool channels, attain this function within a stable form *not* by laying down a uniform pavement of immovable clasts but by an irregular, rapidly varying longitudinal profile (Figure 3).

Traditional analyses of LWD functions have followed a similar approach. The influences are removed from analysis not because they are unimportant but because the presence of LWD is inconvenient. When we exclude that influence in analyzing

an existing channel, however, the representation is incomplete; and when we omit
LWD in a reconstructed channel that would normally include it then the channel
functions are incomplete (and likely inadequate as well).

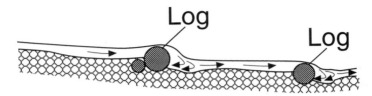

Figure 3. Cross-sectional view of logs in a step-pool channel.

LWD IN URBAN ENVIRONMENTS

The absence of LWD in developed and developing parts of the world is
ubiquitous and long-standing, even where vast forests once blanketed the landscape
(*e.g.*, Wiltshire and Moore, 1983; Petts and others, 1989). Hydraulic considerations,
particularly land drainage and reduction of flood stage, motivated widespread
removal of riparian vegetation and in-stream obstructions. On large rivers any logs
or snags reduced navigability; on small streams mobile debris can be seen to lodge
under bridges and clog culverts, encouraging local sediment deposition and flooding.
In the Pacific Northwest even anadromous fish were thought to suffer from the
migration-blocking effects of LWD accumulations, and so removal was mandated
under commercial forestry permits of the 1970's and early 1980's.

The magnitude of LWD loss in urban streams is best demonstrated by the typical
frequency of logs in their (relatively) undisturbed lowland counterparts. For
example, the number of pieces of well-anchored LWD were tallied in successive 60-
m stream segments of Huge Creek, a 5-m-wide channel draining about 12 km² in an
area of very low urban development in western Washington. On average, log
spacing was less than 7 m (*i.e.*, at least 9 pieces per 60-m segment) and commonly
about equal to the channel width (*i.e.*, about 12 pieces per 60-m segment, or about
200 pieces per km). As we look to progressively more developed watersheds, the
frequency of LWD tends to decrease. This can be displayed most simply by a plot of
percent total impervious area, as a measure of watershed development, against the
number of LWD pieces in a unit distance. Data from Horner and others (this
volume) show abundant scatter but a clear general trend (Figure 4); in extreme yet
all-to-common cases, there is no LWD whatever in an urban channel.

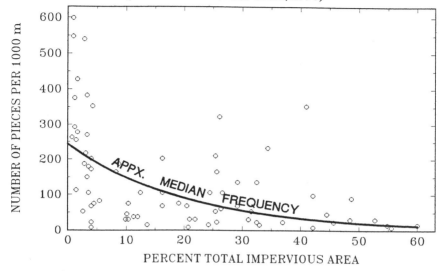

Figure 4. Site-specific measurements and median trend of measured LWD frequency in urban watersheds spanning a range of development intensity.

Loss Mechanisms

Although the processes by which LWD is lost from the active channel are not unique to urban streams, the nature of the changes that accompany urbanization make several removal processes particularly effective.

Washout. Increased discharges are the hallmark of watershed urbanization. By any measure the magnitude of channel flow increases: using the most common measure, the *peak discharges* of annual and multi-year floods increase by typically two- to five-fold (Hollis, 1975), depending on the degree of urban development and the frequency of the flood event. The aggregate *duration of flood flows* may increase more than ten-fold, as once-rare discharges become more commonplace (Barker and others, 1991). Finally, the *frequency of major sediment-transporting events* increases by as much as 50-fold, as discharges with a recurrence of five or ten years in the pre-urban condition can become monthly events after full watershed development (Booth, 1991).

The consequences of these flow increases include a dramatic increase in frequency at which very competent flows, capable of removing normally immobile obstructions, move down the channel. The change in the flow regime also tends to expand the cross-sectional area of the channel through both widening and deepening (Booth, 1990). Local data suggest that the increase in bankfull channel dimensions crudely follows the "downstream hydraulic geometry" relationships of Leopold and others (1964), wherein both width and depth increase in fractional proportion to the increase in bankfull discharge (normally approximated in hydrologic modeling by the 1.5-year or 2-year discharge) raised to a power of about 1.4. A 5-meter-wide channel, responding to a two- or three-fold urban-induced increase in two-year discharge, will therefore expand several meters in width and several tenths of meters in depth, which may expose and undermine most or all of any LWD that was previously well anchored or buried.

Stranding. A second consequence of the increased sediment transport of an urbanized channel affects those streams where the gradient has been established not by a nonerosive bed or a fixed base level, but by the balance between sediment resistance and flow energy, integrated over the suite of flows that has moved down the channel. Where that balance is disrupted by dramatically increased discharges then downcutting may proceed almost unchecked, until a much flatter gradient associated with a deeply incised new channel reduces the competence of the urban discharges (Booth, 1990). The immediate consequences of such a process is a deep and narrow channel that typically has dropped out from under any LWD that once was in contact with the bed. The LWD may be immobile still, but it is ineffectually suspended above the bed. This process is particularly pernicious, for as the incipient incision begins to strand LWD the very source of channel resistance is abandoned and so the rate of downcutting accelerates.

Human Removal. Although we are not aware of any controlled studies of the actions of neighbors on the persistence of LWD in channels, our collective observations suggest a marked correlation between suburban yards and adjacent LWD-free channel reaches that is best explained by intentional removal of debris. Whether motivated by aesthetics or a desire to "improve" the habitat for fish we can only speculate, but we suspect that visual appearances are particularly important.

The few studies on related subjects provide some useful details. Gregory and Davis (1993) asked people to rate the attractiveness of two lowland British streams. They found strong preference for more "natural" channels with clean flowing water and without artificially reinforced or armored banks. However, they also found that woodland channels *without* large woody debris were clearly favored over those with woody debris, despite the ecological advantages that such debris in fact provides. The preferred characteristics of bank vegetation, also among the British population, was investigated in more detail by House and Sangster (1991). They found preferences for (1) trees, ideally a deciduous canopy; (2) a multiplicity of understory vegetation that does not encroach of the channel itself; and (3) some degree of vegetation management (*e.g.* short mown grass rather than long grass). In general, a relatively undisturbed river setting is desirable but by no means mandatory (Mosely, 1989): some highly regarded rivers flow through heavily modified landscapes, and the most truly "natural" settings are not necessarily the most highly valued. Kaplan (1977) found that people already living nearby a stream were more likely to accept more natural, "unkempt" views of a channel than those without that prior experience. These subjective differences become especially important as water-resource design increasingly moves out of the forestland (where it has proceeded almost unnoticed for decades) into restoration and rehabilitation of urban systems (where everybody has an opinion).

Implications for Pervasive LWD Loss

Rarely does the loss of LWD occur in isolation--normally it accompanies upland changes, such as logging or urban development, that initiate significant changes in runoff and sediment-delivery patterns and in riparian vegetation. We can only partly isolate the consequences of LWD loss in channels, because those consequences are amplified by other, concurrent changes as well. Therefore, the mere *replacement* of lost LWD will not reverse all of the stream-channel changes, in large measure because only a fraction of the causal mechanisms are directly addressed by such an action.

Acknowledging the interrelationship between LWD loss and more pervasive watershed changes, we can nevertheless make some judgments about the consequences of that loss by observing the immediate response of channels to LWD removal and by deduction from the observed roles of LWD in undisturbed channels. Those responses include a rapid increase in the rate of channel shifting, both horizontally (typically by widening) and vertically (typically by incision). Changes in the bed morphology are also rapid but depend on the type of channel. Where logs form the "risers" of a step-pool channel their removal leads to immediate disruption of the fundamental morphology of the channel, a morphology that may eventually reform if other hard-to-transport material is available but more commonly will reestablish only after rapid erosion has first increased the channel dimensions and lowered the channel gradient (Booth, 1990). Where logs force pools and riffles in lower gradient streams their removal results in the simple, and relatively rapid, loss of those features.

Sediment discharges also increase following LWD removal or loss. If the change is as a result of riparian clearing then a delay of several years night be expected, as existing logs and jams deteriorate without concurrent replacement (for example, Hedin and others, 1988, report a three-year lag). If instead the LWD is actively removed then the corresponding increase in sediment transport is far more immediate and measurable within a single storm season (*e.g.*, Bilby, 1984; MacDonald and Keller, 1987; Smith and others, 1993).

The concept of *LWD budgets* provides a framework for examining the recruitment, transport, and decay of woody debris in forest stream channels. Analogous to a sediment budget (Dietrich and Dunne, 1978) an LWD budget characterizes any change in the stock of LWD within a channel as the difference between input, typically recruitment through blowdown and landsliding, and output, typically by washout and decay. Yet the dominant mechanisms of LWD input and output differ in urban and forested channels. Clearing of riparian forests and stabilization of channel positions typically dramatically reduce LWD recruitment in urban channels, unless deliberate placement occurs through rehabilitation projects. Even as input decreases the output increases, because of both the higher urban discharges that lead to greater transport rates and more pervasive direct removal of

LWD in urban environments. Decreased recruitment together with enhanced transport inexorably result in lower LWD loadings in urban channels than in their comparable forest counterparts.

REHABILITATION OF URBAN CHANNELS

Principles

Reintroduction of LWD commonly is used in urban channel rehabilitation in an effort to recover lost form and function in the context of massive watershed changes. Whereas complete restoration is often unattainable, the efforts are not entirely misguided for two reasons:

1. Irrespective of what *else* has occurred in the watershed that affects the stream channel, loss of LWD probably has occurred in any once-forested lowland setting, and its functions will need to be replaced at some stage of a comprehensive rehabilitation effort.

2. LWD replacement is commonly the *only* practical stream-restoration activity that can occur in the early stages of a watershed enhancement effort. Other actions that directly address flow changes, sediment-delivery changes, or riparian conditions are much more expensive, contentious, and slow to implement.

The unanswered question, of course, is whether LWD placement *alone* produces any useful long-term effects in urban environments. The evidence from forested environments is not promising (*e.g.,* Frissell and Nawa, 1992; Beschta and others, 1994)--logs can be placed at relatively high densities without any corresponding improvement in the intended targets, typically fish use and production, and most log structures fail in time. A key shortcoming of most in-stream placement of LWD is the lack of geomorphic context and consideration of even the most basic information on channel type and size. The changes to the watershed imposed by urban development are even more disruptive than those resulting from logging, and so we anticipate little or no long-term improvement when LWD replacement is not designed with consideration of these factors.

Extreme increases in flow discharge complicate the use of LWD for channel rehabilitation in urban environments. Concern over local backwater effects and the influence of LWD movement on downstream infrastructure also influences design considerations in urban channels. These limitations are compounded by the fact that most of both what we know and our accumulated experience with reintroduction of LWD to stream channels comes from forest streams. Hence, it is important to consider what is different about specifically *urban* channels and whether these differences matter for channel rehabilitation projects.

Limitations of LWD Placement for Urban Channels

Management Concerns. Even though the long-term consequences of LWD removal are now better recognized among both stream scientists and (some) land managers, the original concerns that first motivated that removal still remain and plague stream-restoration projects that seek to reintroduce LWD into urban channels. Paramount among these concerns are the loss of flood conveyance, the potential for the wood to clog existing channel constrictions, and the possibility of flow diversion causing bank erosion. In the discussion that follows we address the theoretical and empirical basis for these concerns.

Hydrologic Changes. The changes in hydrology that accompany urban development are probably the most severe of the many alterations imposed by such land use. Replacing logs in a channel that experiences a "10-year flow" every month or two (Booth, 1991), and that dries up entirely in the summer, is unlikely to reestablish a robust aquatic-insect or fish population. Bank erosion may be reduced by an increase in flow roughness, local reduction in the water-surface slope by virtue of log steps, and mechanical armoring by careful (or fortuitous) LWD placement, but these factors alone are unlikely to compensate for ten-fold increases in the sediment-transporting capacity of urban channels or the likelihood that accelerated channel changes may abandon or strand the reinserted LWD altogether.

Sediment Fluxes. Related to the hydrologic changes that accompany urbanization, alterations to sediment movement will also compromise the intended functions. In erosional zones, LWD may be left unsupported by lateral channel expansion or stranded by vertical incision. In depositional zones, LWD may be episodically buried and reexcavated as pulses of sediment move down the channel network. In both settings the presence of the LWD may attenuate the response of the reach to the changes in sediment (and water) fluxes, but it cannot entirely mitigate extreme changes.

LWD Budgets in Urban Settings. In contrast to forested settings, opportunities to recruit new LWD are commonly quite poor in urban settings. In combination with concerns over LWD stability, this may result in project designs that seek an unrealistic degree of log stability. In the undisturbed streams that have yielded most of our understanding of the role of LWD, the debris is part of a dynamic system with recognized outputs and balancing inputs. Few urban restoration projects have adequately addressed the realistic need for long-term LWD inputs to maintain the form and function of the restoration effort.

Human Intrusion and Aesthetics. Even if a restoration project is supported by the members of the neighboring community, they may still destroy its functional and biological value in the name of visual improvement. Although the general public appears to value a "derelict but natural" landscape as a reminder of the large natural landscape beyond the urban fringe, people are likely to begin "caring for" these landscape by cleaning out woody debris and other desirable elements or by fashioning homemade retaining walls to stabilize eroding banks. In other words, while people generally like the idea of a stream nearby, they more likely want its appearance to fit into their neighborhood landscape, to look more "manicured" than "scruffy." When public agencies attempt to restore degraded channels, they either complete the manicuring process with smoothed banks having no true rehabilitation value at all, or they build more "ecological" measures that seem unkempt, scruffy, and even more derelict to the nearby homeowner than the original degraded site.

Design Approaches for Urban Channel Rehabilitation

We judge that the rehabilitation of urban channels in the Pacific Northwest must acknowledge the following set of principles:

- In predevelopment time, LWD played a fundamental role in the function of most, but not all, lowland streams in the Pacific Northwest. Any rehabilitation effort must acknowledge those functions and reestablish them in a manner consistent with the geomorphic character of the channel and watershed setting.

- The watershed-scale alterations imposed by urban development cannot be corrected by in-channel, or even near-channel, means alone.

- LWD placement is commonly the most feasible and readily achievable component of a stream-restoration program; that it is not the only necessary action does not negate its value as an early action, as long as long-term rehabilitation targets are closely tied to the implementation of other, *watershed-scale* efforts.

The literature on channel-restoration techniques is replete with designs for "fish-habitat structures" such as single- and double-wing deflectors, notched weirs, and check dams. We find few analogs in undisturbed Pacific Northwest streams for these structures, and so we offer no additional guidelines for their construction. Instead, we focus on the guidelines behind recent projects to reintroduce LWD into urban channels, where the project objectives have been to mimic the character and function of LWD in undisturbed sites. Unfortunately, specific examples are relatively few in number and represent an early stage in our learning: failures are far more prevalent than successes. Nevertheless, they offer some indications for the directions we should be pursuing and should encourage others to follow these directions.

Natural LWD structures have been employed in channel rehabilitation projects using two distinct design philosophies. The first involves placing unanchored debris in the channel and letting high flow events reorganize the debris. The second approach involves the construction of LWD jams patterned after natural analogs (Abbe et al., 1993 and in prep; Abbe and Montgomery, 1996). Examples of the first approach from Washington State illustrate the potential for employing more natural designs in LWD-based stream rehabilitation projects.

Project Experience with LWD Placement in Urban Channels

Project Objectives. The King County Surface Water Management Division has experimented with placement of unanchored LWD in five stream enhancement and stabilization projects. In each case the projects were intended to both decrease sediment discharge and enhance in-stream habitat. The use of unanchored wood represented a distinct change in approach from more conventional channel stabilization projects; here, LWD was considered a dynamic element in the fluvial and riparian system, which had been depleted as a result of human activity. Woody debris was introduced with the explicit expectation that it would be moved, reoriented, and incorporated into the stream system through natural fluvial processes.

Creek Name	Watershed Upstream from Project	Predominate Land Use Upstream from Project	Channel gradient	Bankfull channel width	Predominant bed sediment	Channel Classification
Madsen	5.2 sq km	residential	4.0%	3m	gravel	plane-bed
Boise	22.3 sq km	timber production	3.0%	4 m	gravel	plane-bed
Soosette	14.3 sq km	residential	2.3%	4 m	gravel	plane-bed
Laughing Jacobs	14.6 sq km	residential	2.5%	5 m	gravel	plane-bed
Hollywood Hill	2.2 sq km	residential	7.1%	2 m	gravel	plane-bed

All of the project streams showed evidence of degradation consistent with the combined effects of LWD depletion and changes in basin hydrology as a result of changes in land use. All of the channels exhibited the characteristics of a plane bed channel, although in two cases (Madsen Creek and the Hollywood Hill tributary) the channel gradient exceeded the range typical of such channels. In all cases the project reach was selected to be a substantial distance upstream from the nearest bridge or culvert, in order to minimize the possibility of the debris blocking such a structure.

The size of material placed, and the number of pieces per unit length of stream were based generally on published descriptions of LWD in natural, undisturbed streams in the Pacific Northwest (Bilby and Ward, 1989; Nakamura and Swanson, 1993). Material availability, construction logistics, and budget constraints also significantly affected decisions regarding the type and amount of debris placed. Logs 6 to 12 meters in length in length, lacking both rootwads and branches, were readily available from commercial logging operations and constituted a majority of the pieces used. Rootwads without significant stems were also readily available from land clearing operations. Logs with attached rootwads are much more difficult to acquire and therefore were the used less than either of the bare logs or rootwads alone. All of the woody debris used in these projects was native coniferous species. Log and stump diameters were generally in the range of 0.25 to 0.75 meters.

Creek Name	Year Constructed	Length of Channel Treated (m)	Total # of Woody Pieces	Pieces / 100 Meters	# of Logs w/o Rootwads	# of Logs w/ Rootwads	# of Rootwads	Method of Construction
Madsen	1993	210	51	24	32	9	10	helicopter
Boise	1994	500	93	19	20	51	22	helicopter
Soosette	1994	1600	278	17	278	0	0	helicopter
Laughing Jacobs	1995	300	68	23	22	15	31	crane
Hollywood Hills	1995	80	53	66	17	4	32	crane

Table 1. Summary of LWD Placement Projects

As indicated in Table 1, two different methods were used to place the woody debris in the stream system with minimal disturbance to riparian vegetation. At project sites where access was available, it was possible to place material using a rubber-tired hydraulic crane, which could to reach over stream side vegetation. On the remaining sites, debris was placed by helicopters equipped for long line operation.

These projects have been subject to between one and three years of wet season flows, depending on their construction date. The largest storm to occur in this three year period occurred in February of 1996, subsequent to construction of all five projects. This recurrence interval for this storm (7 day rainfall) varied from 5 to 50 years depending on the project location.

Project Performance. Four of the five projects have achieved only limited success to date in reducing downstream sediment discharge. On these projects, the debris has been relocated by high flows to varying degrees. Much of the wood remains in contact with the channel and is contributing to increased roughness and causing local sediment deposition. In many cases the debris appears to have locally deflected flow, resulting in a more sinuous channel form, and consequently locally lower gradient . On the other hand, flow concentration or deflection by the debris, has caused local areas of bank erosion or channel incision, reducing or negating the net increase in sediment storage resulting from debris placement.

The one project where results to date have been less equivocal is the Hollywood Hill ravine stabilization. Sediment accumulation in the project area has been dramatic. Deposition has occurred throughout the project reach, with local deposition on the order of one-half meter. No significant erosion occurred in this reach. This project was different in several respects from the other four. The contributing basin was the smallest of the five, the number of pieces per unit length was roughly three times greater than the other projects, and a much higher percentage of the pieces used were rootwads rather than logs.

Conclusions. Qualitative evaluation the projects completed to date suggests several conclusions. These conclusions are preliminary, because on the short time period since these projects were completed, and on the lack of consistent, comparable monitoring information:

1) Placement of bare cylindrical logs, with no stems or rootwads seems to encourage streamflow to scour below the log rather than to flow over the top. As a result, such logs do not tend to form "steps" in the channel profile , and therefore do not lead to deposition of an upstream sediment wedge. This seems to be in contrast to natural tree fall, where the resulting debris is often a full length tree, with roughness and complexity provided by both branches and roots.

2) Use of pieces with greater complexity (rootwads, and logs with attached rootwads) and placement of more pieces per unit length both result increased effectiveness. In larger streams, where flow depths are sufficient for debris to become buoyant, rootwads are be more mobile than logs because they are less able to lodge on stream banks, or become wedged against streamside vegetation.

3) The intent of these projects was to simulate the natural input of wood to a channel system. One element which may be critical in this incorporation process is the passage of time. Much of the wood which forms structure in lowland stream channels occurs either as individual pieces which are largely buried in streambed sediment, or as a part of a jam constructed largely of woody debris transported by the stream. In both cases this occurrence implies that the wood was present in the stream for a period of time before it assumed its current role. It may be that it will be necessary to allow for a similar period of time for the debris introduced by these projects to be completely incorporated.

4) A significant effort has been made to evaluate the performance of these projects. Unfortunately the complexity of the physical system has made monitoring their geomorphic effects difficult at best. In addition, the monitoring program between projects has not been consistent, so that the data collected is not readily comparable. For these reasons, it is difficult at this point to draw clear, defensible conclusions about the effectiveness of these projects in meeting their objectives with respect to channel stability and sediment discharge. There is widespread interest in placement of LWD as one element of comprehensive urban stream restoration programs. Before such an approach should be widely adopted, the projects which have been completed to-date should be evaluated through a monitoring program designed to provide a consistent, comparable and meaningful characterization of the effects of such projects, and which yields clear direction for future project designs.

REFERENCES

Abbe, T. B., and Montgomery, D. R., 1986, large woody debris jams, channel hydraulics and habitat formation in large rivers: Regulated Rivers: Research and Management, v. 12, p. 201-221.

Andrus, C. W., Long, B. A., and Froehlich, H. A., 1988, Woody debris and its contribution to pool formation in a coastal stream 50 years after logging: Canadian Journal of Fish and Aquatic Science, v. 45, p. 2080-2086.

Barker, B. L., Nelson, R. D., and Wigmosta, M. S., 1991, Performance of detention ponds designed according to current standards: in Puget Sound Water Quality Authority, Puget Sound Research '91: Conference Proceedings, Seattle, Washington.

Beschta, R. L., Platts, W. S., Kauffman, J. B., and Hill, M. T., 1994, Artificial stream restoration--money well spent or an expensive failure?: Big Sky, Montana, Proceedings of the Universities Council on Water Resources, August 2-5, 1994, p. 76-104.

Bilby, R. E., 1981, Role of organic debris dams in regulating the export of dissolved and particulate matter from a forested watershed: Ecology, v. 62, p. 1234-1243.

Bilby, R. E., 1984, Removal of woody debris may affect stream channel stability: Journal of Forestry, v. 82, p. 609-613.

Booth, D. B., 1990, Stream-channel incision following drainage-basin urbanization: Water Resources Bulletin, v. 26, p. 407-417.

Booth, D. B., 1991, Urbanization and the Natural Drainage System--Impacts, Solutions, and Prognoses: Northwest Environmental Journal, v. 7, p. 93-118.

Bryant, M. D., 1983, The role and management of woody debris in west coast salmonid nursery streams: North American Journal of Fisheries Management, v. 3, p. 322-330.

Buffington, J. M., 1995, Effects of hydraulic roughness and sediment supply on surface textures of gravel-bedded rivers: Seattle, University of Washington, Department of Geological Sciences, M.S. Thesis, 184 p.

Dietrich, W. E., and Smith, J. D., 1983, Influence of the point bar on flow through curved channels: Water Resources Research, v. 19, p. 1173-1192.

Elliot, S. T., 1986, Reduction of a Dolly Varden population and macrobenthos after removal of logging debris: Transactions of the American Fisheries Society, v. 115, p. 392-400.

Frissell, C. A., and Nawa, R. K., 1992, Incidence and causes of physical failure of artificial fish habitat structures in streams of western Oregon and Washington: North American Journal of Fisheries Management, v. 12, p. 182-197,

Gregory, K. J., and Davis, R. J., 1993, The perception of riverscape aesthetics: an example from two Hampshire rivers: Journal of Environmental Management, v. 39, p. 171-185.

Harmon, M. E., Franklin, J. F., Swanson, F. J., and others, 1986, Ecology of coarse woody debris in temperate ecosystems: Advances in ecological research, v. 15, p. 133-302.

Hedin, L. O., Mayer, M. S., and Likens, G. E., 1988, The effect of deforestation on organic debris dams: Verhandlungen der Internationalen Vereinigung für theoretische und angewandte Limnologie, v. 23, p. 1135-1141.

Heede, B. H., 1972a, Influences of a forest on the hydraulic geometry of two mountain streams: Water Resources Bulletin, v. 8, p. 523-530.

Heede, B. H., 1972b, Flow and channel characteristics of two high mountain streams: Fort Collins, Colorado, USDA Forest Service General Technical Report RM-96, Rocky Mountain Forest and Range Experimental Station.

Hollis, G. E., 1975, The effects of urbanization on floods of different recurrence intervals: Water Resources Research, v. 11, p. 431-435.

Horner, R. R., Booth, D. B., Azous, A., and May, C. W., 1996, Watershed determinants of ecosystem functioning: in Roesner, L. A., ed., Effects of watershed development and management on aquatic ecosystems: Engineering, Foundation Conference, Snowbird, Utah, August 4-9, 1996 (this volume).

House, M. R., and Sangster, E. K., 1991, Public perception of river-corridor management: Journal of the Institute of water and Environmental Management, v. 5, p. 312-317.

Kaplan, R., 1977, Preferences and every day nature: method and application perspectives on environment and behavior: in Stokols, D., ed., Theory, research, and application: New York, Plenum.

Keller, E. A., and Swanson, F. J., 1979, Effects of large organic material on channel form and alluvial processes: Earth Surface Processes, v. 4, p. 361-380.

Keller, E. A., and Tally, T., 1979, Effects of large organic debris on channel form and fluvial processes in the coastal redwood environment: in Rhodes, D. D., and Williams, G. P., eds., Adjustments to the Fluvial System, Binghampton, New York, Proceedings of the tenth Annual Geomorphology Symposium, p. 1669-197.

Keller, E. A., MacDonald, A., Tally, T., and Merritt, N. J., 1985, Effects of large organic debris on channel morphology and sediment storage in selected tributaries of Redwood Creek: in Geomorphic Processes and Aquatic Habitat in the Redwood Creek Drainage Basin: U. S. Geological Survey Professional Paper.

Kinerson, D., 1990, Bed surface response to sediment supply: Berkeley, University of California, Department of Geology, M.S. thesis, 420 p.

Kondolf, G. M., 1995, Geomorphological stream channel classification in aquatic habitat restoration: uses and limitations: Aquatic Conservation: Marine and Freshwater Ecosystems, v. 5, p. 127-141.

Leopold, L. B., Wolman, M. G., and Miller, J. P., 1964, Fluvial processes in geomorphology: San Francisco, W. H. Freeman and Co., 522 p.

Lisle, T. E., and Kelsey, H. M., 1982, Effects of large roughness elements on the thalweg course and pool spacing: *in* Leopold, L. B., ed., American Geomorphological Field Group Field Trip Guidebook, Pinedale, Wyoming, 1982 Conference, p. 134-135.

Lobeck, A. K., 1939, Geomorphology: New York, McGraw Hill.

MacDonald, A., and Keller, E. A., 1987, Stream channel response to the removal of large woody debris, Larry Damm Creek, northwestern California: International Association of Hydrologic Sciences Publication no. 165, p. 405-406.

Milhous, R. T., 1973, Sediment transport in a gravel-bottom stream: Corvallis, Oregon, Oregon State University, Ph.D. dissertation, 232 p.

Montgomery, D. R., and Buffington, J. M., 1993, Channel classification, prediction of channel response, and assessment of channel condition: Washington State Department of Natural Resources, Report TFW-SH10-93-002, 84 p.

Montgomery, D. R., Abbe, T. B., Buffington, J. M., Peterson, N. P., Schmidt, K. M., and Stock, J. D., 1996, Distribution of bedrock and alluvial channels in forested mountain drainage basin: Nature (in press).

Mosley, M. P., 1989, Perceptions of New Zealand river scenery: New Zealand Geographer, v. 45, p. 2-13.

Mosley, M. P., 1987, The classification and characterization of rivers: *in* Richards, K., ed., River Channels: Environment and Process: Oxford, Blackwell, p. 294-320.

Petts, G. E., Roux, A. L., and Moller, H., eds., 1989, Historical changes of large alluvial rivers, western Europe: Chichester, John Wiley.

Platts, W. S., 1980, A plea for fishery habitat classification: Fisheries, v. 5, p. 2-6.

Robison, E. G., and Beschta, R. L., 1990, Coarse woody debris and channel morphology interactions for undisturbed streams in southeast Alaska, U.S.A.: Earth Surface Processes and Landforms, v. 15, p. 149-156.

Rosgen, D. L., 1994, A classification of natural rivers: Catena, v. 22, p. 169-199.

Salo, E. O., and Cundy, T. W., 1987, Streamside Management: Forestry and Fishery Interactions: Seattle, University of Washington, College of Forest Resources, Contribution No. 57, 471 p.

Smith, R. D., Sidle, R. C., and Porter, P. E., 1993, Effects on bedload transport of experimental removal of woody debris from a forest gravel bed stream: Earth Surface Processes and Landforms, v. 18, p. 455-468.

Swanson, F. J., Lienkaemper, G. W., and Sedell, J. R., 1976, History, physical effects, and management implications of large organic debris in western Oregon streams: Portland, Oregon, USDA Forest Service General Technical Report PNW-56, Pacific Northwest Forest and Range Experimental Station.

Whittaker, J. G., and Jaeggi, M. N. R., 1982, Origin of step-pool systems in mountain streams: Journal of the Hydraulics Division, Proceedings of the American Society of Civil Engineers, v. 108, p. 758-773.

Wiltshire, P. E. J., and Moore, P. D., 1983, Paleovegetation and paleo hydrology in upland Britain: *in* Gregory, K. J., ed., Background to paleohydrology: Chichester, John Wiley, p. 433-451.

DISCUSSION

The Alluvial Progress of Piedmont Streams
Bruce Ferguson

In the Piedmont region of the southeastern United States, two rehabilitation and
protection projects suggest objectives for healing streams that continue to be
impacted by development.

An equilibrium approach is proposed to reestablish stream stability. Check dams
are grade control structures with intervening pools. They stabilize the base level,
slowing downcutting.

The purpose of grade control is to enable the taking over of the stream corridors by
natural equilibrating mechanisms. All such grade control must include or
accommodate revegetation of the stream corridor.

An equilibrating approach is different from that of sediment basins that are dredged
at intervals of time, and from detention basins that drain after storms are past.
Captured sediment, water, and organic matter are intended to remain largely in
place.

Questions/Comments

Question: Why aren't you dredging the downstream sediment detention pond?
              What about the reduced capacity of the reservoir to retain flow and
              provide the degree of settling that the original pond provided?
Answer: We are trying to reestablish a natural system.

Question: A conclusion of the study is that there was a decrease in sediment
              production upstream of the detention pond because the rate of
              sediment accumulation in the reservoir decreased. Did you measure
              the sediment load in the pond inflow to confirm that sediment load
              had been reduced? Did you measure the sediments in the discharge
              to insure that the sediment was not just washing through pond due to
              the reduced detention time of the runoff in the pond.
Answer: No, sediment sampling of stream flows was not performed in this
              study.

Question: How was it possible to install the flow controls in this stream
              without a permit?
Answer: The local jurisdiction did not require it.

Experience From Morphological Research on Canadian Streams: Is Control of the Two-Year Frequency Runoff Event the Best Basis for Stream Channel Protection
Craig MacRae

Control of the two-year storm alone does not assure the reduction of stream bank erosion. Alternate methods include consideration of flow peak, duration, and stream bank materials. Two methods: 1) "Over Control Approach" is a one-dimensional model that is easy to apply, but does not account for heterogeneous stream bank materials. 2) "Index Approach" is a two-dimensional approach that produces much better control.

Questions/Comments

Question: Can we achieve zero erosion potential in an urbanized watershed?
Answer: It was never possible to reach a design of zero erosion potential; best we could achieve in this study was a 30% reduction in erosion potential.

Question: How do you deal with the Roskin method for establishing design criteria?
Answer: For urban streams, the natural stream methods do not apply, so we need to go to another approach.

Question: To what degree do we try to control erosion of stream channels, what should we leave for natural erosion?
Answer: That is to some extent a social question. Also, there are other techniques for stream stabilization that can be used to complement flow controls.

Effects of Urban Growth on Stream Habitat
Doug Sovern

Stream rehabilitation in urban stream systems can be accomplished to sustain salmon runs as long as the hydraulic changes resulting from urbanization are recognized. These include stream bed stabilization, in stream flow controls, and confining the aquatic habitat channel.

Questions/Comments

Question: How long do these systems constructed with natural materials last?
Answer: We use cedar and large rock and figure it will have a 50-yr life.

Question: How successful have these projects been?

Answer: We have seen some return of salmon, but not very many.

Question: Have federal permits been required?
Answer: Yes, on two streams w/o any trouble, but most projects are done By
                 Fish And Wildlife.

Large Woody Debris in Urban Streams of the Pacific Northwest
   Derek Booth

Large woody debris (LWD) performs key functions in undisturbed streams that
drain lowland forested watersheds, including dissipation of flow energy,
stabilization of bedforms and channel banks, entrapment of sediment, and
formation of pools. Just as LWD is ubiquitous in undisturbed lowland streams of
the Pacific Northwest, it is significantly depleted in urbanized systems where it is
lost through washout, downcutting, and direct removal. We doubt that simple
reintroduction of LWD will fully restore the lost functions of urban streams.
Instead, projects that replace LWD may be best suited to recover a more limited set
of rehabilitation goals.

Derek does not like Doug Sovern's approach because Derek is trying to use logs
and root mats in several applications has had mixed results. Very long debris that
wedges in large material in a relatively narrow channel seems to work better. Bank
stabilization with wood, sandbags, and planted willows (which are the final
stabilization mechanism) seems to be the best. So natural materials are the only
thing in his mind that will ultimately solve the problem.

Questions/Comments

Question: Do you think that you can use large enough material that you don't
                 have to anchor it?
Answer: Yes, and Derek is totally opposed to anchoring

Question: Doesn't adding native material for stabilization cause problems in
                 urban streams?
Answer: There are concerns over the addition of debris and its effect on
                 flooding are largely a concern from the traditional flood manager,
                 but in fact do not impede the actual large storm flows. Also, sewer
                 lines and lack of access to channels makes it difficult to add debris
                 to many urban streams. In these cases you probably can't do too
                 much.

Question: What about citizen opposition?
Answer: It varies, some like LWD restoration, some don't.

Assessing the Condition and Status of Aquatic Life Designated Uses in Urban and Suburban Watersheds

Chris O. Yoder and Edward T. Rankin[1]

Abstract

Ohio EPA employs biological, chemical, and physical monitoring and assessment techniques in biological surveys in order to meet three major objectives: 1) determine the extent to which use designations assigned in the Ohio Water Quality Standards (WQS) are either attained or not attained; 2) determine if use designations assigned to a given water body are appropriate and attainable; and 3) determine if any changes in key ambient biological, chemical, or physical indicators have taken place over time, particularly before and after the implementation of point source pollution controls or best management practices for nonpoint sources. Biological criteria are one of the principal assessment tools by which the status of water bodies is determined in Ohio. The results of biological monitoring in selected small urban Ohio watersheds shows a tendency towards lower biological index scores with an increasing degree of urbanization and allied stressors, becoming more severe as other impact types such as combined sewer overflows (CSOs) and industrial sources coincide. Out of 110 sampling sites examined only 23% exhibited good, very good, or exceptional biological index scores. Of the sites classified as being impacted by urban sources, only two sites (4.5%) attained the applicable biological criteria. Poor or very poor scores occurred at the majority of the urban impacted sites (85%). More than 40% of suburban sites were impaired with many reflecting the impact of new developments for housing and commercial uses. The results demonstrate the degree of degradation which exists in most small urban Ohio watersheds and the difficulties involved in dealing with these multiple and diffuse sources of stress. Well designed biological surveys using standardized methods and calibrated indicators can contribute essential

---

[1] Ohio EPA, Division of Surface Water, Monitoring and Assessment Section, 1685 Westbelt Drive, Columbus, Ohio 43228

information and capacity to urban watershed management. Because the resident biota respond to and integrate all of the various factors that affect a watershed their condition is the cumulative result of what happens within watersheds. It is important issue that ambient monitoring not only be done as part of the overall urban nonpoint source management process, but that it is done correctly in terms of timing, methods, and design.

Introduction

The health and well-being of the aquatic biota in surface waters is an important barometer of how effectively we are achieving the goals of the Clean Water Act, particularly the maintenance and restoration of biological integrity. Simply stated biological integrity is the *combined* result of chemical, physical, and biological processes in the aquatic environment. The interaction of these factors is especially apparent in the effects of nonpoint sources. In order to be successful in achieving Clean Water Act goals, ecological concepts, criteria, and assessment tools need to be better incorporated into the prioritization and evaluation of watershed management efforts (Yoder 1995a).

The monitoring of surface waters and evaluation of the biological integrity goal of the Clean Water Act have historically been predominated by nonbiological measures such as chemical/physical water quality (Karr *et al.* 1986). While this approach may have fostered an impression of empirical validity and legal defensibility it has not sufficiently measured the ecological health and well-being of aquatic resources. An illustration of this point was demonstrated in a comparison of the abilities of chemical water quality criteria and biological criteria to detect aquatic life use impairment in Ohio rivers and streams. Out of 645 water body segments analyzed, biological impairment was evident in 49.8% of the cases where no impairments of chemical water quality criteria based on ambient chemical monitoring were observed (Ohio EPA 1990a). While this discrepancy may at first seem remarkable, the reasons for it are many and lie mostly in the inherent complexity of biological information. Biological communities simultaneously respond to and integrate a wide variety of chemical, physical, and biological factors in the environment whether they are of natural or anthropogenic origin. Simply stated controlling chemical water quality alone does not assure the ecological integrity of water resources (Karr *et al.* 1986).

The health and well-being of surface water resources is the *combined* result of chemical, physical, and biological processes (Figure 1). To be truly successful in attaining biological integrity goals, monitoring and assessment tools are needed that measure both the interaction of chemical, physical, and biological processes and the integrated result of these processes (Karr 1991). This is especially true of nonpoint sources because many of the effects involve the complex and dynamic interaction of these factors. Biological criteria offer a way to measure the end result of watershed level management efforts and successfully accomplish the protection and restoration of aquatic ecological resources. Biological communities respond predictably to gradients of environmental impact which chemical/physical water quality criteria alone cannot adequately discriminate or sometimes even detect. Habitat degradation and sedimentation are two such widespread impacts of nonpoint source origin that simply cannot be measured by chemical/physical assessments alone. As illustrated by Figure 1 it is the cumulative combination of chemical and physical factors that result in aquatic life use impairments from nonpoint sources.

Biological Criteria
Biological criteria are narrative and numerical expressions of the health and well-being of the aquatic biota and are based on measurable attributes of aquatic communities such as fish and macroinvertebrate community structure and function. Ohio EPA adopted numerical biological criteria in the Ohio Water Quality Standards (WQS) regulations in May 1990. Biological criteria are further stratified within a classification system of aquatic life use designations. Numerical biological criteria were derived using a regional reference site approach (Ohio EPA 1987a,b; Ohio EPA 1989a; Yoder 1989; Yoder and Rankin 1995a). Numerical biological criteria, which are expressed as biological indices that represent measurable end-points of aquatic life use designation attainment and non-attainment, are the end-product of an ecologically complex, but structured derivation process. While numerical biological indices have frequently been criticized for potentially oversimplifying complex ecological processes, the need to distill such information to commonly comprehended expressions is both practical and necessary. Numerical biological criteria represent valid ecological end-points so long as the underlying development process is theoretically sound and informationally robust.

The availability of new generation evaluation mechanisms such as the Index of Biotic Integrity (IBI; Karr 1981; Fausch *et al.* 1984; Karr *et al.* 1986), the Index of Well-Being (Iwb; Gammon 1976; Gammon *et al.* 1981), the Invertebrate Community Index (ICI; Ohio EPA 1987b; DeShon 1995), and similar efforts (Plafkin *et al.* 1989; Lyons 1991; Simon 1991; Kerans and Karr 1992; Fore *et al.* 1996; Barbour *et al.* 1996) have satisfied important practical and theoretical gaps not always fulfilled by previously available single dimension indices (Fore *et al.* 1996). Multimetric evaluation mechanisms such as the IBI extract ecologically relevant information from complex biological community data while preserving the opportunity to analyze the data on a multivariate basis. The problem of biological data variability is also addressed within this approach. Variability is controlled by specifying standardized methods and procedures (*e.g.*, Ohio EPA 1989b), compressed through the application of multimetric evaluation mechanisms (*e.g.*, IBI, ICI), and stratified in accordance

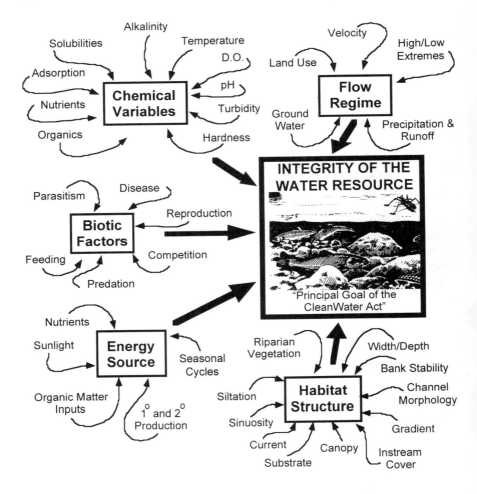

Figure 1. The five principal factors, with some of the important chemical, physical, and biological subcomponents, that influence and determine the integrity of surface water resources (modified from Karr et al. 1986).

with regional and physical variability and potential (*e.g.,* watershed size, ecoregions, tiered aquatic life uses) The result are evaluation mechanisms such as the IBI and ICI that have acceptably low replicate and intra-regional variability (Davis and Lubin 1989; Rankin and Yoder 1990; Stevens and Szczytko 1990).

A few states initially led the effort to establish biological criteria by initiating development and implementation efforts within their own water quality management programs. U.S. EPA effectively endorsed the approach used by some of these states by first issuing national program, policy, and bioassessment guidance (Plafkin *et al.* 1989; U.S. EPA 1990, 1991) and more recently specific guidance for biological criteria development in wadeable streams (U.S. EPA 1995). At the same time more states are undertaking biological criteria development and implementation efforts (*e.g.,* Florida as detailed in Barbour *et al.* 1996). While outstanding and as yet unresolved issues remain surrounding the policy applications of biological criteria (Miner and Borton 1991; Pihfer 1991; Jackson 1992; Ruffier 1992; Schmidt 1992; Schregardus 1992; Yoder 1991a, 1995b), the concept is becoming firmly embedded in emerging state and federal monitoring, assessment, and management initiatives (*e.g.,* environmental indicators, national goals).

Methods and Design
A biological and water quality survey, or "biosurvey", is an interdisciplinary monitoring effort conducted on a water body specific, watershed, or basin/subbasin scale. Biosurveys may be relatively simple, focusing on one or two small streams, one or two principal stressors, and a handful of sampling sites or a much more complex effort including entire drainage basins, multiple and overlapping stressors, and tens of sites. Each year Ohio EPA conducts biosurveys in 10-15 different areas with an aggregate total of 250-350 sampling sites.

Ohio EPA employs biological, chemical, and physical monitoring and assessment techniques in biosurveys in order to meet three major objectives: 1) determine the extent to which use designations assigned in the Ohio Water Quality Standards (WQS) are either attained or not attained; 2) determine if the use designations assigned to a given water body are appropriate and attainable; and 3) determine if any changes in key ambient biological, chemical, or physical indicators have taken place over time, particularly before and after the implementation of point source pollution controls or best management practices for nonpoint sources. The data gathered by a biosurvey is processed, evaluated, and synthesized in a biological and water quality report[2]. Each biological and water quality report contains a summary of major findings and conclusions, recommendations for revisions to use designations, future monitoring

---

2 Approximately 150 of these reports have been produced since 1978.

needs, or other actions which may be needed to resolve impairment(s) of designated uses. While the principal focus of a biosurvey is on the status of aquatic life uses, the status of other uses such as recreation and water supply, as well as human health concerns, are also addressed. These reports are then used to support virtually any Ohio EPA program where the protection of aquatic resources is at issue.

Role of Biological Criteria
There are a number of areas in water resource management in which biological criteria and bioassessment methods can and do play a key role. As a criterion for determining the extent of any aquatic life use impairments, biocriteria have played a central role in the biennial Ohio Water Resource Inventory (305[b] report; Ohio EPA 1994), the Ohio Nonpoint Source Assessment (Ohio EPA 1990b; 1991), generating various priority lists (*e.g.,* 303[d] and 304[l] listings), water quality permit support documents, and comprehensive watershed assessments. Biological criteria represent a measurable and tangible criterion against which the effectiveness of state and federal water pollution abatement and water quality management programs can be judged. However, biological assessments must be accompanied by appropriate program activity measures, ambient chemical/physical measures, measures of pollutant loadings, habitat quality characterizations, land use statistics, and other source information necessary to establish linkages between the activities which impact and degrade aquatic ecosystems (*i.e.,* stressors) and the resultant quality of the ecosystem (as implied by the various exposure and response indicators) to those impacts.

Ohio Water Quality Standards: Designated Aquatic Life Uses
The Ohio Water Quality Standards (WQS; Ohio Administrative Code 3745-1) consist of a classification system of designated uses and chemical, physical, and biological criteria designed to represent measurable properties of the environment that are consistent with the goals specified by each. Use designations consist of two broad groupings, aquatic life and non-aquatic life uses. In applications of the Ohio WQS to the management of water resource issues in Ohio's rivers and streams, the aquatic life use criteria apply to virtually all surface waters regardless of size and frequently result in the most stringent protection and restoration requirements, hence their emphasis in biological and water quality reports. The five different aquatic life uses currently defined in the Ohio WQS are described as follows:

1) *Warmwater Habitat (WWH)* - this use designation defines the "typical" warmwater assemblage of aquatic organisms for Ohio rivers and streams; *this use represents the principal restoration target for the majority of water resource management efforts in Ohio.*

2) *Exceptional Warmwater Habitat (EWH)* - this use designation is reserved for waters which support "unusual and exceptional" assemblages of aquatic organisms which are characterized by a high diversity of species, particularly those which are highly intolerant and/or rare, threatened, endangered, or special status (*i.e.,* declining species); *this designation represents a protection goal for water resource management efforts dealing with Ohio's best water resources.*

3) *Coldwater Habitat (CWH)* - this use is intended for waters which support assemblages of cold water organisms and/or those which are stocked with salmonids with the intent of providing a put-and-take fishery on a year round basis.

4) *Modified Warmwater Habitat (MWH)* - this use applies to streams and rivers which have been subjected to extensive, maintained, and essentially permanent hydromodifications such that the biocriteria for the WWH use are not attainable *and where the activities have been sanctioned and permitted by state or federal law;* the representative aquatic assemblages are generally composed of species which are tolerant to low dissolved oxygen, silt, nutrient enrichment, and poor quality habitat.

5) *Limited Resource Water (LRW)* - this use applies to small streams (usually <3 mi.^{2} drainage area) and other water courses which have been irretrievably altered to the extent that no appreciable assemblage of aquatic life can be supported; such waterways generally include small streams in extensively urbanized areas, those which lie in watersheds with extensive drainage modifications, those which completely lack water on a recurring annual basis (*i.e.,* true ephemeral streams), or other irretrievably altered waterways (*e.g.,* dredged navigation channels, concrete stream channels).

Chemical, physical, and/or biological criteria are generally assigned to each use designation in accordance with the narrative goals defined by each. As such the system of use designations employed in the Ohio WQS constitutes a "tiered" approach in that a gradient of appropriate levels of protection are afforded by each. This hierarchy is especially apparent in the water quality criteria established for parameters such as dissolved oxygen, ammonia-nitrogen, temperature, and the biological criteria. For other parameters such as heavy metals, the technology to construct an equally graduated set of criteria has been lacking. The specified procedure (Stephan *et al.* 1985) has not been able to produce different water quality criteria for the different aquatic life use designations. Thus the same water quality criterion may apply to two or more different use designations. However, we are presently developing a technique for using ambient chemical data and the biological criteria to derive tiered water quality criteria for heavy metals and other parameters.

Determination of Aquatic Life Use Attainment Status

Biological criteria in Ohio are based on two principal organism groups, fish and macroinvertebrates. Numerical biological criteria for rivers and streams were derived by utilizing the results of sampling conducted at more than 350 reference sites that typify the range of "least impacted" conditions within each ecoregion (Ohio EPA 1987b; 1989a). This information was then used within the existing framework of tiered aquatic life uses in the Ohio WQS to establish attainable, baseline biological community performance expectations on a regional basis. Biological criteria vary by ecoregion, aquatic life use designation, site type, and biological index. The resulting array of biological criteria for two of the "fishable, swimmable" use designations, Warmwater Habitat (WWH) and Exceptional Warmwater Habitat (EWH) are shown in Figure 2 which demonstrates the stratification inherent to this process.

The relationship between the aquatic life use designations and narrative ratings of aquatic community condition (termed hereafter as biological community performance) is described in Figure 3. This figure shows the theoretical range of biological integrity (from lowest to highest) compared to the corresponding scale of measurement offered by the multimetric biological indices such as the IBI and ICI. The dual role of biological criteria to serve both as an indicator of aquatic life use status and biological integrity is also demonstrated by Figure 3. For example, the Modified Warmwater Habitat (MWH) use designation is assigned to streams which cannot attain the Warmwater Habitat (WWH) use designation due to circumstances (defined in the WQS; see p. 7) which preclude attainment of the WWH biological criteria[3]. However, the MWH biological criteria, which were derived from a separate set of habitat modified reference sites, reflects only a fair level of aquatic community performance which is not considered to be consistent with the biological integrity goal of the Clean Water Act (CWA). Attainment of the biological criteria for the WWH and Exceptional Warmwater Habitat (EWH) use designations reflect increasingly higher levels of biological integrity which are considered to be consistent with the biological integrity (good and exceptional performance, respectively) goals of the CWA.

---

3 Use designations such as Modified Warmwater Habitat (MWH) and Limited Resource Waters (LRW) do not meet the biological integrity goal of the Clean Water Act and are assigned on a case-by-case basis and must be based on a use attainability analysis which is performed by the state and approved by U.S. EPA.

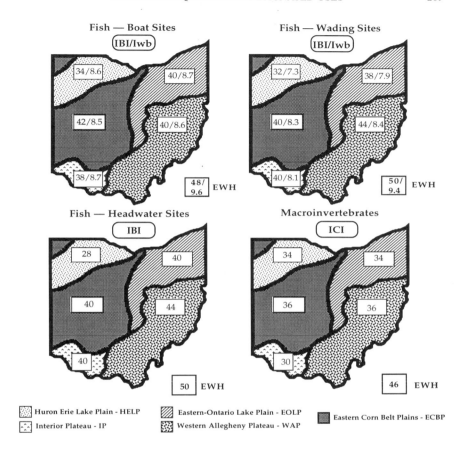

Figure 2. Numerical biological criteria adopted by the Ohio EPA for the Warmwater Habitat (WWH) and Exceptional Warmwater Habitat (EWH) use designations arranged by biological index, site type, and ecoregion. Index values on each map are the WWH biocriteria that vary by ecoregion as follows: IBI/MIwb for Boat Sites (upper left), IBI/MIwb for Wading Sites (upper right), IBI for Headwater Sites (lower left), and the ICI for all sites (lower right). The EWH criterion for each index and site type is located in the boxes adjacent to each map.

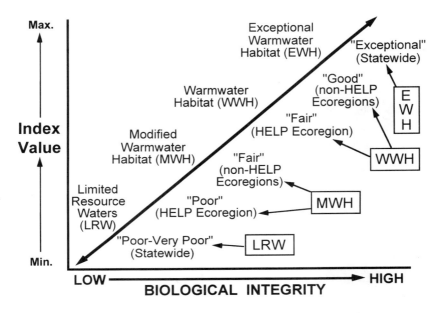

Figure 3. Relationship between the tiered aquatic life uses in the Ohio WQS and narrative evaluations of aquatic community performance and how this corresponds to a theoretical scale of biological integrity and measured biological index values (HELP = Huron/Erie Lake Plain ecoregion).

Procedures for determining the use attainment status of Ohio's lotic surface waters were also developed (Ohio EPA 1987b; Yoder 1991b). Using the numerical biocriteria as defined by the Ohio WQS, use attainment status is determined as follows:

1) FULL - the aquatic life use attainment status is considered to be full if all of the applicable numeric indices exhibit attainment of the respective biological criteria; this means that the aquatic life goals of the Ohio WQS are being attained.

2) PARTIAL - at least one organism group exhibits non-attainment of the numeric biocriteria, but no lower than a narrative rating of fair performance, and the other group exhibits attainment.

3) NON - neither organism group exhibits attainment of the ecoregional biocriteria, or one organism group reflects a narrative performance rating of poor or very poor, even if the other group exhibits attainment.

Following these rules a use attainment table is constructed for a longitudinal stream or river reach organized on a watershed basis. Information included in an attainment table are sampling location (river mile index), biological index scores, the Qualitative Habitat Evaluation Index (QHEI; Rankin 1989, 1995) score, attainment status, and comments about important site specific factors such as proximity to pollution sources. An example is provided by Table 1 for selected small urban and suburban watersheds throughout Ohio. This information may also be graphically portrayed in a classic upstream-to-downstream longitudinal profile comparing the sampling result to longitudinal position in a river or stream or as a scatter plot of the sampling results versus drainage area (an indicator of stream size) at each site. Either technique permits a visual examination of the biological sampling results in terms of position in a water body or watershed and the significance of deviations, if any, from the numerical biological criteria.

Using Biosurveys and Biocriteria to Assess Aquatic Life Use Attainment Status in Urban and Suburban Ohio Watersheds
Biological criteria can play an especially important role in nonpoint source assessment and management since they directly correspond to important environmental goal and regulatory end-points, *i.e.*, the biological integrity goal of the Clean Water Act and aquatic life designated uses in state WQS. Numerous studies have documented the capability of biological assessments to accurately characterize aquatic ecosystem quality and condition in a wide variety of settings. Yoder and Rankin (1995b) described unique combinations of community response variables they termed "biological response signatures" within which different classes of environmental stressors (*e.g.*, toxicity, nutrient enrichment, habitat degradation) can be distinguished. Gammon *et al.* (1983, 1995) documented a "gradient" of compositional and functional shifts in the fish and macroinvertebrate communities of small, agricultural watersheds in central Indiana. Community responses ranged from an increase in biomass with mild nutrient enrichment to complete shifts in community composition and function (*e.g.*, insectivores replaced by omnivores) with increasingly severe impacts. Impacts from animal feedlots had the most pronounced effects. In the latter case the condition of the immediate riparian zone was correlated with the degree of impairment. Other work by Gammon *et al.* (1990) suggested that nonpoint sources are impeding progress in making further biological improvements which have recently been observed in large rivers primarily in response to reduced point source impacts. Bennet *et al.* (1993) used the IBI as an endpoint in a GIS modeling exercise where land use characteristics of agricultural watersheds in Virginia were correlated with the degradation of aquatic communities. Their goal was to develop a method for the most effective use of limited resources in identifying the most critical sources of nonpoint source pollution for changes in land management in order to restore degraded water resources.

Table 1. Aquatic life use (ALU) attainment status for selected headwater stream biological sampling locations in urban/suburban areas of central Ohio. Each line shows sampling location (river mile), index value, the habitat assessment score, use attainment status, and other information about the sampling location and/or watershed area.

RIVER MILE Fish/Invert.	IBI	ICI[a]	QHEI	ALU Attainment Status[b]	Comments
Rose Run (1991)					
E. Corn Belt Plain - WWH Use Designation (Existing)					
0.6/0.6	38ns	MG	72.0	FULL	Suburban dev.
Hamilton Ditch (1992)					
E. Corn Belt Plain - MWH Use Designation (Recommended)					
1.3/0.3	28	8*	40.0	**NON**	Channelized
Rush Run (1994)					
E. Corn Belt Plain - WWH Use Designation (Existing)					
0.2/0.2	26*	4*	69.0	**NON**	Residential, sewage
Trabue Run (1991)					
E. Corn Belt Plain - WWH Use Designation (Existing)					
2.4/2.4	20*	8*	62.0	**NON**	Commercial dev.
0.2/0.2	26*	20*	64.0	**NON**	Light urban, spills
Republican Run (1991)					
E. Corn Belt Plain - WWH Use Designation (Existing)					
0.2/0.2	36ns	--	63.0	[FULL]	Suburban dev.
Eversole Run (1994)					
E. Corn Belt Plain - WWH Use Designation (Existing)					
1.3/1.3	46	F*	70.0	PARTIAL	Rural, intermittent

*   Significant departure from ecoregion biocriterion; poor and very poor results are underlined.

ns Nonsignificant departure from ecoregion biocriterion (≤ 4 IBI or ICI units; ≤ 0.5 MIwb units).

a   The narrative evaluation using the qualitative sample (G-good, MG-marginally good, F-fair, P-poor) is based on best professional judgment utilizing sample attributes such as taxa richness, EPT taxa richness, and community composition and is used in lieu of the ICI when artificial substrate data are not available.

b   Aquatic life use (ALU) attainment status based on one organism group is parenthetically expressed.

Ecoregional Biological Criteria: E. Corn Belt Plain (ECBP)

INDEX - Site Type	WWH	EWH	MWH[c]
IBI - Headwaters	40	50	24
ICI	34	46	22

c - Modified Warmwater Habitat for channelized habitats.

Biological responses to urban nonpoint source impacts have also been documented by numerous investigators. Klein (1979) documented a relationship between increasing urbanization and biological impairment noting that the latter does not become severe until urbanization reaches 30% of the watershed area. Steedman (1988) used a modification of the IBI to demonstrate the influence of urban land use and riparian zone integrity in Lake Ontario tributaries. A model relationship between the IBI and these two environmental factors was developed.

Biological monitoring of nonpoint source impacts and pollution abatement efforts in concert with the more traditional water quality assessment tools (*e.g.*, chemical/physical) can produce the type of evaluation needed to determine where urban nonpoint source management efforts should be focused, what some of the management goals should be, and to evaluate the effectiveness (*i.e.*, end-result) of such efforts (Yoder 1995a). At the same time a well conceived monitoring program can yield multi-purpose information which can be applied to similar situations without the need for site-specific monitoring everywhere. This is best accomplished when a landscape partitioning framework such as ecoregions (Omernik 1987) and their subcomponents are used as an initial step in accounting for natural landscape variability. It is because of landscape variability that uniform and overly simplified approaches to nonpoint source management will fail to produce the desired results (Omernik and Griffith 1991).

Significant uncertainty exists about the link between steady-state water quality criteria applications and ecological indicators, particularly in complex urban settings. In many situations we have failed to detect chemical water quality criteria exceedences at sites where biocriteria impairment is apparent and even severe (Ohio EPA 1990a). Much of the non-attainment that we have observed in urban watersheds is due to non-chemical impacts such as habitat degradation, changes in the flow regime, and sedimentation impacts. However, chemical water quality impacts which frequently escape detection or adequate characterization by the grab sampling approach commonly employed by many local, state, and federal agencies are also thought responsible for a significant portion of the non-attainment (Yoder 1995a). However, reaching this conclusion is made possible only by examining other evidence beyond conventional water column chemical data.

Bioassessments achieve their maximum effective use in the assessment of urban nonpoint sources when a watershed design to sampling and analysis is employed. An example of this design is illustrated by the results of Ohio EPA bioassessments of small urban and suburban watersheds in southwest, central, and northeastern Ohio (Figure 4). The watersheds included in these figures include small, headwater streams that represent a range of land use from largely rural, agricultural settings to intensive

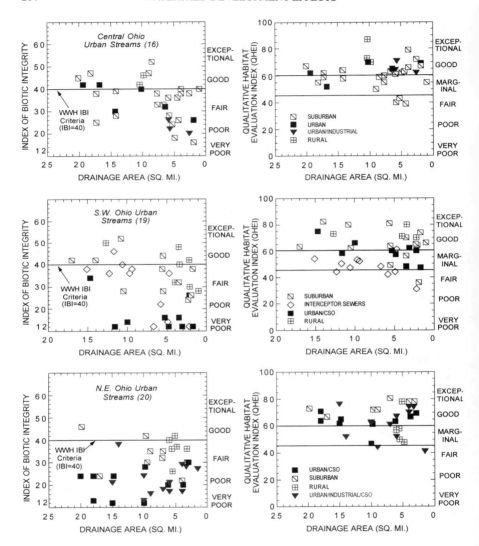

Figure 4. Index of Biotic Integrity (IBI) and Qualitative Habitat Evaluation Index (QHEI) results from 110 biological monitoring locations in small urban and suburban headwater streams in central (upper), southwest (middle), and northeast (lower) Ohio. Narrative ratings of both biological performance and habitat quality are indicated along the Y2 axis. Each location was designated according to the broad land use/impact categories that most influenced each watershed area.

urbanization. An attempt was made to exclude sites which were predominantly impacted by significant point sources. The land use/impact categories used were designated as rural, suburban, urban/industrial, urban with combined sewer overflows (CSO), urban/industrial with CSOs, and interceptor sewer line construction. These categories were assigned to each sampling location based on our general knowledge of the watershed area upstream from the sampling site and is consistent with the assignment of impact types used by Yoder and Rankin (1995b) elsewhere in Ohio.

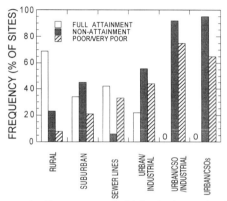

The distribution of IBI scores in these watersheds shows a tendency towards lower IBI scores and a subsequent loss of biological integrity with an increasing degree of urbanization, becoming more severe as other impact types such as CSOs, industrial, or commercial development coincide. Out of the 110 sampling sites examined only 25 exhibited good, very good, or exceptional biological performance which corresponds to meeting the

Figure 5. Frequency of biological monitoring locations in central, southwest, and northeast Ohio headwater streams which were in full attainment or non-attainment of the applicable aquatic life use criteria and the proportion of sites which reflected poor or very poor performance as measured by the IBI. The results are stratified by the broad land use/impact categories that most influenced the watershed area upstream from each sampling location.

WWH (good) or EWH (very good, exceptional) biocriteria for the IBI (Figure 5). An additional 19 sites were marginally good which means the IBI score was in the non-significant range of departure from the WWH IBI biocriterion. Forty-six sites (42%) reflected poor or very poor performance based on IBI values. Of the sites classified as being impacted by urban land use and pollution sources, only two sites attained the applicable IBI biocriterion. Poor or very poor performance was reflected by the majority of the urban impacted sites (85%). More than 40% of the suburban sites were impaired with many of these reflecting the impact of new developments for housing and commercial uses. These results demonstrate the degree of degradation which exists in most small urbanized watersheds and the difficulty thus far in dealing with multiple and diffuse sources of stress. Yoder (1995a) showed that the severity of biological impairments within urban areas was also influenced by stream and river size (as measured by watershed area) with the most severe effects occurring in what we define as headwater streams, *i.e.,* watershed areas less than 20 square miles.

While habitat impacts are responsible for *some* of the observed impairments among the 110 sites, most of the biologically impaired sites offered relatively good instream habitat (Figure 4). Thus factors other than direct habitat deficiencies as measured by the QHEI are likely responsible for the majority of the observed impairment. This includes direct chemical effects from permitted discharges, spills, contaminated runoff, and other releases. CSOs are a major source of impairment in urban watersheds and besides contributing raw sewage, can also include industrial wastewater that is discharged into the sewer system. In many urban settings in Ohio concentrations of chemicals in bottom sediments are frequently elevated compared to concentrations measured at site-specific control or regional reference sites. Contaminated sediments generally result from releases which enter the aquatic environment during regular and episodic releases from point sources (includes CSOs and storm sewers) and/or periodic runoff events from urban nonpoint sources. The correspondence between elevated concentrations of toxic heavy metals and declining aquatic community performance as portrayed by the IBI and ICI which is demonstrated by Figure 6. The results show that increasing levels of seven toxic heavy metals (arsenic, cadmium, chromium, copper, lead, nickel, and zinc) commonly encountered in urban settings corresponded to a much reduced frequency of sites with IBI and ICI scores that meet the typical WWH and EWH values. The frequency of sites meeting the EWH biocriteria declined markedly at levels greater than 150 mg/kg and WWH attainment declined above 200 mg/kg. It is believed that the relationships demonstrated between the indicators of biological integrity and the degree of sediment contamination by heavy metals is an accurate reflection of the history of toxic metals loadings from all sources, something that frequently escapes accurate characterization by the type of chemical grab sampling routinely employed by local and state agencies.

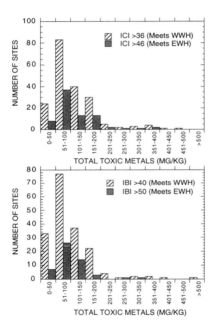

Figure 6. Frequency of biological sampling sites throughout Ohio at which ICI (upper) and IBI (lower) scores consistent with the WWH and EWH biocriteria occurred at corresponding ranges of total toxic heavy metals in sediment.

While much attention is generally paid to toxic substances in urban nonpoint source runoff, evidence suggests that non-toxic impacts are also significant, at least in Ohio and the midwest. Sedimentation (or siltation) resulting from urban and other land use activities is a major impact from urban nonpoint sources and was the second leading cause of impairment (from all sources) identified by the 1994 Ohio Water Resource Inventory (Ohio EPA 1994). Since 1988, this cause category has surpassed ammonia and heavy metals, classes of pollutants most commonly associated with point sources, in rank. Sedimentation is responsible for more impairment (over 1400 miles of stream and rivers and 23,000 acres of lakes, ponds, and reservoirs) than any other category except organic enrichment/dissolved oxygen (D.O.), with which it is closely allied in both urban and agricultural areas.

Watershed impermeability has recently been suggested as an overall indicator of the level of "watershed stress" in terms of being correlated with an increasing degradation of aquatic life (Schueler 1994; Arnold and Gibbons 1996). Imperviousness has been correlated with an increased risk of impairment not only due to adverse effects on watershed hydrology, but as a product of other impacts such as contaminated runoff, more frequent spills, and increasingly severe habitat impacts which correspond to this stressor indicator. In the two papers we reviewed on this subject, watershed imperviousness was negatively correlated with the condition of the aquatic biota with degradation becoming significant at 25-30% within a watershed. While we did not quantify this factor in our Ohio urban/suburban watershed examples (Figures 4 and 5) it seems plausible that imperviousness would be correlated with the results, particularly for small watersheds.

Use Attainability Issues in Urban and Suburban Ohio Watersheds

An emerging issue of increasing importance related to the preceding discussion and to the restoration and management of small urban watersheds is that of use attainability. An important objective of the biosurveys conducted by Ohio EPA is to determine the appropriate and attainable aquatic life use designation. If the results of the sampling and data analysis suggest that an existing use designation is inappropriate (or the stream is presently unclassified) an appropriate use is then recommended. These recommendations are proposed in a WQS rulemaking procedure and adopted after consideration of public input.

The issue of urban and suburban development and the effects of each on aquatic life use attainment in rivers and streams has increased in importance within the surface water programs at Ohio EPA. Small watersheds in established, older urban settings are particularly at issue because of regulatory concerns such as CSOs and stormwater management. As was amply demonstrated by our Ohio examples (Figures 4 and 5), small streams in historically developed urban areas are not only impaired, but severely so. This is generally due to multiple factors including chemical effects, physical habitat modifications, lack of sustained flows during normally recurring dry weather periods, higher peak flows during wet weather periods, and watershed scale modifications of land use characteristics. Overlapping regulatory programs such as NPDES permits for point sources, CSO and sanitary sewer overflow (SSO) control and remediation, stormwater management, and construction site management are commonplace throughout Ohio. The regulatory and/or management requirements associated with each are driven, in part, by the Ohio WQS. In our efforts to develop strategies to protect and restore designated uses the question of use attainability frequently arises. It is widely perceived that the restoration of designated aquatic life uses consistent with the goals of the CWA (*i.e.,* WWH) in intensively urbanized areas is neither practical nor attainable. This in itself can present a premature barrier to the management goal of restoring full use attainment or upgrading use designations for waters now classified for less than CWA goal uses.

The assignment of appropriate and attainable aquatic life uses is a challenge that Ohio EPA has dealt with over the past 20 years. Our approach has relied heavily on experience with observing biological responses to different types of impacts and the habitat assessment provided with the QHEI. Generally speaking if the QHEI reveals that instream habitat is sufficient *on a watershed or reach length scale* to support an assemblage of aquatic life consistent with the WWH use, that use is adopted. Classification of waters to a less than CWA goal use designation such as MWH or LRW requires a showing that the WWH biocriteria are not attained and that habitat is an overriding and precluding factor in the non-attainment. In effect it must be demonstrated that the WWH use is not attainable in the foreseeable future. Rankin (1995) has shown at what point habitat becomes a precluding factor by examining the various attributes of the QHEI which correlate with WWH attainment and non-attainment at sites where non-habitat impacts are minimal. Figure 7 exemplifies this phenomenon by contrasting ranges of IBI values that correspond to the five narrative categories with the ratio of modified:warmwater habitat attributes (as defined by Rankin 1989) which increases as habitat becomes deficient in terms of being able to support an assemblage of aquatic life consistent with the WWH biocriteria. As the predominance of modified habitat attributes increase to a modified:warmwater ratio of greater than 1.0-1.5 the likelihood of having IBI scores consistent with the WWH use declines. This relationship bears out better where the QHEI score and attributes ratios are analyzed on a reach length or watershed scale (Rankin 1995).

The decision to assign a less than CWA goal use (*e.g.,* MWH or LRW) must also meet the conditions prescribed by the U.S. EPA WQS regulations (40 CFR, Part 131) that restoring to a higher designated use would result in widespread, adverse social and economic impacts or the higher use is not attainable due to irretrievable effects of anthropogenic origin or natural conditions. The most frequently used reason for assigning either the MWH or LRW uses in Ohio is due to irretrievable physical effects. For example, the MWH use designation applies in situations of wide-spread stream habitat modifications for agricultural drainage purposes (*e.g.,* channelization) and where that activity is sanctioned by state and/or federal law. Less frequently encountered habitat modifications include run-of-river impoundment by low head dams, or heavy sedimentation due to non-acidic mine drainage and where reclamation activities are not expected. The LRW use applies to cases of severe, watershed-wide drainage modifications and acidic mine drainage where reclamation activities are not expected. With the exception of isolated instances of direct channelization, the most frequently encountered situation with small urban streams is the severe disruption of local habitat such as riparian encroachment and removal, replacement of the natural substrate with artificial materials

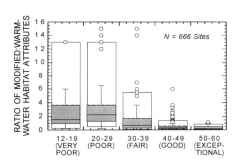

Figure 7. Relationship between the ratio of modified:warmwater habitat attributes and ranges of the IBI corresponding to the five narrative categories of biological community performance. The data is from a set of least impacted and habitat modified reference sites throughout Ohio. This analysis employs a box-and-whisker plot showing the median, interquartile, maximum, minimum, 90th and 10th percentile, and outlier IBI values.

(*e.g.,* concrete, rock-basket gabions), and broad scale watershed modifications. In such cases the QHEI scores are frequently reflective of poor or very poor habitat quality yielding extremely high modified:warmwater habitat ratios (Fig. 7). In such cases flow conditions may also be ephemeral or inadequate to support any except the most tolerant forms of aquatic life, or the stream is virtually eliminated by culverting. Such situations are relatively easy to diagnose and assignment of the LRW use is the result.

The situation is different when the habitat evaluation indicates that sufficient warmwater attributes are present to suggest attainment of WWH is possible. In such cases WWH is viewed as attainable (as the data from several of our small urban/suburban watersheds suggest) even though the aquatic communities only perform in the poor or very poor ranges. As previously mentioned the impairment may be due to sources which theoretically could be abated or sufficiently controlled, thus resulting in the full restoration of the WWH use. The key point here is that uses are based on potential, not the present-day biological attainment status. However, the challenges of managing stressors such as spills, runoff, and CSOs is daunting because of the diffuse nature of these sources and the periodicity of their influence. In some of our urbanized watersheds the attainability of the WWH use has recently come into question even when the QHEI data suggests that WWH is attainable. This issue has become more complicated in light of the recent information about the potential of imperviousness to influence biological performance in urban watersheds (Schueler 1994; Arnold and Gibbons 1996).

Managing CSOs is a growing challenge for Ohio EPA and other local, state, and federal agencies. Current policy involves the establishment of a state-specific strategy and implementation of nine minimum controls by major CSO entities. In some of the major CSO communities of Ohio, questions have been raised about the attainability of the WWH biological criteria and how this might eventually affect CSO abatement strategies. While these questions may have merit in light of the recent literature concerning imperviousness and our own findings about the extent of aquatic life impairment in small urban watersheds, it would be premature to in effect "give up" on WWH attainment without first implementing the nine minimum CSO controls. In addition, resolving this issue will involve an examination of many other factors in addition to imperviousness on a broad geographic scale. Until this type of exploratory research is completed making fundamental changes to the use designation process would be premature.

Applications to the Management of Urban Watersheds
Steedman (1988) observed the IBI to be negatively correlated with urban land use. The land use within the 10-100 km² of a site was the most important in predicting the IBI which suggests that "extraneous" information was likely included if whole watershed land use information was used. Thus, scale will be another important consideration in the assessment of urban watersheds. Steedman (1988) also discovered that the condition of the riparian zone was an important covariate with land use, in addition to other factors such as sedimentation and nutrient enrichment. A model relationship between land use and riparian zone quantity and the IBI was developed. This relationship provided the basis to predict when the IBI would decline below a certain threshold level based on combinations of riparian zone quantity and percent of urbanization. In the Steedman (1988) study the domain of degradation for Toronto area streams ranged from 75% riparian removal at 0% urbanization to 0% riparian removal at 55% urbanization. These results indicate that it is possible to establish the bounds within which the combination of watershed land use and riparian zone quantity must be maintained in order to attain a target level of biological community performance as measured by the IBI. It seems plausible that such relationships could be established for many other watersheds provided the

baseline database is sufficiently developed not only for biological communities, but for land use stressors and riparian condition as well. Additionally including the concept of ecoregions and sub-ecoregions should lead to the development of management criteria for land use, riparian zones, and other important covariates which would assure the maintenance of aquatic life uses in streams and rivers over fairly broad areas without the need to develop a site-specific database everywhere.

Conclusions

Well designed biological surveys and biological criteria can contribute essential information and capacity to urban and suburban watershed management. Because the biota respond to and integrate all of the various factors that affect a particular water body they are essentially the end-product of what happens within watersheds. The important issue is that ambient monitoring be done as part of the overall watershed management and assessment process and be done correctly in terms of timing, methods, and design. However, monitoring alone is not enough. Federal, state, local, and private efforts to remediate impaired watersheds must include an interdisciplinary approach that includes the range of factors responsible for the type of ecosystem degradation that has been documented in urban and suburban watersheds throughout North America. Effective protection and rehabilitation strategies will require the targeting of large areas and individual sites (Schaefer and Brown 1992), as well as the incorporation of ecological concepts in the status quo of land use and water quality management practices and policies.

Urban watershed management and protection strategies will continue to develop as new information is revealed and relationships between instream biological community performance and watershed factors are better developed and understood. However, there are some things which we know now that should become part of our current management strategies. Urban and suburban development must become proactive, *i.e.,* the design of such developments must accommodate the features of the natural landscape and include common sense practices such as minimum widths for riparian zones and the attenuation of peak runoff events. Regulatory agencies also share the responsibility particularly in resolving the difficult use attainability questions. Watersheds which exhibit attainment of aquatic life use biocriteria should be protected to maintain current conditions as new development represents an almost certain threat. Strategies should also include the restoration of degraded watersheds where the potential for recovery actually exists. In systems where the degree of degradation is so severe that the damage is essentially irretrievable, minimal enhancement measures could still be considered even though full recovery is not to be expected. Biocriteria and bioassessments have an important and central role to play in this process now and into the future.

Acknowledgements

Roger Bannerman, Marc Smith, and Jeff DeShon reviewed an original draft and provided helpful comments.

References

Arnold, C.L. And C. J. Gibbons. 1996. Impervious surface coverage: the emergence of a key environmental indicator. J. Am. Planning Assoc. 62(2): 243-258.

Barbour, M.T. and others. 1996. A framework for biological criteria for Florida streams using benthic macroinvertebrates. J. N. Am. Benth. Soc. 15(2): 185-211.

Benke, A.C. 1990. A perspective on America's vanishing streams. J. N. Am. Benth. Soc., 9 (1): 77-88.

Bennet, M.R., J.W. Kleene, and V.O. Shanholtz. 1993. Total maximum daily load nonpoint source allocation pilot project. File Report, Dept. of Agricultural Engineering, Blacksburg, VA. 49 pp.

Davis, W.S. and A. Lubin. 1989. Statistical validation of Ohio EPA's invertebrate community index, pp. 23-32. in Davis, W.S. and T.P. Simon (eds.). Proc. 1989 Midwest Poll. Biol. Mtg., Chicago, Ill. EPA 905/9-89/007.

DeShon, J.D. 1995. Development and application of the invertebrate community index (ICI), pp. 217-243. in W.S. Davis and T. Simon (eds.). Biological Assessment and Criteria: Tools for Risk-based Planning and Decision Making. Lewis Publishers, Boca Raton, FL.

Fausch, K. D., J. R. Karr, and P. R. Yant. 1984. Regional application of an index of biotic integrity based on stream fish communities. Transactions of the American Fishery Society 113:39-55.

Fore, L. S., J.R. Karr, and R.W. Wisseman. 1996. Assessing invertebrate responses to human activities: evaluating alternative approaches. J. N. Am. Benth. Soc. 15(2): 212-231.

Hill, M. T. , W. S. Platts, and R. L. Beschta. 1991. Ecological and geomorphological concepts for instream and out-of-channel flow requirements. Rivers 2: 198-210.

Gammon, J.R., Spacie, A., Hamelink, J.L., and R.L. Kaesler. 1981. Role of electrofishing in assessing environmental quality of the Wabash River, in Ecological assessments of effluent impacts on communities of indigenous aquatic organisms, *in* Bates, J. M. and Weber, C. I., Eds., ASTM STP 730, 307p.

Gammon, J.R., M.D. Johnson, C.E. Mays, D.A. Schiappa, W.L. Fisher, and B.L. Pearman. 1983. Effects of agriculture on stream fauna in central Indiana. EPA-600/S3-83-020, 5 pp.

Gammon, J.R. 1976. The fish populations of the middle 340 km of the Wabash River, Purdue University, Water Resources Res. Cen. Tech. Rep. 86. 73 p.

Gammon, J.R., C.W. Gammon, and M.K. Schmid. 1990. Land use influence on fish communities in central Indiana streams. Proc. 1990 Midwest Poll. Biol. Conf., EPA 905/R-92/003. 111-120.

Gammon, J.R. 1995. An environmental assessment of the streams of Putnam County, Indiana and vicinity with special emphasis on the effects of animal feedlots. Report to Indiana Dept. Nat. Res., DePauw University, Greencastle, IN. 124 pp.

Jackson, S. 1992. Re-examining independent applicability: agency policy and current issues. Water Quality Standards for the 21st Century, Proceedings of the Third National Conference. U.S. EPA, Offc. Science and Technology, Washington, D.C. 823-R-92-009, 135-138.

Karr, J. R. 1981. Assessment of biotic integrity using fish communities. Fisheries 6(6): 21-27.

Karr, J. R. 1991. Biological integrity: A long-neglected aspect of water resource management. Ecological Applications 1(1): 66-84.

Karr, J. R., K. D. Fausch, P. L. Angermeier. P. R. Yant, and I. J. Schlosser. 1986. Assessing biological integrity in running waters: A method and its rationale. Illinois Natural History Survey Special Publication No. 5, 28 pp. Champaign, Illinois

Karr, J. R., L. A. Toth, and D. R. Dudley. 1985. Fish communities of midwest rivers: A history of degradation. BioScience 35 (2): 90-95.

Kelly, M.H. and R.L. Hite. 1984. Evaluation of Illinois stream sediment data: 1974-1980. Illinois EPA, Div. Water Poll. Contr., Springfield, Ill.

Kerans, B. L., and Karr, J. R. 1992. An evaluation of invertebrate attributes and a benthic index of biotic integrity for Tennessee Valley rivers, Proc. 1991 Midwest Poll. Biol. Conf., EPA 905/R-92/003.

Klein, R.D. 1979. Urbanization and stream quality impairment. Water Res. Bull. 15(4): 948-963.

Lyons, J. 1992. Using the index of biotic integrity (IBI) to measure environmental quality in warmwater streams of Wisconsin. Gen. Tech. Rep. NC-149. St. Paul, MN: USDA, Forest Serv., N. Central Forest Exp. Sta. 51 pp.

Miner R. and D. Borton. 1991. Considerations in the development and implementation of biocriteria, Water Quality Standards for the 21st Century,, Washington, D.C., 115.

Ohio Environmental Protection Agency. 1987a. Biological criteria for the protection of aquatic life: Volume I. The role of biological data in water quality assessment. Division of Water Quality Monitoring and Assessment, Surface Water Section, Columbus, Ohio.

Ohio Environmental Protection Agency. 1987b. Biological criteria for the protection of aquatic life: Volume II. Users manual for biological field assessment of Ohio surface waters. Division of Water Quality Monitoring and Assessment, Surface Water Section, Columbus, Ohio.

Ohio Environmental Protection Agency. 1988. Water Quality Inventory - 1988 305(b) report. Volume I. E.T. Rankin, editor. Division of Water Quality Monitoring and Assessment, Surface Water Section, Columbus, Ohio .

Ohio Environmental Protection Agency. 1989a. Biological criteria for the protection of aquatic life: Volume III. Standardized biological field sampling and laboratory methods for assessing fish and macroinvertebrate communities. Division of Water Quality Monitoring and Assessment, Columbus, Ohio.

Ohio Environmental Protection Agency. 1989b. Addendum to Biological criteria for the protection of aquatic life: Volume II. Users manual for biological field assessment of Ohio surface waters. Division of Water Quality Monitoring and Assessment, Surface Water Section, Columbus, Ohio.

Ohio Environmental Protection Agency. 1990a. Ohio Water Resource Inventory, Volume I: Summary, Status and Trends, 1990. E. T. Rankin, C. O. Yoder, and D.S. Mishne, (editors). Division of Water Quality Planning and Assessment, Ecological Assessment Section. Columbus, Ohio.

Ohio Environmental Protection Agency. 1990b. Ohio's nonpoint source pollution assessment. Division of Water Quality Planning and Assessment.. Columbus, Ohio.

Ohio Environmental Protection Agency. 1991. 1991 Ohio nonpoint source assessment. Ohio EPA. Division of Water Quality Planning and Assessment. Columbus, Ohio.

Ohio Environmental Protection Agency. 1994.   1994 Ohio Water Resource Inventory,
    Volume I: Summary, Status and Trends. E. T. Rankin, C. O. Yoder, and D.Mishne,
    (eds). Division of Surface Water, Monitoring & Assessment Section, Columbus,
    Ohio.

Omernik, J. M. 1987. Ecoregions of the conterminous United States. Ann. Assoc. Amer.
    Geogr. 77(1):118-125.

Omernik, J. M., and G. E. Griffith 1991. Ecological regions versus hydrologic units:
    frameworks for managing water quality, Journal of Soil and Water Conservation, 46,
    334.

Pihfer, M. T. 1991. Biocriteria: just when you thought it was safe to go back into the water.
    Environment Reporter, Bureau of National Affairs, 0013-9211.

Plafkin, J. L. and others. 1989. Rapid Bioassessment Protocols for use in rivers and streams:
    benthic macroinvertebrates and fish. EPA/444/4-89-001. U.S. EPA. Washington, D.C.

Rankin, E. T. and C. O.Yoder. 1990. The nature if sampling variability in the Index of Biotic
    Integrity (IBI)in Ohio streams. Pages 9-18. In: W. S. Davis (editor). Proceedings of
    the 1990 Midwest Pollution Control Biologists Conference, U. S. EPA, Region V,
    Env. Sci. Div., Chicago, IL. EPA-905-9-90/005.

Rankin, E. T. 1989. The Qualitative Habitat Evaluation Index (QHEI). Rationale, methods,
    and applications. Division of Water Quality Planning and Assessment, Ecological
    Analysis Section. Columbus, Ohio.

Rankin, E. T. 1995. The use of habitat assessments in water resource management programs,
    pp. 181-208. in W. Davis and T. Simon (eds.). Biological Assessment and Criteria:
    Tools for Water Resource Planning and Decision Making. Lewis Publishers, Boca
    Raton, FL.

Ruffier, P.J. 1992. Re-examining independent applicability: regulatory policy should reflect
    a weight of evidence approach. Water Quality Standards for the 21st Century,
    Proceedings of the Third National Conference, U.S. EPA, Offc. Science and
    Technology, Washington, D.C. 823-R-92-009, 139-147.

Schmidt, W.A. 1992. Water quality protection requires independent application of criteria.
    Water Quality Standards for the 21st Century, Proceedings of the Third National
    Conference, U.S. EPA, Offc. Science and Technology, Washington, D.C. 823-R-92-
    009, 157-164.

Schaefer, J.M. and M.T. Brown. 1992. Designing and protecting river corridors for wildlife. Rivers, 3(1): 14-26.

Schregardus, D.R. 1992. Re-examining independent applicability: biological criteria are the best measure of the integrity of a water body and should control when there is a conflict. Water Quality Standards for the 21st Century, Proceedings of the Third National Conference, U.S. EPA, Offc. Science and Technology, Washington, D.C. 823-R-92-009, 149-156.

Schueler, T.R. 1994. The importance of imperviousness. Watershed Protection Techniques, 1(3): 100-111.

Simon, T. 1991. Development of index of biotic integrity expectations for the ecoregions of Indiana. I. Central corn belt plain. EPA-905/9-91/025. 93 pp.

Steedman, R.J. 1988. Modification and assessment of an index of biotic integrity to quantify stream quality in southern Ontario. Can. J. Fish. Aquatic Sci. 45: 492-501.

Stephan, C. E., and others. 1985. Guidelines for deriving numerical National water quality criteria for the protection of aquatic organisms and their uses. National Technical Information Service, Springfield, VA.

Stevens, J.C. and S.W. Szczytko. 1990. The use and variability of the biotic index to monitor changes in an effluent stream following wastewater treatment plant upgrades, pp. 33-46. in Davis, W.S. (ed.). Proc. 1990 Midwest Poll. Biol. Mtg., Chicago, Ill. EPA-905-9-90/005.

Trautman, M. B. 1981. The fishes of Ohio. (2nd edition). Ohio State University Press. Columbus, Ohio. 782 pp.

U. S. Environmental Protection Agency. 1990. Biological Criteria: national program guidance for surface waters. U. S. EPA, Office of Water Regulations and Standards, Washington, D. C. EPA-440/5-90-004.

U.S. Environmental Protection Agency. 1991. Policy on the use of biological assessments and criteria in the water quality program. Offc. Science and Technology, Washington, D.C.

Yoder, C. O. 1989. The development and use of biocriteria for Ohio surface waters. In: Gretchin H. Flock, editor. Water quality standards for the 21st century. Proceedings of a National Conference, U. S. EPA, Office of Water, Washington, D.C.

Yoder, C. O. 1991a. Answering some concerns about biological criteria based on experiences in Ohio. In: Gretchin H. Flock, editor. Water quality standards for the 21st century. Proceedings of a National Conference, U. S. EPA, Office of Water, Washington, D.C.

Yoder, C. O. 1991b. The integrated biosurvey as a tool for evaluation of aquatic life use attainment and impairment in Ohio surface waters. Biological Criteria: Research and Regulation. Proceedings of a National Conference, U. S. EPA, Office of Water, Washington, D.C.

Yoder, C. O. 1995a. Incorporating ecological concepts and biological criteria in the assessment and management of urban nonpoint source pollution. National Conference on Urban Runoff Management: Enhancing Urban Watershed Management at the Local, County, and State Levels. EPA/625/R-95/003. pp. 183-197.

Yoder, C.O. 1995b. Policy issues and management applications for biological criteria, pp. 327-344. in W. Davis and T. Simon (eds.). Biological Assessment and Criteria: Tools for Water Resource Planning and Decision Making. Lewis Publishers, Boca Raton, FL.

Yoder, C.O. and E.T. Rankin. 1995a. Biological criteria program development and implementation in Ohio, pp. 109-144. in W. Davis and T. Simon (eds.). Biological Assessment and Criteria: Tools for Water Resource Planning and Decision Making. Lewis Publishers, Boca Raton, FL.

Yoder, C.O. and E.T. Rankin. 1995b. Biological response signatures and the area of degradation value: new tools for interpreting multimetric data, pp. 263-286. in W. Davis and T. Simon (eds.). Biological Assessment and Criteria: Tools for Water Resource Planning and Decision Making. Lewis Publishers, Boca Raton, FL.

BIOLOGICAL EFFECTS OF THE BUILD-UP OF CONTAMINANTS IN SEDIMENTS IN URBAN ESTUARIES

D.J. MORRISEY, D.S. ROPER AND R.B. WILLIAMSON[1]

ABSTRACT

Concentrations of many contaminants in sediments near stormwater outfalls in New Zealand often exceed sediment-quality guidelines to protect sediment-dwelling animals, especially in industrial areas. In urbanised estuaries, there is widespread exceedence for the major contaminants Pb, Zn and organochlorines. Surveys of benthic organisms in a number of estuaries near Auckland (New Zealand's largest city) and laboratory toxicity tests on sediments show evidence for stormwater-impacts on aquatic animals through chronic toxic effects. The effects follow those predicted from North American sediment quality guidelines. However, evidence that the observed sub-lethal toxic effect resulted from urban stormwater inputs is circumstantial, and a detailed study on some pollution-gradients in the Auckland region were undertaken to test the hypothesis that the benthic communities were affected along stormwater-derived gradients of contaminant-concentration. These studies are part of a wider programme examining the chemistry, sedimentology, macroinvertebrate biology and ecotoxicity of sediments in New Zealand's urbanised estuaries. The results from the detailed macroinvertebrate survey are discussed in terms of the chemical gradients, other factors such as salinity and sediment texture, and the hypothesised impact.

INTRODUCTION

In 1991, New Zealand adopted the Resource Management Act (RMA), the purpose of which, broadly speaking, was to unify existing environmental legislation within a framework of sustainable management of natural resources. Regulation

---

[1]National Institute of Water and Atmospheric Research Ltd, PO Box 11-115, Hamilton, New Zealand

of human activities under the RMA is effects-based, requiring that adverse effects be avoided, remedied or mitigated.

Stormwater-contamination represents a potential mechanism whereby watershed-development may contravene the objective of sustainable management incorporated in the RMA. Methods of controlling the quality and volume of stormwater are generally designed on the basis of physical and chemical, rather than ecological, parameters. Furthermore, the same criteria are often used to audit the relative success of different methods of control. Because these processes do not incorporate assessments of ecological effects of stormwater on receiving-environments, controls do not necessarily ensure compliance with the requirements of the RMA. Assessment of the ecological effects of stormwater on aquatic receiving-environments is also an essential component of the process of selecting and auditing methods of stormwater-management. Prediction of likely effects of planned developments allows conditions of consent to be formulated and justified.

New Zealand does not have the history of severe industrial pollution of aquatic habitats that many other countries suffer from, particularly in North America and Europe. With a few exceptions, therefore, concentrations of anthropogenic contaminants in these habitats are relatively small. This historical context, coupled with modern methods of control of discharges from point-sources, means that contamination derived from stormwater and other diffuse sources is comparatively important.

Potential ecological effects of stormwater extend from the watersheds themselves, through the streams, rivers and wetlands into which they drain, to the estuaries and adjacent coastal habitats into which waterways eventually discharge. Estuaries and other sheltered coastal environments are sites of deposition of river-borne particulate material, including sediments and the various contaminants which are usually associated with them. This potential for deleterious effects of stormwater on these environments and their perceived value for wildlife, recreation, food-gathering and other commercial activities, has created a need for reliable methods for detecting impacts of contaminants derived from stormwater and other diffuse sources on estuarine and coastal ecosystems.

The detection of impacts of small concentrations of contaminants in estuaries is not a trivial matter. In most coastal environments, the inherent spatial and temporal variability of populations of organisms makes detection of anthropogenic changes difficult at any but extreme levels of impact (Diaz, 1989; Bryan & Langston, 1992). Compounding this, pollutants may cause an increase in the variability in the abundances of organisms at impacted sites (Warwick & Clarke, 1993; Underwood, 1991), making it more difficult (and, perhaps, inappropriate) to distinguish differences in average abundances between impacted and control sites. In estuaries, as in some freshwater environments (Clements & Kiffney, 1995), strong gradients of environmental stressors, such as salinity, sediment texture and exposure to currents and waves, makes detection even more difficult. Natural gradients may impose so strong a signal on the patterns of distribution and abundance of organisms that effects of contamination are all but obscured. Diversity of faunas, for example, is characteristically largest at the seaward and landward ends of the salinity gradient (approaching the diversity of neighbouring marine and freshwater faunas, respectively) and substantially reduced in between, where salinity is low and most variable (McLusky, 1981). Furthermore, those organisms able to tolerate the rigourous conditions of estuaries may also be relatively tolerant of human impacts. Nevertheless, the weight of evidence, if often circumstantial, indicates that accumulated contaminants derived from stormwater do have an effect on estuarine and other coastal ecosystems (e.g., Roper et al., 1988; Diaz, 1989; and from comparison of observed concentrations with the 'effects ranges' of Long et al., 1995 - see below). There is, therefore, a strong need for tools which will enable environmental managers to assess the relative efficiencies of different methods of protecting coastal environments from impacts caused by stormwater.

METHODS OF DETECTION

The various approaches that have been used to detect ecological impacts of runoff in estuaries are briefly described, focusing on impacts on benthic macrofauna. This description is intended to be illustrative rather than exhaustive. We then describe a current study to predict the extent and ecological impact of contaminants derived from stormwater in estuaries in the north of New Zealand.

In addition to the relatively small concentrations of contaminants present, detection of impacts of urban runoff presents an additional challenge in the form of the relatively long time-scales involved. Accumulation of contaminants from diffuse sources probably takes decades to reach environmental concentrations likely to produce detectable impacts. These long time-frames make it difficult or impossible to use some of the better-known and most powerful methods for detecting impacts, such as Before/After-Control/Impact (BACI) designs (Green, 1979; Underwood, 1993). The intensity of sampling required by such a design is high. Sustaining such intensity over the decades before and after the development of an estuarine catchment in order to document consequent environmental effects is beyond the life-expectancy of most researchers, let alone their funding.

Consequently, the most powerful and logically-robust tools for detecting impacts and ascribing cause and effect are not usually available to the environmental scientist attempting to demonstrate ecological impacts of stormwater on estuarine environments. What we are left with are comparisons of current spatial distributions of contaminants with the distributions of various indicators of the ecological integrity of a system (commonly the benthic macrofauna). Selection of control sites for such studies also presents a problem. Internal controls, i.e. uncontaminated locations within the same estuary, are likely to lie at different points on gradients other than that of contamination (salinity, type of sediment, etc.), confounding comparisons. In a simple estuary, it is likely that all sites experiencing the same conditions of salinity and deposition of sediment will also experience the same or similar inputs of contaminants, since these are largely controlled by patterns of settlement of the sediments with which they are associated (Luoma, 1990). A possible solution to this is to compare equivalent points along a contaminated estuarine gradient with those in one or (preferably) more uncontaminated estuaries. Unfortunately, natural variation among estuaries may be such that the power to detect differences between contaminated and uncontaminated is very small, particularly given the probable weakness of the signal from relatively small concentrations of contaminants. An elaboration of this approach has, however, been used successfully in rivers and lakes (see below), and may be useful in future studies of impacts in estuaries.

This correlative approach is made more robust by the additional use of manipulative experiments. The former may provide evidence, in the form of patterns of distribution of contaminants and fauna, *consistent* with an impact. For the purpose of identifying cause and effect, however, such correlations are necessary but not sufficient. In the absence of information on the distribution of contaminants, on the other hand, experimental approaches cannot demonstrate an impact in a particular estuary. Their value lies in testing the causal relationship underlying the correlations.

i. Indices of pollution

The convenience of an index of anthropogenic disturbance which reduces the complexity of benthic communities to a single statistic and, at the same time, removes background noise has led to the proposal of numerous indices based on the relative abundances or biomass of various components of the fauna, including individual species. The attraction of such indices is particularly strong in environments such as estuaries where steep environmental gradients make it difficult to select control-sites with which to compare fauna from putatively contaminated locations (Dauer et al., 1993). Such indices have, however, met with very limited success and have generally been short-lived. One of the most recent and, so far, most successful, is the Abundance Biomass Comparison (ABC) proposed by Warwick (1986) and Warwick & Clarke (1994). The method involves plotting a curve of the cumulative percentage abundance of each species in an assemblage against the rank of that species in terms of its abundance relative to other members of the assemblage (a k-dominance plot). Onto this plot is superimposed an equivalent plot of cumulative percentage of biomass for the same assemblage, also plotted against species-rank (by biomass this time). Empirical studies have shown that assemblages in undisturbed or unpolluted habitats often produce ABC plots in which the biomass-curve lies above the abundance-curve, indicating that the assemblage contains relatively large numbers of species characterised by small abundances of large individuals. Habitats subject to disturbance or pollution generally contain assemblages for which the abundance curve lies above the biomass-curve, indicating that they are dominated by species characterised by large numbers of small individuals. Warwick (1986) has suggested that ABC plots for individual sites can be used as absolute indicators of the degree of disturbance, without the need for comparison with control-sites.

Dauer et al. (1993), however, had mixed success in applying the ABC method to assemblages in contaminated estuarine sediments in Chesapeake Bay. They predicted that assemblages from estuarine areas with low and variable salinity and from areas contaminated with heavy metals and PAH should be classified as highly stressed.   According to Dauer et al., their results were only partially consistent with the predictions of the method. In defence of the method, Warwick & Clarke (1994) pointed out that Dauer et al's suggestion that the ABC method had failed to classify contaminated sites as stressed was based on the assumption that the measured concentrations were likely to cause stress. Absolute concentrations of contaminants, however, are often poor indicators of bioavailability and biological effect (Bryan & Langston, 1992). Further, the 'misclassification' of a supposedly unstressed site was caused by recruitment of large numbers of a small species of polychaete known from other studies to be characteristic of disturbed sites, implying that this site was, in fact, stressed. Overall, Warwick & Clarke (1994) suggest that the ABC method, like most others, be used in combination with other techniques for detecting disturbance that are based on different assumptions.

ii. Correlative approaches

Univariate correlations among concentrations of various contaminants and the abundances of individual species have been used to infer impact in several studies, often in conjunction with other, more sophisticated methods.   These correlations may involved abundances of individual species or indices of community-structure, such as number of species, diversity, evenness or proportions of certain taxa.   In those studies where correlations have been reported, they are generally restricted to areas immediately surrounding outfalls or other sites with relatively large concentrations of contaminants.   Roper (1990), for example, compared abundances of individual species among sites along a gradient of contamination from an estuarine sewage-outfall in New Zealand.   Although abundances of some species were reduced near the outfall, there was little effect on total numbers of taxa or individuals. The impacts on individual species were limited to within 10 m of the outfall.   Roper et al. (1988) found reduced abundances of taxa and individuals at only the most contaminated of several sites on intertidal mudflats in Manukau Harbour, New Zealand.   Each of the sites received runoff from different catchments but varied in concentrations of contaminants present in the sediments. Concentrations of contaminants at most of the sites were very small, consistent with their rural or urbanised catchments. The impacted site, in contrast, received runoff from an industrialised catchment. Diaz (1989) attempted to reduce the confounding effect of the salinity gradient on correlations between biota and contaminants in the James River Estuary, Virginia, by analysing data from different sections of the gradient separately.   In the saline portions of the estuary, abundances and biomass of dominant species were larger immediately below outfalls but effects were not detectable away from outfalls (Diaz, 1989).   In contrast, in the tidal, freshwater part of the estuary, where the confounding influence of salinity and estuarine circulation were absent, the fauna showed more widespread changes which correlated with contamination. Communities showed least diversity in areas where outfall were most abundant. Contrary to the above studies, Schlekat et al. (1994) found no significant correlations among abundances of species or individuals and various sediment-related variables, including organic and heavy-metal contaminants, in their study of tidal rivers around Washington, D.C.

Multivariate techniques have become increasingly common in studies of impacts of contamination on the faunas of estuaries, in part because of their greater ecological relevance. By integrating effects across several taxa, each with its own patterns of response to confounding environmental variables such as salinity and sediment-type, these methods also offer greater power of discrimination of anthropogenic effects from natural variation. The techniques of ordination and classification have been used in numerous estuarine studies to group sampling-sites on the basis of the composition of their fauna, with the aim of identifying differences among areas that correspond to different levels of contamination. Groupings of sites by their fauna may be correlated, formally or informally, with groupings based on environmental variables, including concentrations of contaminants. Some methods of multivariate analysis, such as canonical community ordination (CANOCO: Ter Braak, 1988) and the BIO-ENV procedure described by Clarke (1993: see below) have been designed specifically to allow such correlations.

Roper et al. (1988), for example, in their study of mudflats in the Manukau Harbour, found that the first two axes used to group sites in a canonical discriminant analysis correlated with several sediment-related variables, including the concentrations of hydrocarbons, copper, lead and zinc. In the study of the James River Estuary by Diaz (1989), preliminary cluster-analysis separated sites into groupings corresponding to the salinity-gradient of the estuary. Further classification-analysis of sites within the tidal, freshwater section of the estuary produced associations of sites consistent with analyses of water-quality variables. Schlekat et al. (1994) similarly reported general agreement between groups of sites produced by cluster analyses of faunal data and concentrations of contaminants in sediments, an outcome not seen with univariate analyses.

Bryan et al. (1987) suggested that the absence of certain species of invertebrates, present in nearby uncontaminated estuaries, from contaminated estuaries in southwest Britain was due to accumulations of heavy metals in the sediments. More objective elaborations on this approach have been developed by freshwater ecologists. Systems such as the River Invertebrate Prediction and Classifications System (RIVPACS: Wright, 1995) generate site-specific predictions of the fauna likely to occur at riverine sites in Britain in the absence of environmental disturbance. Predictions are based on several environmental features used to characterise the site and a large database of previously-surveyed, unimpacted (reference) sites, which provides correlations of species with these features. To generate predictions, the multivariate faunal data from existing sites are separated into groups using a classification technique. An ordination technique is then used to determine the combination of environmental variables that separates the groups of sites in the most similar way to the fauna-based classification. The environmental data from a new site (for example, one that is putatively impacted) are ordinated, and the position of the site in the ordination space relative to other sites is used to predict the species to be expected at the new site. Thus, the position of the new site in the ordination places it within a group of existing sites. The proportion of existing sites in the group in which each species occurred then provides an estimate of the probability of that species occurring at the new site. Impacts are identified by comparing actual and predicted faunas.

A similar system, based on RIVPACS, has been developed in Canada for prediction of faunas expected to occur in lake-sediments in the absence of disturbance. The Benthic Assessment of Sediment (BEAST: Reynoldson et al., 1995) focuses on sediment-related environmental variables at the reference sites, but also includes water-quality variables. As with RIVPACS, environmental variables that correlate strongly with the ordination of faunal data are used to predict the species (or higher taxa) likely to occur at new sites. Sites were also classified on the basis of the responses of four species of invertebrate to the sediments in laboratory toxicity tests. This provides a prediction of the toxic response to be expected in the absence of anthropogenic contamination, since the reference sites were chosen on the basis of their distance from such sources. The responses seen with sediments from the references sites are presumed to be due to 'natural' environmental factors such as the type of sediment, background concentrations of metals, etc. If a new site exhibits greater toxicity than the reference sites with which it is grouped, it is presumed to be contaminated.

Systems equivalent to RIVPACS and the BEAST have not yet been developed for estuaries. The principle restriction is the large database required to develop the model. The preliminary phase of development of RIVPACS, for example, included 268 sites and the present total is over 400 (Wright, 1995). Ninety-six sites were used in the development of the BEAST (Reynoldson, 1995). For the time being, therefore, the development and use of such models to detect ecological impacts in estuaries remains a long-term goal. Interest in an 'ESTPACS' model has been shown, however, and at least one study, in southeast Australia, has recently begun (G. Poore, Museum of Victoria, Australia, pers. comm.).

However elaborate, all correlative studies are subject to the limitation that correlation cannot demonstrate cause and effect. The possibility can never be excluded that differences in the distributions of animals are due not to the distributions of contaminants but to some other, perhaps unmeasured, covariable.

Many of the estuarine studies discussed have been done in relatively contaminated estuaries, usually with industrial inputs. In estuaries where the principle source of contamination is stormwater, detection of general impact at the small levels of contamination likely to be present has proved less easy. It is therefore unlikely that correlative methods alone will provide convincing or legally-defensible evidence of ecological effects.

iii. Experimental approaches

As mentioned earlier, demonstration that correlations between the distributions of contaminants and the distributions of organisms are evidence of cause and effect requires an experimental approach. The most widely-used group of methods are laboratory-based toxicity tests (see Rand, 1995). These include spiked-sediment bioassays (e.g. Roper and Hickey, 1994) and microcosm-studies focusing on populations or communities of species (e.g. Moverly et al., 1995). These tests permit the identification of the effects of contaminants under closely controlled experimental conditions. Because they are done in the laboratory, however, they do not reproduce the range of potentially-relevant environmental factors present in nature. Consequently, the actual effects of contaminants in the field may be different from the potential effects identified in the laboratory.

Manipulative field-experiments offer potential for exploiting the advantages of the laboratory- and correlation-based approaches while avoiding many of their short-comings. They also allow assessment of effects over longer time-scales and large spatial scales, since these scales are integrated within many of the processes that structure natural communities. Larval recruitment and predation, for example, may occur over scales of space and time that are impossible to incorporate into even the most elaborate laboratory microcosm.

The field-experimental approach, however, has not been commonly used, largely because of the practical difficulties involved. Studies of the recolonisation of experimentally-defaunated sediments have been done in estuaries (e.g. Zajac & Whitlatch, 1982a and b), including comparisons among sites differing in concentrations of contaminants (Thrush & Roper, 1988). Although there are examples of the use of manipulative experiments to identify effects of contaminants on established communities in sheltered coastal embayments such as Botany Bay, Australia (Morrisey et al, 1996), we are not aware of any in estuaries with strong gradients of salinity.

iv. Weight of evidence approach

Assessments of impact based on weight of evidence draw on evidence from several of the above methods (Chapman et al., 1992). Examples are provided by the 'sediment quality triad' (Long and Chapman, 1985) and the 'effects range' (Long et al., 1995) approaches to assessment of sediment-quality, both of which draw on evidence from the distributions of contaminants and organisms in the field and from laboratory toxicity tests. Schlekat et al. (1994) combined surveys of contaminants and other sediment-related variables and of benthic macrofauna with laboratory toxicity tests using whole sediments to identify ecological effects of contaminants in the upper estuaries of the Potomac and Anacostia Rivers (Washington, D.C.).

ECOLOGICAL EFFECTS OF STORMWATER IN ESTUARIES IN NEW ZEALAND

The objective of our current programme of research is to provide the means to predict the impacts of stormwater on New Zealand's sheltered estuaries and harbours. The first goal is to test and refine a model to predict the extent and rate of accumulation of contaminants in sediments by comparing the predicted and actual spatial variability of contaminants in two urban estuaries. The ecological significance of this chemical contamination will then be assessed through a combination of correlative (biological surveys of the same estuaries) and experimental approaches (laboratory and, in the future, field toxicity studies). We are thus taking a weight of evidence, or integrated, approach by combining comparisons of the distributions of contaminants and fauna with experimental investigation of the mechanisms underlying any correlations detected. The project includes an additional predictive element, represented by the model.

i. Prediction of the extent of accumulation of heavy metals in estuaries

In response to the needs of environmental managers in local government for methods to predict long-term impacts of catchment development, we have developed a model to estimate the accumulation in estuaries of heavy-metal contaminants derived from urban stormwater (Williamson et al., in prep.). Published information on the volumes of runoff from areas of land under different uses, and the concentrations of sediment and contaminants in that runoff, have been combined with information on the historical sequence of development of specific estuarine catchments. The resulting estimates of the cumulative inputs of these variables are combined with estimates of the volume of sediments already present in the estuary, and the concentrations of contaminants in them, to generate predictions of the concentrations of contaminants in the sediments of the estuary at increasing times after the start of catchment-development. The model assumes, furthermore, that 75% of the sediment entering the upper estuary will be deposited in a 'settling-zone' near the freshwater-input and equivalent in area to about 4% of the area of the catchment. This assumption is adapted from principles used in the design of stormwater-treatment ponds (Auckland Regional Council, 1992).

The model was initially developed to predict the accumulation of sediment-associated contaminants in 3 estuaries where proposals had recently been put forward for residential development of the catchment. These estuaries lie within the greater Auckland area, the largest and oldest urban area in New Zealand, and form part of the Waitemata Harbour, an extensive system of main and sub-estuaries.

Testing of the model required making similar predictions for nearby estuaries with catchments that have already been developed. These predictions were then tested by sampling the actual concentrations of contaminants present in those estuaries.

Methods

Two estuaries were chosen for the study, Hellyers/Kaipatiki Creek, in the Upper Waitemata Harbour, and Pakuranga Creek, which forms part of the Tamaki Estuary system, 14 km southeast of Auckland's city centre (Figure 1). Major development of the catchments of both estuaries began in the 1960s. To simplify the development of the model, the estuaries selected were relatively small.

The shore-profile in each estuary consisted of a permanently flooded, low-tide channel flanked by banks or levees of fine mud, generally about 1.5 m high (relative to the bed of the channel). Upshore from the banks, mudflats extended for varying distances (5-50 m) upshore, their lower sections at a slightly lower elevation than the top of the banks and their upper sections up to 2-3 m higher than the bed of the channel. The width of the mudflats generally increased with distance down the estuary. At the upper edge of the mudflats was a zone, roughly 5 m wide, occupied by the pneumatophores of mangroves (*Avicennia marina* var. *resinifera*). Further up the shore again, mangroves extended back to the upper shore of the estuary. The width of the mangrove-zone varied between 10 and more than 100 m.

The sediments in the low-tide channel were very variable in grain-size and invariably coarser than those on the adjacent intertidal areas, as would be expected from the larger rates of movement of water within the channel. The banks and lower flats consisted of mud or fine-sandy mud, while the upper flats, pneumatophore-zone and mangrove-zone contained fine, muddy sediment. Generally, the sediment was of fairly similar texture over the whole area from the banks to the mangrove-zone.

Four locations were selected in each estuary, two within the settling-zone (the area equivalent to 4% of the area of the catchment used in the model: see Williamson et al., in prep.) and two outside it. At each location, the profile of the shore was divided into a number of 'heights' on the basis of topography. Height 1 represented the bed of the low-tide channel, height 2 the top of the bank, heights 3-5 the lower, mid and upper mudflats, respectively, height 6 the pneumatophore-zone and height 7 the mangrove-zone. Heights 3 and 5 were only defined at locations with relatively wide (>10 m) intertidal flats. At each height on the shore, three randomly-located 'sites' were chosen, about 10 m apart. At each of these, two randomly-located core samples were collected, roughly 0.5 m apart. The samples were taken using plastic tubes, 5-cm diameter and 15 cm deep. Metals were extracted with 2 M HCl and analysed by flame AAS. This design of sampling permits differences in concentrations of contaminants and other variables among locations and between locations within and without the settling-zone to be distinguished from variations among heights on the shore and from smaller-scale spatial variation (among sites and among replicate samples).

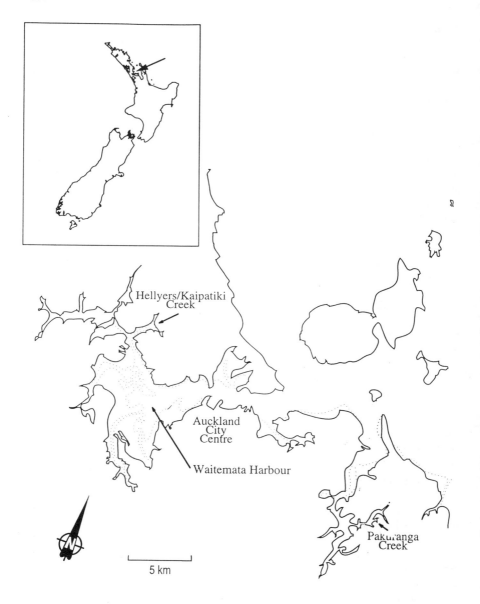

Figure 1. Map of Waitemata Harbour showing
Hellyers/Kaipatiki and Pakuranga Creeks

Results

Concentrations of copper, lead and zinc in Pakuranga Creek were generally larger than in Hellyers/Kaipatiki Creek (Table 1). The maximum concentrations of copper and zinc recorded in Hellyers/Kaipatiki Creek during the present study were smaller than the effects range-low guidelines proposed by Long et al. (1995), representing concentrations at which adverse biological effects are occasionally reported (Table 1). Maximum concentrations of lead in Hellyers/Kaipatiki Creek and of all three metals in Pakuranga Creek, however, exceeded the effects range-low guidelines, indicating a potential for ecological impact of stormwater-derived contaminants on communities of benthic invertebrates in these estuaries. The studies reviewed by Long et al (1995) in deriving their guidelines included various measures of biological effects, ranging from changes in cells and tissues to changes in the structures of benthic communities. None of the values recorded during the present study exceeded the effects range-median concentrations of Long et al. (1995), with which biological effects are frequently associated. Since Long et al. (1995) explicitly excluded from their review any studies which did not use strong-acid digestions, it is perhaps more appropriate to compare their guidelines with maximum concentrations from Hellyers/Kaipatiki and Pakuranga Creeks after applying a correction for strong-acid digestion (Williamson, unpublished data: see Table 1). In this case, all concentrations are still below the effects range-median values, but the corrected concentrations of all three metals from both creeks exceed the effects range-low values. This further accentuates the potential for chronic ecological effects of accumulated contaminants from stormwater.

Correspondences between observed and predicted concentrations were good for Pakuranga Creek but poor for Hellyers/Kaipatiki (Table 1). This was perhaps due to the relatively large intertidal area of the latter, and the longer period of exposure during low tide. During low tide, sediment and associated contaminants are carried down the low-tide channel and outside the settling-zone. Some of this material will probably return on the incoming tide, but the remainder will be dispersed outside the settling-zone, resulting in lower concentrations than predicted within the zone. There were generally differences among locations and among sites within those locations for all three metals in both Hellyers/Kaipatiki and Pakuranga Creeks (Figures 2 and 3; Table 2). Differences among locations followed a trend of increasing concentration from downstream to upstream locations. In Hellyers/Kaipatiki Creek, the concentrations of lead and zinc were also different between locations inside the settling-zone described in the model and those outside (Table 2). There were no clear patterns of variation among heights on the shore.

Table 1. Measured and predicted concentrations of metals in sediments in Hellyers/Kaipatiki and Pakuranga Creeks. Measured concentrations are given as the mean (±SE) of all samples (except low-tide channel) from locations 3 and 4 in Hellyers/Kaipatiki Creek and locations 6 and 11 in Pakuranga Creek, and as the range of values found throughout each estuary. Concentrations corrected for strong acid digestion (see text) are also given. Guidelines refer to the 'effects range-low' (ER-L) and 'effects range-median' (ER-M) concentrations of Long et al. (1995).

		Copper	Lead	Zinc
Hellyers/Kaipatiki	mudflats	19.4 (0.4)	53.8 (1.4)	113.7 (2.1)
	range	8.6-24.2	13.2-70.0	34.3-141.5
	corrected	16.6-36.2	13.2-95.0	58.3-191.5
	predicted	28	78	172
Pakuranga	mudflats	36.5 (1.4)	67.8 (2.2)	232.9 (8.3)
	range	5.7-53.4	22.1-87.1	84.1-294.9
	corrected	13.7-65.4	22.1-112.1	108.1-344.9
	predicted	28	69	225
Guidelines	ER-L	34	46.7	150
	ER-M	270	218	410

Table 2. Summary of results of analyses of variance of the concentrations of metals in sediments at sampling locations in Hellyers/Kaipatiki and Pakuranga Creeks. * $p<0.05$; ** $p<0.01$; *** $p<0.001$.

| | | Hellyers/Kaipatiki | | | | | |
| | | Copper | | Lead | | Zinc | |
Source	df	F	p	F	p	F	p
Zone = Z	1	5.69	ns	39.95	*	61.93	*
Location(Z) = L	2	8.15	**	4.05	*	1.65	ns
Height = H	3	0.45	ns	1.14	ns	1.69	ns
ZxH	3	1.91	ns	2.44	ns	1.27	ns
L(Z)xH	6	1.12	ns	2.10	ns	3.64	**
Site(L(Z)xH)	32	4.22	***	2.44	**	3.46	***
Residual	48						

| | | Pakuranga | | | | | |
| | | Copper | | Lead | | Zinc | |
Source	df	F	p	F	p	F	p
Zone = Z	1	1.96	ns	0.60	ns	1.37	ns
Location(Z) = L	2	39.09	***	66.30	***	54.10	***
Height = H	3	2.21	ns	3.97	ns	1.79	ns
ZxH	3	2.00	ns	7.53	*	2.70	ns
L(Z)xH	6	0.52	ns	0.65	ns	1.24	ns
Site(L(Z)xH)	32	3.48	***	2.03	*	2.18	**
Residual	48						

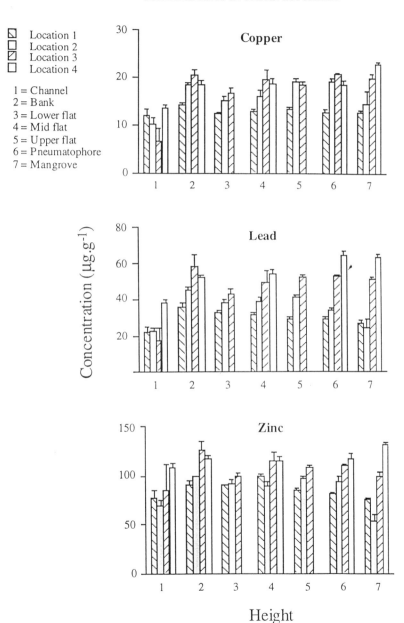

Figure 2. Distribution of heavy metals in Hellyers/Kaipatiki Creek. Values shown are means (± SE) of 2 cores at 3 sites at each combination of height and location. See text for details of design of sampling. Location 1 lies at the downstream end of the estuary and location 4 at the upstream end.

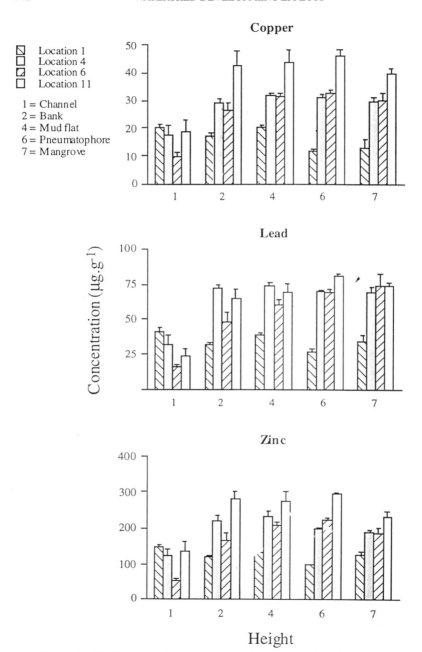

Figure 3. Distribution of heavy metals in Pakuranga Creek. Values shown are means (± SE) of 2 cores at 3 sites at each combination of height and location. See text for details of design of sampling. Location 1 lies at the downstream end of the estuary and location 11 at the upstream end.

ii. Assessment of ecological effects of contaminants

Methods

Samples were collected at six locations along the lengths of each of Hellyers/Kaipatiki and Pakuranga Creeks. At each location, a sampling-area 50 m (parallel to the shore) by 20 m was marked out and samples taken at each of four randomly-located points within the area. The sampling-points were selected using random-numbers, with the restriction that no two points could be closer than 15 m, to ensure distribution of the samples over the area. At each of the four points, a core sample (13-cm diameter, 15-cm deep) was collected for analysis of environmental variables and three for enumeration of the macrofauna in the sediment. One of the macrofaunal cores was taken immediately adjacent to the environmental core, the others approximately 1 m either side.

Faunal cores were sieved (0.5-mm mesh) immediately and fixed in 70% isopropyl alcohol. The animals present were identified to the lowest possible level of taxonomic resolution. The remaining cores were analysed for concentrations of copper, lead, iron, zinc (by flame AAS after extraction with 2 M HCl) and organic matter (combustion at 400 $^{\circ}$C for 4 hours).   Particle-size distribution of the sediments was assessed by sieving and expressed as percentages of gravel, sand and mud. Pore-water was extracted from the sediments by centrifugation and analysed for pH, salinity and concentrations of iron, manganese and zinc (concentrations of copper and lead in pore-water were below detection limits).   Analyses of organochlorines and PAH are in progress.

The multivariate statistical package PRIMER (Clarke, 1993), developed by the Plymouth Marine Laboratory in the United Kingdom, was used to analyse the data. Non-metric multidimensional scaling (nMDS) was used to group samples and sites based on the relative (ranked) similarities of their faunas. Similarities among samples were expressed as Bray-Curtis indices (see Clarke, 1993 for details). Sites were ordinated on the basis of the measured environmental variables using principle components analysis (PCA). To quantify the correspondence between ordinations based on fauna and those based on environmental variables, the ranked Bray-Curtis similarity values among samples were correlated with the Euclidean distance values from the corresponding PCA analysis, using a component of the PRIMER package, BIO-ENV (Clarke & Ainsworth, 1993). Details of these procedures and their rationale are given by Clarke (1993).

Results

A total of 28 taxa were recorded from Hellyers/Kaipatiki Creek, of which 11 were polychaetes, 8 molluscs and 6 crustaceans. The most abundant taxa were capitellid, cossurid, nereid, paraonid and spionid polychaetes and corophiid amphipods. In Pakuranga Creek, a total of 26 taxa were recorded, 11 of them polychaetes, 6 molluscs and 7 crustaceans. The most abundant taxa were the same as in Hellyers/Kaipatiki Creek, although capitellid polychaetes were less numerous. Mean numbers of taxa per sample (133 cm^2) at each location increased down each estuary, ranging from 3.4 (± 0.6 SE, n=12) at location 1 (upstream) to 8.1 (± 0.6) at location 6 (downstream) in Hellyers/Kaipatiki Creek and from 3.9 (± 0.4) at location 1 to 9.3 (± 0.7) at location 6 in Pakuranga Creek. There was considerable patchiness in the distribution of animals within each location (i.e. an area 50 m by 20 m).

For clarity, ordinations of the locations in each creek have been plotted as the sums of each taxon from the three cores at each of the four points within each sampling-area. Ordination of the locations in Hellyers/Kaipatiki Creek (Figure 4.a) did not separate them particularly clearly, although samples from upstream locations tend to lie on the upper-right part of the plot and those from downstream locations in the lower-left part. Locations in Pakuranga Creek separated more clearly (Figure 4.b), with those from upstream locations on the right of the plot and those from downstream on the left. The patchiness of the faunas within each sampling-area is indicated in the plots by the spacing among the four points within each area.

The BIO-ENV analyses were done in two forms. Variables related to pore-water could only be measured on 2 of the cores from each location, while sediment-related variables were measured in all 4 cores. First, therefore, the 3 faunal cores were compared to the sediment-variables, applying the same sediment-variables to all of the 3 faunal cores collected around each environmental core (analysis A in Table 3). Second, each of the 2 environmental cores for which pore-water variables were measured at each location was compared with the single faunal core collected next to it (analysis B in Table 3).

The individual environmental variable that correlates best with the faunal groupings from the Hellyers/Kaipatiki data was the concentration of iron in the sediment (Table 3). The best overall correlation, however, was that provided by a combination of concentration of lead and iron, suggesting that the fauna were responding to the presence of contaminants in the sediments. Of the 3 heavy metals, lead showed the largest concentration relative to its effects range-low value in Hellyers/Kaipatiki sediments (Table 1). When variables related to pore-water were introduced into the analyses, however, the individual variables that correlated best with the faunal groupings were the pH of the pore-water and the concentration of iron in the pore-water. The best overall correlation was provided by a combination of sand and the pH and iron-concentration of the pore-water. This

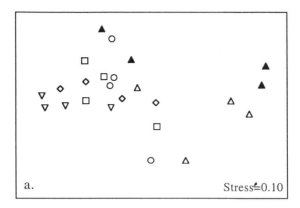

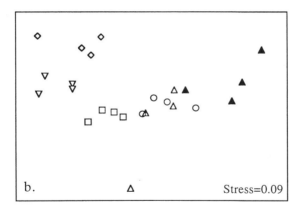

Figure 4. Non-metric MDS ordination of the four sets of samples (totals from 3 cores) at each of the 6 locations in a. Hellyers/Kaipatiki and b. Pakuranga Creeks, based on infaunal communities. Solid triangles - location 1 (upstream); open triangles - location 2; circles - location 3; squares - location 4; diamonds - location 5; inverted triangle - location 6 (downstream). Values of Kruskall's stress are given (see Clarke, 1993).

Table 3. Rank correlation values (unweighted Spearman's coefficients) for combinations of environmental variables yielding the 'best matches' of faunal and environmental similarity matrices. The combination with the largest overall value (in bold) and an illustrative selection of other combinations are shown. See Clarke & Ainsworth (1993) for details of the method. Sediment-related variables are: proportion of gravel (*Gr*), sand (*Sand*) mud (*Mud*) and organic matter (*%Org*) and concentrations of Cu, Pb, Zn and Fe. Pore-water related variables are pH, salinity and concentrations of Fe (*PFe*) and Mn (*PMn*). See text for explanation of *A* and *B*.

Hellyers/Kaipatiki Creek

A

#				
1	Fe 0.375	Pb 0.333	Cu 0.256	Sand 0.133
2	**Pb, Fe** **0.427**	...		
3	Zn, Pb, Fe 0.382	...		

B

#				
1	pH 0.528	PFe 0.514	Pb 0.290	...
2	Sand, pH 0.572	Mud, pH 0.572	Pb, pH 0.569	
3	**Sand, pH, PFe** **0.595**	**Mud, pH, PFe** **0.595**	...	

Pakuranga Creek

A

#				
1	Pb 0.362	Gr 0.344	Mud 0.320	Sand 0.304
2	**Gr, Pb** **0.521**	...		
3	Gr, Mud, Pb 0.509	Gr, %org, Pb 0.509	Gr, Sand, Pb 0.508	...
4	Gr, Mud, %Org, Pb 0.485	Gr, Sand, %Org, Pb 0.483	Gr, Sand, Mud, Pb 0.481	...

B

#					
1	Mud 0.271	Pb 0.252	Sand 0.244	pH 0.239	PMn 0.158
2	**Mud, Pb** **0.359**	...			...
4	Sand, %org, Pb, PMn 0.358	Mud, %org, Pb, PMn 0.358	%org, Pb, pH, PMn 0.354	...	

correlation was much better than the best provided by the sediment-related variables alone (Table 3). These variables may perhaps influence the fauna through their metal-binding properties.

Among samples from Pakuranga Creek, concentration of iron and sediment-texture (proportion of gravel) appeared to be the best 'explanatory' sediment-related variables. When variables related to pore-water were introduced, the best overall correlation was provided by a combination of the concentration of lead and the proportion of mud. The correlation in this case was not as good as with the sediment-related variables alone.

Overall, the BIO-ENV analyses indicated that the predominant 'explanatory' variables among those measured in sediments from Pakuranga Creek were the concentration of lead and those related to sediment-texture. In the case of Hellyers/Kaipatiki Creek, the best 'explanatory' variables were sediment-texture, concentration of lead in the sediment, concentration of iron in sediment and pore-water and the pH of the pore-water. The apparent importance of iron and pH may perhaps relate to their influence on the bioavailability of heavy metals (Luoma, 1990). Alternatively, the correlation with iron may be a consequence of the effects of bioturbation. The improved correlation seen with samples from Hellyers/Kaipatiki Creek when variables related to pore-water were included was not seen with Pakuranga samples. When pore-water salinity alone was added to sediment-related variables, there was no change in the correlations compared with sediment-related variables alone. This suggests that variation in salinity was not exerting a strong influence on the fauna living in the sediments in those parts of both estuaries that were sampled.

iii. Continuing and future work

A similar programme of sampling has also been conducted in two unurbanised estuaries in the Auckland region. This will provide information on concentrations of contaminants in estuaries that have not been subjected to inputs of stormwater and on the relationship between benthic faunal communities and environmental variables under such conditions. Identification of any species present in unurbanised estuaries that are not found in urbanised ones will, in addition to providing circumstantial evidence of impact, suggest suitable candidate-species for laboratory toxicity studies. Development of chronic toxicity tests using indigenous, estuarine species is currently underway (Roper & Hickey, 1994; Nipper et al., 1995; Roper et al., 1995). Other planned future work includes studies of recolonisation of defaunated sediments from contaminated and uncontaminated estuaries and field studies of effects of experimentally-enhanced concentrations of heavy metals on benthic communities (Morrisey et al., 1995, 1996).

ACKNOWLEDGEMENTS

We are grateful to the Aukland Regional Council for partial funding of this study, and for an important impetus to its initiation.

REFERENCES

Auckland Regional Council, 1992. Selection of stormwater treatment volumes for Auckland. Prepared by Beca Carter Hollings & Ferner Ltd for Environment and Planning Division, ARC. Environment and Planning Division Technical Publication No. 4.

Bryan, G.W., Gibbs, P.E., Hummerstone, L.G. & Burt, G.R., 1987. Copper, zinc and organotin as long-term factors governing the distribution of organisms in the Fal Estuary in southwest England. Estuaries, 10: 208-219.

Bryan, G.W. & Langston, W.J., 1992. Bioavailability, accumulation and effects of heavy metals in sediments with special reference to United Kingdom estuaries: a review. Environmental Pollution, 76: 89-131.

Chapman, P.M., Power, E.A. & Burton, Jr, G.A., 1992. Integrative assessments in aquatic ecosystems. In, Burton, Jr, G.A. (ed.), Sediment toxicity assessment. Lewis Publishers, Boca Raton, Florida, pp. 313-340.

Clarke, K.R., 1993. Non-parametric multivariate analyses of changes in community structure. Australian Journal of Ecology, 18: 117-143.

Clarke, K.R. & Ainsworth, M., 1993. A method of linking multivariate community structure to environmental variables. Marine Ecology Progress Series, 92: 205-219.

Clements, W.H. & Kiffney, P.M., 1995. The influence of elevation on benthic community responses to heavy metals in Rocky Mountain streams. Canadian Journal of Fisheries and Aquatic Science, 52: 1966-1977.

Dauer, D.M., Luckenbach, M.W. & Rodi, Jr, A.J., 1993. Abundance biomass comparison (ABC method): effects of an estuarine gradient, anoxic/hypoxic events and contaminated sediments. Marine Biology, 116: 507-518.

Diaz, R.J., 1989. Pollution and tidal benthic communities of the James River Estuary, Virginia. Hydrobiologia, 180: 195-211.

Green, R.H., 1979. Sampling design and statistical methods for environmental biologists. Wiley & Sons, New York, 257pp.

Long, E.R. & Chapman, P.M., 1985. A sediment quality triad: measures of sediment contamination, toxicity and infaunal community composition in Puget Sound. Marine Pollution Bulletin, 16: 405-415.

Long, E.R., MacDonald, D.D., Smith, S.L. & Calder, F.D., 1996. Incidence of adverse biological effects within ranges of chemical concentrations in marine and estuarine sediments. Environmental Management, 19: 81-97.

Luoma, S.N., 1990. Processes affecting metal concentrations in estuarine and coastal marine sediments. In, Furness, R.W. & Rainbow, P.S. (eds), Heavy metals in the marine environment, CRC Press, Boca Raton, Florida, pp. 51-66.

McLusky, D.S., 1981. The estuarine ecosystem. Blackie, Glasgow, 150pp.

Morrisey, D.J., A.J. Underwood & L. Howitt, 1996. Effects of copper on the faunas of marine soft-sediments: a field experimental study. Marine Biology, 125:199-213.

Moverley, J.H., Ritz, D.A. & Garland, C., 1995. Development and testing of a meiobenthic mesocosm system for ecotoxicological experiments. National Pulp Mills Research Program, Technical Report No. 14, Canberra: CSIRO, 117pp.

Nipper, M.G. & Roper, D.S., 1995. Growth of an amphipod and a bivalve in uncontaminated sediments: implications for chronic toxicity assessments. Marine Pollution Bulletin, 31: 424-430.

Rand, G.M. (ed.), 1995. Fundamentals of aquatic toxicology. 2nd edition, Taylor & Francis, Washington, D.C., 1125pp.

Reynoldson, T.B., Bailey, R.C., Day, K.E. & Norris, R.H., 1995. Biological guidelines for freshwater sediment based on BEnthic Assessment of SedimenT (the

BEAST) using a multivariate approach for predicting biological state. Australian Journal of Ecology, 20: 198-219.

Roper, D.S., 1990. Benthos associated with an estuarine outfall, Tauranga Harbour, New Zealand. New Zealand Journal of Marine and Freshwater Research, 24: 487-498.

Roper, D.S. & Hickey, C.W., 1994. Behavioural responses of the marine bivalve *Macomona liliana* exposed to copper- and chlordane-dosed sediments. Marine Biology, 118: 673-680.

Roper, D.S., Nipper, M.G., Hickey, C.W., Martin, M.L. & Weatherhead, M.A., 1995. Burial, crawling and drifting behaviour of the bivalve *Macomona liliana* in response to common sediment contaminants. Marine Pollution Bulletin, 31: 471-478.

Roper, D.S., Thrush, S.F. & Smith, D.G., 1988. The influence of runoff on intertidal mudflat benthic communities. Marine Environmental Research, 26: 1-18.

Schlekat, C.E., McGee, B.L., Boward, D.M., Reinharz, E., Velinsky, D.J. & Wade, T.L., 1994. Tidal river sediments in the Washington, D.C. area. III. Biological effects associated with sediment contamination. Estuaries, 17: 334-344.

Ter Braak, C.J.F., 1988. CANOCO - a FORTRAN program for canonical community ordination by partial detrended canonical correspondence analysis, principle components analysis and redundancy analysis. Groep Landbouwwiskunde, Technical Report LWA-88-02, Wageningen, The Netherlands, 94pp.

Thrush, S.F. & Roper, D.S., 1988. The merits of macrofaunal colonisation of intertidal mudflats for pollution monitoring: a preliminary study. Journal of Experimental Marine Biology and Ecology, 116: 219-233.

Underwood, A.J., 1991. Beyond BACI: experimental designs for detecting human environmental impacts on temporal variations in natural populations. Australian Journal of Marine and Freshwater Research, 42: 569-587.

Watershed Determinants of Ecosystem Functioning

Richard R. Horner[1,2], M. ASCE; Derek B. Booth[2]; Amanda Azous[3]; and
Christopher W. May[2]

Abstract

By the mid-1980s it was clear that urban stormwater runoff was strongly implicated
in the alteration of streams and freshwater wetlands in the Puget Sound Basin of
Washington state to ecosystems lower in biologically diversity and productivity of
the species most valued by society. It was also apparent that the causes of these
modifications were rooted in watershed hydrology and sediment transport as well as
reduced water quality. Recognition of these connections and the rapid pace of
development in the region stimulated research to define the linkages among stream
and wetland habitat structure, conditions in the surrounding landscapes, and the
associated biological responses. One project monitored watershed and riparian
zone conditions, flow, physical habitat characteristics, water quality, benthic
macroinvertebrates, and fish in 31 reaches on 19 low-order streams, representing a
gradient of urbanization, over a three-year period. A second project followed 19
palustrine wetlands during an eight-year period when urbanization began or
increased in the watersheds of about half, while the remainder were essentially
unchanged. Overall, the findings of these projects agree that the effects of modified
hydrology accompanying urbanization exert the earliest and, at least initially, the
strongest deleterious influences on the freshwater ecosystems studied. Furthermore, the results agree that the steepest rates of decline in biological functioning of
both streams and wetlands, and the conditions necessary to support that functioning, occur as urbanization increases total impervious land cover from 0 to about 6
percent, unless mitigated by extensive riparian protection, management efforts, or
both. Thereafter, the decline proceeds at a slower rate as impervious cover
increases further. Functioning at the highest level (e. g., stream benthic index of
biotic integrity, B-IBI, > 35) with very low imperviousness drops by roughly half

[1]230 NW 55th Street, Seattle, WA 98107; [2]Center for Urban Water Resources
Management, University of Washington, Box 352700, Seattle, WA 98195-2700;
[3]P. O. Box 530, Olga, WA 98279.

(e. g., B-IBI < 20) as urbanization progresses to about 45 percent impervious cover. The results suggest that management concentrate on preservation through land use controls where resource values are high, and on prevention of further degradation and enhancement, especially directed at riparian zones, where functioning is impaired but not lost.

Introduction

Ecological Functions and Problems of Pacific Northwest Streams and Wetlands

In the Puget Sound region of the Pacific Northwest, in common with many areas in North America, urban development is rapidly extending into areas containing much of the remaining aquatic resource base. In this region the aquatic ecosystems most directly affected by the prevailing pattern of watershed development are relatively small streams and wetlands in the lowlands surrounding Puget Sound. These streams are critical spawning and rearing locations for seven species of native anadromous and resident salmonid fish of great ecological, cultural, social, and economic importance. Despite its value, this resource is in considerable jeopardy of being lost. More than 100 native salmon and anadromous trout runs in the region are already extinct, and approximately 200 more are seriously threatened (Nehlsen et al. 1991). Wetlands frequently form the headwaters of lowland spawning and rearing streams and comprise substantial portions of their riparian zones. Wetlands also supply other crucial ecological functions and social values, including diverse vegetation; habitat for amphibians, birds, and small mammals; open space; and recreational and educational opportunities.

Ten years ago environmental scientists and managers in the region lacked a foundation to formulate strategies for arresting the decline of these key freshwater resources. Many of them contributed to several research projects over the intervening years to narrow the gaps in knowledge of ecosystem functioning and improve management ability. As the first stage of these efforts is nearing completion, this paper was prepared to summarize the overall approach, key representative results, and implications for managing the region's lowland watersheds to accomplish stream and wetland resource protection goals.

Watershed urbanization effects on streams, in particular, have been fairly well documented and include numerous and extensive changes in the hydrologic regime, channel morphology, riparian corridor, and water quality. These alterations in sum produce a very different habitat structure than the one in which aquatic organisms have evolved, usually resulting in decline or complete loss of at least some populations and substantial modification of biological communities. A full account of the many individual changes is beyond the scope of this paper and may be found elsewhere (e. g., Booth 1990, 1991; Booth and Jackson 1994; Horner et al. 1994).

In analyses of aquatic ecosystem impacts, watershed urbanization has been expressed most commonly in terms of the proportion of watershed area covered by impervious surface. Some past work has identified impervious cover of 10 +/- several percent as the level at which stream ecosystem impairment becomes evident. For example, Klein (1979) used a variety of hydrologic, water quality, biological information to conclude that serious stream quality impairment can be prevented if watershed imperviousness does not exceed 15 percent, or 10 percent for more sensitive ecosystems with self-sustaining trout populations. Booth and Reinelt (1993) found a very similar relationship for Puget Sound Basin lowland streams. Shaver et al. (1995) found decline in macroinvertebrate community indices in Delaware creeks with 8 percent watershed imperviousness and substantially more decrease above 15 percent. They demonstrated that the impact was linked more to physical habitat than water quality degradation.

Toward a Systematic Approach to Managing Watershed and Aquatic Ecosystem Change

With recognition of the web tying together landscapes and aquatic habitats and their inhabitants has come an interest in defining the functional relationships between discharges from watersheds and aquatic resources well enough to avoid or minimize resource losses through management intervention. In this scenario there would be reasonable confidence that a goal to maintain a given organism or community at a specified level could be met by sustaining a certain set of habitat measures, which in turn depend on established values of particular watershed measures. To realize the promise of this approach, of course, it will be necessary to set well conceived goals that set the direction and parameters of the entire management thrust.

Needed for progress in systematically managing watersheds and their aquatic resources are meaningful and convenient measures of watershed conditions, biological "health," and the habitat conditions on which biota depend. Moreover, these measures must be suitable for defining the linkages among these ecosystem components. For expressing watershed conditions, the traditional use of impervious cover must be supplemented by measures that express other facets of the landscape (e. g., riparian zone and drainage system characteristics). Perhaps most critical to progress in describing habitat structure is to identify on an ecoregional scale the variables with the most utility for assessing habitat relationships with watershed circumstances on the one hand and biological responses on the other. The work reported here makes an attempt to advance the measurement and interpretation of both watershed and habitat conditions.

The last 15 years have seen substantial progress in expressing biological community attributes in useful forms for comprehensive watershed studies. Major developments include the index of biotic integrity (IBI; Karr 1981; Karr and Dudley 1981; Karr et al. 1986; Karr 1991; Kerans and Karr 1994, Fore et al. in review), the

U. S. Environmental Protection Agency's (USEPA) Rapid Bioassessment Protocols (Plafkin et al. 1989), and the state of Ohio's biological criteria (Yoder 1991). Further IBI development was performed as part of the study discussed in this paper (Kleindl 1995).

RESEARCH PROGRAM FORMAT

Research Scope

Several research projects have been performed over the last 10 years to identify watershed determinants of freshwater ecosystem character and functioning in the Puget Sound region. These projects have applied the systematic approach outlined in the previous section in which watershed, physical and chemical habitat, and biological variables have been measured and linked with the goal of understanding and learning better how to manage watershed and aquatic ecosystem change. The projects providing most of the data base for this paper are:

- A study of the attributes of lowland streams across a gradient of urbanization, again with an underlying interest in better managing streams, their resources, and the discharges to them from diffuse sources on the landscape (1994-1996). This research is being performed by the University of Washington with a grant from the Washington Department of Ecology's Centennial Clean Water Program and in-kind support from several local governments.

- An investigation of the effects of urbanization, especially urban stormwater discharge, on freshwater wetlands and how to manage wetlands and stormwater for optimal aquatic resource protection (1986-1996). This research is being performed by a cooperative unit involving the King County Natural Resources Department and its predecessor agencies and the University of Washington, with major funding from the Centennial Program, USEPA Region X, and a variety of support from several local governments.

The paper also takes advantage of a series of studies of the habitat and fish resources in lowland streams in the urban and urbanizing area of King County, Washington, performed by King County Surface Water Management Division. This agency has been a participant and contributor of material support and data to both research efforts and has shared and utilized the findings produced by those projects.

Experimental Designs

Stream Research

The design of the stream study was to establish attributes of the watersheds contributing to selected stream reaches and then to make a number of measurements of the riparian, habitat, and biological conditions within those reaches. The concept underlying the study, simply illustrated as follows, was that watershed and riparian characteristics determine habitat conditions, which, in relation to evolved organism preferences and tolerances, set the composition of the biological communities:

$$\text{Watershed and} \quad \Rightarrow \quad \text{Habitat} \quad \Rightarrow \quad \text{Biota}$$
$$\text{riparian characteristics} \qquad\qquad \text{conditions}$$

Data analysis was directed at establishing the linkages represented by the arrows, as well as the less direct connections between watershed characteristics and biota.

Researchers and governmental participants in the research selected streams and study reaches from over 100 candidates that best met 15 criteria expressing such factors as representativeness of Puget Sound lowland stream conditions and resources, ability to cover a range of watershed urbanization from near zero to more than 50 percent imperviousness, and data availability. All study reaches are of the same general scale, being of first, second, or third order. They have mean annual discharges in the approximate range 0.1-1.7 m^3/s (3.5-60 ft^3/second); lie at altitudes of less than 150 meters above sea level; and have average stream gradients under 5 percent.

Site selection emphasized contributing catchments in the early stages of urbanization, when it was hypothesized that the most rapid change in ecological conditions occurs, or with very little to no urbanization. These latter catchments represent what is considered to be the maximum attainable ecological function in Puget Sound lowland streams. Until recently, these "reference" sites have experienced relatively little human activity after being logged approximately 100 years ago. Watersheds contributing to the 31 study reaches monitored since 1994 exhibit the following distributions of TIA: 0-5%--8 sites, 5-10%--7 sites, 10-20%--3 sites, 20-30%--3 sites, 30-50% 6 sites, and > 50%--4 sites.

The study watersheds cover two distinctly different geologic conditions. All but two are underlain by glacial till, in which continental glaciation (15,000 years b. p.) deposited a dense hardpan, now approximately 1 meter beneath the surface and overlain by a loamy soil. The other watersheds are underlain by glacial outwash, with surface coarse material providing direct hydraulic communication between the surface and the regional aquifer. In the undisturbed forested condition glacial till catchments are capable of providing precipitation storage on the order of 15 cm in the overlying soil, and the outwash areas much more (Wigmosta et al. 1994).

Suburban development of routine density and style with the stormwater management standards prevailing in King County between 1979 and 1990 reduced storage by roughly 90 percent. New standards taking effect in 1990 in that county and elsewhere in the study area would recover more storage but still provide no more than 25 percent of the original amount (Barker et al. 1991; Booth and Jackson 1994). Because the new standards affected little new development between 1990 and 1994, conditions in this study more reflect the pre-1990 situation; i. e., little effective stormwater management.

Monitoring occurred over three years in 31 reaches on 19 streams and consisted of determination of watershed land cover and other attributes, riparian zone dimensions and characteristics, a host of physical habitat measurements, flow and water quality under storm and base flow conditions, bed sediment quality, benthic macroinvertebrate community composition, and salmonid fish usage. Habitat and riparian zone assessments were conducted on 120 segments, representative of local physiographic and land use conditions and approximately 1 km in length, extending between study reaches and headwaters. Additional details on study design and the specific monitoring methods are given by Olthof (1994), Bryant (1995), Kleindl (1995), and May (1996).

Wetlands Research

The wetland and stream study designs were similar, with one exception. While the stream research was confined within a fairly short stretch of time, during which little land use change occurred, the wetland study extended long enough to observe the effects of land use changes in some of the contributing watersheds. Those that did not change were regarded as references. As with the stream study, wetland research sites were chosen according to specific criteria, in this case from approximately 150 candidates. The complete set of activities over all years of monitoring involves 19 wetlands. An additional seven wetlands served for special studies. All lie at altitudes of less than 150 meters above sea level. Wetland areas range from 0.61 to 12.55 ha. Contributing watersheds of the 19 principal study sites exhibited the following TIA distributions at the beginning and end of the project: 0-5%--8 sites in 1989, 6 in 1995; 5-10%--3 sites in 1989, 2 in 1995; 10-20%--4 sites in 1989, 5 in 1995; 20-30%--2 sites in 1989, 3 in 1995; 30-50%--1 site in 1989, 2 in 1995; and > 50%--1 site in 1989, 1 in 1995.

Monitoring encompassed determination of watershed land cover and other attributes; wetland morphology; water level fluctuation patterns; water and sediment quality; and insect, amphibian, bird, and small mammal community characteristics during 5 years over the period 1986-1995. Like the stream research, the wetlands project investigated associations among those system components. Additional details on study design and the specific monitoring methods are given by King County Resource Planning Section (1987, 1988), Reinelt and Horner (1990, 1991), Cooke (1991), Taylor (1993), and Richter and Azous (1995).

Research Results and Interpretations

Stream Research

Stream Biology in Relation to Watershed and Riparian Characteristics

This discussion considers observed linkages between the source and response components of the system, letting aside for the moment the intermediate habitat component:

Watershed and	$\Rightarrow$	Habitat	$\Rightarrow$	**Biota**
riparian characteristics		conditions		

Throughout this paper watershed condition is characterized by impervious cover. Analyses demonstrated that the relationships discussed are very similar if watershed condition is alternatively expressed by road density (km/km^2 of watershed area, May 1996).

The ecological condition and functioning of the benthic macroinvertebrate community was expressed in terms of a Benthic Index of Biotic Integrity (B-IBI), which accounts for the relative presence in the respective reaches of certain taxa and trophic groups with varying tolerances to stress (Kleindl 1995). Figure 1 portrays the relationship between B-IBI values computed in 1994 and 1995 and urbanization level, expressed by total impervious area as a percentage of the total watershed (% TIA). Only reaches with TIA < 3.9 percent exhibited an index of 35 or greater. All B-IBI values of at least 25 were associated with watersheds having no more than 11 percent impervious, with eight notable exceptions. These eight points (B-IBI = 25-31) were computed for reaches on two streams having contributing catchments with 25-34 percent TIA. Despite the moderately high level of urbanization in these cases, these streams have more of their riparian areas in intact wetlands (17.2-21.5 percent) than all but one of the 19 streams and have overall wider riparian zones (~ 60 percent > 30 meters wide) than most cases of similar urbanization level. These observations give an indication that maintenance of the adjacent stream buffer zone may help ameliorate the effects of more distant urbanization.

Setting aside these eight points, the general shape of the curve in Figure 1 indicates a relatively rapid rate of decline in biological function as impervious area increases to about 8 percent, following which the rate of decline appears to slow as urbanization increases further. It appears from these data to be probable that Puget Sound lowland streams would have a B-IBI of 15 or less with more than 45 percent imperviousness.

Among the salmonid fish, the research concentrated on the coho salmon (*Oncorhynchus kisutch*), the species arguably most vulnerable to urban runoff effects. Coho especially rely on small lowland creeks and is the only Northwest salmon species that over-winters in fresh water as a juvenile (Nickelson et al. 1992), dwelling primarily in pools (Bisson et al. 1988). In urban streams coho suffer from loss of pools through increased sedimentation and less large woody debris (LWD). On the community level, urbanization seems to alter the relationship between coho salmon and cutthroat trout (*Oncorhynchus clarki*), another salmonid species with far less social and economic recognition. These two species are often in competition for food and habitat space, and adult cutthroat prey on juvenile coho. Coho tend to dominate in less urbanized watersheds, while the advantage shifts to cutthroat with increasing development (Perkins 1982, Richey 1982, Steward 1983, Scott et al. 1986).

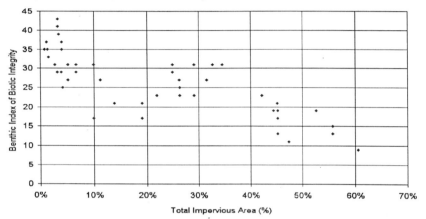

Figure 1. — Benthic Index of Biotic Integrity in Puget Sound Lowland Streams Over a Gradient of Watershed Impervious Land Cover

Figure 2 shows salmonid fish data available for a subset of the study streams, expressed in terms of the ratio of coho salmon to cutthroat trout abundance. The less tolerant coho dominate only with a small amount of urbanization, the level of which can not be fixed because of a gap in the data between 4 and 10 percent TIA. They appear to have a slight dominance in the 10-15 percent TIA range, following which dominance switches to the trout. Lucchetti and Fuerstenberg (1993) earlier found a similar trend of decreasing coho abundance with increasing imperviousness in a partially overlapping set of King County creeks.

Habitat Conditions in Relation to Watershed and Riparian Characteristics

This discussion considers the direct linkages between the source and physical and chemical habitat components of the system:

| **Watershed and** | ⇒ | **Habitat** | ⇒ | Biota |
| **riparian characteristics** | | **conditions** | | |

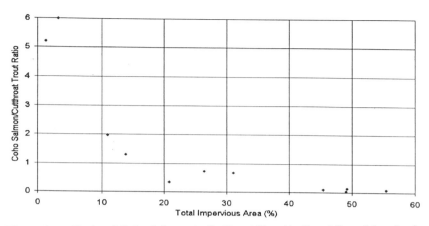

Figure 2. — Ratio of Coho Salmon to Cutthroat Trout in Puget Sound Lowland Streams Over a Gradient of Watershed Impervious Land Cover

Water quality was examined in wet and dry season base flow and during runoff from small (0.6-1.2 cm precipitation), medium (1.2-1.8 cm), and large (> 1.8 cm) rain storms. Figure 3 illustrates a representative result, showing total zinc concentrations versus TIA for all three storm sizes. Also shown are acute and chronic criteria for protection of aquatic life for the hardness prevailing in these streams. The concentration was well below both criteria under all conditions until TIA rose above 40 percent. The gaps between measured concentrations and regulatory criteria were even greater for other metals. The distributions of storm flow concentrations were similar for total suspended solids and other contaminants. Base flow concentrations were generally lower. It does not appear that measured water quality effects are strongly associated with the biological responses seen with rather small impervious proportions in Figures 1 and 2.

Figure 4 plots sediment zinc concentrations over the % TIA gradient. All were below the "lowest effect threshold" of the Washington Department of Ecology (1991) freshwater sediment criteria and far below the "severe effect threshold." The low measured concentrations relative to advisory or regulatory criteria were found with other metals as well. As with water quality, it appears that sediment quality does not change appreciably until urbanization reaches the vicinity of 50 percent impervious. Again, there is no sign of a strong association between sediment quality and the biological changes that occur much earlier during the onset of urbanization.

In contrast to the absence of associations between water and sediment quality with relatively low and moderate levels of urbanization, increasing hydrologic fluctuation seems to be an early harbinger of rising impervious cover. Discharge

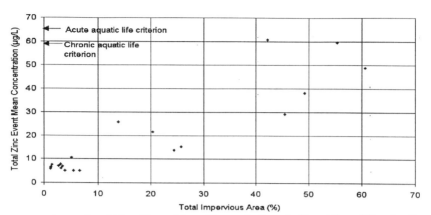

Figure 3. — Total Zinc Event Mean Concentrations in Puget Sound Lowland Streams During Storm Runoff Over a Gradient of Watershed Impervious Land Cover

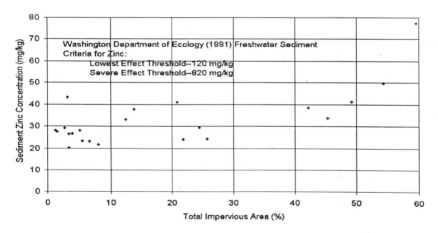

Figure 4. — Zinc Concentrations in Puget Sound Lowland Stream Sediments Over a Gradient of Watershed Impervious Land Cover

records were modeled using either: (1) the Hydrologic Simulation Program-FORTRAN (HSPF) continuous simulation computer program; (2) "runoff files" generated from regional data and HSPF runs (King County Surface Water Management Division 1995); or (3) multiple regression analyses to determine an unknown discharge from a known one in terms of relative catchment areas, % TIA, mean annual precipitation, and proportion of catchment with till soils (Cooper 1996). Modeling results were compared to values from stream gauge records where available. Figure 5 illustrates the available results in the form of a plot of the ratio of the 2-year frequency peak to the winter (wet season) base flow versus % TIA.

The ratio expresses the relative stream power, and thus physical stress on habitats and biota, exerted by storm runoff in relation to interevent conditions. The ratio is consistently < 20 with TIA < 15 percent, mostly 20-30 with TIA = 20-40 percent, and usually above 40 thereafter.

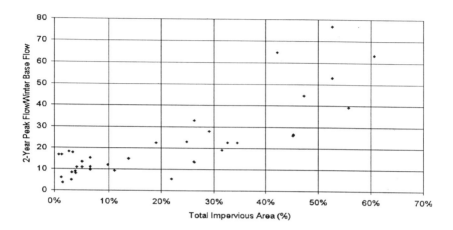

Figure 5. — Ratio of 2-Year Peak Flow to Winter Base Flow in Puget Sound Lowland Stream Sediments Over a Gradient of Watershed Impervious Land Cover

Several other habitat features have been related to urbanization as expressed by impervious cover. Figure 6 shows, for example, the results for large woody debris quantity. This feature is particularly important in Northwest streams, providing roughness that regulates velocities, cover to fish, and aid in forming pools in which fish feed and rest (McMahon and Hartman 1989). LWD more numerous than 300 pieces/km never occurred when TIA was more than 9 percent, with two exceptions where, respectively, wood was added to the channel in a habitat improvement project and a culvert prevents wood movement downstream. Numbers were always below 100/km with TIA > 40 percent. As a second example, not illustrated, the proportion of fines (< 0.85 mm) in the stream bed surface (top 10 cm) did not exceed 16 percent until TIA went above 20 percent. Values were in the range 22-27 percent fines with TIA at 45 percent or above, except where flushed by high flows.

A major reason, probably the leading one, for the death of salmonid embryos in the egg is ineffective dissolved oxygen (DO) interchange through fines that cover the spawning grounds. Measurement of both water column and intragravel DO gave the opportunity to assess this risk in relation to urbanization, as pictured in Figure 7. The ratio of intragravel to water column DO was above 90 percent in about half of the cases where TIA was 5 percent or less and above 80 percent for all but one of the remainder. Values generally fell below 70 percent above 10 percent TIA, but some reaches with more riparian protection exhibited more effective DO interchange. At TIA above 40 percent all ratios were below 70 and fell as low as 30 percent.

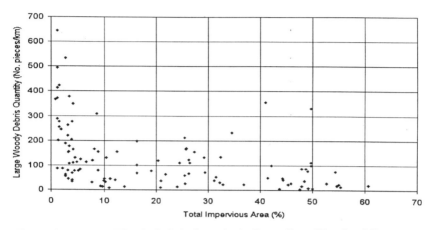

Figure 6. — Large Woody Debris Quantity in Puget Sound Lowland Stream Sediments Over a Gradient of Watershed Impervious Land Cover

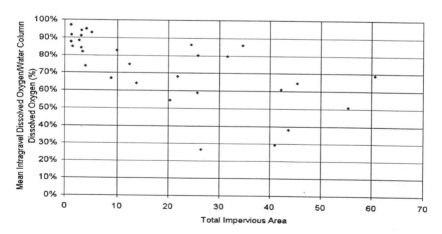

Figure 7. — Ratio of Intragravel to Water Column Dissolved Oxygen in Puget Sound Lowland Stream Sediments Over a Gradient of Watershed Impervious Land Cover

<u>Stream Biology in Relation to Habitat Conditions</u>

Given that relationships were identified between biological communities and contributing catchment conditions and between those conditions and habitat attributes, it is reasonable to suppose that associations of similar form exist between habitat conditions and biology. This discussion explores those linkages:

Watershed and	$\Rightarrow$	**Habitat**	$\Rightarrow$	**Biota**
riparian characteristics		**conditions**		

Figure 8 illustrates how B-IBI varies in relation to the ratio of 2-year peak flow to winter base flow. All indices of 35 or above were found in reaches where the hydrologic ratio was below 20. With ratios of 39 and above all but one B-IBI value was less than 20.

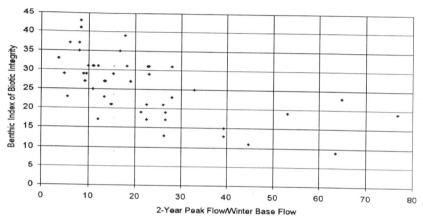

Figure 8. — Benthic Index of Biotic Integrity in Relation to Ratio of 2-Year Peak Flow to Winter Base Flow in Puget Sound Lowland Streams

Figure 9 shows the influence of the riparian zone on the biotic index. The highest indices (>/= 35) were all found where at least 60 percent of the riparian buffer zone upstream of the sampling point was at least 30 meters wide, and the lowest (< 20) where less than 50 percent of the buffer was that wide.

Also investigated, but not shown, was B-IBI versus percent fines (< 0.85 mm) in the bed substratum. All indices >/= 35 were consistent with no more than 15 percent fines, whereas reaches with 22 percent or more fines all had B-IBI < 20. The benthos thus appear to be quite sensitive to a relatively small alteration of the substratum.

Finally, Figure 10 shows B-IBI and coho salmon/cutthroat trout ratio in relation to a Qualitative Habitat Index. This index was derived by assigning scores of 1-4 to 15 attributes and summing (May 1996). Coho salmon dominance is consistent with B-IBI > 35 and a habitat index > 50. At the other extreme, cutthroat trout dominance is absolute with B-IBI under 20 and habitat index less than 40.

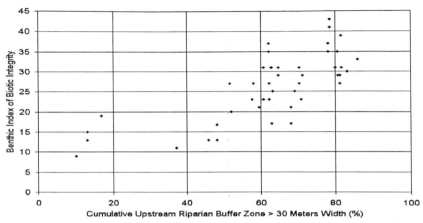

Figure 9. — Benthic Index of Biotic Integrity in Relation to Cumulative Upstream Riparian Zone Width in Puget Sound Lowland Streams

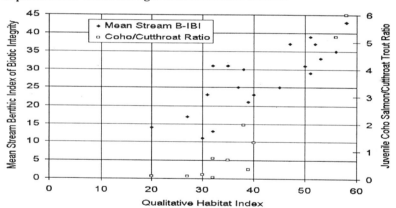

Figure 10. — Benthic Index of Biotic Integrity and Coho/Cutthroat Trout Ratio in Relation to Qualitative Habitat Index in Puget Sound Lowland Streams

WETLANDS RESEARCH

<u>Wetland Biology in Relation to Habitat Conditions</u>

This discussion considers observed linkages between WLF and key elements of wetland biological communities:

Watershed and riparian characteristics	⇒	**Habitat conditions**	⇒	**Biota**

As in the stream study, water pollutant concentrations were generally well below levels known to have direct biological effects, thus raising the relative significance of hydrology further (Reinelt and Horner 1990). The widely recognized primacy of hydroperiod (inundation pattern, specifically its depth, frequency, and duration) as a determinant of wetland ecosystem character (Mitsch and Gosselink 1993) produced an interest in finding a convenient means of measuring and expressing hydroperiod and investigating its relationship to biological communities on the one hand and attributes of contributing watersheds on the other. Instantaneous and crest (maximum since preceding measurement) water level readings were taken and used in computing water level fluctuation (WLF) as the difference between the crest and the average instantaneous depth in a time period (Azous 1990, Taylor 1993).

Plant species richness (number of species represented) was found to decline with increased mean annual WLF in the emergent (Figure 11) and scrub-shrub zones (not shown). Richness differed significantly (Mann-Whitney test, $p < 0.002$) between areas with mean annual WLF above and below 22 cm in both zones. No emergent areas had more than 14 species if WLF exceeded 22 cm, whereas 30 percent of the areas with WLF < 22 cm had at least 16 species. In the higher WLF group 33 percent had five or less species, in contrast to only 8 percent in the group subject to less fluctuation. Very similar distinctions pertained to scrub-shrub zones.

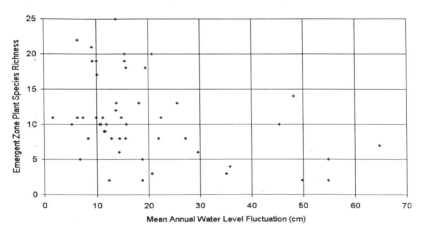

Figure 11. — Emergent Zone Plant Species Richness in Relation to Mean Annual Water Level Fluctuation in Puget Sound Lowland Wetlands

Being restricted to wetlands for reproduction, amphibians represent a biological indicator useful in assessing wetland condition. Figure 12 graphs richness in this community during the baseline years of the study versus mean annual WLF and presents a picture similar to the plant community. Of the wetlands with WLF < 22 cm, 62 percent had five or more species, and all had at least three. Of the wetlands in the group with WLF >/= 22 cm, 83 percent had four or less species. In later years amphibians declined in all wetlands, whether affected by urbanization or not. The reason for this decline is not known but may be the result of natural variation, meteorological effects, or human-induced stress other than urbanization. Over all years, 83 percent of wetlands with mean annual WLF < 22 cm averaged three or more species, while 73 percent with WLF at or above this level had two or fewer species.

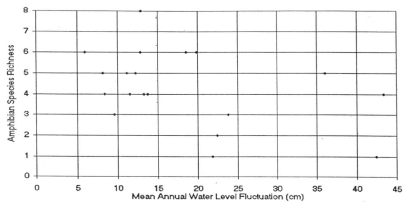

Figure 12. — Amphibian Species Richness in Relation to Mean Annual Water Level Fluctuation in Puget Sound Lowland Wetlands

The patterns of decline of species richness for both plants and amphibians exhibit a continuous rather than threshold response to water level fluctuation. Still, the point at about WLF = 22 cm seems to represent a boundary of tolerance in both communities, above which substantially fewer species tolerate the more dynamically fluctuating environment.

Wetland Habitat Conditions and Biology in Relation to Watershed Conditions

With respect to linkages between habitat and watershed conditions:

| **Watershed characteristics** | $\Rightarrow$ | **Habitat conditions** | $\Rightarrow$ | Biota |

WLF was found to have a relationship to urbanization. All watersheds with < 5.5 percent TIA had mean WLF in the two periods 1988-1990 and 1993-1995 no higher than 21 cm. With TIA > 21 percent, WLF exceeded this level in 89 percent of the cases.

Not surprisingly in light of these results and those reported earlier, watershed and biological connections were also established:

| **Watershed characteristics** | $\Rightarrow$ | Habitat conditions | $\Rightarrow$ | **Biota** |

Amphibians responded negatively to urbanization in a pattern similar to the variation of WLF with imperviousness (Figure 13). Of the instances of highest amphibian richness (5 or more species), all were in watersheds with 22 percent or less impervious area, and 78 percent had watersheds with TIA < 6 percent. Another observation for this group of wetlands, although not graphed, was that watershed forest cover was at least 35 percent in 88 percent of the cases.

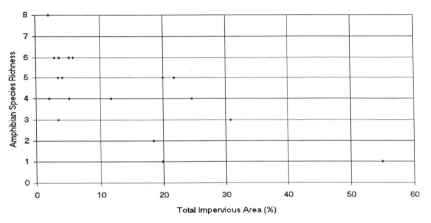

Figure 13. — Amphibian Species Richness in Puget Sound Lowland Wetlands Over a Gradient of Watershed Impervious Land Cover

Conclusions and Implications for Watershed Management

Research in representative sets of Puget Sound lowland streams and wetlands has shown that a host of physical habitat and biological characteristics change with increasing urbanization in a continuous rather than threshold fashion. Although the patterns of change differ among the attributes studied and are more strongly evidenced for some than for others, physical and biological measures generally were seen to change most rapidly from levels lightly affected by urbanization as total impervious area increased to 5-8 percent. With greater urbanization, the rate of alteration of habitat and biology usually slowed. There was direct evidence in both stream and wetland cases that altered watershed hydrology was at the source of the overall changes observed.

Water quality measures and concentrations of metals in sediments did not follow this pattern. They did not change much over the urbanization gradient until imperviousness approached 50 percent. Even then water column concentrations did not surpass aquatic life criteria, and sediment concentrations remained far below freshwater sediment criteria.

Thus, physical and biological change were seen to start early, almost immediately, in the urbanization process. Chemical pollutants did not exhibit a role in these early stages. They may not even have a large role at the intermediate levels represented in these studies, still being below regulatory criteria. Of course, for various reasons these criteria may not provide a realistic basis for judgment in the dynamic urban runoff environment. As urbanization increases above the 60 percent impervious level, with pollutant concentrations rising rapidly at that point, it is likely that the role of water and sediment chemistry will become more important biologically.

It is clear that biological community alterations in urban streams and wetlands are functions of many variables representing conditions in the immediate and more remote surroundings. In addition to urbanization level, a key determinant of biological condition appears to be the integrity of the riparian area available to buffer the aquatic community, in some measure, from negative influences in the watershed. The involvement of these numerous variables throughout the wider system shows the wisdom of moving to the watershed level to consider how to manage aquatic systems.

From the research results there appears to be a set of necessary, though not by themselves sufficient, conditions for the highest level of biological functioning. If maintenance of that level is an adopted goal, then this set of conditions constitutes standards that must be achieved if the goal is to be met. Using Puget Sound area lowland streams as an example, if that level is taken to be a B-IBI of at least 35, TIA must be < 5 percent, unless mitigated by extensive riparian protection, management efforts, or both. Some of the specific conditions that must be met would be:

- 2-year peak flow/winter baseflow < 20;

- > 60 percent of upstream buffer > 30 meters wide; and

- < 15 percent of surface bed material < 0.85 mm.

This set will be enlarged and refined as data analysis proceeds. Other sets of conditions could, and will, be similarly established for other biological goals.

To assess what these observations mean for watershed management, it may be useful to generalize the observed pattern of physical habitat and biological change with urbanization and consider what management objectives and challenges would most likely apply in different situations (Figure 14). In the broad region on the right side of the graph, some of the original resource base has been lost. Attempting to recover it, even if possible, would require large reductions in or compensations for urban land use. Thus, dramatic ecological restoration would be extremely challenging. The principal management objective in this situation should properly be stemming decline. Fortunately, with the relatively slow rate of change here, the system is likely to be more "forgiving" of change than in the region to the left. Holding the line can be successful if changes with negative tendencies are compensated by actions that maintain the overall status in about the same position. Another management objective should be to use any available opportunities to enhance the ecosystem structure and functioning. One of the leading strategies, on the basis of both its feasibility and probably effectiveness, should be to protect the riparian zone and, whenever possible, acquire riparian property and return it to a natural condition.

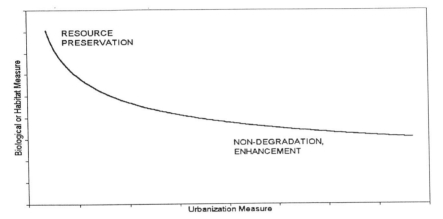

Figure 14. — General Pattern of Physical Habitat Characteristics in Puget Sound Lowland Streams and Wetlands Over a Gradient of Urbanization, with Regions Representing Different Watershed Management Objectives and Challenges

In the region on the left of the graph, resources have not been substantially affected and are likely to be highly valued. Here, a relatively small modification outside of the aquatic system is likely to create a proportionately bigger change within. The management objective would probably be to preserve the resources. Accomplishing this objective would require intense management to control the amount and location of impervious surfaces, maintain existing runoff storage capacity, prevent disruption of riparian areas, and the like, a very difficult challenge technically, and perhaps even more so, politically. The only way to meet such an objective is through strong preservation and regulation that allows very little and only certain types of watershed alteration. Implementing this program may be possible only with extensive property-owner incentives, government purchase of a large portion of the contributing catchment, transfer of development rights to an already highly degraded watershed, or some combination. Obviously, government could not move in such a fashion, nor justify the political and economic costs, without a strong case for the value of the resource thereby saved.

References

Alley, W. A. and Veenhuis, J. E., "Effective Impervious Area in Urban Runoff Modeling," *Journal of Hydrological Engineering, ASCE* 109(2):313-319, 1983.

Barker, B. L., Nelson, R. D., and Wigmosta, M. S., "Performance of Detention Ponds Designed According to Current Standards," *Proceedings of the Puget Sound Research '91 Conference*, Puget Sound Water Quality Authority, held at Seattle, WA, 1991, pp. 64-72.

Bisson, P. A., Sullivan, K., and Nielsen, J. L., "Channel Hydraulics, Habitat Use, and Body Form of Juvenile Coho Salmon Steelhead, and Cutthroat Trout in Streams," *Transactions of the American Fisheries Society* 117:262-273, 1988.

Booth, D. B., "Stream Channel Incision Following Drainage Basin Urbanization," *Water Resources Bulletin* 26(3):407-417, 1990.

Booth, D. B., "Urbanization and the Natural Drainage System--Impacts, Solutions, and Prognosis," *Northwest Environmental Journal* 7:93-118, 1991.

Booth, D. B. and Jackson, C. R., "Urbanization of Aquatic Systems--Thresholds and the Limits of Mitigation," *Proceedings of the Symposium on Effects of Human-Induced Changes in Hydrologic Systems*, held at Jackson Hole, WY, 1994, pp. 425-434.

Booth, D. B. and Reinelt, L. E., "Consequences of Urbanization on Aquatic Systems--Measured Effects, Degradation Thresholds, and Corrective Strategies," *Proceedings of the Watersheds '93 Conference*, U. S. Environmental Protection Agency, held at Alexandria, VA, 1994, pp. 545-550.

Bryant, J., "The Effects of Urbanization on Water Quality in Puget Sound Lowland Streams," thesis presented to the University of Washington, at Seattle, WA, in 1995, in partial fulfillment of the requirements for the degree of Masters of Science in Civil Engineering.

Cooke, S. S., "The Effects of Urban Stormwater on Wetland Vegetation and Soils," *Proceedings of the Puget Sound Research '91 Conference*, Puget Sound Water Quality Authority, held at Seattle, WA, 1991, pp. 43-51.

Cooper, C., Untitled thesis to be presented to the University of Washington, Seattle, WA, in 1996, in partial fulfillment of the requirements for the degree of Masters of Science.

Fore, L. S., Karr, J. R. and Wisseman, R., "A Benthic Index of Biotic Integrity for Streams in the Pacific Northwest," *Journal of the North American Benthological Society*, in review.

Horner, R. R., Skupien, J. J., Livingston, E. H., and Shaver, H. E., *Fundamentals of Urban Runoff Management: Technical and Institutional Issues*, Terrene Institute, Washington, D. C., 1994.

Karr, J. R., "Assessment of Biotic Integrity Using Fish Communities," *Fisheries* 6(6), 1981.

Karr, J. R., "Biological Integrity: A Long Neglected Aspect of Water Resources Management," *Ecological Applications* 1(1):66-84, 1991.

Karr, J. R. and Dudley, D. R., "Ecological Perspective on Water Quality Goals," *Environmental Management* 5:55-68, 1981.

Karr, J. R., Fausch, K. D., Angermier, P. L., Yant, P. R., and Schlosser, I. J., "Assessing Biological Integrity in Running Waters: A Method and Its Rationale," Special Publication 5, Illinois Natural History Survey, Urbana, IL, 1986.

Kerans, B. L. and Karr, J. R., "A Benthic Index of Biotic Integrity (B-IBI) for Rivers of the Tennessee Valley," *Ecological Applications* 4(4):768-785, 1994.

King County Resource Planning Section, "Detailed Planning of the Puget Sound Wetlands and Stormwater Management Research Program," King County Resource Planning Section, Seattle, WA, 1987.

King County Resource Planning Section, "Puget Sound Wetlands and Stormwater Management Research Program: Initial Year of Comprehensive Research," King County Resource Planning Section, Seattle, WA, 1988.

King County Surface Water Management Division, "King County Runoff Time Series (KCRTS), Version 3.0, Computer Software Reference Manual," King County Department of Public Works, Seattle, WA, 1995.

Klein, R. D., "Urbanization and Stream Quality Impairment," *Water Resources Bulletin* 15:948-963, 1979.

Kleindl, W., "A Benthic Index of Biotic Integrity for Puget Sound Lowland Streams," thesis presented to the University of Washington, Seattle, WA, in 1995, in partial fulfillment of the requirements for the degree of Masters of Science.

Lucchetti, G. and Fuerstenberg, R., "Relative Fish Use in Urban and Non-urban Streams," *Proceedings of the Conference on Wild Salmon*, Vancouver, B. C., Canada, 1993.

May, C., "Assessment of Cumulative Effects of Urbanization on Small Streams in the Puget Sound Lowland Ecoregion: Implications for Salmonid Resource Management," dissertation to be presented to the University of Washington, Seattle, WA, in 1996, in partial fulfillment of the requirements for the degree of Doctor of Philosophy.

McMahon, T. E. and Hartman, G. F., "Influence of Cover Complexity and Current Velocity on Winter Habitat Use by Juvenile Coho Salmon," *Canadian Journal of Fisheries and Aquatic Sciences* 46:1551-1557, 1989.

Mitsch, W. J. and Gosselink, J. G., *Wetlands*, 2nd Ed., Van Nostrand Reinhold, New York, NY, 722 pages.

Nehlsen, W., Williams, J., and Lichotowish, J., "Pacific Salmon at the Crossroads: Stocks at Risk from California, Oregon, Idaho, and Washington," *Fisheries* 16(2):4-21, 1991.

Nickelson, T. E., Rodgers, J. D., Johnson, S. L., and Solazzi, M. L., "Seasonal Changes in Habitat Use by Coho Salmon in Oregon Coastal Streams," *Canadian Journal of Fisheries and Aquatic Sciences* 49:783-789, 1992.

Olthof, J. "Puget Sound Lowland Stream Habitat and Relations to Basin Urbanization," thesis presented to the University of Washington, Seattle, WA, in 1994, in partial fulfillment of the requirements for the degree of Masters of Science.

Perkins, M. A., "An Evaluation of Instream Ecological Effects Associated with Urban Runoff to a Lowland Stream in Western Washington, U. S. Environmental Protection Agency, Washington, D. C., 1982.

Plafkin, J. L., Barbour, M. T., Porter, K. D., Gross, S. K., and Hughes, R. M., "Rapid Bioassessment Protocols for Use in Streams and Rivers: Benthic Macroinvertebrates and Fish," EPA 440-4-89-001, U. S. Environmental Protection Agency, Washington, D. C., 1989.

Prych, E. A. and Ebbert, J. C., "Quantity and Quality of Storm Runoff from Three Urban Catchments in Bellevue, Washington," Water Resources Investigations Report 86-4000, U. S. Geological Survey, Tacoma, WA, 1986.

Reinelt, L. E. and Horner, R. R., "Characterization of the Hydrology and Water Quality of Palustrine Wetlands Affected by Urban Stormwater," King County Resource Planning Section, Seattle, WA, 1990.

Reinelt, L. E. and Horner, R. R., "Urban Stormwater Impacts on the Hydrology and Water Quality of Palustrine Wetlands in the Puget Sound Region," *Proceedings of the Puget Sound Research '91 Conference*, Puget Sound Water Quality Authority, held at Seattle, WA, 1991, pp. 33-42.

Richey, J. S., "Effects of Urbanization on a Lowland Stream in Western Washington," dissertation presented to the University of Washington, Seattle, WA, in 1982, in partial fulfillment of the requirements for the degree of Doctor of Philosophy.

Richter, K. O. and Azous, A. L., "Amphibian Occurrence and Wetland Characteristics in the Puget Sound Basin, " *Wetlands* 15(3):305-312, 1995.

Scott, J. B., Steward, C. R., and Stober, Q. J., "Effects of Urban Development on Fish Population Dynamics in Kelsey Creek, Washington," *Transactions of the American Fisheries Society* 115:555-567, 1986.

Shaver, E. Maxted, J., Curtis, G., and Carter, D., "Watershed Protection Using and Integrated Approach," *Stormwater NPDES Related Monitoring Needs*, H. C. Torno, ed., American Society of Civil Engineers, New York, NY, 1995, pp. 435-459.

Steward, C. R., "Salmonid Populations in an Urban Stream Environment," thesis presented to the University of Washington, Seattle, WA, in 1983, in partial fulfillment of the requirements for the degree of Masters of Science.

Taylor, B. L., "The Influence of Wetland and Watershed Morphological Characteristics on Wetland Hydrology and Relationships to Wetland Vegetation Communities," thesis presented to the University of Washington, Seattle, WA, in 1993, in partial fulfillment of the requirements for the degree of Masters of Science in Civil Engineering.

Washington Department of Ecology, "Summary of Criteria and Guidelines for Contaminated Freshwater Sediments," Washington Department of Ecology, Olympia, WA, 1991.

Wigmosta, M. S., Burges, S. J., and Meena, J. M., "Modeling and Monitoring to Predict Spatial and Temporal Hydrologic Characteristics in Small Catchments," U. S. Geological Survey Water Resources Technical Report Number 137, 1994, 223 pp.

Yoder, C., "The Integrated Biosurvey as a Tool for Evaluation of Aquatic Life Use Attainment and Impairment in Ohio Surface Waters," *Biological Criteria: Research and Regulation*, U. S. Environmental Protection Agency, Washington, D. C., 1991, pp. 110-122.

SESSION 2B: Impacts of Watershed Development on Aquatic Biota, Round 1

DISCUSSION

Assessing the Condition and Status of Aquatic Life Designated Uses in Urban and Suburban Runoff
Chris Yoder

An important objective of the bioassessments conducted by Ohio EPA is to determine the appropriate and attainable aquatic live use designation.

Biological criteria are narrative and numerical expressions of the health and well-being of the aquatic biota and are based on measurable attributes of aquatic communities such as fish and macroinvertebrate community structure and function. The numerical biological criteria, expressed as biological indices which represent measurable end-points of use designation attainment and non-attainment, are essentially the end-product of an ecologically complex, but structured derivation process.

Questions/Comments

Question:	Are you looking at sediment control programs?
Answer:	Yes we are trying to get a hold on this, but we need more than silt fences; we need biologic setbacks.
Question:	Where are you going with IBI improvement?
Answer:	We're pretty satisfied with the protocol we have, we are working on developing more subregions.
Question:	With low IBIs do you know what the problem is?
Answer:	Yes, it seems like sediment is a culprit, but it may be that the IBI does not include hydrologic effects to the extent that it should, in CSO areas however, we are not certain that this is a real problem.
Question:	How do you make certain that you get effects of CSOs accounted for when the effect may be many miles downstream?
Answer:	We of course do mathematical modeling which helps us know where these impacts will be.

Biological Effects of the Build-up of Contaminants in Sediments in Urban Estuaries
Don Morrisey

The objective of our current programme of research is to provide the means to predict the impacts of stormwater on New Zealand's sheltered estuaries and harbours. The first goal is to test and refine a model to predict the extent and rate of accumulation of contaminants in sediments. The ecological significance of this chemical contamination will then be assessed through a combination of correlative and experimental approaches.

We have developed a model to estimate the accumulation in estuaries of heavy-metal contaminants derived from urban stormwater. Published information on the volumes of runoff from areas of land under different uses, and the concentrations of sediment and contaminants in that runoff, have been combined with information on the historical sequence of development of specific estuarine catchments.

Two estuaries were chosen for the study, Hellyers/Kaipatiki Creek, in the Upper Waitemata Harbour, and Pakuranga Creek, 14 km southeast of Auckland's city centre. Four chemical sampling locations were selected in each estuary, two within the settling-zone and two outside it.

Correspondences between observed and predicted concentrations were good for Pakuranga Creek but poor for Hellyers/Kaipatiki (see Table 1 of paper). This was perhaps due to the relatively large intertidal area of the latter, and the longer period of exposure during low tide. Faunel core samples were collected at six locations along the lengths of each of Hellyers/Kaipatiki and Pakuranga Creeks.

BIO-ENV (a statistical analysis program) analyses were done in two forms. First, the 3 faunal cores were compared to the sediment-variables, applying the same sediment-variables to all of the 3 faunal cores collected around each environmental core. Second, each of the 2 environmental cores for which pore-water variables were measured at each location was compared with the single faunal core collected next to it.

Overall, the BIO-ENV analyses indicated that the predominant "explanatory" variables among those measured in sediments from Pakuranga Creek were the concentration of lead and those related to sediment-texture. In the case of Hellyers/Kaipatiki Creek, the best "explanatory" variables were sediment-texture, concentration of lead in the sediment, concentration of iron in sediment and pore-water and the pH of the pore-water.

Questions/Comments

Question: Did you consider sampling metals in sediments upstream and
              correlate them to metals in the estuaries?

Answer: We have done some of that, but also we took deep core samples that tell us what the sediment was like before urbanization.

Watershed Determinants of Ecosystem Functioning
    Rich Horner

By the mid-1980s it was clear that urban stormwater runoff was strongly implicated in the alteration of streams and freshwater wetlands in the Puget Sound Basin of Washington state. Recognition of these connections and the rapid pace of development in the region stimulated research to define the linkages among stream and wetland habitat structure. One project monitored watershed and riparian zone conditions, flow, physical habitat characteristics, water quality, benthic macroinvertebrates, and fish. A second project followed 19 palustrine wetlands during a period when urbanization began or increased in the watersheds of about half, while the remainder were essentially unchanged. Overall, the findings of these projects agree that the effects of modified hydrology accompanying urbanization, exert the earliest and, at least initially, the strongest deleterious influences on the freshwater ecosystems studied. The steepest rates of decline in biological functioning of both streams and wetlands occur as urbanization increases total impervious land cover from 0 to about 6 percent. The decline proceeds at a slower rate as impervious cover increases further.

Questions/Comments

Question: Given what you know, how do we control the aquatic ecological effects of urban development?

Answer: We have to "bite the bullet". The bite-the-bullet answer is "Now that we know what to do technically, let's figure out how we get it done at the institutional level."

Riparian Buffer Widths at Rocky Mountain Resorts[1]

Edward W. Brown, Jonathan E. Jones, Jane K. Clary, and Jonathan M. Kelly[2]

Abstract

Development in the Rocky Mountains of Colorado is occurring rapidly and is concentrated on valley floors around streams, rivers, wetlands, and lakes. Without adequate planning and management, development around these water bodies can degrade water quality and riparian habitat. Many Rocky Mountain towns are addressing protection of riparian areas with more scrutiny as they realize the multiple values of these areas. As is the case throughout much of the country, development and protection of environmental resources may often conflict. However, in the Rocky Mountains, where much of the economy is based on the recreational opportunities and the aesthetics of the mountains, development and preservation of environmental resources go hand-in-hand. This paper explores the nature of regulations and policies at resort areas in Colorado involving riparian corridors. In addition, different approaches to preserving riparian corridors will be explored.

I.   Introduction

Land development in the Rocky Mountains is occurring rapidly. The mountainous terrain dictates that development occur along valley floors, in the vicinity of streams, rivers, wetlands, and lakes. Without adequate planning and management, development around these water features can adversely impact water quality, water quantity, and riparian habitat. Preserving undeveloped riparian corridors in high mountain valleys is one of the more significant commitments that can be made to protect water quality and aquatic life.

As is the case throughout much of the country, development and protection of environmental resources can conflict. However, in the Rocky Mountains, where much of the economy is based on tourism due in part to the recreational opportunities

---

[1] Presented at the August 1996 Engineering Foundation Conference in Snowbird, Utah, *"Effects of Watershed Development and Management on Aquatic Ecosystems."*
[2] Members of Wright Water Engineers Inc., Denver, Colorado.

and the aesthetics of mountain streams and their surrounding environment, a strong majority of resorts and city/county governments recognized that land development and preservation of environmental resources must go hand-in-hand. For example, although nearly all resort towns in the Rocky Mountains have flood control and traditional "setback" regulations, some towns are now extending that commitment by providing additional protection of riparian areas. Similarly, ski resorts themselves (in addition to the adjoining municipalities) often designate wide, undeveloped riparian corridors.

This paper focuses on the value of preserving riparian corridors and the different approaches to preserving this resource. Riparian corridors serve a variety of valuable functions which are summarized in this paper. Depending on the riparian functions present, various buffer widths may be appropriate to ensure preservation. An overview of current protective measures in place in Colorado resort areas is provided in addition to a brief literature review of similar regulations and approaches in other parts of the country. Based on this review, additional suggestions to enhance riparian corridor preservation in the Rocky Mountains are presented.

II.   Functions of Riparian Corridors

Just as there are multiple functions of riparian corridors, there are also multiple definitions of what exactly constitutes a riparian corridor. Basically, these definitions can range from the broad "stream banks" to a more complex ecological definition which includes considerations such as soils, hydrology, vegetation and species distribution (Johnson et al. 1981; USFS 1995; BLM 1986). For purposes of this paper, the Federal Bureau of Land Management (BLM) definition identified in the 1986 "Draft BLM Riparian Area Management Policy," will be assumed. The BLM defines riparian ecosystems as "land transitional between aquatic and upland habitats that is characterized by hydric soil and distinctive vegetation requiring free or unbound water."

Riparian corridors serve a variety of functions which can be categorized into four general areas: water quality, wildlife/aquatic life, water quantity, and aesthetics. Each of these categories has several sub-functions which are briefly described below (Johnson and Ryba 1992; Castelle et al. 1991; Matthews and Moore 1989). Riparian corridors may serve some or all of the following functions to varying degrees.

Water Quality

- Stabilizing streambanks and limiting erosion;
- Filtering or otherwise immobilizing suspended solids, nutrients, fecal coliform and harmful or toxic substances (wide buffer zones are particularly valuable during the construction of ski runs and lifts);

- Moderating impacts of stormwater runoff;
- Nutrient cycling;

Wildlife/Aquatic Life

- Supporting and protecting aquatic and wildlife species (including benthic communities);
- Moderating the microclimate of a riparian system (i.e., cold water trout streams);
- Providing migration corridors (i.e., mammals and birds);
- Maintaining and enhancing habitat diversity and integrity;

Water Quantity

- Groundwater recharge/discharge;
- Flood flow/stormwater runoff attenuation (flood losses can be reduced to insignificant levels if riparian zones are left undeveloped);
- Maintenance of the current surface water and/or groundwater regime;

Aesthetics

- Aesthetic and scenic values of natural features;
- Noise reduction;
- Recreation.

Several of these functions are particularly vital to resorts of the Rocky Mountains. One of the major attractions to many resorts is trout fishing. To have a healthy trout population, it is important to preserve healthy benthic communities, maintain cold water temperature, reduce pollutant loads from urban runoff and nonpoint sources, and strictly limit increased sediment loads. (Increased sediment loads can result from inadequate erosion and sediment control practices and channel erosion accompanying increased streamflows caused by land development in the absence of proper detention practices.) Riparian corridor protection promotes all of these objectives.

Similarly, the aesthetics of the stream and stream corridor are important for other tourist activities such as rafting and kayaking. Preservation of migration corridors is important to maintain healthy elk and deer populations, which support hunting and wildlife viewing. In Colorado, preservation of high-quality potable water sources and water rights has been and will continue to be a vital issue. Finally, to minimize direct and indirect adverse impacts to wetlands, it is often desirable to preserve riparian corridors.

III.   Buffer Zone Width Determination

The main strategy used to protect and preserve riparian corridors is to create regulations which establish buffer zones where development is prohibited or restricted. There are two general methods to define or measure buffer zone widths. One way to measure the buffer zone is to start from the *streambank* out to some distance. The second way is to measure the buffer zone from the edge of the *riparian* boundary (as determined by vegetation, hydrology, and soils) out to some distance. The first approach applies to a buffer zone for the *stream* which limits development in the riparian area, while the second approach applies to a buffer zone for the *entire* riparian area. For purposes of this paper, riparian buffer width is defined as distance from the streambank, unless otherwise noted. Developing rules for the determination of buffer widths can be difficult, as it requires weighing the political/administrative desire to minimize the complexity and interpretive aspects of the issue with the more time and resource intensive methodology of implementing site-specific ecological criteria. Each of these is discussed below.

Political/Administrative Considerations

Two alternative approaches are used for delineating buffer zones: *fixed* or *variable*. The first approach employs a static distance from the streambank regardless of site-specific conditions, while the second approach enables regulatory bodies to take into consideration site-specific conditions. Although the fixed approach is easier to administer and enforce, it may have little ecological and/or environmental protection correlation to the site conditions (Matthews and Moore 1989). The main advantage with variable width buffers is greater flexibility for varying site conditions and land management practices. The variable buffer width approach is more ecologically reasonable in most cases; however, the difficulty in administration makes it cost prohibitive for many regulatory agencies, particularly local governments.

In general, counties in the Rocky Mountain regions of Colorado defer to the fixed-width policy. On ski resort projects, however, there is generally extensive regulatory review at the local, state and federal levels and project proponents spend large amounts of money on permitting. Consequently, there is more of an opportunity to utilize the variable-width approach.

To facilitate the delineation of riparian corridor boundaries using the variable-width approach, it is desirable to present the following information on map "overlays":

- Existing and projected land use and general nature of land ownership;

- Utilities and easements;

- Regulatory boundaries based on existing local regulations such as minimum construction setbacks from a water body;

- Vegetation, soils, and topography;

- Hydraulic information on stability aspects of the stream on a reach by reach basis;

- Information of significance gained from aerial photo review;

- Floodplain information for the 100-year return frequency flood and potentially others;

- Aquatic or terrestrial habitat of special importance;

- Wetlands, with emphasis on wetlands of special importance and/or functions;

- Other site-specific characteristics.

Ecological Considerations

The following factors are important ecological considerations for determining buffer widths:

- **Regional Geology: Erosion/Sedimentation Potential.** Regional geology, including soil characteristics, plays a significant role in the susceptibility of a riparian area to erosion and sedimentation problems. Matthews and Moore (1989) note that the University of Florida Center for Wetlands recommends that a buffer zone should "abate destructive water velocities and quantities of pollutants carried by surface runoff from uplands that may have a negative impact on downstream water quality, flora, and fauna." The Center recommends an empirical equation dependent on slope, soil type and erodibility as the basis for determining water quality buffer width. This formula is $B_w = S^{1/2}/E$ where B_w is the buffer width (ft); S is the average slope of the land in feet per 100 feet; and E is the erodibility factor based on Natural Resources Conservation Service (NRCS, formerly the Soil Conservation Service) erosion factors (K) (Matthews and Moore 1989). The NRCS also emphasizes the importance of slope and recommends buffers of 3 to 8 meters next to pastures with 0-30 percent slopes, 8 to 46 meters near logging operations in areas of 0-70 percent slopes, and 23 to 92 meters for livestock feedlots and liquid waste treatment on 2-6 percent slopes (Johnson and Ryba 1992).

- **Vegetation/Canopy.** The composition, age, and condition of buffer vegetation are important for determining buffer widths (Johnson and Ryba 1992). One of the functions of canopy preservation is maintenance of stream temperature to maintain healthy trout populations. Barton et al. (1985) studied 40 sites on 38 southern Ontario trout streams to determine the relationship between riparian land use and environmental parameters that define the suitability of the streams for trout. As a result of weekly observations of maximum and minimum temperature, coarse and fine suspended matter, and discharge during the summer of 1980, they determined that maximum temperature appeared to be the most critical variable in determining the suitability of a stream for trout. Regression analysis relating maximum weekly temperature to buffer strip length and width accounted for 90 percent of the observed variation in water temperature for these sites. The study concluded that control of temperature, and to a lesser extent turbidity and stability of discharge, can be achieved through establishment or maintenance of forested riparian buffer strips.

Other studies by Brazier and Brown (1973) found that canopy density along the path of incoming solar radiation in Oregon best described the shading capacity of a buffer strip and suggested that such strips need to be wide enough to include trees contributing to the canopy cover. In general, wider streams require taller trees. Given these considerations, Barton et al. (1985) further suggested that minimum buffer width depends on the age and species composition of the riparian vegetation and its resulting height and foliage density.

- **Adjacent Land Usage.** The type of activity being conducted adjacent to the stream is important in determining buffer widths because some land uses cause greater disturbance than others to the riparian area. For example, improper livestock grazing practices significantly impact stream corridors by causing the decline of streambank vegetation, increases in turbidity, nutrients and bacteria, and general loss of streambank stability and structure (White 1991; Johnson and Ryba 1992; EPA 1990).

During construction activities, it is beneficial to separate construction projects from receiving waters by wide buffer strips, in the event that erosion and sediment control measures function improperly. Many resorts in Colorado face very strict water quality requirements for both point and non-point sources; therefore, preservation of a wide buffer width promotes water quality protection.

- **Stream Characteristics.** Considerations include stream order, slope, the 100-year floodplain and ambient water quality (Schueler 1994). For example, a sensitive, small tributary will likely benefit more from a large buffer zone than a large river (Johnson and Ryba 1992).

- **Species Present.** Considerations include the presence of threatened or endangered animals and plants, large mammal migration corridor requirements, and aquatic life requirements such as cold water fishery needs.

- **Natural Hazards.** Protecting riparian corridors can provide the added benefit of reducing the risks of natural hazards including catastrophic floods, debris flows, avalanches, rockfall (in some cases), and slope instability. These factors are often of great significance at Rocky Mountain resorts.

- **Existing Riparian Condition.** Considerations include quality, size, and sensitivity to disturbance of the area being buffered.

- **Land Planning and Public Recreation Considerations.** Normally, planners and landscape architects should be consulted about the most appropriate width(s) for a proposed buffer zone. Considerations related to land planning, zoning, density, public recreation, trails, etc., often play a significant role in defining the width of the protected riparian zone.

- **Public Participation.** Members of the public frequently provide valuable insight into factors that should influence the nature of the buffer zone. Public interest in most resort proposals is significant, and developers often find that public involvement is mutually beneficial and facilitates project review and approval.

IV. Additional Considerations

Even with established riparian corridor buffer widths, there will be times when disturbances may be necessary, such as construction or maintenance of bridge crossings, utilities, and irrigation structures. Certain restrictions and guidelines on these activities need to be specified to protect the established riparian boundaries to the maximum practicable extent. Examples include:

- Strictly limit the amount of vegetation removal during construction;

- Use structural and non-structural BMPs to minimize inflow of sediments and other pollutants during and after construction;

- Minimize pesticide, herbicide and fertilizer usage, and implement stringent selection, application, and monitoring practices;

- Restrict and/or control activities such as grazing, pumphouses, sewage treatment plants, golf courses, campgrounds, timber harvesting, hydropower, roads/bridges, athletic fields, playground equipment;

- Enforce leash laws/pet controls;

- Locate nature trails or bike paths to minimize potential impacts;

- Provide designated access points for recreational activities such as fishing, canoeing, and kayaking;

- Construct educational signs to inform users about sensitivity and importance of riparian habitat, the importance of staying on designated routes and keeping pets on the leash;

- Revegetate any temporarily disturbed sites with native species.

V.   Problems Associated with Establishing and Maintaining Buffer Widths

Significant challenges exist with regard to the issues of the "invisibility of stream/wetland buffers" and difficulties in enforcing these areas. Schueler (1994) summarized two recent studies which suggested that buffers might have limited usefulness as a watershed protection tool as they are currently enforced. The key problem is that buffer boundaries are often invisible to property owners, contractors, and even local governments. Without defined boundaries, urban buffers face enormous pressure from encroachment, disturbance, and other incompatible uses. A study by Heraty (1993) on buffer zones administration in 36 jurisdictions throughout the country identified the following issues:

- Most developers were required to delineate a stream buffer on concept or final plans for purposes of development review. However, only half the jurisdictions required that buffer boundaries be clearly delineated on the plans for clearing/grading and sediment control.

- Local governments often did not record buffer zone boundaries on their own official maps, making it difficult to inspect or enforce these boundaries.

- Nearly 90 percent of most buffer zones were on private lands. For most property owners, these buffer zones were particularly invisible due to little education by local authorities. Usually, the only notification given to property owners about buffer limits was a one-time legal disclosure, such as notes on the deed of sale, language in a homeowners' association charter, or prescribed notice upon property resale.

- Staffing was inadequate to verify delineated buffer zone boundaries. Post-construction field walks were rarely conducted.

- Many local government buffer ordinances were too simplistic and lacked a clear vegetative goal. For example, the goal of maintenance of pre-vegetative cover should be to protect mature riparian forest cover.

- Pressures for various uses within the buffer zone existed, as did difficulties in defining acceptable uses.

Another study by Cooke (1991) of 21 wetland buffers in Seattle, Washington, showed that of the original buffers defined in development plans, 95 percent had been altered. Forty percent had been so altered that their capability to protect the adjacent wetland had been severely compromised. Buffer disturbances included tree removal, conversion into lawns and foot trails, filling encroachment, dumping of yard wastes, and erosion by stormwater runoff. Narrow lots in residential areas were particularly susceptible.

The message of the Cooke and Heraty studies is that local governments must do more than merely require buffers during development review. They must also make the effort to manage buffers after they become established, which includes education of the community. Further, private entities must commit to abiding by the regulations. Four season resorts in Colorado, for example, generally play an active role in promoting the preservation of riparian areas and are an integral part of the community education process.

VI.   Colorado Regulations for Preserving Riparian Corridors

The State of Colorado does not have regulations for preserving riparian corridors or establishing buffer zones; however, many Colorado counties have regulations which require setbacks from streams and wetlands. Often these regulations may be established for other specific purposes such as floodplain zoning or wetland protection, but riparian corridors are protected nonetheless, especially if riverine wetlands are located along the streambanks. In addition, most counties qualitatively address the value of preserving riparian areas.

Most of the Colorado counties in resort areas currently have minimum fixed buffer width regulations, with several counties requiring additional buffer widths under certain circumstances. Similarly, watershed management plans are being developed in some counties to address the importance of proper protection of riparian areas and may lead to increased consideration of variable buffer widths in the future. In addition, much of the land in the resort areas is owned by the U.S. Forest Service (USFS) and must be protected under its regulations and policies. USFS staff respond favorably to the preservation of wide riparian buffer zones and recognize that this is a necessary practice where receiving waters are subject to anti-degradation review.

County and USFS regulations for representative mountain areas in Colorado are discussed below. Many of the factors identified in the literature as being important to determining buffer widths have been incorporated into these county and USFS regulations. In addition, four season resorts will often take added measures to protect the riparian corridor, as they have realized the benefits from establishing these buffers.

Counties

The Vail, Beaver Creek, Arrowhead and proposed Adam's Rib Recreational Area ski areas are located in Eagle County, Colorado. The county is also a popular summer time destination for outdoor activities such as whitewater rafting, kayaking, camping and mountain biking. Gore Creek flows through the center of the Town of Vail and into the Eagle River. The Vail Valley is a relatively narrow area with steep mountainous terrain bounding both sides of the valley. Development is concentrated along the valley bottom in the vicinity of the Gore Creek and Eagle River.

The Eagle County regulations require that a 15 meter strip of land or the 100-year floodplain, whichever is the greater distance measured horizontally from the natural identifiable high water mark on each side of any live stream, be protected in a "natural state" with the exception of footpaths, irrigation structures, flood control, and erosion protection devices. A greater setback, up to 46 meters, may be required when: the slope equals or exceeds 30 percent; highly erodible soils are present; or the proposed use of the property presents a special hazard to water quality, e.g., storage or handling of hazardous or toxic materials.

The Aspen, Buttermilk and Snowmass ski areas are located in Pitkin County, Colorado. Similar resort activities to those in Eagle County take place in Pitkin County. It is the policy of the County to preserve its natural, rural scenery for the benefit of its residents and the continued viability of its resort economy. This includes maintaining natural vegetative buffers along its surface water such that the surface water and groundwater of the area are not encroached upon by land uses or other human activities which could cause deterioration of water quality or impair natural treatment processes provided by wetlands and meadows. Typical buffers range from 6 to 46 meters.

Summit County includes the towns and ski resorts of Keystone, Frisco, Arapahoe Basin, Breckenridge, and Copper Mountain. Although the valley area is wider than that of Vail, development is still centered around waterbodies including Dillon and Green Mountain Reservoirs, the Snake River, and Ten Mile Creek. Much of the county's economy is tied to the ski industry in the wintertime, and conference and convention business, outdoor recreation and construction in the summertime. Riparian corridors are protected under several zoning regulations which address preservation of wildlife habitat, flood control, water quality protection, and an interim wetland regulation.

Summit County recognizes that "preservation of wildlife habitat provides economic, recreational, and environmental benefits to the residents and visitors of Summit County." The county has developed wildlife overlay districts, which protect wildlife habitat, defined as "those regions or environments containing those elements necessary for the survival and health of wildlife species, and consisting of principal feeding areas, winter range, summer range, shelter areas, concentration areas, production areas, buffer zones, areas providing essential minerals and water, and species habitat needs." From the perspective of riparian corridors, wildlife habitat includes shoreline vegetation, which is defined as "plant life which grows along streambanks and the shorelines of lakes and ponds."

The Telluride ski area is located in San Miguel County, Colorado. San Miguel County Land Use regulations state that development is not allowed in mapped and known wetlands, a 46-meter buffer zone surrounding them, or the waters of San Miguel County without undergoing a special review process. The definition of wetland in this regulation includes all waters of San Miguel County and all riparian areas in the county. "Buffer Zone" is defined as all areas where development could impact wetland areas, extending at least 30 meters around the wetland areas.

The Eldora ski area is located in Boulder County, Colorado. Riparian areas in Boulder County are protected by a combination of regulations restricting development in floodplains and by wetland protection regulations. Based on a comprehensive county wetland inventory, it was determined that approximately 90 percent of the naturally occurring wetland areas were located in riparian corridors. The protection provided to riparian corridors under the wetland regulations is generally more restrictive than that required under the floodplain regulations.

The USFS is driven by regulations under the National Forest Management Act of 1976, where forest officers are directed to:

> "Give special attention to land and vegetation up to approximately 100 feet from the edges of all perennial streams, lakes, and other bodies of water. This distance shall correspond to at least the recognizable area dominated by riparian vegetation (36 CFR 219.27e). Give special attention to adjacent terrestrial areas to assure adequate protection for the riparian dependent resources. Inventory,

analysis, and monitoring of the riparian areas are to be conducted as part of the continuing forest planning process." (Schmidt 1987).

Resort and Community Measures

Some developers in Eagle County have chosen to protect buffer zones larger than those required by the regulations. For example, in the proposed Adam's Rib Recreational Area, considerable field evaluation was conducted to identify and quantify wetland functions in a proposed development area known as Vassar Meadow. Field evaluation showed that the riverine wetlands provided the highest overall functional capabilities, particularly for flood flow attenuation, sediment retention, nutrient removal/transformation, and aquatic life functions. Consequently, the developer agreed to: impose no development on the riverine wetlands; use special "minimization" measures for utilities and road crossings; and evaluate indirect hydrologic impacts caused by upgradient disturbances.

Further, the developer committed to implement a conservative and innovative variable width buffer zone determination approach along nearly 15 kilometers of perennial streams in the project area. A set of four criteria were established and evaluated, with the most conservative of them (i.e., the farthest from the stream) serving as the buffer zone boundary delineation. The four criteria are:

- the riparian vegetation boundary as determined by aerial photography review;

- the jurisdictional wetland boundaries;

- the 100-year floodplain boundary; and

- the Eagle County 50-foot setback requirement.

The criteria were applied on a reach by reach basis throughout the project area. Figure 1 is an example of how the criteria were applied. The resulting buffer zone boundary is represented by the heavy black line.

These protection measures result in average buffer widths ranging from 82 to 137 meters (depending on the stream) from the stream channel (Adam's Rib Recreational Area 1996). The wide riparian corridors at Adam's Rib provide multiple planning and engineering benefits and are instrumental in helping the resort to demonstrate to regulators that the development can occur without significant adverse impacts to receiving water quality.

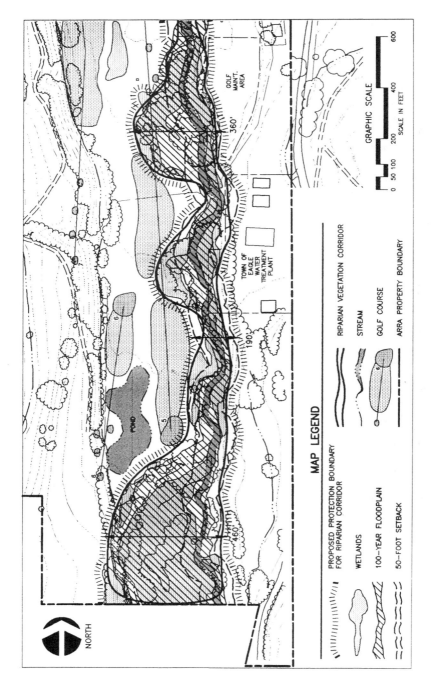

Figure 1. Schematic Mapping of Proposed Protection Boundary for Riparian Corridor at ARRA, Eagle, Colorado

Two watershed management plans recently developed in Eagle County, which address the Eagle River and Gore Creek, also recognize the value of preserving riparian corridors. Vail Associates is playing an active role in the development of the comprehensive Gore Creek Watershed Management Plan. These two plans are likely to provide an additional mechanism for protecting riparian corridors as they are further developed and implemented.

The Town of Snowmass Village is currently undergoing an extensive stream restoration and riparian enhancement project which incorporates stream bank stabilization and habitat improvement with "visual enhancement" that integrates the improvements with the surrounding landscape. Community education and the development of an overall watershed management plan are also being emphasized by the Town. In addition, the Aspen Skiing Company frequently establishes buffer zones along streams that go beyond applicable regulations.

The Summit County regulations are closely tied to the need to strictly limit phosphorus loads from new development projects to Lake Dillon, which has a chlorophyll-a standard of 7.4 µg/L. To meet this standard, "pound for pound" mitigation is required. Developers such as Keystone and Keystone-Intrawest recognize that their ability to attain this objective is inextricably linked to the width of the stream "setbacks" that they utilize. As a result, the riparian buffer zones typically seen on submittals from Keystone and Intrawest go beyond relevant requirements.

Keystone Resort and the base area developer, Keystone-Intrawest, have prepared a master drainage plan for a major expansion of the base area, including a new golf course. Nearly all of the existing major drainageways will be undisturbed during the development process, and buffer zones will exceed county requirements. The golf course will be designed to strictly minimize stormwater runoff into the primary receiving stream, the Snake River. An important design question facing Keystone/Intrawest and other resorts is the extent to which portions of a golf course can encroach into the riparian zone, and under what conditions.

VII.   Buffer Zone Recommendations in Other States

Johnson and Ryba (1992) conducted a literature review of buffer zones recommended by 38 separate investigators to maintain seven major riparian functions. These buffer widths ranged from 3 meters to 200 meters. These large ranges were attributed to the various focuses of those conducting the scientific research. With the exception of grazing, the range centered around 15 to 50 meters. They concluded that buffers less than 10 meters provide little if any maintenance of various riparian functions. Buffers of approximately 15 to 30 meters appeared adequate for most functions. They recommended a minimum buffer distance of 15 to

30 meters, depending on the riparian function to be maintained. A selected summary of some of Johnson and Ryba's (1992) findings according to riparian function to be maintained follows:

- Water Quality: The efficacy of vegetated buffers in maintaining water quality, including sediment removal, fecal coliform reduction, nutrient reduction, and stormwater runoff management generally increases with increasing buffer width. Most investigators recommended buffer widths of 30 to 120 meters.

- Sediment Control: The widest range of recommended widths was for buffers to filter suspended sediments. Four sources recommended a range of 30 to 38 meters. However, recommendations ranged from 3 meters for filtering sand up to 88 meters for filtering clay.

- Temperature Control: The relative degree of shading provided by a buffer strip depends on a range of factors such as species composition, age of stand, density of vegetation. Buffer strips with widths of 30 meters or more generally provided the same level of shading as that of an old-growth stand.

- Wildlife Habitat Protection: Recommended buffer widths for protecting wildlife habitat ranged from 30 meters for salmonid, 67-93 meters for small mammals, 75-200 meters for some birds during the breeding season, and 100 meters for large mammals.

- Benthic Communities: Approximately 30 meters is recommended to protect benthic communities.

VIII. Conclusion

Preservation of riparian corridors is widely recognized as vital to protecting water quality, wildlife/aquatic life, water quantity and aesthetic values along streams throughout the county. In Colorado resorts, and adjoining towns, protection of riparian corridors is particularly important because of the economic dependence of these communities on tourism. However, because of the limited terrain available for development, pressure exists to develop portions of these riparian areas. While regulations exist for minimum setbacks of 7.5 to 15 meters from streams and other waterbodies in most counties hosting resorts, several counties may also require additional setbacks under some circumstances. None of the counties surveyed had specific quantitative requirements specifically designated for the purpose of protecting "riparian corridors"; however, these areas are protected under a combination of flood control, wildlife habitat preservation, and wetland protection regulations. Most counties qualitatively address the value of preserving riparian corridors or riparian vegetation. Further, several counties are exploring additional approaches and/or regulations which may further protect riparian areas while still working to reach cooperative resolutions with development interests. Aside from county requirements, many four season resorts are voluntarily committing to preserving wide riparian corridors because this practice has multiple engineering, economic, social, environmental, planning and other benefits.

REFERENCES

Adam's Rib Recreational Area, 1996. "Proposed Adam's Rib Recreational Area Stream and Riparian Corridor Protection Zone Commitments."

Barton, David R., Taylor William D., and R.M. Biette, 1985. "Dimensions of Riparian Buffer Strips Required to Maintain Habitat in Southern Ontario Streams," *North American Journal of Fisheries Management*, 5:364-378.

Castelle, A.J., C. Conolly, M. Emers, F.D. Metz, S. Meyer, M. Witter, S. Mauermann, T. Erickson, and S.S. Cooke, 1991. *Wetland Buffers: Use and Effectiveness*. Adolfson Associates, Inc. for Shorelands and Coastal Zone Management Program, Washington Dept. Ecology, Olympia, WA. Publ. No. 92-10.

City of Boulder Planning Department and County of Boulder Land Use Department, 1990. *The Boulder Valley Comprehensive Plan*. Revised through December 18, 1990.

Eagle County, Colorado Zoning Regulations, Article 2, Section 2.23.25 "Streams."

EPA, 1990. *Livestock Grazing on Western Riparian Areas*. Prepared by the Northwest Resource Information Center, Inc., Eagle, Idaho.

Johnson, Alan W. and Diane M. Ryba, 1992. "A Literature Review of Recommended Buffer Widths to Maintain Various Functions of Stream Riparian Areas." Prepared for King County Surface Water Management Division. February.

Johnson, R. R., S. W. Carothers, and J. M. Simpson, 1981. "A Riparian Classification System," Paper presented at the California Riparian Systems Conference, University of California, Davis, September 17-19.

Matthews, Frank E. and Kathleen E. Moore, 1989. The Wekiva River Buffer Experience: Land Use in the Fourth Dimension," in *Proceedings of the International Wetland Symposium Wetlands and River Corridor Management*, July 5-9, 1989, Charleston, South Carolina, The Association of Wetland Managers, Inc.

Pitkin County Land Use Code, 1995. Prepared by the Aspen/Pitkin County Planning Department.

San Miguel County, 1992. *San Miguel County Land Use Code.* Amended through October 1, 1992. Prepared by the Board of County Commissioners, Planning Commission, and Planning Department.

Schmidt, Larry J., 1987. "Recognizing and Improving Riparian Values: the Forest Service Approach to Riparian Management," in *Proceedings of the Society of Wetland Scientists' Eighth Annual Meeting,* May 26-29, 1987, Seattle, Washington.

Schueler, Tom, 1994. "Technical Note 7, Invisibility of Stream/Wetland Buffers: Can Their Integrity Be Maintained?," in *Watershed Protection Techniques.* Vol., No. 1, February.

Summit County Community Development Division, 1996. Letter to the Countywide Planning Commission Regarding #95-16: Interim Wetland Regulations. January 30.

Summit County, 1988. *Summit County Land Use and Development Code.* Adopted September 12, 1988 (Resolution #88-52). Effective October 1, 1988.

U.S. Bureau of Land Management (1986) "Draft BLM Riparian Area Management Policy," Paper presented at BLM Riparian Management Workshop. Helena, MT. June 3-4, 1986.

U.S. Forest Service, 1995. "No Surface Occupancy Stipulation Wetlands/Floodplains/Riparian Areas," Serial No. C-56447, Revised March 9, 1995.

White, R.J., 1991. "Objectives Should Dictate Methods in Managing Stream Habitat for Fish;" *American Fisheries Symposium,* 10:44-52.

Development and Application of the Rapid Stream Assessment Technique (RSAT) in the Maryland Piedmont

John Galli[1]

Abstract

In response to a growing demand for both quickly identifying channel erosion problem areas and systematically characterizing stream quality conditions in the Washington Metropolitan Area (WMA), a rapid stream assessment technique was developed by COG in 1992. RSAT represents a synthesis of several well-known stream survey techniques, such as EPA's Rapid Bioassessment Protocols (RBP), together with COG's stream sleuthing experience in the mid-Atlantic Piedmont. RSAT employs both a reference stream and an integrated numerical scoring and verbal ranking approach. Major abiotic and biotic factors that influence overall stream quality have been streamlined, weighed and placed into the six following general RSAT evaluation categories: 1.) channel stability, 2.) channel scouring/sediment deposition, 3.) physical aquatic habitat, 4.) water quality, 5.) riparian habitat conditions and 6.) biological indicators. Under RSAT, the stream, including its channel network, is surveyed in its entirety. Current applications include: watershed-wide stream quality reconnaissance, rapid screening of general stormwater BMP performance, providing supplemental information for EPA RBP level II-, III- and V-type studies, elucidating general watershed land use/stream quality relationships and providing baseline data for incorporation into a regional stream data base. RSAT's intended use and applicability is presently limited to small- to medium-size Piedmont streams.

---

[1] Senior Environmental Engineer, Metropolitan Washington Council of Governments (COG), 777 North Capitol Street, NE, Washington, DC 20002.

Introduction

Recognizing a growing need to both quickly identify existing channel erosion problem areas and systematically evaluate general stream quality conditions on a watershed-wide scale, the Montgomery County Department of Environmental Protection (DEP) contracted the Metropolitan Washington Council of Governments (COG) to develop a set of rapid stream assessment protocols. In response, the Rapid Stream Assessment Technique (RSAT) was developed for Montgomery County, Maryland by COG in 1992. RSAT has been intentionally designed to provide a simple, rapid, reconnaissance-level assessment of stream quality conditions.

The RSAT system represents a synthesis of US Environmental Protection Agency's Rapid Bioassessment Protocols (Plafkin, et al. 1989), the Izaak Walton League and Save Our Streams stream survey techniques (Kellog, 1992), US Department of Agriculture, Water Quality Indicators Guide: Surface Waters (Terrell and Perfetti, 1989), together with COG staff's many years of Piedmont stream survey experience. Presently, the intended use and applicability of RSAT is limited to non-limestone, first through fourth order Piedmont streams with drainage areas less than approximately 100-150 square miles. RSAT employs both a reference stream and an integrated numerical scoring and verbal ranking approach. Major abiotic and biotic factors which influence overall stream quality have been streamlined, weighed and placed into the six following general RSAT evaluation categories: 1.) channel stability, 2.) channel scouring/sediment deposition, 3.) physical instream habitat, 4.) water quality, 5.) riparian habitat conditions and 6.) biological indicators.

In order to provide a quantitative measurement of the six preceding evaluation factors, the RSAT system employs a rigorous field evaluation protocol in which over 30 physical, chemical and biological parameters are measured at approximately 400-foot intervals along the stream (typically, 12-13 riffle transects per stream mile for smaller streams).[2] Data is first recorded via field survey sheets and later transferred into a spreadsheet data base. Transect locations, the presence of storm drain outfalls, fish barriers, stream channel erosion problem areas and other noteworthy observations are additionally mapped onto topographic maps (preferably 1" = 200' horizontal scale or larger). In addition, necessary general reference condition descriptor adjustments have also been incorporated into RSAT to account for differences associated with increased stream/catchment area size (Galli, 1996). Last, photographic information (35mm color slide format) is catalogued so as to provide a permanent historical reference for stream areas surveyed.

RSAT employs a riffle transect-based assessment approach for two main reasons: 1.) in Piedmont stream systems riffles are the principal macroinvertebrate-producing areas and 2.) riffles are both prominent and relatively permanent geomorphological features of a stream; thereby facilitating repeatable and comparable future stream assessment studies.

---

[2] RSAT stream survey work is performed during baseflow conditions, only.

Unlike most bioassessment techniques, under the RSAT system, the stream, including its channel network, is surveyed in its entirety in an upstream-downstream fashion. As implied by its name, RSAT is designed to provide a quick yet accurate assessment of stream conditions. An experienced two-person monitoring team can, under normal field conditions, generally survey 1.0-1.25 stream miles per day (roughly equivalent to 12-15 transects). Although a two-person team approach is strongly recommended, the RSAT survey can be performed satisfactorily by a single investigator.

I.   RSAT Evaluation Categories, Scoring System and Example Survey Form

An example of the 0-50 point RSAT scoring system, as applied in Montgomery County, Maryland to the Brooke Manor Country Club Tributary of Rock Creek, has been included as Table 1. As seen in Table 1, the channel stability evaluation category is weighed slightly more heavily than the other five categories. This was intentionally done to reflect the major influence which the stream flow regime exerts on all six evaluation categories.

Table 1.   Example of RSAT Scoring for Brooke Manor Country Club Tributary

RSAT Evaluation Category	General Verbal Rating Categories and Associated Point Range				
	Excellent	Good	Fair	Poor	Points
1. Channel Stability	9-11	6-8	3-5	0-2	7
2. Channel Scouring/Deposition	7-8	5-6	3-4	0-2	4
3. Physical Instream Habitat	7-8	5-6	3-4	0-2	6
4. Water Quality	7-8	5-6	3-4	0-2	6
5. Riparian Habitat Conditions	6-7	4-5	2-3	0-2	5
6. Biological Indicators	7-8	5-6	3-4	0-2	8

Total Points   Verbal Ranking   Good   Total Score   36

42-50   Excellent
30-41   Good
16-29   Fair
< 16   Poor

An example of a completed RSAT stream survey form has been included as Table 2. As seen in Table 2, upon survey completion stream parameters/conditions are averaged over the entire stream segment length surveyed. These averaged results, together with the RSAT general representative stream characteristics and the investigator's professional experience and judgement, are used in assigning a summary condition score for each of the six RSAT evaluation categories. Within the point range of each of the four verbal rating/assessment categories (i.e., excellent, good, fair and/or poor) discretion is used to add or subtract points based on representative stream conditions, taking into account stream segments observed between transect station locations.

Table 2. Example RSAT Stream Survey Form (modified from Galli, 1996)

Watershed ___Rock Creek (North Branch)___ MC Street Map Book Coordinates ___MC 24: K-6, J-6, H-6___

Tributary Name and Number ___Brooke Manor Country Club Tributary___ D.A. (ac) __498__ Watershed Imperviousness (%) __7.0__ Stream Gradient (%) __1.9__

MDE Class __III__ Date __12/21/92__ Time __0900 hrs.__ Investigators __J. Galli & P. Trieu__ Current Weather Conditions __cloudy/cool in am: sunny/cool in pm__

Last Precipitation Event __12/19/92__ Survey Reach __Brooke Manor golf course to confluence with North Branch (Rock Creek)__

Survey Length (ft.) __5800__ Riffle/Pool Ratio __0.83:1__ Baseflow (cfs) __1.03 @ x-13__ General Accessibility __good (upstream of Emory La.) / poor (below Emory La.)__

Photo No.s __14 (large upper golf course pond), 15 (piping), 16 (x-1), 17 (x-3), 18, 19, 20 (eroding bank & house), 21, 22 (x-13)__

Avg. Canopy Coverage (%) __71%__ (0, 95, 95, 90, 30 85, 90, 60, 90, 75, 70, 80, 65)

Total No. of Tree Falls __2__ No. of Recent Tree Falls __0__

Relative Amount of Trash Present __none - very light__

No. of Observed Fish Barriers: Partial __0__ Complete __1 (@ lower g.c. pond)__

No. of Exposed Sewer Lines __0__

Transect No.	Reach Length (ft.)	Top Channel Width (ft.)	Bottom Channel Width (ft.)	Wetted Perimeter (ft.)	Avg. Riffle Depth (in.)	Avg. Bank Height (ft.)		Bank Stability (%)			Bank Material Type		Riffle Substrate Material Composition	Embeddedness (%)	Substrate Fouling Level (%)	Riparian Vegetation (Type)		Buffer Width (ft.)		Pool Habitat	
						R	L	R	L	Avg.	R	L				R	L	R	L	Max. Depth (in.)	Quality
1	50	10.0	8.0	5.0	1.5	1.0	1.0	100	100	100	S/C	S/SI	R,B,Cb,G,S	95	25	F	F	200	100	none	n/a
2	50	7.5	4.0	4.0	1.5	2.0	2.0	90	90	90	S/L	S/L	S,G,Cb,R,B	100	30	F	F	50	200	20	v. good
3	50	11.0	9.0	4.0	2.0	2.5	1.5	70	85	79	Si/L	S	R,Cb,G,S,B	85	25	F	F	100	200	17	excel.
...																					
N=13																					
Avg.		12.9	10.3	6.5(63%)	2.6	2.0	2.0			77.3				61.5	31.7			100	140	18.5	

Key Abbreviations: C = clay, Sl = silt, S = sand, G = gravel, Cb= cobble, R = rubble, B = boulder, F = forest, Grs= grass, * = highly erodable bank material present,
5 = abundant, 4 = common/abundant, 3 = common, 2 = scarce/common, 1 = scarce/common, Ra = rare, Bd = bedrock, Sap = saprolite, Conc = concrete, Gab = gabion, Sh = shrub

Water Quality Conditions

Tran-sect No.	Time (hr)	Temp (°C) Air	Temp (°C) Water	DO (mg/L)	pH	TDS (mg/L)	Cond. (ms/cm)	Turb. (NTU)	Water Color	Odor
4	1047	3.0	6.0	10.62	7.52	70	0.116	6	clear	none
11	1506	8.0	6.4	11.58	7.13	70	0.126	2	clear	none

Miscellaneous Conditions TDS readings: x-1 = 30, x-2 = 90, x-3 = 70, x-5 = 60, x-6 = 70, x-7 = 70, x-8 = 60, x-13 = 60

Biological Conditions   **Representative Taxa - Abundance**

Stoneflies - 2	Mayflies - 3/2	Caddisflies - 3/3	Midgeflies - 2	Craneflies - 3	Water Pennies - 1	Helgrammites - 3	Dragonflies	Damselflies	Beetles - 2	Segmented Worms - 2	Unsegmented Worms	Snails - 1	Crayfish - 2	Amphipods	Isopods - 1	Leeches - 1	Black Nose Dace	Creek Chub	Darters	Rosyside Dace	Sunfish/Bass	Suckers/Sculpins
×	×	×		×	×	×			×	×	×	×	×		×	×	×	×		×	×	×

Miscellaneous Conditions 1.) stoneflies, mayflies, caddis all common/abundant @ most transects; 2.) both macroinvertebrate and fish diversity are excellent, very high macroinvertebrate biomass; 3.) riffles contain excellent substrate mix; 4.) recommend MdDNR survey for trout introduction potential in lower half of stream

Channel Modification/Storm Drain Outfalls:
Channel Modification R_____ ft.   Gabions _____   Rip-Rap _____   Other _____
                                L _____ ft.   _____   _____   _____
Storm Drain Outfall(s)   24" CMP @ Sycamore La

RSAT Stream Evaluation Summary

Category	Excellent	Good	Fair	Poor	Points
1. Channel Stability	9-11	6-8	3-5	0-2	7
2. Channel Scouring/Deposition	7-8	5-6	3-4	0-2	4
3. Physical Instream Habitat	7-8	5-6	3-4	0-2	6
4. Water Quality	7-8	5-6	3-4	0-2	6
5. Riparian Habitat Conditions	6-7	4-5	2-3	0-1	5
6. Biological Indicators	7-8	5-6	3-4	0-2	8

Total Points   VERBAL RANKING: Good   TOTAL SCORE: 36
42-50 = Excellent
30-41 = Good
16-29 = Fair
<16 = Poor

II.   RSAT Applications

A.   Stream Channel Stability and Cross-Sectional Characterization

As previously stated, one distinguishing feature of RSAT is that the stream, including its channel network, is surveyed in its entirety; thereby, providing a comprehensive picture of overall stream conditions. Depending on stream size, either a 50- or 100-foot long channel reach is selected at each riffle transect and carefully examined for signs of instability, such as bank sloughing/slumping, recently exposed non-woody tree roots, the general absence of any vegetation within the bottom one-third portion of the bank, recent tree falls, etc. which are noted. After the total length of stable area for each bank has been determined, it is converted into a percentage and recorded along with the average percent stability for the entire section surveyed. In addition, general bank stability conditions between transect stations are visually rated and mapped.

As the stream channel is walked, particularly close attention is paid for evidence of major channel downcutting or degradation. Average bank heights provide a good approximate indication of the extent of downcutting along most streams. For example, bank heights which average five feet for a small headwater stream would suggest that downcutting on the order of two to three feet has probably occurred. Other reliable indicators include the presence of nickpoints and exposed concrete footers for retaining walls, weirs, culverts and other man-made instream structures. In urban streams, the presence of exposed sewer lines provide yet another good measure of channel degradation.

One of the more practical applications of the RSAT data is for generating representative channel cross-sections of the surveyed stream. This allows the investigator to: 1.) quickly calculate the average cross-sectional area of the channel, 2.) determine the gross level of channel widening and degradation which the stream has experienced and 3.) importantly, create a historical reference baseline for future follow-up surveys.

Figure 1 provides an example of the use of RSAT for characterizing flow regime-related impacts, associated with urbanization, on stream channel morphology. The stream channel cross-sectional area presented in Figure 1 is illustrative of the major common differences in channel widening and degradation frequently observed between relatively lightly developed headwater streams and older, highly urbanized ones lacking any form of stormwater control.

B.   Bank Material Characterization (Soil Sampling)

At each riffle transect area, the general soil texture of material located in the lower one-third of each bank is classified by feel. Depending on the degree of homogeneity of the bank material, one to three, two- to three-inch deep samples per bank are taken from an exposed soil area with a short soil sampling tube, small hand trowel or equivalent. The dominant textural class is then recorded for each bank on the survey form.

Figure 1.   Average Stream Channel Cross-Section: Brooke Manor Country Club Tributary versus Stoney Creek (Galli, et al., 1996)

Brooke Manor C.C. Tributary (DA: 0.79 mi²; Imperviousness: 7.0%. X-Sect.: 23 ft²)

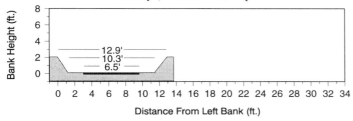

Stoney Creek (DA: 0.68 mi²; Imperviousness: 37.0%. X-Sect.: 122 ft²)

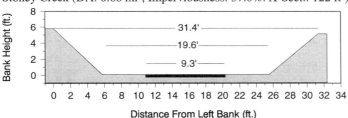

The preceding soil textural information is used for 1.) quick screening of the relative potential erodibility of the stream bank network and 2.) providing insight into both potential in-channel sources of sandy material and possible future susceptibility to high embeddedness levels.

C.   Water Quality

As part of the RSAT survey work, baseflow water quality grab sampling is conducted to provide a snap-shot picture of stream water quality conditions.   The following 10 parameters are measured at regular well-defined intervals along the stream: air temperature, water temperature, dissolved oxygen (DO), pH, conductivity, turbidity, total dissolved solids (TDS), water color and odor, and substrate fouling. Of the preceding 10 parameters, TDS and substrate fouling are used for direct stream to stream comparisons. The three principal reasons for their selection are as follows: 1.) while qualitative, they provide a good indirect measure of overall long-term water quality conditions in a stream, 2.) neither one is readily influenced by prevailing daily or seasonal meteorological/ climatological conditions, and 3.) each one is strongly correlated with   watershed imperviousness levels, macroinvertebrate community condition and overall stream quality.

For example, TDS levels often increase in response to the introduction of a variety of pollutants such as sewage from septic field/sanitary sewer line leakage, road salts, fertilizers, etc.   Substrate fouling level, which for RSAT purposes is defined as "the percentage of the underside surface area of a cobble-sized stone (or larger), lying free on the streambed, which is coated with a biological film or growth, " can similarly provide a qualitative measure of chronic nutrient (primarily nitrogen) and organic carbon loadings in a stream. In relatively clean, non-limestone, Piedmont streams, baseflow TDS levels are typically 50 mg/L or less and substrate fouling levels are normally on the order of 10 percent or less (Galli, 1995).

Figure 2 has been included to illustrate the general relationship typically observed between average substrate fouling levels and watershed imperviousness.

Figure 2. Rock Creek Tributaries: Average Substrate Fouling Versus Imperviousness (modified from Galli, et al., 1996)

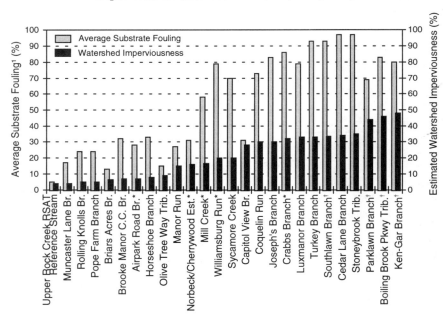

Substrate fouling rating scale: 0-10% = Excellent, 11-20% = Good, 21-50% = Fair, >50% = Poor.
* Major SWM controls.
† Predominantly commercial/industrial land uses.

D.   Macroinvertebrate Community Characterization

As part of the RSAT evaluation, a screening level biosurvey of the stream's riffle macroinvertebrate community is performed. The primary purpose of the biosurvey is to characterize macroinvertebrate community composition and relative abundance of major representative taxonomic groups, so as to shed additional light on overall stream quality/level of impairment. It should be noted that in Piedmont stream systems riffles are the primary macroinvertebrate-producing areas and the majority of the fish in these streams rely upon macroinvertebrates for food.

The standard RSAT biosurvey protocol involves both turning over 10 cobble-size stones (or larger), as well as, taking a minimum of three one-square foot, 30-second kick samples per riffle transect station. Kick sampling of smaller streams is performed using a minimum six-inch wide, fine-meshed dip net. For larger streams, a 12-inch wide D-net is used. Macroinvertebrate identification is performed at each riffle transect site via visual examination. Individuals are identified to taxonomic order and, whenever possible, to either the family or genus level. Representative individuals are captured and placed into a voucher collection for either further identification or future reference. General pollution tolerance for major taxonomic groups is per Bode et al., 1991 and Lenat, 1993. Macroinvertebrate relative abundance categories used in the biosurvey are comparable to EPA's RBP level I and are as follows: 1.) absent/no group found, 2.) scarce, 3.) scarce/common, 4.) common, 5.) common/abundant and 6.) abundant. Relative abundance ratings are made based on the investigators experience and judgement.

Figure 3 provides but one example of how RSAT can be used for both characterizing general macroinvertebrate community conditions along a stream continuum and for identifying stream quality problem areas. The stream depicted in Figure 3 originates in a semi-rural setting, flows through a small multi-purpose lake and then through an older, predominantly residential suburban area.

Figure 3.   Rock Creek Mainstem: General Macroinvertebrate Community Condition (modified from Galli, et al., 1996)

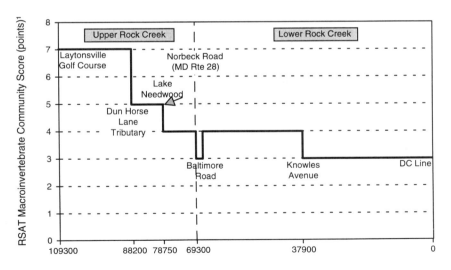

Distance in Feet along Mainstem from DC Line (0 ft.) to Headwaters (109300 ft.)

[1] Macroinvertebrate rating scale: 7-8 pts. = Excellent, 5-6 pts. = Good, 3-4 pts. = Fair, 0-2 pts = Poor.

E.   BMP Performance Monitoring

Another RSAT application is in the area of BMP performance monitoring. RSAT may be employed to provide a screening-level analysis and/or to supplement more rigorous monitoring studies. Figure 4 shows how various individual parameters of interest can be examined via RSAT. As seen in Figure 4, before and after substrate embeddedness levels associated with the construction of a flow diversion-parallel pipe storm drain system, designed to improve physical aquatic habitat conditions in a small urban stream, were tracked using RSAT.

**Figure 4.   Flora Lane Tributary Embeddedness[1]
1993-1995 RSAT Survey Averages (Galli and Corish, 1996)**

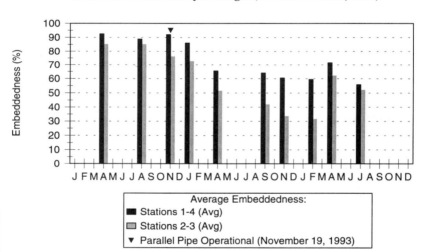

[1] Embeddedness (0-25% = Excellent, 26-50% = Good, 51-75% = Fair, > 75% = Poor)

III.   Future Direction

While RSAT has already been applied to over 140 Piedmont stream miles, RSAT methodology will continue to be evaluated, updated and where deemed necessary, revised. Additional short- and long-range goals include: 1.) expanding the reference stream data base to include limestone streams and more agricultural and forested watershed areas; 2.) creating a regional stream channel morphology data base for the WMA; 3.) developing an RSAT training course; 4.) integration with both local and regional GIS systems and 5.) incorporation of innovative technical improvements, such as the use of digital photography.

References

Bode, R.W., M.A. Novak, and L.E. Abele, 1991. Methods for Rapid Biological Assessment of Streams. NYS Dept. of Environ. Cons., Albany, NY.

Galli, F.J., 1995. Water Quality Grab Sampling of Streams in Montgomery and Prince George's County, Maryland - Unpublished Notes. Metropolitan Washington Council of Govts., Wash. DC.

Galli, F.J., 1996. Technical Memorandum: Rapid Stream Assessment Technique (RSAT) Field Methods. Prepared for Montgomery County Department of Environmental Protection, Mont. Co., MD. 36 pp.

Galli, F.J., K. Corish, J. Lawson, P. Trieu, 1996. Rapid Stream Assessment Technique (RSAT) Survey of the Rock Creek Watershed. Prepared for Montgomery County Department of Environmental Protection, Mont. Co., MD. 86 pp.

Galli, F.J., K. Corish, 1996. Draft - Rapid Stream Assessment Technique (RSAT) Monitoring Results of Wheaton Branch, Crabbs Branch and Flora Lane Tributary - In Preparation. Metropolitan Washington Council of Govts., Wash., DC.

Kellog, L.L., 1992. Save Our Streams - Monitor's Guide to Aquatic Macroinvertebrates. Izaak Walton League of America, Arlington, VA. 46 pp.

Lenat, D.R., 1993. A Biotic Index for the Southeastern United States: Derivation and List of Tolerance Values, with Criteria for Assigning Water-Quality Ratings. J. N. Am. Benthol. Soc. 12:279-290.

Plafkin, J.L., M.T. Barbour, K.D. Porter, S.K. Gross and R.M. Hughes, 1989. Rapid Bioassessment Protocols for Use in Streams and Rivers: Benthic Macroinvertebrates and Fish. U.S. EPA, Off. of Water. EPA/444(440)/4-39-001, Wash. DC.

Terrel, C. and B. Perfetti, 1989. Water Quality Indicators Guide: Surface Waters. USDA Soil Conservation Service, P.O. Box 2890, Washington, DC 20013. SCS-TP-161. 129 pp.

Cumulative Impacts of Watershed-Scale Development
on Stream Morphology

Michelle A. Girts[1], William R. Blosser[1], and Thomas T. Ogee[2]

Abstract

NEPA requires that the cumulative effects of impacts from projects be considered in addition to direct and indirect impacts. For most projects, cumulative impacts are not a major consideration, but in the case of linear infrastructure (highway, railroad, pipeline), cumulative impacts can be significant and multiple watersheds involved. In the Pacific Northwest, many of the streams along which transportation corridors run support special status fish species.

Union Pacific Railroad proposed to add a second track in nineteen 2-mile segments over approximately 150 miles through the Blue Mountains of eastern Oregon. CH2M HILL developed a watershed-based methodology to evaluate cumulative impacts associated with the project. Following assembly of regional background information, field reconnaissance, review of the literature regarding cumulative impacts, and extensive meetings with the permitting agencies, the methodology was applied to determine cumulative impacts. Cumulative impacts attributable to Union Pacific Railroad activities were identified in two watersheds with narrow valleys and steep stream gradients: Meacham Creek and Dry Creek in the Umatilla River and Grande Ronde River drainages, respectively.

[1] Project Manager and Sr. Project Manager, CH2M HILL, 825 NE Multnomah, Suite 1300, Portland, Oregon 97232

[2] Project Manager, Engineering Division, Union Pacific Railroad, 1416 Dodge Street, Room 1030, Omaha, Nebraska 68179

Project Description

Since 1986, Union Pacific Railroad (UPRR) train growth through the Pacific Northwest increased annually by about 5 percent. From 1993 to 1998, the company anticipates annual growth rates from 1.5 to 4.5 percent, which would further expand its traffic base. However, there are physical constraints to increased traffic in Oregon, primarily in the Blue Mountains.

Throughout much of the Pacific Northwest, UPRR typically handles 35 to 40 trains per day, with peak days approaching 45 to 50 trains per day. UPRR's main line in Oregon currently carries an average of 26 freight trains per day as well as two passenger trains. Both the number of trains on UPRR's main line and the average number of cars in each train have increased in recent years. Bottlenecks have developed on the line where long, bulk-transport trains are forced to travel at low speeds because of steep track grades and sharp curvatures. They occupy the main line track for long periods, thereby preventing high-speed, time-sensitive, intermodal trains from passing. It is these bottlenecks that UPRR will correct with its proposed Blue Mountains project.

UPRR's main line enters Oregon at Huntington and continues east through Baker City, La Grande, the Blue Mountains, and Pendleton before reaching the Columbia Gorge and continuing on to Portland. To relieve the constraints affecting track capacity on this line, UPRR will spend more than $120 million over 10 years, starting in 1993, to install a second main-line track at 19 locations in the Blue Mountains. These capital improvements will enable UPRR to better meet the demands of its customers for timely freight and passenger service, as well as to serve new customers.

In undertaking improvements to the 19 project segments, UPRR focused on using existing rights-of-way and minimizing disruption to the surrounding environment. The first theoretical issue facing the team preparing a Cumulative Impact Assessment (CIA) was to define "the project" against which impacts would be assessed. Defining the project is difficult because construction of the project segments is staged over 10 years and detailed engineering for these segments was not done at the time of the CIA. UPRR prepared a theoretical or prototypical project description that could be used as a basis for assessing impacts. Because operational impacts can also be significant, operations and maintenance protocols were also developed.

Cumulative Impacts Assessment Methodology

NEPA is the legal basis for requiring a CIA. Under this act, the Council on Environmental Quality (CEQ) prepared regulations calling for the consideration of cumulative impacts. CEQ requires environmental impact statements to "anticipate a cumulatively significant impact on the environment from Federal action." The term "federal action" has been interpreted by the courts and CEQ to include any action regulated by the federal government.

Cumulative impacts are defined by CEQ as follows:

> ...the impact on the environment which results from the incremental impact of the action when added to other past, present, and reasonably foreseeable future actions regardless of what agency (Federal or non-Federal) or person undertakes such actions. Cumulative impacts can result from individually minor but collectively significant actions taking place over a period of time.

While the assessment of cumulative effects is typically a minor portion of an agency's NEPA review, it has gained greater significance in this project. The U.S. Army Corps of Engineers (USCOE), the lead federal agency, made an initial determination that construction of project segments can avoid significant environmental impact through appropriate mitigation. This initial determination is tested through the permit review process required for each project segment. However, the agencies and the tribes expressed concern about the project's cumulative effects. As a result, the process of assessing the cumulative effects of the entire project has been separated out from environmental review of the individual project segments. This represents a unique approach to the consideration of cumulative effects and, accordingly, required development of a new methodology for their measurement.

Development of Methodology

The methodology for this CIA was developed and discussed in several meetings with the UPRR Blue Mountains Project Technical Working Group, which included representatives of state and federal resource agencies, Confederated Tribes of the Umatilla Reservation (CTUIR), UPRR, and CH2M HILL.

It was agreed by this group that the following items are the critical components of a CIA:

- Identifying the actions (activities that generate, induce, or contribute to cumulative effects) creating the impacts
- Identifying the impacts (actual damages or benefits to a system) of the projects on the surrounding environment
- Identifying the affected system (entire ecosystem or particular resources within an ecosystem; Eckberg, 1986)

To determine the actions, the number, location, size, and type of projects proposed are considered. The assumptions outlined in the project description, the prototypical designs and construction standards, and the operations and maintenance protocols define the standards for project actions. The impacts are measured according to a resource's abundance, viability, type, overall landscape integrity, and overall value (Hirsch, 1988). The system is defined in terms of space (impact area) and time (duration of the impact).

Before a determination of the significance of impacts can be made, goals are established by which to measure cumulative effects. These goals identify the environmental conditions that are preferred in any given area.

Blue Mountains Project Actions. The UPRR Blue Mountains project contains the following anticipated actions that may contribute to cumulative impacts.

- Excavation (for example, widening right-of-way, creating drainage channels, and constructing structures)
- Filling (for example, widening roadbed, constructing roads, and disposing of waste material)
- Blasting (for example, widening roadbed)
- Construction of structures in and adjacent to jurisdictional waters and aquatic habitats (for example, bridges, culverts, and rail line)
- Maintenance (for example, replacing ties, removing snow from access roads, repairing culverts and channels, maintaining ballast, repairing fence, and controlling vegetation)
- Ongoing operation of railroad (for example, traffic delays, noise, wildlife collisions, runoff and erosion, trash and debris disposal, contaminant spills, and stockpiling of ballast and other materials)

Potential Cumulative Impacts. On the basis of preliminary discussions with resource agencies and an understanding of project actions and the general environment in which they will occur, this analysis focuses on measuring the cumulative contribution of the proposed projects to the following types of impacts:

- Loss of floodplain area and function
- Stream realignment
- Loss of riparian vegetation
- Water quality impacts
- Floodplain erosion and bank stability
- Reduction of fish habitat and fish passage
- Reduction of large organic woody debris recruitment in streams
- Wildlife and fish mortality and obstructions to movement
- Population and habitat changes of sensitive plant, fish, and wildlife species
- Intrusion of weedy nonnative species

Only impacts related to stream morphology and fish habitat are addressed in this paper.

Systems to Be Studied. Defining the system or area in which the analysis will be performed is one of the most challenging aspects of a CIA. As noted earlier, the system needs to be defined in terms of both space and time.

Obviously, the boundary of the study area must fully enclose the system functions to be evaluated. There is considerable case law on the issue of what constitutes a proper study area. Principally, the study area selected for each resource must be appropriate to the type of impact that may affect that resource. Thus, for some resources, the area of potential impact may be as large as a river basin; for others, it may be as small as the railroad right-of-way. River basins, or hydrologic units, are usually the most appropriate systems in which to study fish, water quality, riparian vegetation, wildlife, recreation, and cultural values.

On the basis of these considerations, two study areas were defined. The first is the immediate project area, which is defined as an area extending 0.25 mile on either side of the centerline of the existing UPRR tracks and approximately 0.5 mile beyond the end of each segment, because of the possibility of construction staging impacts. There are 19 segments, each averaging 2.7 miles in length. The total length of double tracking is approximately 52 miles, distributed along 147 miles of railroad alignment.

The second study area, the extended project area, generally is defined as the watershed from the farthest downstream point of the farthest downstream segment to the headwaters. However, an exception to this rule was made for the Meacham watershed because potential UPRR cumulative impacts were thought by the UPRR Blue Mountains Project Technical Working Group to extend to the confluence of Meacham Creek and the Umatilla River. Five watersheds were identified:

- Meacham Creek upstream of its confluence with the Umatilla River
- Grande Ronde River above La Grande
- Pyles Creek above its confluence with Catherine Creek (also a subbasin of the greater Grande Ronde watershed)
- Powder River upstream of the point approximately 1 mile below its confluence with the North Powder River
- Burnt River watershed above Huntington (including Huntington; less than 1 mile from the confluence of the Burnt River and the Snake River)

Use of this definition of the extended project area ensures that all upstream factors contributing to the conditions in a particular segment are considered. Also, the impacts of construction downstream of the segment can be evaluated apart from the non-project-related factors, such as land management or road construction. Impacts along the entire line, such as those related to increased frequency of train traffic, will also occur in the individual segments and therefore can be assessed.

All actions proposed in the UPRR Blue Mountains project are included in the CIA because they are "reasonably foreseeable future actions." The time frame needs to be sufficiently long to include recovery from impacts (Lee and Gosselink, 1988). Some identified resources may recover from disturbance within hours or days, while others may respond over years.

The temporal scale for evaluation of cumulative impacts for the UPRR project extends back to the time of exploration and settlement of the project area, the early 19th century, as previously described. The analysis includes impacts expected to occur in the future until 2014, which is 10 years past the date of the last planned project. This extended time frame will allow the impacts and mitigations of the earliest project action to be well documented, which, in turn, will make it possible to determine the proper cumulative impacts that could extend beyond 2014. The following assumptions have been made with respect to the future impacts of non-railroad activities in the immediate project area and extended project area:

1.   Logging, ranching, mining, and agricultural activities will continue at approximately the same rate as in 1994.

2.   The methods used in implementing these activities may have less of an impact on the study area than methods used currently and historically. The UPRR Blue Mountains Project Technical Working Group debated how to define that lessening of impact. The subject of what are appropriate management practices for logging, ranching, mining, and agriculture is being broadly debated at present. Adopted federal land management plans offer a some direction on what the policies are and will be, but most of these plans are being revised and/or are in litigation. Forecasting management practices on private lands is even more uncertain, because there is no clear process in place for adopting management regulations. Thus, for purposes of this CIA, we decided to assume that the policies embodied in existing federal land management plans would be followed in the future and that there would be no substantial changes to present practices on private lands.

3.   No new major roadways or transportation corridors will be constructed.

Goal Setting (Desired Future Conditions and Threshold Establishment). Cumulative impacts cannot be evaluated unless there are goals against which to measure the impacts. That is, a system may be able to tolerate change (impact) and remain viable up to a certain point. Beyond that point, however, it may deteriorate rapidly. Two types of goals can be set. The first, called a threshold, is the absolute minimum acceptable environmental condition that will still allow a system to survive, albeit at levels of performance considerably below natural conditions. The second, called a desired future condition, is a goal, usually set by public resource agencies, for a level of environmental condition that is above the threshold. It is a measure or description of the condition that is acceptable or desirable for continued viability of the resource.

The definition of goals allows agencies to judge individual permit applications in terms of cumulative impacts. As impacts on one or more parameters approach the limits set by the goals, projects may have to be modified to reduce cumulative impacts, or permits may be denied. If specific project cumulative impacts are minor and do not cause parameters to approach the limiting values of the goals, permits can be processed without additional steps (assuming that project-related direct impacts are adequately mitigated).

Unfortunately, scientific information and methods needed to support quantitative assessment of cumulative impacts are lacking (Hirsch, 1988). There have been significant advances in recent years in certain methodologies to assess the cumulative impacts on certain resource parameters, but such advances have not been observed in other fields. However, the greatest limitation to a quantitative assessment of cumulative impacts is not methodology as much as it is lack of data. In preparing this CIA, we discovered very early that data for many key potential measures, such as existing or past areas of riparian vegetation, do not exist. In consequence, this CIA relies on professional judgment where quantitative information was not available.

The agencies and UPRR discussed at length the issue of whether a methodology using nonquantitative methods could fulfill the goals of NEPA and meet the needs of the agencies to identify significant negative cumulative impacts needing mitigation. After reviewing a draft of the CIA, the group concluded that the methodology adequately identified the areas where significant negative cumulative impacts had occurred or would occur with the addition of the proposed projects. The next step, which necessarily would need to be more quantitative and location-specific, would be to develop a mitigation plan for the affected areas

A decision concerning the tradeoff between theoretical and applied cumulative impacts assessment methodology was also reached. Although the concept of thresholds is discussed in the literature, researchers and agencies have rarely defined and applied thresholds. This infrequency of application is logical, because the object of most agency work is not to determine the point at which an ecosystem collapses, but rather to identify the point at which it can remain viable. Thus, the agencies' effort has been put into identifying goals or desired future conditions, not thresholds. Consequently, the concept of thresholds was dropped from this CIA.

The following primary sources were used for setting desired future conditions:
- Corps of Engineers: floodplain regulations.
- Oregon Department of Fish and Wildlife (ODFW): Letter from Tim Bailey, February 17, 1994.
- Umatilla National Forest Working Group: Draft Proposal on Desired Future Conditions (Umatilla National Forest Working Group, 1994) referred to herein as the Umatilla National Forest Working Group proposal. It should be noted that the Umatilla National Forest Working Group revised its proposals during the time this CIA was being prepared. The project team used the most recent revisions, but the reader should be

aware that considerable further revisions are possible before the proposals is finally adopted.

- Oregon Department of Environmental Quality: 1988 Statewide Assessment of Nonpoint Sources Water Pollution (NPS Assessment; Oregon Department of Environmental Quality, 1988); and State Water Quality Standards.
- Environmental Protection Agency: Endangered Species Act (1977).

Determination of Impact. The methods used to rate potential impacts are largely subjective, because most of the analyses did not lend themselves to quantified methods, and site-specific construction techniques and designs are unknown for all but three segments (constructed in 1993). Potential impact ratings are ranked on a six-point scale from "major" (maximum impact) to "none" (no impact), based on indicators, criteria, and best professional judgment as identified in each analysis (Table 1). "No impact" ratings were reserved for cases where a floodplain, stream channel, or wetland was so far away from the UPRR right-of-way that widening would have a very low probability of affecting an adjacent waterbody.

Table 1. Impact Rating Scale

Criteria	Major	Substantial	Moderate	Minor	Negligible
Floodplain area	Decreased >25%	Decreased 10-25%	Decreased <10%	No change in active channel	Unlikely noticeable change
Alteration of channel morphology	Fill into active channel	Fill adjacent but not into active channel	No hydraulic changes	No change	Unlikely noticeable change
Sedimentation	Likely >100 ids downstream	Likely >100 yds downstream	Localized and temporary	Localized and temporary	Unlikely noticeable change
Loss of riparian vegetation	Entire	Entire	Less than 25 feet of riparian corridor remaining	At least 25 feet of riparian corridor	Unlikely noticeable change
Potential for revegetation	Minimal	Localized	Expected over time	Expected over 1-2 years	Unlikely noticeable change
Soil erosion	Likely >100 yds downstream	Likely >100 yds downstream	Localized and temporary	Localized during high flow events	Unlikely noticeable change
Loss of non-riparian vegetation communities	Entire	Partial	Partial	Single season partial disturbance	Unlikely noticeable change

Resulting Applied Methodology. To make the methodology and assumptions described above explicit, each potentially affected resource has been analyzed in accordance with the following steps:

1. Resource Function: a description of the resource and its function within the ecosystem.

2. Potential Impacts: the project actions that could adversely affect the resource function.

3. Indicators: observable indicators of resource viability.

4. Historical Impacts on Indicators: historical conditions in the extended project area that have modified indicators.

5. Status of Indicators Before UPRR Project: a description of the present condition of the resource as measured by the indicators prior to the construction of the new project segments.

6. Desired Future Conditions: ideal indicator and resource status and function in the future, as expressed in terms of agency standards or guidance. Where no such standards or guidance exist, they were proposed by the project team.

7. Status of Indicators after UPRR Project (Project Impacts): a comparison of the identified indicators after project construction with the existing conditions and desired future conditions (item 6 above). The discussion of project impacts (direct and indirect) normally would not occur within a discussion of cumulative impacts. However, in this case it seems necessary to include it because this document is being issued separately from an environmental assessment or environmental impact statement, and the reader thus would have no context to judge the project's cumulative impacts without knowing the expected project impacts.

8. Significance of Project Impacts: a discussion of the significance of project impacts (direct and indirect, but excluding cumulative), based on existing conditions and desired future conditions.

9. Significance of Cumulative Impacts: a discussion of the significance of the impacts of the specific projects in conjunction with the cumulative impacts of previous, proposed, and known future actions.

Historical Setting

The railroad has existed in this corridor since the 1870s, when the first transcontinental railroads were being constructed. Although the line today is largely in the same place in which it was constructed more than 120 years ago, realignments have straightened it to increase its safety and capacity. What exists today is the result of a large initial construction project and many smaller projects occurring over more than a century.

The following is a review of the historical record of the entire impact area to help the reader understand the context in which the railroad was constructed and operates. While this review is not quantitative, it will provide a clear qualitative picture of the types of human activity that have occurred within the corridor and their likely effects. The historical record is not sufficiently detailed to provide answers to some fascinating questions about the relationship between a certain actions and impacts those actions may have caused. For example, did construction of the railroad decrease the overall impacts in the corridor by replacing wagon traffic that was very destructive to the countryside, or did it increase the overall impacts by enabling more settlers to come to the area and extract more resources? Causes and effects are so intertwined that untangling them is a matter of much historical conjecture.

Prehistory. The area from Pendleton to Huntington has been occupied by humans since the last glaciation as a result of immigration from Asia to North America, at least 13,000 years ago. One site a few miles south of La Grande has been dated to as early as 8,000 years ago. Ethnographic studies identify two cultural areas: the Southern Columbia Plateau and the Northern Great Basin. Native American plateau cultures within the study area included the Umatilla, the Cayuse, the Walla Walla, and the Nez Perce, all of whom were semi-sedentary hunters and gatherers. Anthropologists estimate that 41 bands of Nez Perce occupied the Grande Ronde Basin, or about 6,000 people (Guthrie, 1994). The Paiutes, of the Northern Great Basin culture, were food gatherers more than hunters and ranged over a large area, including the Blue Mountains, to supplement the less dependable resources of the Northern Great Basin. By the 1700s, annual trading events took place between the various Native American groups. The largest were in the Grande Ronde Valley and at Wallowa Lake, and trails interconnected the entire area. The plateau cultures most closely associated with the Blue Mountains raised horses, which they acquired in the early 1700s and grazed in the river valleys; as many as 5,000 horses were counted by mountain man Joe Meek in one Grande Ronde herd. It is possible that overgrazing occurred where these horses were corralled in large numbers, particularly in the drier areas (Guthrie, 1994). Native Americans also burned areas of underbrush, particularly adjacent to streams, to increase hunting success. Annual salmon catches by the tribes in the Columbia Basin are estimated as being equivalent to catches obtained through capital-intensive means in the early 1900s, the peak of historical fish harvest. Vegetables apparently were cultivated on a small scale.

Exploration and Settlement. Sporadic visits to Oregon and Washington from the sea were made by the Spanish, Russians, British, and Americans prior to the 1800s. Tales of bounty, primarily of furs that could be sold in the Far East, drew attention to the potential for wealth from Pacific Northwest resources. Exploration of the Pacific Northwest by people from other lands began in earnest with overland expeditions in the early 1800s, including the well-known trips by Lewis and Clark and John Jacob Astor.

The Oregon Trail (which was never a single trail, but, rather, a loose web of trails generally following the same route) through the Blue Mountains was used during this early period of exploration for both westward movement and eastward return journeys. With the establishment of the fur trade by the North West Company and the Hudson's Bay Company, furbearing animal populations were hunted and trapped almost to extinction; by 1829, the sea otter had been all but exterminated, and by the 1830s, a lack of beaver caused beaver trappers to shift to muskrat.

Along with fur trapping came ancillary activities: increased fishing and hunting for food; logging for structures; fires for clearing, warmth, and cooking; and clearing for trails and building. Missionaries, doctors, and teachers began to arrive in noticeable numbers in the Pacific Northwest in about 1834. The most influential early missionaries were the Whitmans and the Spaldings, who set up a mission near Walla Walla that served as a major provision point for the settlers who followed. With the missionaries came more logging and clearing for building, fires, and farming. As wagons became a major mode of transportation, trails widened into roads with associated dust and erosion.

News reached the East that emigrant travel to Oregon was feasible, and the first wagon trains were organized in 1841. While the route varied, the most frequent route used for travelers to the Pacific Northwest started in Independence, Missouri, and wound through what is now Kansas, Nebraska, Wyoming, and Idaho, crossing the Snake River at Farewell Bend about 4 miles south of Huntington (Evans, 1990). The Oregon Trail is adjacent to or within 8 miles of the current UPRR right-of-way from Huntington to Pendleton; a side route taken by Whitman and a few subsequent groups descended directly through the Meacham Creek Canyon. Diaries of the emigrants document the resources found in the area in detail, and the environmental effects of the emigration can be traced over the years (Evans, 1990). Early emigrants found their way through the mountains by driving their wagons up streambeds full of brush and trees, probably willow and alder. They observed dense forest, park-like savannahs, and open grasslands devoid of any shrubs or trees. They observed fish running in the streams and noted the smoky haze that often hung over the mountains in the fall along with other signs of frequent fire (most emigrants arrived at Farewell Bend in September).

By 1849, after some 11,600 people had traveled the Oregon Trail, conditions had changed. Writers commented on garbage and animal carcasses littering the trail, the wide, dusty track, and the lack of wood for fires, which lead to extensive tree cutting at some distance from the trail. In many areas, the track was wide enough for four wagons to travel side by side. Many of the oxen and cattle died from lack of food; not enough grass was available for grazing, and the streams and rivers ran muddy, with carcasses lying midstream. Fire scars were frequent along the mountainous portions of the route. Emigrants hunted and fished to augment their provisions. Also because of their lack of provisions, economic incentives developed among the earlier settlers and the tribes to farm, fish, and hunt for trading or sale to the emigrants. Even at the time of the earliest wagon trains, Native American tribes traded potatoes, corn, peas, beans, squash, and horses for cattle and tools. Emigrants jettisoned plant materials and seeds, which, in addition to the seed sources that were inadvertently carried on wagons and livestock, and in combination with extensive soil disturbance, undoubtedly introduced many nonnative species along the trail.

Economic Development. Gold was discovered in California in 1848, and by 1849 more than 24,000 people traveled to California to find their fortune. Gold was discovered in Oregon by the Meek emigrant party of 1845, but was not relocated until prospectors found placer gold in the Burnt and John Day Rivers in the 1850s. The major gold-producing area in northeastern Oregon includes nearly all of the Powder and Burnt River drainages, and a small portion of the southern headwaters of the Grande Ronde River, John Day River, and Malheur River. Mining of placer gold deposits began in northeastern Oregon in 1862, as a result of the movement of prospectors from California through all the mountain areas of the West (Brooks and Ramp, 1968). Development of most of the important underground mines had been initiated by the 1880s, but they never contributed as much to gold production as placer mining. By 1865, placer deposits were being worked in many of the drainages within the study area, including a tributary of the Powder River a few miles south of Baker.

The first mining settlement in the region was established in Baker in early 1862; by the end of the year, the size of the camp was estimated at 5,000 to 6,000 persons. Ditches of great length were dug and water was diverted from streams and rivers to support sluicing and hydraulic mining (parts of some of these ditches continue to be used, as part of the Baker City water system and for irrigation). Supplies and equipment were hauled from The Dalles, and networks of rough roads and trails were developed. Placer mining was done with hand-operated equipment until the early 1900s, when dredges came into use. The first floating bucketline dredge was operating on Burnt River near Durkee in 1900. Sumpter Valley, in the Powder River watershed, is the largest dredge field in the state (approximately 8 miles of the mile-wide valley below Sumpter are mined). A dredge began working there in 1913 and removed enough gold to double the state's 1912 gold output. Elsewhere in the Powder River drainage-for example, on the east flanks of Elkhorn Ridge-nearly every stream and gulch shows evidence of early placer operations. These operations took their toll on entire aquatic systems, moving sediment into the streams and decimating the aquatic food chain (algae, insect larvae, clams, etc.). Stream flows were forced under and through the gravel in many reaches that had been worked. Since the early 1900s, gold production in the area has varied with the price of gold. Gold production dropped in the 1920s, picked up during the Depression of the 1930s, and virtually ceased with the curtailment of gold mining during World War II. Placer mining operations in northeastern Oregon have continued to lead the region in gold production since World War II.

The rush to the gold fields encouraged grazing in the project area. In the first 8 months of 1862, 46,000 head of cattle were shipped by steamship to The Dalles and driven overland to eastern ranches (Lavender, 1958). Horses, mules, hogs, and sheep were also shipped by steamship in large numbers. The heavy use of livestock for packing and freighting in the mining areas created a demand for hay, and farms sprang up in the lower valleys to fill this demand. Produce farms were established in Baker on the Powder River and along the Grande Ronde and Umatilla Rivers. Also associated with the mining boom were clearing and logging for construction materials and fuel. Hunting, trapping, and fishing augmented food shipped in from The Dalles. Human-caused fires were common, both in the forests and in towns. Roads and clearing brought erosion and impaired water quality. Because of the continuous movement of people and goods into and out of this region via wagon and boat, nonnative plant species were undoubtedly spread throughout the area.

Early indications are that commercial logging began, on at least a moderate scale, in the 1880s (McIntosh et al., 1994). In the late 1880s to 1919, the Grande Ronde River was a major log-driving river. Splash dams were built on the Grande Ronde River at Perry, at Vey Meadows, and on several tributaries to provide flow to move logs year-round. These splash dams and the associated log drives caused considerable damage to the stream channel and aquatic habitat. It is not clear whether such practices also occurred in the Meacham Creek drainage or in the forested tributaries of the Powder and Burnt Rivers, but logging did occur in these areas to some degree.

Farming became established throughout the overall project area in the late 1800s. The Homestead Act of 1862 promoted establishment of farms. The Desert Land Act of 1877 allowed a settler 640 acres (in contrast to the normal homestead allotment of 160 acres) if 80 of the acres were irrigated (Lavender, 1958), which stimulated irrigation diversions. Private irrigation companies were founded before the 1880s to boost the value of land. The favored dryland crop was wheat, which flourished in the project area. The vision of extensive wheat fields was promoted by the Oregon Steam Navigation Company (OSN), so that downriver steamboat traffic might equal the volume and profits of upriver traffic. Early records of live-stock grazing in the upper Grande Ronde Basin suggest that the area had been overgrazed by the 1880s (Skovlin, 1991).

As of 1979, between 50 and 75 percent of the land in Umatilla County is in farming and grazing, while 25 to 50 percent of the land in Union and Baker Counties is in farming and grazing (Highsmith and Kimerling, 1979). Hay, alfalfa, and sweet corn have become important products in the southern sections of the project area, while a variety of vegetable and feed crops are grown in Umatilla County under irrigation. Cattle and sheep are the major grazing stock. Surveys in the 1970s estimated more than 50,000 acres of grazing land in Baker County, 30,000 acres in Union County, and 25,000 acres in Umatilla County (Highsmith and Kimerling, 1979).

Diversion of stream flows for irrigation and stock watering accompanied the development of farming and grazing. Additional clearing and removal of riparian vegetation triggered erosion, and wildfires were common. In later years, with the advent of farm machinery, streams were channelized to create more easily farmed parcels. The introduction of foreign seeds and soil disturbance from cultivation and grazing helped nonnative vegetation to increase its foothold in the area.

Transportation System Development. Mining, logging, and agriculture required tools, machinery, and movement of goods for distribution and sale. The major sources of tools were St. Louis or San Francisco. From San Francisco, the major trade routes were the old Overland Trail through Nevada, across the mountains from the upper Sacramento Valley, or by coastal vessels to the Columbia River. From Astoria, goods were shipped by OSN to The Dalles or Umatilla, then hauled along the Oregon Trail to Boise, Fort Hall, and the northern Rockies (Lavender, 1958). The Oregon Trail continued to move people and supplies through the region long after the early emigrants had passed.

OSN was a conglomerate of portages, steamers, barges, landing facilities, and roads that had been formed in 1861 (Asay, 1991). Oregon's first railroad was built in 1862 along the Columbia River rapids known as the Cascades to replace a mule-powered wooden tramway, on the north side of the river, that had been damaged in winter floods. Crews then built a railroad from The Dalles to Cellilo Falls, to move goods around The Dalles rapids. The Northern Pacific Railroad was chartered in 1864 with the intention of building a rail line on the south side of the Columbia River. OSN sold controlling interest in the company to the Northern Pacific Railroad in 1870, but in 1873, Northern Pacific folded. Meanwhile, the Oregon and California Railroad Company proposed a line from San Francisco to Portland, making an east-west connection to Portland more lucrative. A railroad from Walla Walla to the OSN landing in Wallula began construction in 1872 to move the wheat downriver that was being grown in the Walla Walla area; OSN bought this railroad in 1878, and the line was completed in 1879. The Oregon Railway and Navigation Company (OR&N) was formed in 1879, and in 1881, under the leadership of Henry Hilgard Villard, the OR&N and Northern Pacific Railroad merged to form the Oregon and Transcontinental. Construction to fill the gaps along the Columbia between Portland and Walla Walla began in 1880, and the OR&N planned to install a line to Huntington, Oregon, to meet the Oregon Short Line, a subsidiary of UPRR, which was laying tracks from points east through Idaho. The line from Umatilla toward Huntington was surveyed by both UPRR and OSN; grading began in 1881. The Huntington line reached Meacham in 1883 and Huntington in 1884. UPRR leased the OR&N in 1887 and purchased half of its capital stock in 1889. Spur branches north from La Grande to Joseph were completed in 1908, and a branch to Sumpter Valley was built. No other branch lines were owned or operated by the UPRR although a number of private spur lines existed to provide access to logging and mining areas.

Government Land Surveys for the sections between Pendleton and Huntington through the Blue Mountains were conducted in the 1860s, 1870s, and 1880s; a few sections were not surveyed until the 1890s. The survey notes documented an extensive network of roads, such as the Boise City stage road and Old Emigrant Road; villages; houses; stores, such as the Martin and Swift store in Pendleton; agency buildings; grist mills; and sawmills, as well as natural features, such as vegetation, streams, wetlands, and open fields. Meacham Creek was described in 1882 as a dry streambed in numerous locations, such as at about Milepost (MP) 261, MP251, MP252, and MP248, and as a flowing stream in others. Cottonwood ("balm"), alder, birch, and willows are described along all the stream beds. The Powder River and Burnt River areas are described as having good bunch grass and thick willow brush along the river.

The surveyors described the township where project segments B-2 and B-3 (Grande Ronde drainage) are located as mountainous. They said it had many fine springs but that the creeks dried up when warm weather came. The land's chief value was thought to be for timber, much of which had already been removed by 1895 (Lavender, 1958). The timber was primarily tamarack, black pine, fir, and spruce. The area between North Powder and Huntington was generally described as open grassland, with some thick areas of sage; timber was confined to the stream banks and consisted primarily of birch and alder; willows were thick along the rivers.

The construction of the railroad through the Blue Mountains required more logging, clearing, and fill. Tie-cutting camps were located near the construction. Railroad grades often followed streams because they generally provided the gentlest overall grades. However, the valleys were often so narrow that extensive filling, riparian vegetation removal, and channelization of the streams were required. Given the reported high frequency of fires already in the area, it is not apparent that the steam engines increased the incidence of range and forest fires in the area.

After construction of the railroad line from Pendleton to Huntington, highways continued to be used for local travel through the mountains. U.S. 30 was constructed on the grade of the Old Emigrant Road and is now called the Old Oregon Trail Highway. I-84, a four-lane freeway with a median, was constructed in the same corridor for high-speed road traffic in the 1970s. Local county paved roads, dirt roads, and U.S. Forest Service (USFS) roads are also spread throughout the study area.

Significance of Cumulative Impacts - Floodplain Area and Function

Overall cumulative impacts on flood and bank storage, including proposed mitigated project impacts, are summarized in Table 2.

Table 2 Floodplain Area and Function Cumulative Impacts (CIs)						
Watershed	Pre-Project Indicator Status	Post-Project Indicator Status	Historical CIs	UPRR Contribution to Historical CIs	Mitigated Project Impacts	CIs from All Sources
Meacham	-	-	Substantial	-	0	Substantial
Grande Ronde	-	-	Moderate	0*	0	Moderate
Pyles	0	0	Neglig.	0	0	Neglig.
Powder	0	0	Neglig.	0	0	Neglig.
Burnt	0	0	Neglig.	0	0	Neglig.

Note:
- Indicates significant negative impact.
+ Indicates significant positive impact.
0 Indicates no significant impact.

Scale of impact evaluation, from least to most: negligible, minor, moderate, substantial, major.

* Except for the floodplains of Dry Creek, where there is greater possibility of diminished flood and bank storage although actual events have not been documented.

Meacham Creek. Historical documents do not show that early road and trail building in the Meacham extended project area altered the flood and bank storage aspects of Meacham Creek. The UPRR tracks, constructed in the 1880s, parallel Meacham Creek along approximately 85 percent of the river's total length. UPRR embankments have constrained an estimated 13 percent (based on documented constraints in project areas) of this 85 percent, or approximately 11 percent of the length of Meacham Creek's active channel. Historical floodplain encroachment is assumed to represent 35 percent of the length of railroad grade proximity. Of the total length of Meacham Creek, then, an estimated 30 percent has some fill by UPRR in the historical floodplain. The desired future conditions indicate that a natural channel configuration controlled by only geological constraints is desired.

No documentation of historical flooding conditions (elevation or extent) along Meacham Creek was found. Project investigations did not document indicators of diminished flood storage in Meacham Creek, although historical fill within the channel could be expected to have reduced flood storage, particularly in the area of the finger dikes and where the Meacham Creek Canyon is narrow. The volume of flow displaced by historical fills may have raised water elevations more than 1 foot in certain upstream reaches; no documentation is available to determine conclusively whether this displacement effect occurred. No increased flooding has been reported.

Stream channels are constrained by bedrock along much of the upper reaches of Meacham Creek. Erosion appears to be related to increased stream energy in the headwaters and lower reaches. Loss of riparian vegetation also appears to be related to such factors as grazing, fire, logging, and fills. Because dry reaches of Meacham Creek were observed by emigrants and surveyors before UPRR construction, they cannot be attributed to lack of bank storage as a result of encroachment.

On the basis of the status of these indicators and the desired future conditions, historical cumulative impacts from all factors in the watershed to flood and bank storage are substantial. UPRR actions and structures have been the primary source of floodplain constraint in the Meacham drainage and, hence, of the flood and bank storage impacts. UPRR's contribution historical cumulative impacts have thus been significant in the Meacham extended project area.

Grande Ronde. Early historical impacts on flood and bank storage in forested areas of the Grande Ronde extended project area were primarily a result of splash dams built to provide flow for movement of logs to mills. Placer mining also occurred in upper Grande Ronde tributaries, with diversions of streams and accumulation of tailings debris within stream channels. Water is diverted through diversion and control structures for agriculture in the lower watershed and in agricultural areas; many of these structures are temporary.

In the late 1800s, UPRR located the track bed adjacent to streams in the Grande Ronde extended project area. The UPRR right-of-way parallels Dry Creek, Pelican Creek, Five Points Creek, and the Grande Ronde River along approximately 100 percent, 20 percent, 9 percent, and 13 percent of their respective total lengths within the project drainage (3 percent of the length of the Grande Ronde River if the greater drainage is considered). Of the active channels, approximately 11 percent of Dry Creek, 2 percent of Pelican Creek, 1 percent of Five Points Creek, and 1 percent of the Grande Ronde River active channel are estimated to have been constrained as a result of historical UPRR activities. The desired future conditions indicate that a natural channel configuration controlled by only geological constraints is an objective.

Subsequent to UPRR construction, I-84 and U.S. 30 were built, confining reaches of Pelican Creek and the Grande Ronde River where they were previously unconstrained or where they were constrained on only one side by the railroad. An estimated 1 percent of Dry Creek, 1 percent of Pelican Creek, and 2 percent of the Grande Ronde River length are constrained by roads. A network of logging roads was also built in the upper reaches of tributaries to the Grande Ronde, but the extent of constraint of streams is not known.

Identified indicators of diminished flood and bank storage have not been documented in the Grande Ronde extended project area, although historical fills within the channel could be expected to have reduced flood storage. Fills in the historical floodplain related to UPRR activities affect 14 percent, 1.5 percent, 0 percent, and 10 percent of the respective stream lengths within the Dry Creek, Pelican Creek, Five Points Creek, and Grande Ronde project watersheds (2 percent of the length of the Grande Ronde River). The volume of flow displaced by permanent historical fills may have raised floodwater elevations in upstream reaches of Dry Creek; no documentation is available to determine conclusively whether this displacement effect occurred. No upstream flooding has been reported. Erosion appears to be related to fire, logging, grazing and scouring of stream banks due to increased flow velocities. Dry reaches of Dry Creek were observed historically (hence the name); they probably are related to bedrock controls. Cutoff reaches, however, have been created by UPRR embankments, where the active channel is isolated from the historic channel; in these active channels, bank storage and low-flow recharge are probably minimal.

On the basis of the existing and desired future conditions, historical cumulative impacts on flood and bank storage are moderate in Dry Creek and minor in the Grande Ronde project reaches. Logging, mining, and road-building have had a greater impact on conditions in the Grande Ronde extended project area than UPRR has had. UPRR's contribution to historical cumulative impacts has thus been insignificant in the Grande Ronde extended project area.

Impacts on floodplains from the proposed project include embankment fill along 520 yards of the Grande Ronde River (less than 1 percent of stream length within the watershed area). This isolated fill will not eliminate stream meanders or create cutoff reaches. The impact of this channel fill on flood storage is expected to be insignificant.

Pyles Creek. Historical impacts on flood storage in the Pyles Creek extended project area are related primarily to channel realignments and irrigation diversions, both due to agriculture. Irrigation diversions cause frequent modification of stream channels as a result of the reconstruction of diversions at 1- or 2-year intervals. Local roadways, highways (State Route 237), and modifications by landowners have redefined the channel in places. The UPRR tracks, built in the 1880s, parallel Pyles Creek along approximately 55 percent of its total length, primarily in the lower reaches of the creek. The desired future conditions indicate that a natural channel configuration controlled by only geological constraints is desired.

Indicators of diminished flood storage have not been documented, because most flooding is now controlled in the upper reaches of the extended project area by a storage reservoir. No documentation of historical flooding was found. Thunderstorm events are probably the primary cause of floodwaters, because precipitation is low and in the small drainage area and the presence of seeps that support the base flow of the stream. Historical fill and modification of Pyles Creek may have raised flood elevations during storm events; however, no documentation of this rise is available. Upstream flooding generally has not been reported to occur.

Stream channels have a relatively low gradient, except in the headwaters, and erosion appears to be related more to grazing impacts than to increased stream energy. Loss of riparian vegetation also appears to be related primarily to grazing effects. Bank storage is not a critical element for providing base flow in this stream, because of the year-round groundwater discharge to the stream and because of the upstream reservoir.
On the basis of the status of these indicators and the desired future conditions, historical cumulative impacts on flood storage are negligible. Flood storage has increased in an upstream reservoir rather than in the stream channel, and site observations indicate no changes to bank storage. Some loss of flood storage within the stream channel may have resulted from realignments and diversions. UPRR actions and structures have had minimal effect on bank and off-stream storage. Therefore, UPRR's contribution to historical cumulative impacts is insignificant in the Pyles Creek extended project area.

Powder River. Historical impacts on flood storage in the Powder River extended project area are related primarily to mining and irrigation diversions. Mining, particularly dredging, recontoured stream channels and created cutoff reaches in many headwater streams of the Powder River. Irrigation diversions have caused frequent modification of stream channels because the diversions are reconstructed at 1- or 2-year intervals. Local roadways, highways, and modification by landowners have redefined channels in places. The UPRR tracks, built in the 1880s, parallel an unnamed tributary along approximately 20 percent of the tributary's total length, along Antelope Creek for approximately 33 percent of its length, and along the Powder River for approximately 7 percent of its length within the project watershed and approximately 3 percent of its total length. The UPRR right-of-way also parallels Sutton Creek along approximately 50 percent of its length in the southern portion of the Powder River watershed. The desired future conditions indicate that a natural channel configuration controlled by only geological constraints is an objective.

Indicators of diminished flood storage have not been documented, as most flooding is now controlled in the lower reaches of the extended project area by Phillips Lake and other reservoirs. Documentation of historical flooding was not found. Historical fill and modification of the Powder River and tributary channels may have raised flood elevations during storm events; however, no documentation of this rise is available. Upstream flooding has not generally been reported. Stream channels have a relatively low gradient, except in the headwaters, and erosion appears to be related more to grazing impacts than to increased stream energy. Loss of riparian vegetation also appears to be related primarily to grazing effects. Bank storage has become less of a critical element in base flow downstream of the storage reservoirs.

On the basis of the status of these indicators and the desired future conditions, historical cumulative impacts to flood storage are negligible. Flood storage has increased in reservoirs, rather than in the stream channels, and site observations indicate no changes to bank storage. UPRR actions and structures have had minimal effect on bank and off-stream storage. Therefore, UPRR's contribution to historical cumulative impacts is insignificant in the Powder River extended project area.

Project activity in 1993 relocated 0.35 mile of the unnamed tributary (approximately 7 percent) and could encroach on 0.35 mile of Antelope Creek (3.5 percent). Project impacts on flood and bank storage are expected to be insignificant because only a short length of stream is affected and downstream flood storage is available.

Burnt River. Historical impacts on flood storage in the Burnt River extended project area are primarily related to agricultural channel relocations, irrigation diversions, and road construction. Ranches and farms have moved and channelized drainages for ease of fencing and access. Irrigation diversions have caused frequent modification of stream channels because they are reconstructed at 1- or 2-year intervals. Local roadways and highways have redefined channels. I-84 and the Old Oregon Trail Highway run adjacent to the UPRR right-of-way for its entire length in this extended project area, except for the Oxman segment (D-3) and near Huntington where the roads turn south and the UPRR right-of-way follows the river. Roadway channel constrictions are 66 percent of total channel constrictions from all sources. The UPRR tracks, built in the 1880s, parallel Alder Creek along approximately 40 percent of its total length, Pritchard Creek along approximately 40 percent of its length, and the Burnt River along approximately 26 percent of its length within the project watershed. Two percent of Burnt River has been constrained by the UPRR and I-84 grades. The desired future conditions indicate that a natural channel configuration controlled by only geological constraints is an objective.

Identified indicators of diminished flood storage have not been documented, as most flooding is now controlled in the lower reaches of the extended project area by Unity Reservoir. No documentation of historical flooding was found. Historical fill and modification of the Burnt River and tributary channels may have raised flood elevations during storm events; however, no documentation of this rise is available. Upstream flooding has not generally been reported. Stream-channel gradients are relatively low, except in the headwaters, and erosion appears to be related more to grazing impacts than to increased stream energy. Loss of riparian vegetation also appears to be related primarily to grazing effects. Bank storage has become less of a critical element in base flow downstream of the storage reservoirs.

On the basis of the status of these indicators and the desired future conditions, historical cumulative impacts on flood storage are negligible. Flood storage has increased in reservoirs, rather than in the stream channels, and site observations indicate no changes to bank storage. UPRR actions and structures have had minimal effect on bank and off-stream storage. Therefore, UPRR's contribution to historical cumulative impacts is insignificant in the Burnt River extended project area.

Proposed activities could constrain an additional 80 yards of the Alder Creek channel (less than 1 percent), no reaches of Pritchard Creek, and 420 yards (less than 1 percent) of Burnt River. Project impacts on flood and bank storage are expected to be insignificant because only the short lengths of streams are affected and because upstream flood storage is available.

Conclusions

In conclusion, the CIA process was a useful screening tool to identify the watersheds where most historical impacts to stream morphology occurred, which could be related to Union Pacific Railroad construction, operations, and maintenance: the Meacham Creek and Grande Ronde River watersheds. The technical studies and analysis performed to support the CIA will assist in identifying and prioritizing restoration and rehabilitation projects in these watersheds. As a result of the mitigation obligations and as a gesture of good will, Union Pacific Railroad has established a trust fund to be managed by the regulatory agencies, the CTUIR, and UPRR, the proceeds of which are dedicated to stream restoration and research projects within these watersheds.

References

Asay, Jeff. 1991. *Union Pacific Northwest: The Oregon-Washington Railroad & Navigation Company.* Edmonds, Washington: Pacific Fast Mail.

Brooks, Howard C., and Len Ramp. 1968 (reprinted 1980). *Gold and Silver in Oregon.* Bulletin 61. State of Oregon Department of Geology and Mineral Industries.

CH2M HILL. 1993. *Union Pacific Blue Mountains Project Cumulative Impacts Assessment: Background Information.*

Confederated Tribes of the Umatilla Indian Reservation (CTUIR). 1993. *Nonpoint Sources of Water Pollution Assessment and Management Program Umatilla Water Basin-Public Review Draft.* CTUIR. December 1993.

Eckberg, D. K. 1986. Cumulative Effects of Hydropower Development Under NEPA. *Environmental Law* 16:673-703.

Evans, John W. 1990. *Powerful Rockey: The Blue Mountains and the Oregon Trail.* Enterprise, Oregon: Pika Press.

Guthrie, Roger. 1994. Presentation at 1994 Oregon Wildlife Association Meeting, Bend, Oregon.

Highsmith, Richard M., and A. Jon Kimerling, Eds. 1979. *Atlas of the Pacific Northwest,* 6th edition. Corvallis, Oregon: Oregon State University Press.

Hirsch, A. 1988. Regulatory Context for Cumulative Impact Research. *Environmental Management* 12(5):715-723.

Lavender, David. 1958. *Land of Giants: The Drive to the Pacific Northwest.* Garden City, New York: Doubleday & Company, Inc.

Leckenby, D.A. 1984. *Elk Use and Availability of Cover and Forage Habitat Components in the Blue Mountains, Northeast Oregon, 1976-1982.* Oregon Department of Fish and Wildlife, Research and Development, Wildlife Research Report Number 14. Portland, Oregon. 40 p.

Lee, L. C. and J. G. Gosselink. 1988. Cumulative Impacts on Wetlands: Linking Scientific Assessments and Regulatory Alternatives. *Environmental Management* 12(5):591-602.

McIntosh, Bruce A. 1992. "Historical Changes in Anadromous Fish Habitat in the Upper Grande Ronde River, Oregon, 1941-1990." Master's thesis, Oregon State University.

McIntosh, Bruce A., James R. Sedell, Jeanette E. Smith, Robert C. Wissmar, Sharon E. Clarke, Gordon H. Reeves, and Lisa A. Brown. 1994. *Management History of Eastside Ecosystems: Changes in Fish Habitat over 50 Years, 1935 to 1992.* General Technical Report PNW-GTR-321. Portland, Oregon: U.S. Department of Agriculture, Forest Service, Pacific Northwest Research Station.

Northwest Power Planning Council. 1993. Smolt Density Model Database.

Oregon Department of Environmental Quality. 1988. *1988 Oregon Statewide Assessment of Nonpoint Sources of Water Pollution.* Prepared by the Planning and Monitoring Section, Water Quality Division, Oregon Department of Environmental Quality.

Oregon Department of Fish and Wildlife, Umatilla Tribe, Nez Perce Tribe, Washington Department of Fisheries, and Washington Department of Wildlife. 1990. *Columbia Basin System Planning. Salmonid and Steelhead Production Plan. Grande Ronde River Subbasin.* Northwest Power Planning Council. September 1990.

Oregon Department of Fish and Wildlife. 1990. *Mule Deer Plan*. December 5, 1990. Portland, Oregon. 49 p.

Oregon Department of Fish and Wildlife. 1992. *Oregon's Elk Management Plan*. July 1992. Portland, Oregon. 63p.

Oregon Department of Fish and Wildlife. 1993. *1993 Big Game Statistics*. July 1993. Portland, Oregon. 53 p.

Skovlin, J. M. 1991. *Fifty Years of Research Progress: a Historical Document on the Starkey Experimental Forest and Range*. General Technical Report PNW-GTR-266. Portland, OR: U.S. Department of Agriculture, Forest Service, Pacific Northwest Research Station.

Strahler, A. N. 1964. "Quantitative Geomorphology of Drainage Basins and Channel Networks." Sec. 4-II in V. T. Chow (ed.): *Handbook of Applied Hydrology*. New York: McGraw-Hill.

Umatilla National Forest Working Group. 1994. *Draft Proposal on Desired Future Conditions*. Umatilla National Forest, USFS, Pendleton, Oregon.

FANNO CREEK WATERSHED PLANNING & ENHANCEMENT

Kendra L. Smith[1]

Abstract

The Fanno Creek watershed, located in the Portland Metropolitan area, is one of the oldest urbanized watersheds of the Tualatin River Basin. Modifications to the landscape alter the hydrology, soil conditions, native plant communities, and habitats of the watershed. The urban flow regime causes excessive stream bed and bank erosion and favors the invasion of non-native plant species. Polluted urban runoff, impervious surfaces, and stream/floodplain encroachment also contribute to the degradation of the water quality and habitats of Fanno Creek. However, despite all the abuse, Fanno Creek continues to support native fish and wildlife. The remaining greenspaces of the watershed offer considerable opportunities to enhance the quality of the watershed and the health of its streams for both animals and people.

Formulating a holistic community supported Watershed Plan, whose recommended enhancement options address a variety of water quality and flood management problems, is the critical first step towards effective watershed management. Extensive public involvement, watershed analysis, and management options evaluation, are needed to develop action-oriented watershed management projects. This paper discusses Fanno Creek's watershed planning process and some of the management actions that may be used to improve water quality and aquatic/wildlife habitat, provide flood management, and enhance the overall livability of the watershed.

[1] Environmental Scientist, Kurahashi & Associates, Inc. 12600 SW 72nd Tigard, OR 97223.

Acknowledgments

The author gratefully acknowledges contributions to this working paper from consultant team members Greg Koonce and Bill Norris of Inter-Fluve, Inc.; David Smith of Wildlife Dynamics, Inc.; Dennis O'Connor, Restoration Ecologist; Roger Sutherland and Seth Jelen of Kurahashi & Associates, Inc.; and the client Lori Faha of the Unified Sewerage Agency. Their expertise and insight are invaluable to the project and this paper.

Introduction

Fanno Creek watershed consists of approximately 20,510 acres of land that drains water from the Tualatin Mountain Range, Sexton Mountain, and Bull Mountain down to the Tualatin River. The waters of the Tualatin flow into the Willamette River, then on to the Columbia River and Pacific Ocean. There are approximately 117 miles of streams in the Fanno watershed, including two major tributaries (Ash and Summer Creeks) and twelve smaller tributaries (Figure 1). There are also three counties, five cities, two park districts, and two surface water management agencies that govern various aspects of the watershed.

Regional Issues and Management Objectives

Growth management in the Portland metropolitan area is coordinated by Metro, the regional land use planning agency. The 2040 regional land use planning process found that citizens favor intensified urban densities within the "urban growth boundary" as a mechanism for reducing detrimental urban/suburban sprawl. However, citizens also strongly support the protection of a green infrastructure within the urban environment. These greenspaces are vital to the protection of surface water quality and provide flood benefits, wildlife habitat, and passive recreation opportunities. Watershed management plans are being developed to support these regional land use and natural resource conservation goals.

Fanno Creeks' watershed planning is funded by the Unified Sewerage Agency of Washington County (USA). They manage surface and storm water in the urbanized areas of the county, which is located entirely within the Tualatin River Basin. The Oregon State Department of Environmental Quality (DEQ) has mandated USA to improve the water quality of the Tualatin River and its tributaries in order to protect beneficial uses (fish/wildlife habitat, recreation, health and safety) of these systems. The Tualatin River and Fanno Creek are water quality limited with established Total Maximum Daily Loads (TMDL's) and water quality standards. The Watershed Plan will identify prioritized management actions that will improve water quality and flood management, and involve the watershed community in its implementation.

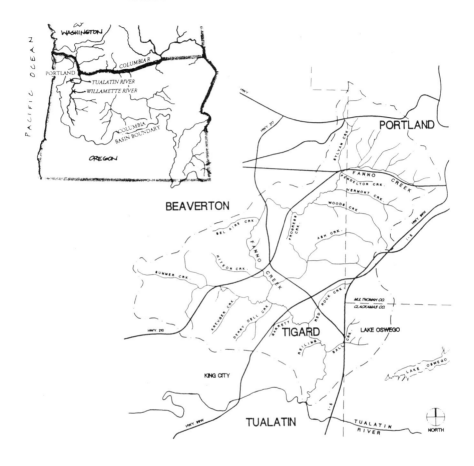

Figure 1: Fanno Creek Watershed

The Watershed Planning Process

The watershed planning process for Fanno Creek is summarized in Figure 2. Each planning stage is guided by a diverse group of watershed stakeholders that work with the technical consultants, establish management priorities, and will eventually decide on watershed management actions. The participants that make up this Project Committee represent cities, the county, state regulators, the park district, businesses, citizens, USA, and the local stream group FANS of Fanno Creek.

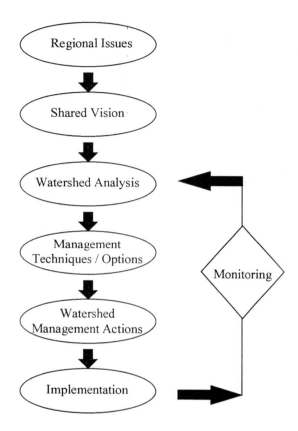

Figure 2: The Watershed Planning Process

Shared Vision

A shared vision is the beginning point of watershed planning. For Fanno Creek, the visioning process occurred through the Project Committee. As a first assignment, these stakeholders photographed the good and bad qualities of the watershed in order to identify the major issues. Watershed issues were also gathered from the community-at-large through newsletters, open houses, and one-on-one communication with the USA and the consultant team; these issues were presented to the Project Committee. Six outcomes or goals for the Plan identified through this visioning process, include:

Improve and protect water quality
- Reduce flood impacts
- Preserve open spaces, floodplains, and parks
- Enhance fisheries and wildlife habitat
- Provide greater public education and involvement
- Evaluate / improve regulatory usage (new and existing)

Public involvement is a cornerstone of this planning process and will be a driving force behind the decisions made for the Plan. With a clear vision of expected outcomes, the technical consultants began the watershed analysis.

Watershed Analysis

Watershed analysis began with an understanding of the landscape's history. Since current conditions are manifestations of the watershed's past, key features of the landscape were evaluated in a historic context. How did the watershed evolve over time? What influences played a role in shaping the system as it exists? Can/should some of those influences be restored or is a substitute necessary given the watershed conditions? What can be realistically achieved in the present and future urban environment? These questions will play a vital role in identifying recommended management actions that will achieve the goals set forth by the Project Committee.

Historic Context

The prehistoric landscape of the Fanno creek watershed was formed by re-occurring floods. The floodwaters that created the Columbia River Gorge, filled up and drained from the Tualatin River Basin several times during the retreat of the ice age. The massive force of these floods created geologic features such as Lake Oswego and the Tounqin Scablands in Tualatin. Centuries later, Fanno creek was altered by the beaver that lived in the wetlands of the Basin. They provided habitat for a variety of wildlife by damning the creek with large trees and encouraging complex channel configurations. The upland forests and prairies of the watershed were influenced by natural fires, insect infestations, and wind throw. Flooding of the riparian corridor provided a consistent flushing of the system. These changes in the landscape where natural, and the environment was self-sustaining and healthy.

A small population of Native Americans utilized the resources of the watershed for subsistence living until European settler's moved to the area in the mid-1800's. The health of the watershed began to decline as new settlers engaged in agricultural and forestry practices that modified the landscape. Settlers suppressed wildfires, drained wetlands to reclaim more land, and ponded creeks with earthen berms for irrigation. The forests and agricultural lands of the watershed were gradually replaced with residential development throughout the 1900's. Recognizing that ecosystem stress increases with deviations from historic disturbance regimes, what can we expect to find in today's urbanizing landscape?

Existing Conditions

In general, the urbanization of the Fanno Creek watershed has reduced the stream's health and it's ability to repair itself. The hydrology, water quality, channel stability, and fish/wildlife habitat have changed dramatically due to our presence. However, the system still contains areas of good habitat that provide water quality and flood reduction benefits.

The existing watershed landscape was evaluated for: land use, soils, hydrology, geomorphology, vegetation, fish/wildlife habitat, water quality, and analogs or "representative good sites". The consultant team gathered information for each subwatershed using existing reports and monitoring data, GIS, computer modeling, and in-field surveys. The key findings of the watershed analysis are briefly discussed below:

Land Use

The watershed land uses include: commercial business, light industrial, residential homes of different densities, parks, schools, golf courses, transportation corridors, and creek/floodplain/wetland areas. The existing total effective impervious area for the watershed is 21% and unconnected impervious area is 11.60%. Pervious uplands are 63.5% and wetland/stream areas are 3.9% of the total watershed coverage.

Soils

The watershed soils are highly consolidated silts and clays, with relatively low infiltration rates. Even "pervious" areas have a moderate runoff rate if the landscape lacks a dense vegetative layer to absorb the water. Excessive soil erosion is a problem throughout the watershed due to loss of vegetative cover, soil compaction, stream/floodplain fill, and altered hydrology. A majority of the soils in the watershed contain moderate to high background levels of phosphorous, a water quality parameter of concern. Combined high soil erosion potential and high phosphorous availability areas are predominately in the headwater areas of the creek and its tributaries.

Hydrology

Historically, water storage capacity was contained in the forested canopies, the dense litter layers of forests and prairies, and in the complex wetland and stream systems of the watershed. The increase in urbanization decreased water storage and infiltration, causing intensified rainfall to runoff rates. The hydrographs (flow versus time relationships) for the watershed show the classic steep "spike," which represents direct runoff from the urban environment (Figure 3). The consequences of this flashy flow regime permeate the system dynamics from geomorphology to habitat to water quality. Flooding problems in the watershed primarily result from undersized or clogged culverts, channel modifications (armorment and damning) and creek/floodplain encroachment.

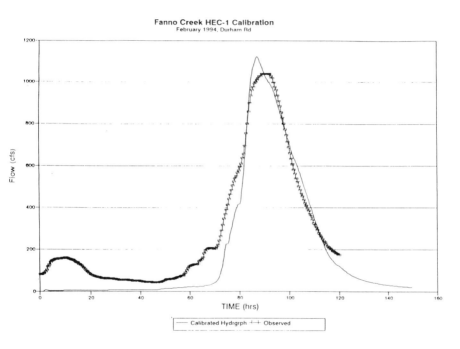

Figure 3: Example of a Fanno Creek Hydrograph

Geomorphic Conditions

    Geomorphic changes in the stream channel are a reflection of the hydrologic and landscape (soil) conditions of the watershed. The stable bends and bars, backwater areas, undercut banks, and large woody debris that dominate healthy streams of the Basin are limited on Fanno Creek. The geomorphic processes resulting from urban hydrology encourage a less complex stream system, excessive bank erosion, and channel incision. Sewer lines, utilities, bridges, culverts, weirs, and other structures in and over the creek also disrupt geomorphic progression.

Vegetation

    Riparian vegetation is affected by hydrology and the geomorphic changes of the channel. Flashy flood events dislodge native trees and shrubs, stripping the banks of needed root structure and vegetative protection. Stream bed incision and reduced upland storage has lowered the water elevation of summer flows, leaving native riparian species with inadequate amounts of water during the dry season. The riparian buffer width per bank averages approximately 6 feet. However, the range of extremes from 0 to 100 feet is more representative of various stream segments.

The stressed conditions of the watershed threaten the historic native plant communities. Non-native species, which establish quickly following disturbance and out-compete native plants, dominate a majority of the landscape. The non-native plants typically grow as a monoculture, reducing vegetative diversity and wildlife habitat value. Landscaping around homes, businesses, parks, and some mitigation sites also contribute non-native seed to the natural areas via birds, small mammals, and high flow events. The healthy native plant communities that remain are typically associated with relatively large habitat patches.

Wildlife Habitat
Significant wildlife habitats covering approximately 75 un-contiguous acres were identified for protection of water quality and wildlife. The five habitat blocks contain relatively mature trees mixed with a diversity of native plant assemblages, dead wood (snags and logs) and clean water. Reduction in the size of these parcels, alteration of the vegetative layers, or other human disturbances will diminish the functions and value the sites provide for water quality and wildlife. Creating linkages and increasing the vegetative structure will improve habitat suitability for many wildlife species that have co-evolved with the native plants.

Overall, there is suitable habitat within the watershed for species that can tolerate at least moderate human disturbance and fragmented vegetated areas. The habitat complexity of the watershed has been simplified and a majority of the remaining natural areas favor generalist species that can adapt to changing environmental conditions and can tolerate non-native plant food sources.

Fish Habitat
Understanding the relationship between fish assemblages and the physical habitat in the creek is important, if the effects of urbanization are to be recognized from the standpoint of stream health. Oregon Department of Fish and Wildlife (ODFW) studied fish and crayfish distribution and habitat availability in the Tualatin River Basin from 1993-1995 for USA, in order to establish baseline information on the health of the various streams of the Basin. Siltation, bank erosion, lack of woody debris, and insufficient overhead cover were noted in the ODFW report as significant factors affecting fish habitat in the streams. A summary of the dominant habitat characteristics and fish assemblages for the nine sample sites in Fanno, Ash and Summer Creeks is provided in Table 1.

The dominant species in the watershed are reticulate sculpin and redside shiner. Both species are intermediate to tolerant of poor habitat and water quality conditions. Intolerant species such as salmon are found only rarely in the healthy habitat area near the confluence with the Tualatin River.

Table 1: Fish Habitat and Character in Fanno Creek

Dominant Habitat Characteristics			Fish Assemblage Statistics	
	Range	Average		Total
Glides	0-99%	79.40%	IBI Score**	12-38
Eroding Banks	50-100%	77.50%	Tolerant fish species***	75.90%
Shade	21-71%	51.40%	Intermediate fish species	21.90%
Gradient	.6-3.6%	1.20%	Intolerant fish species	1.70%
Undercut Bank	0-14%	3.50%	Introduced fish species	6.10%
Wood Index*	1-2.8	1.7	Crayfish (# range)	0-187

* Wood index: 1=very low, 5=abundant
** Index of Biotic Integrity reflects the functional and structural characteristics of fish communities. Total range of scores throughout the Basin was 12 to 44
*** Tolerance relates to fishes ability to survive in existing habitat and water quality conditions. Intolerant species such as salmon are not found in areas of poor habitat and water quality.

Water Quality

Fanno Creek is designated as a water quality limited stream (one of 900 in the State) because greater than 10% of the water quality samples taken in-stream, exceed the set TMDL standard for total phosphorous and at least two exceedences of this standard occur between May and October. Fanno creek and its major tributaries are also in varying degrees of noncompliance for the dissolved oxygen, temperature, and bacteria water quality standards. Other water quality parameters of concern include pH, chlorophyll a, ammonia, metals and total suspended solids. Point source pollution, such as illegal discharges and/or dumping, into the creeks has declined as surface water agencies have strongly promoted recycling and fines for illegal discharges of contaminating materials.

A majority of the water quality problems in Fanno Creek are related to non-point source pollution such as urban runoff and erosion. Sediment is an important transport mechanism for some of the water quality parameters of interest (phosphorous, bacteria, TKN, and metals). Understanding the source of sediment is critical to formulating appropriate management options for the each stream reach. In-stream water quality and flow data combined with urban runoff modeling were used to evaluate the extent of upland and in-stream sediment contributions.

Utilizing the SIMPTM urban storm water quality model, the total mass loading of sediment and its associated pollutants (phosphorous, TKN, metals, COD, BOD) from the urban landscape were simulated for an average or representative rainfall year. The loadings calculated by SIMPTM were then compared to estimated total mass loadings developed from over 200 in-stream water quality samples collected at a gauge station. The information generated provides an estimate of:

- How much sediment is transported on an annual basis during dry weather flow
- How much of the sediment is coming from the urban landscape versus in-channel sediment during wet weather events

Sediment estimates for other reaches of the creek with different upland slope, channel grade, and land uses will be made using the flows from HEC-1 and the water quality data collected along the streams by USA and DEQ over the last nine years. Documenting in-stream erosion versus urban washoff is a critical step towards understanding the appropriateness of certain management options in given areas. The significance of bioengineered stream enhancement on water quality improvement is of particular interest.

<u>Analog Sites</u>
In order to evaluate how to enhance the ecological functions that are damaged throughout the system, a search for "analogs" (sites representative of how the system used to be) was conducted. Few of these sites actually exist in the watershed. In most cases, a site would have remnant parts of a healthy system (i.e. historic plant communities, stable channel characteristics, etc.) but lacked the all the elements needed to maintain a healthy state. However, these sites provided insight to the stages of system response, particularly to changing hydrology. Combinations of the analog elements will be used in formulating conceptual enhancements for the stream reaches as part of the recommended Watershed Plan.

The results of the watershed analysis were presented to the Project Committee using photographs, slides, and color maps. Problems such as high soil erodibility and soil phosphorous availability, channel erosion, flooding of structures, degraded upland and riparian area habitats, disconnected floodplains, unshaded water, and excessive impervious cover were documented on a watershed map. The Project Committee will ultimately decide if, and in what priority, the problems will be solved.

Management Techniques / Options Identification

Opportunities to enhance the watershed are based on the available management techniques (toolbox of ideas) which are developed into options that are technically feasible and social acceptable. The pro's and con's of the techniques identified below were presented to the Project Committee, who reviewed and approved the list for options development.

Flood Management

- Increase conveyance or capacity
- Increase detention storage locally and regionally
- Increase in-stream and wetland storage
- Divert flows around flooding problem
- Acquire vulnerable (flood prone) properties
- Decrease / disconnect impervious cover

Water Quality Improvement

- Increase public education and involvement
- Increase tree planting (upland and riparian)
- Increase and diversify vegetated buffer widths
- Stabilize streambanks (bioengineering)
- Stabilize stream bed (grade control)

- Increase floodplain function (in-stream storage)
- Enhance wetlands for passive treatment
- Install sedimentation devices in storm water catchments
- Install bioswales and on-site detention
- Enhance storm water filtration and treatment (leaf compost, peat-sand filter)
- Disconnect impervious surfaces roof drains and porous pavement
- Increase street and storm system catchment cleaning
- Reduce mandatory parking spaces
- Reduce residential street widths
- Enhance and enforce erosion control regulations
- Reduce debris dumping and promote enhanced recycling

The consultant is currently formulating options for water quality and flood management for site specific problems, each stream reach, and the overall watershed. Each option developed by the consultant must be accessible, constructable (if in-field), technically feasible, and meet the goals and objectives of USA by providing water quality and flood management benefits. Options must meet a majority of specific technical criteria including:

Reduce in-stream temperatures
- Provide sediment management
- Reduce nutrient and bacterial loading
- Maintain or increase dissolved oxygen levels
- Reduce petroleum pollution
- Maintain or enhance channel stability
- Moderate peak flows
- Reduce negative flooding impacts
- Minimize the migration of problems upstream and downstream

For sites or problems that have more than one management option, the Project Committee will balance the technical information with additional evaluation criteria that may be weighted based on relative significance or numerically. The non-technical evaluation criteria includes:

- Social acceptability (aesthetics, livability, risk/hazard)
- Habitat benefits (fish, herptiles, birds, mammals)
- Relative capital, operations, & maintenance costs
- Recreation opportunities (passive and active)
- Public awareness and education opportunities (school involvement, adult outreach)
- Partnership opportunities
- Permitability (ease and cost)
- Land use compatibility

Watershed Management Actions

The final stage in the watershed planning process will require the Project Committee to recommend specific watershed management actions based on the evaluation of the proposed options. They will define what the creek will be managed for and how much money may be spent to protect its beneficial uses. Key elements and issues associated with Fanno Creek's management include:

- Multiple use of open space for water quality treatment, habitat, and recreation
- Background nutrient loading from native soils
- Planned densities and open space preservation
- Public use and private rights of landowners
- Urban storm system retrofitting and regional water quality treatment
- Individual pretreatment areas and linear stream enhancement
- Prevention of sensitive lands development and wetland mitigation
- Long term monitoring and maintenance of projects
- Public participation and acceptance
- Short term and long term funding

Generalized watershed management actions being considered by the consultant are listed below. More specific enhancement actions and conceptual designs for each stream reach and problem area will be developed following the initial approval of the options by the Project Committee. *Note: None of the following actions have been presented to or endorsed by the Project Committee.*

1. Link and increase the protection of open spaces that provide water quality benefits (floodplains, existing healthy wetlands, stream bordered uplands, and headwater areas where soil phosphorous availability and soil erosion potential are high). Expand buffer requirements, acquire conservation easements, and strengthen design/construction standards in and around these sensitive areas.

2. Reduce the negative effects of urban hydrology through a combination of stream bed grade control (rock and large woody debris), peak flow storage on the floodplain, bioengineered streambank stabilization, localized detention, and reduction of urban runoff from upland sites.

3. Re-establish or enhance native riparian and floodplain plant communities in order to increase vegetative complexity. Native vegetation will help to improve fish and wildlife habitat, trap sediments that carry pollutants, decrease water temperatures and bacteria levels, and increase dissolved oxygen levels.

4. Enhance in-stream ponds by diversifying or creating an emergent plant zone, increasing vegetated pond edge length, removing non-native plant and animal species, improving habitat structure, and providing public education. If the opportunity is present, restore creek conditions so that low flows will bypass the ponds.

5. Encourage appropriate lawn care and create incentives for owners to establish minimum natural area buffers at schools, golf courses, commercial and residential areas throughout the watershed.

6. Manage urban "hot spots" by providing incentives for older areas to retrofit and provide on-site detention and water quality treatment using bioswales and leaf-compost filtration (new development already has these requirements). Provide economic incentives for the maintenance of these on-site facilities.

7. Establish and enforce tighter construction standards for grading, erosion control, limited land clearing, and soil compaction.

8. Reduce impervious surfaces on new, redeveloping, and existing areas by: decreasing parking requirements per unit, encouraging cooperative parking, allowing skinny streets and shared driveways, promoting cluster development and taller buildings, and modifying sidewalks and parking areas to drain to vegetated strips or gravel catchments (City of Olympia, 1995)

9. Promote an un-paralleled level of stewardship of the urban natural areas through public education and involvement, in order to maintain a healthy and livable environment.

Conclusion

The goal of the watershed management process is to achieve the healthiest environment possible given the land use constraints. In the Portland Metropolitan area, citizens have chosen to increase the density of the urban environment in order to protect lands outside the urban growth boundary. However, they also favor the protection and enhancement of greenway linkages along stream corridors for water quality, passive recreation, and fish/wildlife habitat benefits.

Stream health is linked to the land use, hydrology, geomorphology, vegetation, and water quality of the watershed. Improving watershed conditions on both public and private lands is needed in order to improve the overall health of the stream. "Hot spot" management of problem water quality/quantity contributions from the urban landscape can reduce the impacts of urbanization. Stream stewardship programs that encourage active participation in the enhancement of the system will be the cornerstone of a successful plan implementation. Finding a balance between the watershed's social, political, physical, and biological components will lead to holistic watershed management that will improve overall stream health.

References

City of Olympia and Washington State Department of Ecology. Impervious Surface Reduction Study, Final Report. May 1995.

Oregon Department of Fish and Wildlife. Distribution of Fish and Crayfish and Measurement of Available Habitat in the Tualatin River Basin. Final Report of Research. June 1995.

DISCUSSION

Riparian Buffer Widths at Rocky Mountain Resorts
  Jon Jones

In the Rocky Mountains, where much of the economy is based on the recreational
opportunities and the aesthetics of the mountains, development and preservation of
environmental resources go hand-in-hand.

The State of Colorado does not have regulations for preserving riparian corridors or
establishing buffer zones; however, many Colorado counties have regulations
which require setbacks from streams and wetlands.

In Colorado resorts, and adjoining towns, protection of riparian corridors is
particularly important because of the economic dependence of these communities
on tourism. However, because of the limited terrain available for development,
pressure exists to develop portions of these riparian areas. Aside from county
requirements, many four seasons resorts are voluntarily committing to preserving
wide riparian corridors because this practice has multiple engineering, economic,
social, environmental, planning and other benefits.

Questions/Comments

Question;	Why is the 100 yr flood plain limit included as a protection boundary?
Answer:	Just a conservative measure to help insure no development encroaches on the stream corridor.
Question:	How are you able to do this in Colorado, when we can't do it in the Pacific NW?
Answer:	This is really just implemented for ski areas, where there is a concerted effort to protect the streams. In urbanized areas this is not the case.
Question:	Who manages these corridor areas?
Answer:	The landowner is the responsible party.
Question:	Is there a relationship to stream width to buffer width?
Answer:	No, it is just the application of the criteria as shown in the talk.

Development and Application of the Rapid Stream Assessment Technique (RSAT) in the Maryland Piedmont
John Galli

In response to a growing demand for both quickly identifying channel erosion problem areas and systematically characterizing stream quality conditions in the Washington Metropolitan Area, a rapid stream assessment technique (RSAT) was developed by COG (Council of Governments). RSAT represents a synthesis of several well-known stream survey techniques, such as EPA's Rapid Bioassessment Protocols, together with COG's stream sleuthing experience in the Mid-Atlantic Piedmont. Under RSAT, the stream, including its channel network, is surveyed in its entirety. RSAT's intended use and applicability is presently limited to small- to medium-size Piedmont streams.

Questions/Comments

Comment: Galli's information indicates that runoff controls in Maryland are in fact retarding the rate of stream degradation.

Comment: The 10 - 20 percent impervious range seems to be a threshold for uncontrolled runoff after which the stream system seems to significantly degrade with respect to integrity of the invertebrate community.

Question: How about cross-correlation of parameters to eliminate some and increase the speed of the survey?
Answer: Shape of the channel is one, and the number of cross sections for mainstream areas, but not number of cross sections for small tributaries.

Question: What about doing isolated reaches rather that walk the entire streams?
Answer: You could but we don't, I don't think there is a characteristic area or reach.

Cumulative Impacts of Watershed-Scale Development on Stream Morphology
Michelle Girth

For most projects, cumulative impacts are not a major consideration, but in the case of linear infrastructure (highway, railroad, pipeline), cumulative impacts can be significant and multiple watersheds involved. In the Pacific northwest, many of the streams along which transportation corridors run support special status fish species.

Cumulative impacts attributable to Union Pacific Railroad (UPRR) activities were identified in two watersheds with narrow valleys and steep stream gradients.

In undertaking improvements to project segments, UPRR focused on using existing rights-of-way and minimizing disruption to the surrounding environment. The first issue facing the team preparing a Cumulative Impact Assessment (CIA) was to define "the project" against which impacts would be assessed.

It was agreed that the following items are the critical components of a CIA:

- Identifying the actions (activities that generate, induce, or contribute to cumulative effects) creating the impacts
- Identifying the impacts (actual damages or benefits to a system) of the projects on the surrounding environment
- Identifying the affected system (entire ecosystem or particular resources within an ecosystem.

Fanno Creek Watershed Planning and Enhancements
Kendra Smith

The Fanno Creek watershed, in the Portland Metropolitan area, is one of the oldest urbanized watersheds. Polluted urban runoff, impervious surfaces, and stream/floodplain encroachment contribute to the degradation of the water quality and habitats of Fanno Creek. However, Fanno Creek continues to support native fish and wildlife. The remaining greenspaces of the watershed offer considerable opportunities to enhance the quality of the watershed and the health of its streams for both animals and people.

The goal of the watershed management process is to achieve the healthiest environment possible given the land use constraints.

Stream stewardship programs that encourage active participation in the enhancement of the system will be the cornerstone of a successful plan implementation. Finding a balance between the watershed's social, political, physical, and biological components will lead to holistic watershed management that will improve overall stream health.

Questions/Comments

Question: Are you using geomorphological computations to design your projects?

Answer: No these are early action demo projects of bioengineering to demonstrate to public what we are trying to do; will use engineering data to help design rehabilitation of future projects.

Question: How did you get such support out of your sewerage agency?
Answer: Individual initiative of the agency director to DEQ setting of
               standards for the Tualatin River. DEQ feels that the agency is
               moving in the right direction.

Question: What is the cost of this program?
Answer: The Planning process is $300,000. Implementation is projected to
               be $2M/yr over a number of years.

CAN ENVIRONMENTAL IMPACTS OF WATERSHED DEVELOPMENT BE MEASURED

by W. J. Snodgrass, B. W. Kilgour, M. Jones, J. Parish and K. Reid
respectively of Ministry of Transportation, University of Waterloo, Ministry of
Natural Resources, Ortech International, and BCI International

Abstract

The objectives of this paper are to evaluate whether impacts upon aquatic
ecosystems, particularly at a watershed scale can be measured, or whether the state-
of-the-practice is one of being able to measure only a few effects using indicators,
inferences and simulation models.

Four approaches are presented which have the potential to provide the analytical
and quantitative structure for demonstrating measurable effects: (i) relationships
between watershed development and aquatic ecosystem properties, (ii) a stress-
response framework which ultimately may provide the integrative methodology for
measuring effects, (iii) a systems analysis approach for studying component parts of
the ecosystem. and (iv) a synthesis of various ecological (physical, biophysical,
biochemical) indicators for a range of stream ecosystems. As a conclusion, an
assertion is made that the field is at least half a decade away from being able to
quantify the "stress-response" relationships as a predictive tool for impact assessment, and that the immediate future will depend upon relationships and synthesis of
models and experience.

1. Introduction

1.1 Scope of this Paper

This conference addresses the effects of watershed development upon aquatic
ecosystems. Watersheds may be dominated by forested lands, grasslands, deserts,
agriculture, urbanlands, or a combination of these land uses. The principle aquatic

receiving waters may be streams (rivers), lakes, wetlands, estuaries, or oceans. The receiving water ecosystems may be dominated by discharges from municipal or industrial wastewater discharges, combined sewer overflows, stormwater runoff (from the rural or urban catchment) or a combination of these sources.

The objective of this paper is to provide a foundation for determining whether the effects (impacts) of watershed development upon aquatic ecosystems can be measured. To use a database which is as thorough and widely understood as possible, the focus of this paper is two fold:

(i) it centers upon the impact of urbanization upon aquatic ecosystems largely contained within streams and rivers. The natural boundary is the surface water based, watershed. The other types of ecosystem boundaries which need to be considered are outlined below.

(ii) the impacts of urbanization evaluated, are those caused by stormwater (a) runoff to streams and (b) infiltration to groundwater. Other impacts such as those associated with point source releases of pollutants or stream alteration due to infrastructure is discussed only in brief detail.

1.2 Urbanization and Receiving Water Issues in a Watershed Context

(a) Overview of Impacts of Urbanization

The impacts of urbanization upon environmental systems include:

i) direct, spatial effects, involving loss of lands;

ii) indirect impacts through pathways, especially hydrological and ecological alterations; and

iii) cumulative effects.

(i) Direct Spatial Effects - Loss of lands and direct alteration of stream channels (direct spatial effects) have historically been one of the largest effects of urbanization upon surface waters and their contained ecosystems. In fact, a permanent feature of urban development is the loss of natural features, valued land forms, groundwater recharge/discharge areas and wildlife habitats if development occurs on lands occupied by these land uses. In some literature, these effects are called "form effects" or "spatial impacts" in the Geographical Information System (GIS) sense of "spatial analysis".

In general, spatial impacts due to urbanization include: reconstruction, encroachment, and intrusion by humans. For receiving environments such as stream corridors and valleys, direct effects have historically been caused by enclosing streams in pipes or open channels, channel realignments and stream crossings by infrastructure.

(ii)   Pathways - Pathways involve the movement of water or contaminants from industrial, residential, commercial or agricultural areas through atmospheric, surface water or groundwater routes (Indirect Impacts Through Pathways) to impact biota in the various ecosystems where they live, reside, breed and raise their young. These ecosystems units are called habitats.

The major impacts of urbanization upon hydrological pathways include: (i) increases in surface water flow, (ii) reduction in infiltration to groundwater, (iii) increased risk of water quality degradation, (iv) increases in flooding and erosion potential, and (v) habitat degradation and even destruction.

The major impacts of biological pathways include the bioaccumulation of toxic elements through the food chain. The major effects on air pathways include increased emissions of various contaminants into air from fuel consumption for heating residences, industrial and commercial buildings, transportation, and industrial operations.

(iii)   Cumulative Effects - The current system for reviewing planning applications, particularly in rural areas within the urbanizing fringe, is primarily oriented to site-specific analysis and, therefore, does not anticipate the broader, longer term environmental implications of permitting many individual sites to be developed. In some planning circles, these are known as cumulative environmental effects. (See Constant and Wiggins, 1991 for additional discussion.)

1.3.   Proposed Analytical Frameworks

This paper is principally concerned with the effect of urban stormwater runoff upon surface waters, groundwater, and their contained ecosystems, impacts which occur through hydrological pathways.

Four analytical frameworks, for evaluating whether the effects of watershed scale development can be measured, are presented in this paper:
(i)   relationships between per cent imperviousness and characteristics of the watershed;
(ii)   a stress response framework for the impacts of development upon hydrological pathways;
(iii)   a systems analysis framework for studying "an ecosystem"; and
(iv)   a synthesis of models, literature; and field measurements/experience.

2. Relationships between degree of development and impacts on hydrological pathways or receiving waters

2.1 Literature relationships

Relationships between land use changes (forested to agriculture or urban; agriculture to urban; regeneration of forests) or infrastructure and urban density (using indicators such as per cent impervious) and receiving water characteristics provide one approach for describing and potentially quantifying the stress that watershed development has on surface water ecosystems. This section evaluates the ability of the field to measure the impact of watershed development, using a few representative relationships published previous to the publication of this conference. The characteristics include:

1. Watershed imperviousness and hydrological Characteristics
   a) Watershed imperviousness and Storm runoff Coefficient
   b) Hydrological Coefficient Coefficients for Meadow and Parking Lot
   c) Effect of Urbanization on Distribution of May to November Water Budget (Ontario) for Forested Areas and Urban Areas
   d) Urbanization and Baseflow
2. Watershed imperviousness, hydrology and fluvial geomorphology
   a) Channel stability as a function of Imperviousness
   b) Channel width, relative to changes in channel slope
3. Watershed Imperviousness and Macroinvertebrate Density Index

In presenting the relationships for urban areas, the amount of imperviousness on a site or in a watershed is used as a unifying theme. Imperviousness is the per cent (or decimal fraction) of the total catchment covered by the sum of roads, parking lots, sidewalks, rooftops, and other impermeable surfaces of the urban landscape. Operationally for mature urban areas, watershed imperviousness can be defined as the fraction of watershed area that is unvegetated.

The relationships presented in this section do not sub-divide imperviousness. It is composed of two primary components - the *rooftops* under which humans live, work and shop, and the *transport* system (roads, driveways, and parking lots). The transport component now often exceeds the rooftop component in terms of total impervious area created (Schueler, 1994b). Two key aspects which may influence our ability to measure the impacts of development upon aquatic ecosystems are a matter of active debate in the field: (i) does consideration of these primary components provide a better measurement approach? (ii) how have different studies calculated percent imperviousness?

a) **Watershed imperviousness and Hydrological Characteristics**

Changes in the site runoff coefficient, hydrological model coefficients, and watershed water budget as a result of site imperviousness are three indicators of the impacts of development upon hydrological runoff pathways.

(i)   Watershed Imperviousness and Storm runoff Coefficient The runoff coefficient which ranges from zero to one, expresses the fraction of rainfall volume that is actually converted into storm runoff volume. The data, from over 40 runoff monitoring sites across the USA (see Fig 1.1, WEF, 1996), show that the runoff coefficient closely tracks percent impervious cover, except at low levels where soils and slope factors also become important. In practical terms this means that the total runoff volume for a one-acre parking lot (Rv=0.95) is about 16 times that produced by an undeveloped meadow (Rv = 0.06). (Schueler, 1994b).

(ii)   Hydrological Coefficient Coefficients for Meadow and Parking Lot - The changes in the coefficients used in the HYMO class of models for simulating the different hydrological characteristics of a meadow and a parking lot (see Table 1.5 of WEF, 1996) can be used as an index for quantifying the effects of urbanization upon watershed runoff characteristics.

(iii)   Effect of Urbanization on Distribution of May to November Water Budget (Ontario) for Forested Areas and Urban Areas - The calculated impact on a catchment's hydrology due to a land use change from a forested area to a typical single family residential area, (see Table 1.6, WEF, 1996) provides an approach for estimating the potential forcing functions upon a stream due to urbanization.

b) **Effect of Urbanization Upon on Groundwater and Baseflow**

Because infiltration is reduced in impervious areas, one would expect groundwater recharge to be proportionately reduced (see section 3c in this paper). This, in turn, should cause lower dry weather stream flows. Actual data, however, that demonstrate this effect are rare (see WEF, 1996) - some studies show a decrease, some an increase and some no change in baseflow after urbanization.

c) **Watershed imperviousness, hydrology and fluvial geomorphology**

The graph of Booth and Reinelt (1993: between imperviousness (X-axis) and the ratio of flows (10-yr forested /2 yr current) on the Y-axis; see also Fig 1.3, WEF, 1996) is significant in the field for suggesting that a threshold for urban stream stability exists at about 10% imperviousness of a watershed - watershed development beyond this threshold consistently resulted in unstable and eroding channels.

d) Watershed Imperviousness and Macroinvertebrate Index

There are a variety of studies (e.g., Shaver et al, 1995) of macroinvertebrate index in a stream versus % imperviousness which suggest that a major change in the relationship (i.e., a threshold) occurs in watersheds having 10-20% imperviousness, and is consistent with relationships observed by other investigators.

2.2   Discussion of these and Other Relationships

The following points are made:

1. There are many other published pieces of data which could be used infer the effects of watershed development on aquatic ecosystems. For example, relationships (see Figure 4, Simon, 1995) between changes in "hydraulic geometry characteristics of a stream" (channel width to depth ratio) and channel gradient after a disturbance (such as urbanization), suggest that moderately steep channels (channel gradient of 1%) experience a greater increase in width than depth (especially for coarse-grained systems).

2. Relationships change as more data are accumulated. For example, the hypotheses of Shaver et al (1995) that (i) there is a threshold effect and (ii) that installation of BMPs in urban watersheds may protect the stream and change the threshold of imperviousness from a value of 10-25% to a value of 30 -50% (see dotted line in Figure 1.4, WEF, 1996) have changed. In their recent update (see Figure 4 Maxted and Shaver, 1996, respectively) (i) there is not an apparent threshold because the impervious relationship is continuous and (ii) BMP's do not seem change the macroinvertebrate - imperviousness relationship.

3. This conference represents a significant advance in the professional field due to the number of studies which evaluated relationships between indicators of watershed development and aquatic ecosystem characteristics.

3. Stress-Response Framework As An Analytical Tool

A stress - response framework is second method for measuring the impact of watershed development upon aquatic ecosystems. A stress-response (cause-effect) framework attempts to link specific changes in the ecosystem (the response, effect) to the primary causes of the change (the cause, the stress). For the purpose of state-of-the-environment reporting, indicators of changes are classified (Hunsacker and Carpenter, 1990; US EPA, 1992) as:

i)   STRESSOR, caused by pollutants, pesticides, climate change, forest harvesting, etc. STRESSOR indicators are mostly associated with human activities, but can also include natural sources (e.g. forest fire, hurricane, volcanic eruption, etc.);

ii)   EXPOSURE, of ecosystems to STRESSORs. Exposure is the intensity of stress experienced at a point of time, as well as the accumulated doses over time. Examples are concentrations or accumulations of toxic substances, rate of habitat loss, etc.; and

iii)   RESPONSE, of organisms and/or ecosystems to EXPOSURE to STRESS-ORs, for example, mortality, change in productivity, change in species occurrence or diversity, etc.

Urban development, highway activities and other landuse changes cause a variety of direct and indirect effects. Understanding the exposure route: the biophysical and atmospheric pathways by which ecosystem degradation occurs, allows the analyst to place the effect of development into its proper context and magnitude. This approach allows the practicing scientist to use common sense to determine the relative role of physical, chemical, and biological aspects in protecting environmental quality.

In terms of the stress-response terminology, the key stressor of concern to this paper is urbanization. The exposure pathway is hydrological, from the urban imperviousness to the receiving stream. The response of concern is the aquatic ecosystems, largely contained within the stream's waters.

To apply a stress response framework to the effects of urbanization on receiving water ecosystems, one needs to have
-   a mathematical model or a quantitative methodology of relationships between stressor and response;
-   indicators or measurements of stressor and response; and
-   protection objectives,

This paper now presents a proposed model in qualitative terms to illustrate the use-of-a stress - response framework for measuring the impact of watershed development upon aquatic ecosystems. The model can be entitled a "**Proposed Stress Response Framework for Hydrological Pathways to Aquatic Ecosystems**". Comments on protection objectives and indicators are provided at the end of the paper.

The framework is presented as a sequential set of hypotheses of effect (Figure 1) linking urbanization of the watershed and the consequential changes to the receiving water ecosystem. The effects are described sequentially under the headings:
-   urban watershed hydrology,
-   stream hydrology,
-   groundwater hydrology,
-   stream geomorphology,
-   stream water quality,

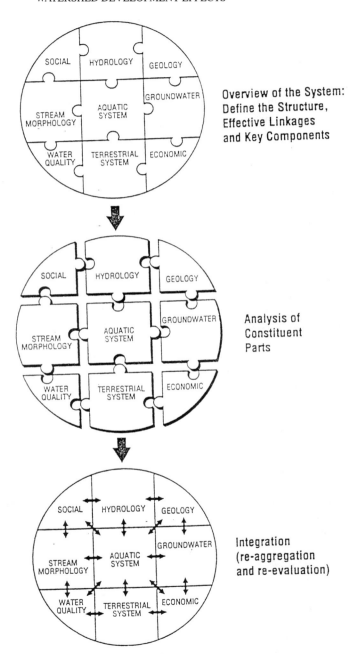

Figure 1: Stress - Response Framework for Impacts of Urbanization upon
Aquatic Ecosystems Through Hydrological Pathways

- instream habitat
- direct effects on riparian canopy, and
- instream ecosystem.

The impacts are described in this section in narrative form for communication purposes. This structure is an adaptation of Schueler (1992), who loosely grouped the impacts on urban streams into four categories: changes to stream hydrology, fluvial geomorphology, water quality and aquatic ecology. Because the extent of an alteration to a stream is a function of the climatic regime (wet, dry) and impervious land cover, the reader should be cognizant that the narrative in this section may be representative of the urbanization of a forested headwater watershed in a relatively wet (90 cm), temperate Eastern Seaboard area. The reader should consider the hydrological characteristics of other climatic areas, and their stream's morphological setting to assess the potential impact of hydrological alterations on their receiving streams.

(a) Watershed Changes Caused by Urbanization

Urbanization may change the hydrologic cycle by:
(i)   increasing the volume of runoff and reducing the degree of infiltration due to the hardening of surfaces;
(ii)   reducing the amount of depression storage due to regrading, or reducing evapotranspiration due to removal of vegetative cover; and
(iii)   reducing the time of travel to the receiving body of water due to the construction of efficient sewer systems. In the extreme, small streams have been completely replaced by pipes and open channels after urbanization, resulting in streams and open channels which are completely dry between storms.

(b) Changes in Stream Hydrology

The net effect of development is a dramatic change in the hydrologic regime of the urban stream. The effects include:

1. There is an increase in the magnitude and frequency of severe flood events.
2. More of the stream's annual flow is delivered as surface storm runoff rather than baseflow or interflow.
3. The velocity of flow during storms becomes more rapid.

(c) Changes in Groundwater Infiltration and Baseflow Recharge

The cover of a water shed provided by buildings, pavement and concrete within an urban area is expected to

1. Reduce the volume of water available recharge.
2. Lower the groundwater table
3 Reduce stream baseflow, provided by groundwater discharge.

The relative magnitude of these effects are dependent upon the watershed topography, and soils - if flat sandy soils are covered by pavement, the reduction in infiltration will be much more dramatic than if clay soils are covered with urban surfaces.

(d) Changes in Urban Stream Morphology

Stream channels in urban areas must respond and adjust to the altered hydrologic regime that accompanies urbanization. The severity and extent of stream adjustment is a function of the degree of watershed imperviousness. The adjustments include the following:

1. The primary adjustment to the increased stormflow is increasing the stream's cross-sectional area to accommodate the higher flows.
2. Sediment loads in the stream increase due to increased streambank erosion and upland construction site runoff.
3. Together, the increased sediment load and channel widening produce a major change in the morphology, and in the riffle - pool - run - sequences and portions of urban streams.
4. The nature of the streambed is also modified by the urbanization process. Typically, the grain size of the channel sediments shifts from coarse grained particles towards a mixture of fine and coarse grained particles.
5. In intensively urbanized areas, many streams are totally modified by man to "improve" drainage and reduce flooding risks. Headwater streams tend to suffer disproportionately from enclosure.
6. Another inevitable consequence of urbanization is stream crossings by roads and pipelines. These structures must be heavily armored to withstand the down-cutting power of stormwater. Such armoring exports effects to downstream reaches.

(e) Changes in Stream Water Quality

Changes in storm water quality are associated with two major phases of urbanization. During the initial phase of development, an urban stream receives a significant pulse of sediment eroded from upland construction sites, even if erosion and sediment controls are used. Sediment contributions from the land surface usually decline once upland development stabilizes and are replaced by increased streambank erosion. In the second phase of urbanization, the dominant source is the washoff of accumulated deposits from impervious areas during storms (Schueler, 1992).

In general, the constituent concentrations in urban streams are one to two orders of magnitude greater than those reported in forested watersheds. Their degree of loading has been shown to be a direct function of the percentage of watershed imperviousness (Schueler, 1987). In urban streams, the higher loadings translate into water quality problems, such as:

1. turbid water,
2. nutrient enrichment,
3. bacterial contamination,
4. organic matter loads,
5. toxic compounds,
6. temperature increase, and
7. trash/debris.

(f) Changes in Riparian Zone Stream Habitat

Riparian Zone Stream Habitat is the vegetation (grasses, overhanging trees) on the stream banks which protects the stream banks from the erosive forces of fast flowing waters and shades the stream from the hot sun, and other vegetation in the flood plain where fish (such as pike) and other organisms spawn during periods of flood flow. Physical changes in riparian vegetation is caused by:

1. Tree removal for infrastructure crossings, golf courses, and other flood plain land uses.
2. Erosion of stream banks, due to the increased work done by stream waters and the changes in the meandering stream.

(g) Changes in In-Stream Habitat and Ecology

Physical changes in instream habitat are shaped by the changes in the stream's hydrology and geomorphology. Other habitat characteristics are shaped by mats of algae and other biological phenomena, and sediments that cover habitat features. Of the four major habitat features (In-stream cover; benthic invertebrates as an indicator of water quality; channel morphology - width/depth ratio, bank stability, sediment transport and substrate characteristics: and thermal conditions) presently being measured to adapt the Habitat Suitability Index model to Ontario's coldwater - cool water streams, the following changes may occur:

1. <u>In-stream cover components</u>
   i) The riffle-pool structure associated with a stream's meanders is altered as the sinuosity, stream length along its thalweg and stream gradient are changed.
   ii) In-stream cover provided by bank undercuts is increased at points of bank-toe erosion

iii)   Access to sub-pavement materials for over-wintering habitat is more
         difficult where inter-bedding of the substrate is caused by finer grained
         sediments

2. Benthic Invertebrates
    i)   The relative abundance of various taxa change as the nutrient status at a
            site becomes more eutrophic (for example, certain taxa such as
            oligochaetes and dipterans are very abundant at enriched sites, while
            mayflies and stoneflies are more common at oligotrophic and colder
            sites) and the thermal regime changes.

3. Channel Morphology
    i)   The width/depth ratio increases if the channel widens, but decreases if
            the channel downcuts. (The process of widening or deepening is a
            function of the geological terrain and other channel characteristics.)
    ii)   Bank stability decreases as a result of geomorphic changes, leading to
            changes in fish habitat in undercuts etc,
    iii)   Increased sediment transport and the altered grain size distribution
            provides a source of material for interbedding the substrate; and
    iv)   An alteration of substrate characteristics due to interbedding or changes
            in pavement - subpavement ratios reduced the extent of fish habitat for
            critical life-stages, depending upon the fish species
    v)   Armoring reduces or eliminates stream bank - related fish habitat.

Of the four characteristics of channel morphology being evaluated as a part of
fish habitat, the linkages to life cycle requirements are now being measured (put
into next section).

4. Thermal Conditions

Removal of riparian canopy causes the stream waters to heat up, causing shifts
from cold water habitat to cool or warm water habitat conditions.

(h) Changes in the Stream's Ecosystem

The ecology of urban streams is shaped and molded by the extreme shifts in
hydrology, geomorphology and water quality that accompany the development
process. The stresses on the aquatic community of urban streams are both subtle
and profound, and are often manifested in the following ways.

1. Shift from external (leaf matter etc.) to internal (algal organic matter) stream
    production.
2. Reduction in diversity in the stream community.
3. Destruction of freshwater wetlands, riparian buffers and springs.
4. Changes from cold water communities to cool or warm water communities.

(i) Structure of these Changes

These changes: first "stream hydrology" and "groundwater hydrology", then "fluvial morphology", then "water quality", followed by "stream habitat" and the "aquatic ecosystem", provide a structure for the reader to develop an integrated picture of impacts.

This sequence of changes (Figure 1) is called "impact flow" in this paper. The "flow" of impacts occurs first through physical pathways and then chemical, biophysical or biochemical pathways, dependent upon which part of the sequence that one is considering. For example for the sequence of

   i) hydrology
   ii) hydrogeology
   iii) fluvial geomorphology
   iv) surface water quality
   v) fish habitat (instream, riparian canopy)
   vi) aquatic ecosystem

The first three are physical, the fourth is chemical/biochemical while the last two are physical/biophysical in nature.

(j) Application of Stress-Response Framework to Measurement Questions

There are a variety of questions to be considered in applying the stress response framework to the impacts of urbanization. One set of questions, such as the following, involve watershed scale analysis:

1. What plan form changes will occur in streams after urbanization?
2. What will be site level, reach level, and subwatershed scale changes expected in habitat?

The analytical methodology for "hydrological impacts" addresses part of these questions but does not address all plan form changes - these changes need to apply the broader knowledge base of geomorphology. The "impact flow" structure is limited to addressing only "indirect effects in plan form changes due to changes in hydrological pathways". It does not include "direct form effects" from infrastructure crossing the stream, infrastructure running longitudinally and parallel to the stream in the valley, and other valleyland uses such as golf course. In addition, it addresses surface aquatic systems; it does not address changes in valleyland terrestrial systems, nor tableland terrestrial systems which result from urbanization.

3. A Systems Study Process for Ecosystem Analysis

(i) The need for the study process
The term "ecosystem approach to management" has evolved to link and to integrate human objectives with ecological objectives in environmental decision-making. Several descriptions of the term "ecosystem approach" exist in the literature including:

- everything is connected to everything else;
- human beings are part of nature and not separate from nature;
- human beings are responsible for their actions and associated impacts; and
- economic health and environmental health are mutually inclusive.

One of the more vexing aspects of ecosystem analysis is the apparent non-quantitative characteristics of relating various component parts to an understanding of how the overall ecosystem will respond to various stresses such as urbanization. A key conceptual process used in any measurement methodology addresses such questions as "How do you study the ecosystem? How do you tear it apart? How do you reassemble it?." Since "an ecosystem approach" is defined as "everything is connected to everything else," a measurement methodology should include the "whole", the "component parts" and "the linkages and connections".

Another need is that of an integrative approach. Why is there a need for integration in watershed management? In simple terms, the ecosystem approach forces us to look, not only at air, water, land and living things, but also at the interrelationships among these components of a watershed. Traditionally, science has taken a reductionist approach, studying the individual components of a problem. In the context of a watershed, this would mean studying the streams, the groundwater, the aquatic communities, and other components of interest. Ecological planning tells us that we must do much more than just study the parts of the watershed: we need to understand how these components interact, and we need to study the whole as well. Integration is the study of the complete system (the watershed, its components, and their interrelationships). Integration in watershed planning is both a process and an evolving discipline — a science that is cross-disciplinary in nature. As a discipline, integration is a very young science.

(ii) Systems Study Process
The need to integrate forces a change in how analysis and management is carried out. How can we take a holistic approach to studying a watershed? An integrative approach to watershed analysis can be considered to have a number of discrete steps (see Figure 2):

1) Overview the system (identify what resources, habitats and human uses exist and where);

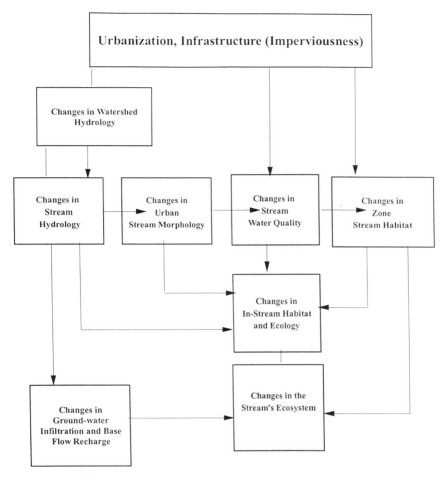

Figure 2: Systems Study Process for Watershed Analysis

2) Define the structure, the effective linkages of the system, and key components to be studied (i.e. functions);

3) Reduce the system to its constituent components for the purpose of scientific study;

4) Study the component parts (hydrology, groundwater, aquatic systems, etc.);

5) Reaggregate the system; and

6) Re-evaluate the overview focusing on interrelationships and the whole system.

A cartoon illustrating this process for the 9 component parts of watershed systems is presented in Figure 2. The system is still disaggregated for the study of its component parts, however, this traditional approach is augmented by the process of integration which looks at interrelationships and the whole system (including the interrelationships between component parts).

Care must be taken to ensure that the system overview is properly formulated upfront (i.e. that the problem is properly defined). The major difference between the integrative approach outlined above and traditional forms of resource planning is the emphasis on linkages in Step 2 (including external influences from outside the watershed boundaries), and the reaggregation and re-evaluation of the system in Steps 5 and 6. The study of the component parts in Step 4 is carried out with a respect for linkages (e.g. by having constant communication between study team members as they carry out their tasks).

Conceptually, the science of integration is synonymous with "systems analysis". The use of systems analysis in watershed management has been slowed by a reliance on the traditional reductionist approach to scientific research, and by administrative and jurisdictional barriers (in particular, the fragmented distribution of responsibilities for environmental management). Systems analysis has long been applied to water management questions such as reservoir optimization to arrive at recommendations on how multiple, often competing objectives, can best be met.

The greatest technical challenges in integration lie in having the tools and techniques to allow reaggregation and re-evaluation. Some of the tools available (Kidd et al, 1995) for integration are:

- Spatial Analysis (GIS)
- Models for Pathways Analysis
  - Flood Flow
  - Water Balance
  - Water Quality Mass Balance
  - Habitat Suitability Index (HSI), Index of Biotic Integrity (IBI)
- Time Sequence of Change
- Hierarchical Frameworks (Scale Dependent)
- Trade-off Analysis (McHarg-Overlaying Matrix)

(iii)   Measurements of Effect Using Stream Component Data

Site scale measurements have been initiated in Ontario which will provide data for examining relationships between discretized components of the ecosystem. Two components outlined in this section are: 1. Fish habitat measurement and its relationships to fish biomass; and 2. Relationships between macro-invertebrate communities and fish communities.

a) Fish habitat relationships with fish biomass

A significant new applied research effort is underway to adapt the Habitat Suitability Index Models to Southern Ontario Cold and Cool Water Streams. (Stanfield et al, 1996). The four major measurement components (see section 4 (g)) are: 1. In-stream cover components, 2. Benthic Invertebrates, 3. Channel Morphology, and 4. Thermal Conditions.

The habitat measurement procedures have shifted from using a visual method to using a quantitative point transect method for evaluating habitat. In the visual method (similar to the technique of Galli [see paper this volume]), the portion of habitat which was in pools, riffles, runs etc was visually assessed and the percent of area for each feature was estimated. In the point transect method, the depth of water, substrate composition (Pavement, subpavement), stream point velocity etc is measured at a number of points across a transect and the geometry of each bank at either end of the transect is measured. A sufficient number of transects are measured to provide at least 60 point measurements. Then based upon a dichothomous key, the point data are ranked into quantitative estimates of the portions of pools, riffles, runs etc.

The new procedure is more repeatable by different crews than the older visual method, while still meeting criteria such as efficient, cost-effective, and scientifically based. The older method was less repeatable between visits due to scale influences (how large does a feature have to be before it is counted), difficulties in interpreting criteria (what exactly is a pool), and the fact that streams are transitory by nature, which defies clear definitions of habitat features.

Present research is focusing on using this measurement methodology for impact assessment and long term monitoring.

b) Associations between Invertebrate Communities and Fish Communities

Although benthos and fish are often used to characterize the condition of aquatic resources, fish and benthos monitoring data are difficult to make decisions with because we currently have no decision criteria in Canada. Historically, any demonstrated impact on either of these groups has been used as evidence of a significant biological impact. However, Kilgour et al. (1996b) demonstrates that such criteria are faulty because the detection of a difference is dependent on sample size. As a result, we need to specify non-zero differences that we consider relevant for detection.

In Canada, the federal *Fisheries Act* has a general goal of maintaining the productive potential of fish and fish habitat. Although several ecological measurements such as toxicity, habitat surveys and benthic surveys have been historically used to demonstrate non-compliance with the act, only surveys of fish community composition have any obvious relationship to the goals of the act itself (Kilgour et al., 1996a). Using the terminology outlined by Cairns et al. (1993), one then might consider the fish community to be a compliance indicator, while the benthic community might be considered an early warning indicator if it can be demonstrated that it can predict when fish communities will be in non-compliance. Kilgour and Barton's (1996) work in southern Ontario streams confirms that fish and benthos do respond to similar suites of environmental variables, although benthos respond more quickly because of shorter life cycles.

For ecosystem endpoints that are considered worth protection (such as fish community composition), Kilgour et al. (199b) propose that impacts in ecological responses are of little interest unless they exceed the normal range of variation in reference areas. The normal range for any response variable is defined by the region encompassing 95% of the possible reference-area observations, which for single variables is approximately ± 2 x the standard deviation (þ) of reference-area observations. DFO & EC (1995) and the Ontario Ministry of Environment (K. Somers, pers. comm.) have incorporated such suggestions into their aquatic environmental effects monitoring (AEEM) programs. Given that we can specify a magnitude of change that is acceptable, it follows that a statistical test that determines the probability that such an event has not occurred (i.e., that the measured response is in fact different by more than ±2 þ) should be conducted. The usually applied test of differences is not appropriate (McBride et al., 1993; Kilgour et al., 1996). Rather, one should test the null hypothesis $H_0:|\mu_1-\mu_2|\leq 2þ$, versus $H_A:|\mu_1-\mu_2|>2þ$, where μ_1 and μ_2 are the means of two areas (a reference and test site) with within-location variation $þ^2$. Specifying a non-zero difference in the null hypothesis requires that test statistics be judged against critical values derived from non-central t or F distributions (McBride et al., 1993; Kilgour et al., 1996). Use of such a generic approach to derivation of decision criteria has several advantages including:

(i)   applicability to any ecosystem parameter considered worth protection, not just freshwater fish;

(ii)   only a sample of the background variation needs to be described. This contrasts with most other approaches that require literally hundreds of reference samples before there is reasonable comfort that the reference condition has been properly described (Reynoldson et al., 1996);

(iii)   the approach can be applied to survey data that incorporates either regional-reference (Hughes, 1995) or site-specific-reference locations (Green, 1979), or temporal controls (Green, 1979). Other decision criteria are generally derived based on regional-reference criteria that do not allow for determining whether a point-source alone has ecologically relevant effects on a system.

Benthic invertebrate community surveys have many advantages for evaluating the condition of stream ecosystems (Rosenberg and Resh, 1993), but are usually rationalized on the basis of presumed relationships between the condition of invertebrates and the condition of fish (DFO & EC, 1995, Hodson et al., 1996). Benthos are also favored because they have shorter life cycles (and therefore quicker responses to stress), and can be collected without any obvious effects on fish. Since their use is rationalized on the basis of a relationship with fish, it is apparent that impacts on invertebrates are acceptable (by many ecologists and legislators) if impacts on fish are within an acceptable range, whereas impacts on invertebrates would be unacceptable when they coincide with unacceptable impacts on fish. Using data from wadeable streams in southern Ontario, we have demonstrated that fish and benthos respond to similar suites of environmental variables and that impacts measured using benthic invertebrates can be used to predict when impacts on fish community composition exceed 2 þ (Kilgour and Barton, unpublished data). Such relationships between fish and benthos can, therefore be used to derive critical benthic effects that themselves can be used in hypothesis testing as described above.

5. Synthesis of Models, Literature, and Field experience

A synthesis of representative models (USFWS, 1981), literature and field experience is a fourth method for assessing how ecosystems and their biophysical requirements change if the hydrological system changes due to urbanization. It provides a method for calibrating the stress-response framework of this paper, in the absence of the a comprehensive measurements in one stream.

This synthesis is presented in the form of an ecological framework (Table 1 and Table 2) which summarizes the characteristics (stream types, stream hydrology, water quality) in a watershed necessary for particular aquatic habitats (Reid et al, 1996). The two tables are:

Table 1 Biological Communities Representative of Alternative Habitat Units; and

Table 2 Biophysical Characteristics of Each Habitat Unit

a) Biological Communities Representative of Habitat Units

The two major biological entities used in this framework as indicators of ecological communities are fish and benthic organisms. Fisheries have traditionally formed one of the basis for aquatic resources management plans in North America, particularly due their commercial and recreational value. More recently, fisheries ecologists have focused on additional concepts such as biotic integrity in the management of aquatic ecosystems (e.g., Karr *et. al.*, 1986).

Benthic-based assessments of aquatic systems have also been proven to provide reliable measures of ecosystem health. Benthic macroinvertebrates, which are relatively immobile, must endure any stress or disturbance to their environment in order to survive. Unlike higher trophic level organisms, they generally cannot move to more suitable habitat conditions when a stress is applied.

The potential aquatic habitat management units presented in Table 1 range from the relatively pristine condition of a native coldwater fish community dominated by brook trout, sculpins or brook lamprey, to degraded conditions represented by a tolerant, low-diversity warmwater community, or even the absence of any aquatic community.

In a qualitative sense, one can describe a shift toward the right side of the framework (Table 1), i.e., a transformation of an existing aquatic community to one which indicates a more degraded condition, as the result expected empirically if a forested watershed is transformed to an agricultural stream, or to an urban stream. On the other hand, a shift to the left side of Table 1 would lead to less degraded aquatic communities and represent management goals of enhancement, and restoration.

The different aquatic ecosystems presented in Table 1 at the end of this paper are representative of the range of habitat conditions which presently or potentially may exist in Ontario. They are as follows:

- Type IA - a self-sustaining native coldwater fish community;
- Type IB - a self-sustaining non-native coldwater fish community;
- Type IIA - a highly diverse native warmwater fish community;
- Type IIB - a diverse warmwater fish community;
- Type III - a moderately tolerant warmwater community;
- Type IVA - tolerant warmwater fish community;
- Type IVB - highly tolerant warmwater fish community; and
- Type V - no fish community present on a self-sustaining basis.

Potential benthic macroinvertebrate community characteristics for these habitat units are described in detail in the appendix. In summary they include :
- Type I - stable coldwater community (including a minimum of four sensitive coldwater taxa);
- Type II - stable warmwater community (at least five sensitive warmwater taxa);
- Type III - unstable warmwater community (at least six moderately tolerant warmwater taxa);

- Type IV - impaired warmwater community (at least four highly tolerant taxa); and
- Type V - severely impaired community (dominated by at least five of the most tolerant taxa).

b) Biophysical Characteristics (Table 2)

The following categories of information are used in Table 2 at the end of this paper to characterize the habitats:

- hydrological targets;
- water quality targets;
- channel morphology targets;
- in-stream cover targets; and
- substrate targets.

This information is based upon models (largely Habitat Suitability Index Models), literature and field observations used to develop these targets. Empirically, the following impacts of watershed development are captured in the Table 2 if one scans the table from left (Type 1a -Intolerant Coldwater Community) to right (Type V - no aquatic community):

- hydrological targets, - the portion of baseflow needed by an aquatic community decreases as the watershed develops and can no longer support a coldwater community;
- water quality targets, - the maximum summer water temperature, a prime determinant of cold water habitat, increases as the degree of forest cover id decreased and replaced by agriculture or urban land uses; the requirements for dissolved oxygen become less severe allowing communities to develop which can tolerate lower levels of oxygen due to allochthonous inputs of carbon;
- channel morphology targets, - the changes as urbanization proceeds in parameters such as e.g., percent pool area, percent riffle area, mean pool depth, are a function of watershed development, but also upstream physiography;
- in-stream cover targets, for parameters such, percent area as undercut banks, percent bottom cover in pools/backwaters become less strict as streams respond to altered sediment regimes die to urbanization; and
- substrate targets, such as dominant substrate type in riffle-run areas, percent fines in riffle-run areas, etc. become less strict for certain species, but also dependent upon the basic provision of sediment supply to the stream from physiography influences of the watershed.

c) Other Influences on Habitat Units

Implicit in this framework are the influences of physiography and the degree and type of cover (vegetation in terms of % wooded, length of continuous intact riparian canopy upstream of the habitat unit) in the watershed which provide these reach level habitat characteristics. The characteristics are completely implicit in Table 1 and somewhat more explicit in Table 2. Such relationships are available from recent research in Ontario where relationships (i) between % forested and % of annual flow in the form of baseflow are available, or (ii) between % forested and presence of Brook Trout community, are being developed. Such relationships have presently being incorporated into the table, but are incomplete at the time of writing this paper. The quantitative background for such relationships will improve in the near future from the active research of the fields of biogeography, landscape architecture, and conservation biology.

6. Discussion

This section discusses four aspects: state-of-the-practice re measurement of impacts, concept of threshold, protection provided by urban BMPs, and environmental indicators.

(i)   Measurement of Effect. The major question addressed in this paper is: "what analytical structure will provide quantitative data for relating the stresses of urbanization to impacts on aquatic ecosystems". The above four methods present alternative methods for attempting to measure the effects of watershed-scale development upon aquatic ecosystems. It is likely that all four methods, and others, will ultimately be needed to measure effects/impacts. Two methods (#1 - relationships; #4 - synthesis) provide immediately useful approaches and information for characterizing the linkage of development and effects, but are not rigorous in a cause - effect sense. Two methods (#2 - stress - response; #3 - systems analysis) provide the potential to scientifically demonstrate cause-effect, but the state-of-the practice of measurement appears to be at least half a decade to a decade from being able to actually provide the needed measurements.

It is the assertion of this paper that:
(a) The state-of-the art is: i) to document the big picture relationships from empirical data sets (e.g., from S Ontario and other examples) ii) to use the word narrative of impact flow to provide the systems analysis type construct, and iii) to develop relationships of specific linkages of the disaggregated system.
(b) The state-of-the art is some time away from having all the necessary data sets and relationships from dissaggregated system at the same site or in the same aquatic system which would permit one to reaggregate the system and evaluate the systems model for the "development question."

(c) The integrative template of "impact flow" may be quantifiable, or may simply provide a narrative word equation to show the logical method for quantitative impact calculations, and measurements;

(ii)   Concept of Threshold. The concept of a threshold effect between percent impervious and aquatic system response has evolved and is in widespread use in the field. Inspection of the data used, suggests that there is not a threshold in the classical sense of defining the term - threshold, but rather a continuous relationship. But the concept is still useful for communication purposes because there are distinct zones of change in these relationships. Hence when defined as representing zones of change, the concept has utility and a reasonable basis-in-fact.

(iii)   Influence of BMPs upon Aquatic Systems. The influence of urban BMPs such as wet ponds upon stream systems is complex and requires several more years of research and field measurements to discern their explicit benefits. In view of the impact flow structure presented in this paper, physical rather that biological effects may be more readily measurable.

The BMP Protection hypothesis of Shaver et al (1995) has merit if the major pertubation caused by urbanization is linked to physical changes (Hydrological, morphological) since a similar threshold is suggested by Booth and Reinhold (1993) for essentially physical changes in the stream as a result of urbanization. But how clearly are invertebrate communities linked to physical changes in a stream as urbanization occurs?

This BMP hypothesis also has merit for macroinvertebrate communities because BMPs are particularly designed as water quality control ponds to remove nutrients - a typical value of 50-60% phosphorus removal would decrease the degree of eutrophy in a stream. Since macroinvertebrate communities are a useful indicator of water quality, particularly trophic conditions, a change in threshold may be expected. But the question that is unclear is whether a significant enough change may be expected, since reduction in stormwater EMC's for total phosphorus from 200ug/L to 100 ug/l would still leave the stream eutrophic, assuming that 30 ug/l is a reason estimate of the transition from oligotrophic to mesotrophic conditions, and that 50 ug/L is a reasonable estimate of the transition from mesotrophic to eutrophic conditions in streams.

Actually the effect on invertebrate communities may be due to both physical and biochemical (nutrient) changes in water quality. In this case, a causal relationship which uses both physical and biological stresses as its underpinnings should be sought to determine whether either a physical or biochemical effect is more dominant. Such relationships might be discernable at a watershed scale with appropriate site selection, but it is more probable that relationships from site scale data (as illustrated above in section 5) and site selection based upon experimental design principles will be the measurement methodology needed to evaluate this question.

(iv) Environmental Indicators and Principles for their Selection. A substantial body of literature is evolving on environmental measurement methodologies. Two terms should be distinguished: "Mass Concentration" and "Environmental Indicator". The term measurement of mass is used to describe an environmental entity whose total mass is measured (e.g., total phosphorus, total fish biomass, total macroinvertebrate biomass), and can be modelled using mass balance principles. The term, indicator, is used to denote a part of the total mass (e.g., soluble phosphorus concentration, fish catch per unit effort, macroinvertebrate index) and an entity which still provides a used index of environmental quality, but which is only gathered occasionally. The field is evolving toward using environmental indicators rather than mass measurements in many measurement efforts.

Of the rapidly expanding literature on the selection of Environmental Indicators, two recent papers (Cairns et al, 1993; Noss 1995) provide representative examples of the rapidly evolving state-of-the-art. Pertinent points from these papers include:

1. Know what you want to do or protect. For example, the Clean Water Act has provided one of the major legislative bases in the US for aquatic ecosystem management (e.g., one key Wetlands function stressed is the "water-cleansing ability" of the Wetland), while the Canadian Federal Fisheries Act provides a mayor national impetus in Canada (under the act, fish, fish habitat and the chemical and biological components of the ecosystem necessary to sustain fish, are protected from anthropogenic influences). Recent watershed studies are providing a forum for significant human input into defining the direction of management objectives, since measurement and modelling methodologies may not be adequate for establishing quantitative relationships between urban stresses and ecosystem response.

2. Different Type of Indicators are needed for different purposes. Three types of indicators for management purposes are
   - compliance indicators (used to judge the attainment of ecosystem objectives)
   - diagnostic indicators (used to judge the cause of ecosystem deterioration, where the cause is not obvious of simple to determine)
   - Early warning indicators (because the first two types of indicators are reactive indicators, the authors advocate the development of early warning ecosystem indicators, analogous to 'leading economic indicators', to allow for the implementation of management actions before conditions have deteriorated to the point where compliance indicators are).

3. Tradeoffs between desirable characteristics, costs, and quality of information is inevitable when choosing indicators for management use.

4. There is a tautology in the field, which provides a most useful perspective on selection of indicators "Everything is an indicator of something, but nothing is an indicator of everything".

5. As indicators of ecological integrity, indicators should be selected at several ecosystem scales. For example for reserve networks, measurable indicators of ecological integrity are provided for landscape-regional levels, community ecosystems levels, and species levels.

(v) Impact Prediction Methodologies for Watershed Scale Development. It is suggested that predictions of changes due to watershed development are possible for hydrology, but that predictions, based upon a quantitative impact assessment methodology, are not presently possible for biological entities such as benthos and fish. However the general direction of response can be estimated - the major uncertainty is given a large variance structure in environmental data, how large a change is necessary to quantify a change in direction.

7. Summary

The objective of this paper is to provide a foundation for determining whether the effects (impacts) of watershed upon aquatic ecosystems can be measured.

As a summary, an assertion is made that the field is at least half a decade away from being able to quantify the "stress-response" relationships as a predictive tool for impact assessment in order to provide tools for evaluating issues associated with watershed development.

8. References

Booth, D. and L. Reinelt. (1993). "Consequences of Urbanization on Aquatic Systems, measured effects, degradation thresholds, and corrective strategies". pp. 545-550 in *Proceedings Watershed Management*. March 21-24, 1993. Alexandria, Virginia.

Cairns, J., P.V. McCormick, & B.R. Niederlehner "A proposed framework for developing indicators of Ecosystem Health" Hydrobiologia, 263, 1-44.

Constant C.K., and L.L. Wiggins (1991) "Defining and Analyzing Cumulative Environmental Impacts." *Environ. Impact. Assess. Rev.*, 11, 297 -309.

Department of Fisheries and Oceans & Environment Canada (DFO & EC). 1995. Further guidance for the invertebrate community survey for aquatic environmental effects monitoring related to federal *Fisheries Act* requirements. EEM 2, February 1995.

Galli, J. (1996) "Development and Application of the Rapid Stream Assessment Technique (RSAT) in the Maryland Piedmont." In *Effects of Watershed Development and Management on Aquatic Ecosystems*. This volume.

Green, R.H. 1979. *Sampling Design and Statistical Methods for Environmental Biologists*. John Wiley & Sons, Toronto. 257 pp.

Hodson, P.V., K.R. Munkittrick, R. Stevens and A. Colodey. 1996. A tier-testing strategy for managing programs of environmental effects monitoring. *Water Quality Journal of Canada*, 31:215-224.

Hunsacker, C.T. and D.E. Carpenter (eds.). 1990. *Ecological Indicators for the Environmental Monitoring and Assessment Program*. EPA 600/3-90/060. U.S. Environmental Protection Agency, Office of Research and Development, Research Triangle Park, N.C.

Hughes, R.M. 1995. Defining acceptable biological status by comparing with reference conditions. In, W.S. Davis and T.P. Simon (eds), *Biological Assessment and Criteria, Tools for Water Resource Planning and Decision Making*. Lewis Publishers, Boca Raton, Florida. PP 31-48.

Karr, J.R., K.D. Fausch, P.L. Angermeier, P.R. Yant, and I.J. Schlosser. 1986. Assessing Biological Integrity in Running Waters: A Method and its Rationale. Special Publication 5. Illinois Natural History Survey, Champaign, Illinois. 28 p.

Kidd J and multiple Co-authors (1995) State-of-the Science, State-of the Art, State-of-the-Practice of Watershed Analysis, Management and Planning in Ontario. Ontario Ministries of Natural Resources and Environment Publication.

Kilgour, B.W. and D.R. Barton. 1996. Fish-benthos-environment correlations in southern Ontario streams. Manuscript.

Kilgour, B.W., D.R. Barton, D.G. Dixon, H.C. Duthie, D.G. Fitzgerald, B. Goebel, R.P. Lanno, E. Rott and J.G. Winter. 1996a. Quantifying stream ecosystem condition: new approaches to developing decision criteria and diagnostic tools. *Canadian Water Resources Journal*, in press.

Kilgour, B.W., K.M. Somers and D.E. Matthews. 1996b. A 95% rule as a criterion for ecological significance in environmental monitoring and assessment. *Ecology*, in review.

Maxted J. R., and H.E. Shaver (1996). The Use of Retention Basins to Mitigate Stormwater Impacts of Aquatic Life. In *Effects of Watershed Development and Management on Aquatic Ecosystems*. This volume.

McBride, G.B., J.C. Loftis and N.C. Adkins. 1993. What do significance tests really tell us about the environment? *Environmental Management*, 17:423-432.

Noss, R (1995) Maintaining Ecological integrity in Representative Reserve Networks. World Wildlife Fund - Canada; United States Discussion Paper 90 Eglington Ave E.,Suite 504,Toronto, Canada, M4P 2Z7.

Reid et al (1995). Development of a Dichotomous key for the Selection, planning and design of Fish Habitat Mitigation and Compensation Measures for Linear Corridors. Ontario MOT Report RD - 95 - 22.

Reynoldson, T.F., R.C. Bailey, K.E. Day and R.H. Norris. 1995. Biological guidelines for freshwater sediment based on Benthic Assessment of SedimenT (the BEAST) using a multivariate approach for predicting biological state. *Australian Journal of Ecology*, Volume 20.

Rosenberg, D.M. and V.H. Resh. 1993. *Freshwater Biomonitoring and Benthic Macroinvertebrates*. Chapman and Hall, New York.

Schueler, T.R. (1987). *Controlling Urban Runoff: A Practical Manual for Planning and Designing Urban Best Management Practices*. Metropolitan Washington Council of Governments.

Schueler, T. (1992). "Mitigating the Impacts of Urbanization" in *Implementation of Water Pollution Control Measures in Ontario*. (W.J. Snodgrass and J.C. P'Ng edit) pp. 219 to 242, University of Toronto Press. ISBN-0-7729-9964-3.

Schueler, T. (1994a). "Watershed Protection Techniques", *A Quarterly Bulletin on Urban Watershed Restoration and Protection Tools,*Vol. 1, No. 1, Feb. 1994. The Center for Watershed Protection,8630 Fenton St., Suite 910, Silver Spring, MD., 20910.

Schueler, T. (1994b). The Importance of Imperviousness. Watershed Protection Techniques. Volume 1 (No. 3) pp. 100-111 ISSN: 1073-9610. Available from Editor; 8630 Fenton Street, Suite 910, Silver Spring MD 20910.

Simons A., (1995). Adjustment and Recovery of Unstable Alluvial Channels: Identification and Approaches for Engineering Management. Earth Surface Processes and Landforms vol 20, pp 611- 628.

Shaver, E., J. Maxted, G. Curtis, and D. Carter (1995). "Watershed Protection Using An Integrated Approach". In Stormwater NPDES Related Monitoring Needs (H.C. Torno ed.). *Engineering Foundation, American Society of Civil Engineers, 345 E 47th Street, New York N.Y. 10017.* ISBN-0-7844-09665-2. pp 435 - 459.

Stanfield L., M.L. Jones, B. Kilgor, and J. Parrish (1996) Point - Transect Method for Fish Habitat Assessments - A Field Manual and Protocol. Ontario Ministry of Natural Resources Publication.

U.S. EPA (1992) "Framework for Ecological Risk Assessment" report published by *Risk Assessment Forum,* U.S. EPA, Washington, D.C., 20460. EPA/630/R-92-001.

USFWS. 1981. U.S. Fish and Wildlife Service. Ecological services manual - standards for the development of habitat suitability index models. ESM 103. Washington, DC: U.S. Department of the Interior, Fish and Wildlife Service. (Chapters are paginated individually).

Warren-Hicks, W., B.R. Parkhurst and S.S. Baker (1988) *Ecological Assessments of Hazardous Waste Sites: A Field and Laboratory Reference Document.* Report prepared by Kilkelly Environmental Associates, Highway 70 West. The Water Garden, Raleigh, N.C., 27622.

Water Environment Federation and American Society of Civil Engineers (WEF, 1996) Urban Runoff Quality Management. WEF Manual of Practice No 23. ASCE Manual and Report of Engineering Practice 87. Chapter 1 (in press).

TABLE 1: Biological Communities Representative of Habitat Units
in Ontarion Streams and Rivers

Type 1A Intolerant Coldwater Community	Type IB Tolerant Coldwater Community	Type IIA Highly Diverse Warmwater Community	Type IIB Diverse Warmwater Community
Fisheries			
Minimum of one of the following fish species: • brook trout • sculpin • brook lamprey	Minimum of one of the following fish species: • rainbow trout • chinook salmon • brown trout	Minimum of 18 fish species, including at least 6 of the following: • northern hog sucker • pike • smallmouth bass • Iowa darter • longear sunfish • yellow perch • walleye • intolerant minnows[1] • stonecat	Minimum of 14 fish species, including at least 4 of the following: • northern hog sucker • pike • smallmouth bass • Iowa darter • longear sunfish • yellow perch • walleye • intolerant minnows[1] • stonecat
Type III Moderately Tolerant Warmwater Community	Type IVA Tolerant Warmwater Community	Type IVB Highly Tolerant Warmwater Community	Type V No Aquatic Community
Fisheries			
Minimum of 10 fish species, including at least 2 of the following: • pumpkinseed/ bluegill • largemouth bass • blackside darter • greenside darter • redhorses • central stoneroller • insectivorous minnows[2]	Minimum of 4 fish species, including at least 1 of the following: •Green sunfish • rock bass • black crappie • white sucker • gizzard shad • johnny darter • omnivorous minnows[3]	Minimum of 1 of the following fish species: • carp • goldfish • brown bullhead • brook stickleback • central mudminnow	No fish present

TABLE 1: continued

Macroinvertebrates		
Type I Stable Coldwater Community	Type II Stable Warmwater Community	Type III Unstable Warmwater Community
WQI > 13 EPT ≥ 15	WQI > 12 EPT ≥ 10	WQI 10-12 EPT ≥ 5
At least four of the following:	At least five of the following:	At least six of the following:
• *Amphinemura* • *Leuctra* • *Haploperla* • *Ectopria* • *Heterotrissocladius* • *Eukiefferiella* • *Rhyacophila*	• *Acroneuria* • *Isoperla* • *Taeniopteryx* • *Paraleptophlebia* • *Serratella* • *Chimarra* • *Rhyacophila* • *Diamesa* • *Lumbriculus variegatus* • Turbellaria • *Eukiefferiella*	• Turbellaria • *Baetis* • *Caenis* • *Stenacron* • *Tricoythrodes* • *Cheumatopsyche* • *Hydropsyche* • *Neophylax* • *Optioservus* • *Stenelmis* • *Micropsectra* • Simulidae
Type IV Impaired Warmwater Community	Type V Severely Impaired Community	
WQI 6-10 EPT 2-5	WQI <6 EPT ≤ 2	
At least four of the following:	At least five of the following:	
• *Sialis* • *Berosus* • *Cheumatopsyche* • *Hydropsyche* • *Dubiraphia* • *Probezzia* • *Cryptochironomus* • *Paratanytarsus* • *Rheotanytarsus* • *Chaetocladius* • *Hemerodromia* • *Helobdella*	• *Nais pardalis/bretscheri* • *Limnodrilus hoffmeisteri* • *L. Claparedianus* • *Tubifex tubifex* • *Sparganophilus* • *Berosus* • *Probezzia* • *Chironomus* • *Physella*	

[1]Blacknose shiner, sand shiner, rosyface shiner, river chub.

[2]Hornyhead chub, emerald shiner, common shiner, blacknose shiner, striped shiner, spottail shiner, rosyface shiner, spotfin shiner, sand shiner, redfin shiner, blacknose dace, longnose dace, mimic shiner.

[3] Fathead minnow, northern redbelly, bluntnose minnow, goldfish, creek chub, brassy minnow, golden shiner.

TABLE 2: Biophysical Characteristic of Each Aquatic Habitat Unit

Aquatic Performance Standard	AQUATIC ECOSYSTEM OBJECTIVE			
	Type IA	Type IB	Type II	
Hydrology				
• 7Q2 or 3Q2 flow as proportion of average daily flow April to September	• minimum 30%	• minimum 30%	• minimum 10-20%	
Channel Morphology				
• channel stability	• dynamically stable channels with natural features	• dynamically stable channels with natural features	• dynamically stable channels with natural features	
• average pool area as % of total surface area at low flow	• >12%	• >12%	• >20%	
• average riffle area as % of total surface area at low flow	• >12%	• >12%	• >15%	
• average minimum summer pool depth	• 0.5 m	• 0.5 m	• 0.5 m	
In-Stream Cover	• minimum total in-stream cover 30-40% by surface area • woody debris present up to 10% of surface area	• minimum total in-stream cover 25-30% by surface area • minimum 5% of surface area with overhead cover	• minimum 15% of surface area during low flow • 20-80% of pool/backwater habitat with cover over bottom	

| | AQUATIC ECOSYSTEM OBJECTIVE | | |
Aquatic Performance Standard	Type IA	Type IB	Type II
In-Stream Cover Continued	• minimum 15% of surface area with overhead cover • minimum 5% of surface area with overhead cover		• minimum 10% area as submerged macrophytes • cover at stream margins critical for juvenile fish
Substrate	• well-sorted riffle zones • maximum 25% fines in spawning substrates • maximum 30% fines in riffle zones • upwelling conditions required • minimum 50% of riffles composed of cobble, rubble, small boulder	• well-sorted riffles • maximum 25% fines in spawning areas • maximum 305 fines in riffle areas • no upwelling conditions required • minimum 30% of riffle/ substrate composed of cobble, rubble, small boulders	• D50 in pool areas generally exceeds 80 mm • maximum fines in riffle zones <50% • pool and backwater substrates silt to gravel, broken rock
Riparian Habitat • shaded during 1000-1400 hours	• minimum 35%	• minimum 20-30%	• minimum 10% • maximum 30%
• woody debris	• important component of in-stream cover and roughness	• important for roughness and in-stream cover	• very important for roughness and refuge during peak flows

AQUATIC ECOSYSTEM OBJECTIVE			
Aquatic Performance Standard	Type IA	Type IB	Type II
Water Quality			
• maximum annual water temperature	22° C	26° C	28° C
• average annual total suspended solids	<20	<40	<150
• dissolved oxygen (ppm)	>5	>5	>3-4
• spills	none	none	none
Barriers	• remove as feasible	• remove as feasible	• remove as feasible
• Typical • Rosgen • Stream types	• C-type and E-type	B-type C-type E-type	C-type Da-type
Example	tributary to Cavan Creek	Da-type Duffin Creek	Thames River

AQUATIC ECOSYSTEM OBJECTIVE		
Aquatic Performance Standard	Type III	Type IV	Type V
Hydrology			
• 7Q2 or 3Q2 flow as proportion of average daily flow April to September	• minimum 10%	• minimum 5% or sufficient to maintain isolated pools	• no requirement

AQUATIC ECOSYSTEM OBJECTIVE				
Aquatic Performance Standard	Type III	Type IV	Type V
Channel Morphology			
• channel stability	• dynamically stable channels with natural features	• no requirement	• no requirement
• average pool area as % of total surface area at low flow	• >4%	• >4%	• no requirement
• average riffle area as % of total surface area at low flow	• >10%	• generally >5%	• no requirement
• average minimum summer pool depth	• 0.3 m	• 0.2 m	• no requirement
In-Stream Cover	• minimum 10% of stream area during low flow • 10-20% of bottom of pool/ backwater habitats covered by logs, vegetation, woody debris and boulder • cover at stream margins critical for juvenile fish	• minimum 5% of stream area during low flow • <10% of bottom of pools/ breakwaters covered by logs, vegetation, woody debris and boulder	• no requirement
Substrate	• D50 in pools generally <80 mm • fines in riffle zones moderate to low (<50%)	• D50 in pools generally <40 mm • more fines in riffle zones, generally >50%	• no requirement

Aquatic Performance Standard	AQUATIC ECOSYSTEM OBJECTIVE		
	Type III	Type IV	Type V
Riparian Habitat			
• shaded during 1000-1400 hours	• minimum 0% • maximum 50-75%	• minimum 0% • maximum 75%	• no requirement
• woody debris	• important for roughness and refuge during peak flows	• woody debris less important	• no requirement
Water Quality			
• maximum annual water temperature	30° C	31-35° C	35° C
• average annual total suspended solids	<150	<200	<400
• dissolved oxygen (ppm)	>3-4	>2	>2
• spills	none	none	none
Barriers	• minimize as feasible	• minimize as feasible	• minimize as feasible
• Typical • Rosgen • Stream types	• virtually any	• virtually any	• virtually any
Example	West Humber River	tributary to Crow Lake	rural or urban drains

The Impact of Urban Stormwater Runoff on Freshwater Wetlands and the Role of Aquatic Invertebrate Bioassessment

Anna L. Hicks[1]
Joseph S. Larson[2]

Abstract

The impact of urbanization on freshwater wetlands is a rapidly growing area of scientific research. One of the key landscape indicators of urbanization is the amount of impervious surface within the local drainage basin of the wetland. The connection between impervious surface and the quality and quantity of stormwater runoff has already been documented by many authors, and there is a consensus that degradation of aquatic ecosystems occurs at a threshold of 10-20% imperviousness. The evaluation of impact to wetlands through water, soil and habitat analysis can be a difficult and expensive process for management agencies. These measurements often do not indicate the response of the biological community to the impact of nonpoint pollution. Wetlands, whilst primarily driven by hydrology, are essentially biological systems with important water purifying, ecological and wildlife sustaining functions. The Clean Water Act mandates the development of wetland water quality standards, and EPA has determined to focus on biological criteria to supports this effort. Invertebrates are becoming increasingly important as a measuring tool to monitor ecological integrity of water bodies and have proven value in evaluating the health of streams and rivers.

Only recently has research commenced on the application of an aquatic invertebrate bioassessment protocol suitable for wetland conditions. The preliminary results of the research conducted by the Environmental Institute, University of Massachusetts for EPA suggests that ecological integrity of wetlands is affected by the amount of impervious surface in the watershed, and that aquatic macroinvertebrate communities do serve as indicators of wetland condition.

---

1 Project Wetland Scientist, The Environmental Institute, University of Massachusetts, Blaisdell House, Amherst MA 01003-0820

2 Director, The Environmental Institute, University of Massachusetts, Blaisdell House, Amherst MA 01003-0820

Background

Prior to the last quarter of this century wetlands were perceived as nuisance environments with low economic value. They were inevitably targeted for drainage, filling and eventual development. With the change in attitudes over the last two or so decades, accompanied by the growing realization of wetland functional, societal and economic values (Table 1), planning authorities are now concerned about the continued destruction and degradation of this resource.

HYDROLOGY	WATER QUALITY
Flood Storage Flood Desynchronization Groundwater Recharge Groundwater Discharge Shoreline Stabilization	Sediment Trapping Toxicant Trapping and Transformation Nutrient Retention, Removal and Transformation
ECOLOGICAL INTEGRITY	**SOCIO-CULTURAL VALUE**
Wildlife Habitat Finfish and Shellfish Habitat Production Export	Open Space Noise Buffer Recreation Visual/Aesthetics Education Economic Return

Table 1 Functions and Values of Wetlands adapted from Hicks, 1993.

Not all wetlands can provide all of the above services (Larson, 1995). Determining factors are: appropriate location, characteristics of the landscape setting, opportunity, size, and health of the wetland. In this day of harsh economic reality appraisal of the value of wetlands needs a new approach. It can be argued that the land value of property incorporating a wetland is low, and therefore development of that wetland provides cost savings to the developer. Development will provide goods, services, employment and land tax, all desirable. What is not factored into the account are the costs to the community, either passed on directly by the developer, or through increased local government taxes for: alternative site, design and construction costs for the mandated replacement of the wetland destroyed or impacted; engineered drainage and flood control measures due to increase stormwater runoff; treatment mechanisms for the increased nonpoint pollution; possible loss of ground water supply; loss of open space, aesthetics, and wildlife habitat; and, resulting reduction in property values to the local landowners. Harder to evaluate in terms of costs are degraded ecosystem stability and loss of biodiversity (Stevens, 1992; Stevens et al., 1995).

Impact of Urban Stormwater Runoff on Nontidal Wetlands

The impact of urbanization on wetland systems, as summarized in Figure 1 and Table 2, is a rapidly developing area of research. One of the key landscape indicators of urbanization is the amount of impervious surface within the local drainage basin of a water body. The connection between impervious surface and the quality and quantity of stormwater runoff has already been documented by many authors (Hicks, 1995; Klein, 1979; O'Brien, 1990; Office of Wetlands, Oceans, and Watersheds, 1991; Schueler, 1987; Schueler, 1992; Shaver et al., 1995; Woodward-Clyde Consultants, 1990).

URBAN LAND USES
Industry, Commerce, Transportation, Administration, Residential, Open Space, and Recreation

CHARACTERISTICS
Imperviousness
High Population Density
Intensive Land Use
High Resource and Energy
  Consumption
High Pollution Generation

ACTIVITIES
Land Clearing
Excavation
Filling
Channelization and Drainage
Construction
Manufacturing and Retailing
Transporting
Waste disposal

CUMULATIVE IMPACTS

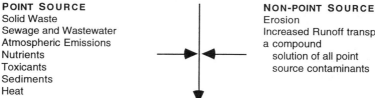

POINT SOURCE
Solid Waste
Sewage and Wastewater
Atmospheric Emissions
Nutrients
Toxicants
Sediments
Heat

NON-POINT SOURCE
Erosion
Increased Runoff transporting
a compound
  solution of all point
  source contaminants

WETLAND STRESSES

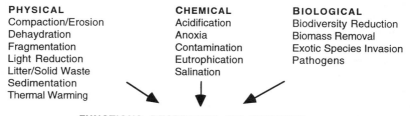

PHYSICAL
Compaction/Erosion
Dehaydration
Fragmentation
Light Reduction
Litter/Solid Waste
Sedimentation
Thermal Warming

CHEMICAL
Acidification
Anoxia
Contamination
Eutrophication
Salination

BIOLOGICAL
Biodiversity Reduction
Biomass Removal
Exotic Species Invasion
Pathogens

FUNCTIONS DESTROYED OR IMPAIRED
Flood Control
Water Quality Improvement
Ecological Integrity
Wildlife Habitat
Aesthetics
Other Societal Values

Figure 1 Urban Land Uses and Linkage to Wetland Impairment

URBAN IMPACTS	Acidification	Anoxia/DO Fluctuation	Biodiversity Reduction	Biomass Reduction	Compaction/Erosion	Contamination/Toxicity	Dehydration	Eutrophication	Exotic Species Invasion	Habitat Fragmentation	Inundation	Light Reduction	Litter/Solid Waste	Pathogens	Salination	Sedimentation	Thermal Warming
STRESSES																	
Urbanization and Imperviousness	■	■	■	■	■	■	■	■	■	■	■	■	■	■	■	■	■
LAND USES																	
Industry	■		■	■	■	■		■		■		■				■	■
Commerce			■	■		■		■		■				■		■	■
Transportation	■	■		■		■			■	■		■			■	■	■
Administration & Public Service				■		■			■	■		■				■	■
Residential		■	■	■		■		■	■	■		■	■	■		■	■
Recreational		■		■	■	■			■	■		■	■		■		■
Open & Public Space			■						■	■			■				
ACTIVITIES																	
Domestic & Hazardous Waste Disposal						■							■			■	
Septic & Waste Water Treatment		■	■			■		■				■	■	■		■	
Channelization & Water Treatment				■						■							■
Drainage			■	■												■	■
Filling			■	■				■	■		■	■					■
Excavation											■						■
Impoundment/Outlet Widening							■				■						■
Land Clearing			■	■	■			■	■	■		■				■	■
Road Construction	■		■		■				■	■			■			■	■
Ground Water Extraction			■	■			■										

Table 2 Impacts and stresses on wetlands caused by urbanization.

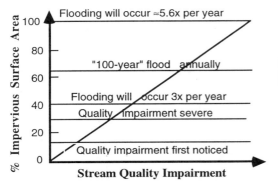

Figure 2 Thresholds for impervious surface impact. Taken from Klein, 1979.

Role of Impervious Surface

There is strong justification for taking the percent imperviousness of the wetland's drainage basin as an integrating indicator of the cumulative impacts of urbanization. The study of ecological indicators along urban gradients is important for the improved management of urbanized landscapes (Limburg and Schmidt, 1990). Data provided by geographic information systems (GIS), land use maps, or direct satellite imagery to be overlaid onto topographic maps with delineated watershed boundaries of associated wetlands facilitates the derivation of calculations of percent imperviousness and provide such a gradient. Even though differing methods may produce a range of results, they can be used to determine if responses along a gradient are linear, or if a threshold is discernible. Schueler (1994), reported thresholds from a number of studies, and there appears to be a consensus that biota are impacted between 10 and 15%, with severe problems after imperviousness exceeds 30% (Pitt et al., 1995). Habitat is impacted once 15% is exceeded. Figure 2 has been drawn from early work on imperviousness by Klein (1979). Whilst local drainage basin imperviousness is a powerful tool as an indicator of impact, the location of the imperviousness in relation to the water body is also a determining factor. For example, a wetland bordered by sheets of imperviousness in the nature of parking lots and roadways with no intervening buffer of vegetation is more at risk than a wetland immediately surrounded by a scrub-shrub or treed landscape.

An Integrated Landscape Approach to Assessing Wetland Health

Measuring and controlling impact from cumulative nonpoint pollution and hydrological alterations requires an integrated watershed approach - examining the wetland within a larger ecosystem, not just as a "black box":

Watershed assessment focuses on those elements that affect beneficial uses. Generally, the elements of beneficial uses are contained in the hydrological, biologic, and topographic features of the watershed. These elements can characterize or determine water flow sediment flow, vegetation change, wildlife habitat, and pollution sources. Assessments must also consider social factors of ownership and political boundaries. Together these elements give a picture of a watershed as a hydrological unit and as an ecosystem with human influences (Euphrat and Warkentin,1994).

In an integrated approach, three sets of factors must be considered:

1. The source of the impact - the drainage basin and immediate surrounding landscape
2. The nature of the sink - the characteristics of the wetland, and how effectively they can mitigate impact
3. The logical geographic scale for planning and management - the assessment unit, whether local, sub, or whole drainage basin.

These factors and the concept of a watershed approach are well illustrated in Figure 3. In seeking indicators for the rapid assessment of urban wetlands using a watershed approach, Hicks and Larson (1995) listed those found in Table 3.

Biomonitoring and Application to Water Quality BMP's

The Clean Water Act (1972) was introduced to "restore and maintain physical, chemical, and biological integrity of the nation's waters". The Watershed Protection Approach (WPA) Framework Document (U.S. EPA, 1991) promoted a holistic approach to address nonpoint source pollution (Euphrat and Warkentin, 1994) as past emphasis on water quality parameters did not bring about the required improvements to aquatic ecosystems (Karr and Kerans, 1992). Bioassessment to measure biological integrity is now recognized as a powerful tool for planners and managers seeking to restore and protect freshwater ecosystems (Davis and Simon, 1995).

Bioassessment and ecological integrity is particularly important for wetlands. Although the driving force is hydrology, it is the biological components of the wetland that perform functions of flood desynchronization, sediment trapping, nutrient and toxicant trapping and recycling. BMP's for water quality improvement are implementing the construction of wetlands to further their aims (U.S. EPA, 1993; Hammer, 1989; Moshiri, 1993). Ecological integrity and biological components are closely bound to other wetland functions: wildlife habitat, fish habitat, aesthetics, recreation and education.

Biological monitoring can be cost effective, especially if a well organized and a quality-controlled network of volunteers is utilized. The River Watch Network and the Massachusetts Water Watch Partnership are two volunteer organizations using aquatic invertebrates incorporating a series of metrics and indices to assess degrees of impairment to streams and rivers of New England, and many states now utilize aquatic invertebrate biomonitoring - Ohio, Delaware, Oregon, Maine, just to name a few. Protocols and criteria vary, although most have a foundation drawn from Plafkin et al., 1989.

3a & 3b Ecological Management Boundary

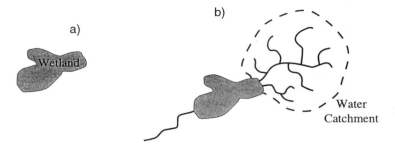

a)

b)

Wetland

Water Catchment

3c Administrative Management Boundary

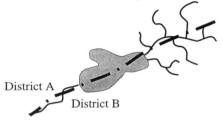

District A

District B

3d Comprehensive Management Boundary

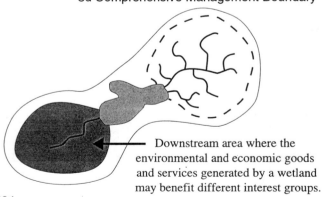

Downstream area where the environmental and economic goods and services generated by a wetland may benefit different interest groups.

Burbridge, 1994

Figure 3 Levels of Integrated Planning and Management

HYDROGEOMORPHIC INDICATORS	(from Brinson, 1993)
Landscape	Stream Order
	Gradient of Topographic Slope
	Soil Classification
	Hydrological Modifications within the Drainage Basin
Wetland	Size of Wetland
	Ratio Wetland/Local Drainage Basin Area
	Juxtaposition of other Water Bodies
	Shoreline Configuration and Island Inclusions
	Hydrological Regime of the Wetland
	Hydrological Modifications within the Wetland
	Rates of Sedimentation
BIOLOGICAL /PHYSICAL/CHEMICAL/INDICATORS	
Landscape	Upland Vegetation Classes as a % of the Drainage Basin
	Distance from Wetland to Nearest Forested Habitat
Wetland	Presence/width of Buffer Zone Surrounding Wetland
	Number of Wetland Classes (Cowardin et al., 1979)
	% Wetland Vegetated
	Sources of Food for Various Trophic Groups
	Nature of the Substrate
	Quality of Water and Sediments
	Aquatic Invertebrate Community Composition
ANTHROPIC INDICATORS	
Landscape	Current and Zoned Land Uses within 500' of Wetland
	% Imperviousness
	Rate of Development
	Potential Sources of Pollution
Wetland	Nature of Inlet/Outlet
	Degree of Draining, Dredging, Filling
	Degree of Vegetation Removal
	Other Human Impacts (recreation, litter, etc.)

Table 3. Landscape and Wetland Indicators for Risk Assessment of Urbanized Wetlands

Publications by Rosenberg and Resh (1993), Davis and Simon (1995), and Hicks (1996) provide full coverage of the application of aquatic invertebrates for biomonitoring. To date no set protocol has been designed specifically for wetlands, although research is now under way (Ludwa, 1994; Hicks, 1995).

Case Study: An Integrated Approach to Assess Impact of Urbanization on Wetlands in the Middlesex County, Connecticut

A co-operative agreement, involving The Environmental Institute, University of Massachusetts, U.S. Environmental Protection Agency, Region 1 and Corvallis, and the University of Connecticut, funded and assisted the research outlined below. In order to assess the impact of urban stormwater nonpoint pollution on permanently flooded palustrine wetlands it was necessary to develop a rapid assessment methodology for wetlands based on Plafkin et al. (1989)'s Protocol II for streams and rivers, and then determine whether aquatic invertebrate communities responded to a gradient of local drainage basin imperviousness.

Eleven wetland sites representing a gradient of urban impact were selected in South-central Connecticut, in Middlesex County. The choice was facilitated by the use of GIS derived land use intensity maps. The flow chart in Figure 4 outlines the methodology procedures. Landscape and wetland characteristics were assessed and scored in accordance with Appendix A, and used to derive a Habitat Assessment score. As shown in Figure 5, when graphed against local drainage basin imperviousness, a negative correlation was found.

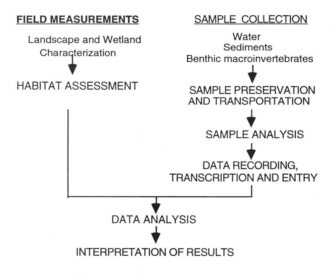

Figure 4 Steps for sampling program

The aquatic invertebrate sampling was conducted in May and August 1995. Field techniques and analysis procedures closely followed Plafkin et al.'s Protocol II (1989), chosen due to its suitability for possible volunteer application. Sampling was performed using a combination of D-net and sediment core sampler in aquatic bed vegetation, within 30' if each wetland outlet, and for the most urbanized wetlands within 30' of the inlet also. Retrieved samples were preserved in 70% alcohol. Sorting, identification to family level and enumeration were completed under laboratory conditions. Raw data were converted into selected metrics, indicated in Table 4, which, in turn, were summarized into an Invertebrate Community Index (ICI) score. The data from the three least impaired wetlands were averaged to provide reference standards against which the remaining sites were compared.

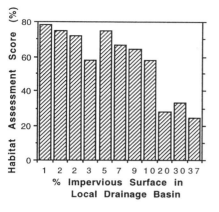

Figure 5 Correlation between impervious surface and habitat assessment.

Figure 6 Seasonal variation of Invertebrate Community Index.

Figure 6 combines ICI results for both seasons, plotted against imperviousness. May's ICI scores declined with increasing imperviousness with only one exception. August results were not so responsive. There was a marked improvement in biotic condition in August compared to May, especially in the case of the most urbanized wetland - a large water body surrounded by 37% impervious surface. The explanation is as follows:

a) Aquatic invertebrate monitoring is designed specifically to measure the impact of cultural eutrophication which, in urban communities, arrives in the water body via stormwater runoff.

b) Connecticut in the summer of 1995 experienced a severe drought with little, if any, stormwater runoff prior to sampling. Thus the urbanized wetlands were relieved of their normal loading of nonpoint source pollution between May and August, and the ICI procedure was sufficiently sensitive to reflect the circumstances.

Total Organisms
EOT Richness
Total Taxa Richness
EOT / Chrionomidae Ratio
Family Biotic Index
%Contribution Dominant Family
% Composition:
* Ephemeroptera
* Odonata
* Trichoptera
+ Chrionomidae
+ Diptera other than Chrionomidae
+ Coleoptera
+ Hemiptera
* Amphipoda (Hyalellidae)
* Isopoda
* Pelycypoda (Sphaeriidae)
+ Oligochaeta
+ Nematoda (Mermithidae)
+ Hirundinea (Glossiphoniidae)
Community Similarity Index
Invertebrate Community Index
* **Sensitive** to urban eutrophication.
+ **Tolerant** of urban eutrophication.
EOT=Ephemeroptera+Odonata
+Trichoptera

Table 4 Aquatic Invertebrate Metrics and Indices

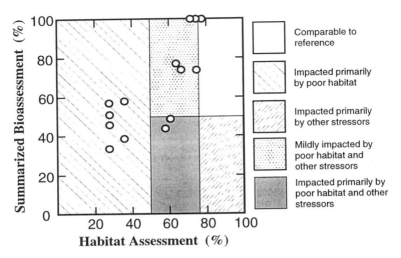

Figure 7 Integrated Assessment Summary.

The Habitat Assessment and Invertebrate Community Index scores for May are drawn together in Figure 7. This summary graph, following the pattern set by Shaver *et al.*, 1994, is a powerful tool with significant planning applications. It provides a means of measuring and interpreting the degree of impairment that characterize each wetland. All wetlands with more than 20% local drainage basin imperviousness were moderately to severely impaired, primarily due to poor habitat, i.e. the impact of urbanization. Wetlands with less than 5% local drainage basin imperviousness were close to or comparable to reference conditions, and if impaired at all, this was due to poor habitat combined with other stressors.

Conclusion

Research results suggest that ecological integrity of palustrine wetlands is affected by the amount and quality of stormwater runoff from impervious surfaces of urbanized landscapes, and that the aquatic invertebrate community does respond to the associated stresses. In adapting Plafkin et al.,'s Protocol II (1989) to wetland conditions a new design of Habitat Assessment with accompanying scoring criteria was required to incorporate a watershed approach. The choice of metrics to derive an Invertebrate Community Index depends to a large extent on the ecoregion, the wetland system, the selected habitat type within the wetland being sampled, and the results obtained by the chosen sampling device/s.

The integrated approach described above and summarized with a simple direct visual provides a scientific, rapid assessment tool easily implemented by environmental engineers, and applicable to volunteer programs under supervision of a qualified coordinator. Results can be presented to town planners, environmental commissioners and members of the community, to pinpoint those impaired wetlands requiring remedial action, those that may be at risk from increased urban development, and finally, those wetlands that provide high quality ecological integrity and wildlife support functions, requiring special protective measures.

Appendix

HABITAT ASSESSMENT PROTOCOL

INDICATOR	HABITAT CONDITION SCORING CRITERIA			
	0	2	4	6
Landscape:				
Dominant land use	Commercial, industrial, transportation	High-medium density residential, agriculture	Low-density residential, grazing	Forestry and open Space
% Impervious surface of local drainage basin	> 15	10 - 15	4 - 9	<4
% Forested	<10	10 - 29	30 - 50	> 50
Ratio wetland/local drainage basin	< 2%	2 - 5%	6 - 10%	> 10%
Possible major sources of pollution	Industrial/commercial effluent, urban storm water run-off	Nutrients and pesticides from gardens, golf courses, agriculture. Sediments and de-icing salts from roads	Septic sewage leakage	No discernible source of pollution
Wetland:				
Water level fluctuation	Water level fluctuations extreme and unseasonal due to upstream dam releases or dehydration /inundation as a result of high % imperviousness within the landscape	Water Level controlled by damming of the outlet	Some modification to natural hydrology through artificial controls	Water level fluctuation (if any) due to natural seasonality
Outlet restriction (Culvert, sluice, spillway, beaver dam, etc.)	Wetland control length <5'	Wetland control length between 5 - 30'	Wetland control length > 30'	No outlet constriction
Nature of sediments	Sediments composed almost entirely of rocks, cobbles, gravel and sand, with no organic matter	Sediments predominantly gravel, sand, with some silt/mud and organic matter	Sediments consist equitably of gravel, sand, silt/mud and organic debris	Sediments predominantly silt/mud and organic debris
Vegetation diversity	< 2 Cowardin classes	2 - 3 Cowardin classes	4 Cowardin classes	> 4 Cowardin classes
% Presence of a buffer (vegetated with trees, scrub/shrub, open space herbs) of 200' width	< 20	20 - 49	50 - 80	> 80
Food Sources	No macrophytes, algae or periphyton. Little, if any, CPOM and FPOM	Some algae and periphyton, CPOM and FPOM	Some macrophytes. Moderate levels of algae, periphyton, CPOM and FPOM	Abundant sources of macrophytes, algae, periphyton, CPOM and FPOM
Degree of impact from human activities in wetland: (litter, trails, roads, vegetation removal/trampling, boating, swimming, shoreline modification, etc.	High level	Moderate level	Low level with minimal impact	No human impact discernible

CPOM = Course particulate organic matter
FPOM = Fine particulate organic matter

Literature Cited

Brinson, M.M., Krucsynski, W., Lee, L.C., Nutter, W.L., Smith, R.D. and D.F. Whigham. 1993. Developing an Approach for Assessing the Functions of Wetlands. In, W.J. Mitsch and R.E. Turner (Eds.), Wetlands of the World: Biogeochemistry, Ecological Engineering, Modeling and Management. Elsevier Publishers, Amsterdam.

Burbridge, P.R. 1994. Integrated Planning and Management of Freshwater Habitats, Including Wetlands. Hydrobiologia 285:311-322.

Cowardin, L.M., Carter,V., Golet, F.C. and E.T. LaRoe. 1979. Classification of Wetlands and Deepwater Habitats of the United States. FWS/OBS-79/31. U.S. Fish and Wildlife Service, Washington, D.C. 193 pages.

Davis, W.S. and T.P. Simon. 1995. Biological Assessment and Criteria: Tools for Water Resource Planning and Decision Making. Lewis Publishers, Boca Raton, FL. 415 pages.

Euphrat, F.D. and B.P. Warkentin. 1994. A Watershed Assessment Primer. Oregon Water Resources Research Institute, Oregon State University. U. S . EPA 910/B-94/005, Seattle, WA. 94 pages plus appendices.

Hammer, D.A. 1989. Constructed Wetlands for Wastewater Treatment: Municipal, Industrial and Agricultural. Lewis Publishers, Boca Raton, FL. 831 pages.

Hicks, A.L. 1993. Environmental Indicators of Wetland Attributes in Urban Landscapes: Function, Sensitivity to Impact, Replacement Potential: A Literature Review. A Report submitted to Wetland Research Program, U.S. EPA Corvallis, OR. 124 pages plus appendices.

Hicks, A.L. 1995. Impervious Surface Area and Benthic Macroinvertebrate Response as an Indicator of Impact from Urbanization on Freshwater Wetlands: Results and comparison of two seasons. Report submitted to U.S. EPA Corvallis, OR. 50 pages plus appendices.

Hicks, A.L. 1996. Aquatic Invertebrates and Wetlands: Ecology, Biomonitoring and Impact from Urbanization. A Literature Review. The Environmental Institute, University of Massachusetts, Amherst. MA. 177 pages

Hicks, A.L. and J.S. Larson. 1995. Environmental Indicators of Wetland Attributes in Urban Landscapes: Function, Sensitivity to Impact, Replacement Potential: Final Report submitted to Wetland Research Program, U.S. EPA Corvallis, OR. 37 pages plus appendices.

Karr, J.R. and B.L. Kerans. 1992. Components of Biological Integrity: Their Definition and Use in Development of an Invertebrate IBI. pp 1-16. In, T.P. Simon and W.S. Davis (Eds.) Proceedings of the 1991 Midwest Pollution Control Biologists Meeting. Environmental Indicators: Measurements and Assessment Endpoints. EPA 905/R92/003. Environmental Protection Agency, Region V, Instream Biocriteria and Ecological Assessments Committee, Chicago, IL.

Klein, R.D. 1979. Urbanization and Stream Quality Impairment. American Water Resources Association, Water Resources Bulletin 15(4):948-963

Larson, J.S. 1995. Identifying the Functions and Values of Freshwater Wetlands. pp 75-90. In, B. Gopal (Ed.), Handbook of Wetland Management. Wetlands Division, WWF India, New Delhi, India. 307 pages.

Limburg, K.E. and R.E. Schmidt. 1990. Patterns of Fish Spawning in the Hudson River Tributaries: Response to an Urban Gradient? Ecology 7(4): 1238-1245.

Ludwa, K.A. 1994. Urbanization Effects on Palustrine Wetlands: Empirical Water Quality Models and Development of a Macroinvertebrate Community Based Biological Index. MS Thesis. Department of Civil Engineering, University of Washington, Seattle, WA. 131 pages.

Moshiri, G.A. 1993. Constructed Wetlands for Water Quality Improvement. Lewis Publishers, Boca Raton, FL. 560 pages.

Plafkin, J.L., Barbour, M.T., Porter, K.D., Gross, S.K., and R.M. Hughes. 1989. Rapid Bioassessment Protocols for Use in Streams and Rivers: Benthic Macroinvertebrates and Fish. U.S. EPA, Assessment and Watershed Protection Division, Washington, D.C. Variously paged.

O'Brien, A.L. 1990. Hydrological Aspects of Stormwater Detention in Wetlands. In, Wetland Law and Policy: What You Need to Know. Resource Education Institute Co., Northboro, MA.

Office of Wetlands, Oceans, and Watersheds. 1993. Natural Wetlands and Urban Stormwater: Potential Impacts and Management. Wetlands Division, U.S. EPA, Washington, DC. 76 pages.

Pitt, R., Field, R., Lalor, M. and M. Brown. 1995. Urban Stormwater Toxic Pollutants: Assessment, Sources, and Treatability. Water Environment Research 67(3):260-275.

Rosenberg, D.M. and V.H. Resh. 1993. Freshwater Biomonitoring and Benthic Macroinvertebrates. Chapman and Hall, New York. 488 pages.

Schueler, T.R. 1987. Controlling Urban Runoff: A Practical Manual for Planning and Designing Urban BMP's. Metropolitan Washington Council of Governments, Washington, DC. 275 pages.

Schueler, T.R. 1992. Mitigating the Adverse Impacts of Urbanization on Streams: A Comprehensive Strategy for Local Government. In, P. Kumble and T. Schueler, (Eds.) Watershed Restoration Source Book: Collected Papers Presented at the Conference, "Restoring Our Home River: Water Quality and Habitat in the Anacostia," Publication No. 9271 of the Metropolitan Washington Council of Governments, Washington, DC. 19 pages.

Schueler, T.R. 1994. The Importance of Imperviousness. Watershed Protection Techniques 1(3): 100-111.

Shaver, E., Maxted, J., Curtis, G. and D. Carter. 1994. Watershed Protection Using an Integrated Approach. In, C. Torno (Ed.), Stormwater NPDES Related Monitoring Needs. Proceedings of an Engineering Foundation Conference. American Society of Civil Engineers, New York.

Stevens, T.H. 1992. Economic Valuation Procedures for Wetland Development Permit Decisions. Draft for publication.

Stevens, T.H., Benin, S. and J.S. Larson. 1995. Public Attitudes and Economic Values for Wetland Preservation in New England. Wetlands 15(3):226-231.

U.S. Environmental Protection Agency. 1993. Created and Natural Wetlands for Controlling Nonpoint Source Pollution. Office of Research and Development, Corvallis, OR. and Office of Wetlands, Oceans and Watersheds, Washington, DC.

U.S. Environmental Protection Agency, Office of Water. 1991. The Watershed Approach: An Overview. EPA 503/9-92-001. Washington, D.C.

Woodward-Clyde Consultants. 1990. Urban Targeting and BMP Selection: An Information and Guidance Manual for State Nonpoint Source Program Staff, Engineers and Managers. Prepared for the U.S. EPA, Region V, Water Division, Chicago, IL and the Office of Water enforcement and Permits, Office of Wetland Protection, U. S. EPA, Washington, DC.

Bioassessment of BMP Effectiveness
in Mitigating Stormwater Impacts on Aquatic Biota

R. Christian Jones, Allyson Via-Norton, and Donald R. Morgan[1]

Abstract

Stormwater and associated nonpoint pollution are powerful agents in the degradation of suburban aquatic ecosystems. Best management practices (BMP's) have been utilized to dampen erosive flow pulses and trap pollutants, but their ability to mitigate impacts on suburban stream ecosystems remains unproven. Macroinvertebrate and fish communities were used to assess the ability of urban BMP's to mitigate stormwater impacts in a suburban watershed. A total of eight practices were assessed including wet ponds, dry ponds, a retrofitted culvert, and a riparian park. Results suggest that appropriately designed and properly sited BMP's can provide some mitigation of stormwater impacts on stream communities. However, the resulting communities differ greatly from those in undeveloped watersheds and reflect a fundamental alteration in stream biotic diversity, structure, and function.

Introduction

The management of nonpoint source pollution (NPS) in urban and suburban watersheds is becoming an increasing focus of strategies designed to improve water quality in streams, rivers, lakes, and estuaries. Best management practices (BMP's) are a major tool for the mitigation of NPS and stormwater impacts in these watersheds. Originally designed to diminish downstream flooding

[1]Respectively, Associate Professor, Graduate Assistant, and Graduate Assistant, Department of Biology, George Mason University, Fairfax, VA 22030.

and stream erosion, BMP's have been modified in recent years to trap nutrients and other pollutants by increasing retention time. All of these practices result in some modification of the pre-development stream ecosystem. Designing an appropriate mix of on-site and regional facilities to maximize stream ecosystem integrity will require substantial knowledge of impacts of individual practices on stream biotic communities.

Ideally, NPS effects should be controlled on-site by eliminating sources and/or trapping pollutants before they leave the site. Since they are located high in the watershed network, effective on-site controls minimize the portion of downstream watercourses impacted by NPS pollution. However, substantial obstacles exist to total reliance on on-site treatment of NPS pollution. Among these are poor performance and maintenance problems associated with on-site methods as well as the high cost of retrofitting these facilities in built-up areas. Thus, some have called for regional facilities which are located some distance downstream and handle NPS pollution from larger areas. While these facilities, mostly regional ponds, may provide economies-of-scale in construction, in maintenance, and in the potential for good pollutant removal, upstream lotic habitats go unprotected and substantial areas of valuable riparian wetland may be inundated by the impoundments. Furthermore, regional facilities are large enough to have their own impacts on downstream habitats such as increased temperature and altered trophic webs.

Numerous constituents of urban NPS may have deleterious or undesirable impacts on freshwater communities. Suspended sediment levels are markedly enhanced in urban and suburban streams, particularly following storm events. Suspended sediments may interfere with respiration and feeding of stream invertebrates (Lemly 1982) and fish (Gardner 1981). The increased levels of toxic contaminants such as heavy metals, petroleum hydrocarbons, and pesticides found in urban NPS pollution contribute to its deleterious effect. Road de-icing salts can have a major negative impact on stream invertebrates (Crowther and Hynes 1977). Although less important in flowing waters than in lakes and ponds, nitrogen and phosphorus stimulate the growth of nuisance algae which can alter stream food webs. Temperature is a critical factor controlling the life cycles of many aquatic organisms (Vannote and Sweeney 1980). Urbanization alters the temperature regime of streams by decreasing riparian vegetation and base flows (Galli and Dubose 1990). This results in the elimination of cold-water animals such as stonefly nymphs and trout.

Habitat alteration is another factor resulting from suburbanization which may have major impacts on stream ecosystems. Increased frequency and magnitude of storm flows result in undercutting and downcutting, altering channel structure and producing large amounts of fine sediments. Deposited fine sediments embed the open spaces around larger, preferred substrates eliminating required habitat of many invertebrates as well as fish nesting areas. Decreases in riparian vegetation alter the base of the stream food web.

Knowing the potential for urban NPS constituents to impact freshwater organisms, it should not be surprising to find that most watershed studies to date indicate substantial degradation of the fauna of urban streams. Jones and Clark (1987) found that watershed urbanization had a major impact on benthic insect communities even in the absence of point source discharges. Watershed development had little impact on total insect numbers, but shifted the taxonomic composition markedly. Chironomids increased while mayflies, stoneflies, beetles, and dobsonflies decreased. Other studies report similar results (Benke et al. 1981, Pitt and Bozeman 1982, Duda et al. 1982, DiGiano et al. 1975).

Little data exist to allow evaluation of the impacts of specific BMP's on stream communities. Galli (1988) examined the response of downstream communities to discharge from a small wet pond BMP. The benthic invertebrate community was substantially degraded downstream due to adverse water quality impacts and changes in organic matter supply. Impoundments generally harbor substantial quantities of plankton which can alter downstream food webs. Trophic structure downstream of impoundments often shifts to filter-feeding invertebrates which utilize the plankton being flushed from the impoundment (Herlong and Mallin 1985, Mackay and Waters 1986, Robinson and Minshall 1990). Robinson and Minshall concluded that impoundments also interfere with colonization of downstream reaches by drift and that reestablishment of the native stream community may not occur for some distance downstream. Deeper impoundments also stratify thermally which may dramatically alter temperature regimes and chemical concentrations in downstream reaches. Dams restrict the movement of fishes which would eliminate recolonization of upstream reaches.

In this paper we report the results to date of an on-going study designed to determine the efficacy of stormwater BMP's in maintaining and enhancing stream biotic integrity. Both benthic macroinvertebrate and fish bioassessment are employed and habitat analysis data are collected to help determine the mechanisms for stream improvement or degradation.

Study Sites

The study sites were located in Prince William County, Virginia, a rapidly growing suburban jurisdiction in the Washington, D.C. metropolitan area. The eastern portion of the county along the Washington-Richmond corridor has undergone substantial development in the past 20 years predominantly in the northernmost Neabsco Creek watershed. Since the mid-1980's this development has been accompanied by the use of BMP's to control stormwater impacts.

Table 1

Site Characterizations

Site	Type	Watershed Area (km^2)	HD	MD	LD
				Land Use[1]	
Parkway Pond	Wet Pond				
PPUL		0.53	19.0	0.0	81.0
PPUR		0.40	20.8	0.0	79.2
PPD1		1.30	38.7	0.0	61.3
PPD2		1.47	46.7	0.0	53.3
PPD3		1.64			
PPD4		1.82			
Galinsky West (GW)	Wet Pond	0.21	35.1	0.0	64.9
Galinsky East (GE)	Wet Pond	0.23	59.6	0.0	40.4
Daleview Manor (DM)	Dry Pond	1.00	7.3	56.0	36.7
Wexford-Highbridge (WH)	Wet Pond	0.34	16.2	33.2	50.6
Dale Blvd. (DB)	Retrofit Culvert	2.01	7.9	13.7	78.4
Minnieville Elem. (ME)	Storm Sewer	4.84	5.5	53.3	41.1
Andrew Leach Park	Riparian Park				
N@P (upstream end)		3.06	6.2	13.1	80.7
N@L (downstream end)		8.65	3.6	42.4	54.0
Mary Bird Br. (MB@S)	Reference	1.71	0.0	0.0	100.0
Mary Bird Br. (MB@11)	Reference	0.94	0.0	0.0	100.0
Mary Bird Br. (MB@7)	Reference	0.23	0.0	0.0	100.0
Quantico Trib. A (QA@O)	Reference	0.39	0.0	0.0	100.0
Quantico Trib. A (QA@7)	Reference	0.70	0.0	0.0	100.0
Quantico Trib. A (QA@S)	Reference	1.25	0.0	0.0	100.0
Quantico Trib. B (QB@S)	Reference	3.47	0.0	0.0	100.0

[1]Land Use: HD=high density (commercial, townhouse and apartments, highways of 4 or more lanes), MD=medium density (detached residences with lots size less than 1 acre), LD (detached residences with lot size greater than 1 acre, open space, and forest).

Individual BMP's examined included wet ponds, retrofitted culverts, dry ponds, and riparian park land, all located in the Neabsco Creek watershed (Table 1). The Parkway Pond and the Galinsky Ponds were designed to control 2- and 10-year storms with some additional pool size for water quality enhancement. The Wexford-Highbridge site contained two sequential wet ponds designed to control the 10-year storm. Sampling stations were established below and, where possible, above each BMP site. In a number of cases, it was impossible to sample above the BMP since runoff entered only through small, normally dry stream channels or directly from storm drains. Reference sites were located on Quantico Creek, also in Prince William County, whose watershed is occupied by Prince William Forest Park, a unit of the National Park System, and the Quantico Marine Base. The Quantico Creek watershed is almost entirely forested and contains few or no residences. Land use characteristics for the catchment draining to each station are included in Table 1.

Methods

A modification of EPA Rapid Bioassessment Protocol (RBP) II was used as the basic tool for macroinvertebrate bioassessment (Plafkin et al. 1989). RBP II utilizes semiquantitative field collections in riffle/run and leaf habitats to determine the values of eight metrics which characterize the status of the benthic macroinvertebrate community. The protocol allows for the modification of metrics and the use of alternative metrics depending on regional conditions. Based on previous work in these watersheds (Jones and Kelso 1994), we deleted the scrapers/filter collectors metric. The shredders/total number metric could not be used since CPOM was not available at many sites. Sorensen's index was used for community similarity.

Macroinvertebrates were sampled once at each station in late spring-early summer 1994, fall 1994, and late spring-early summer 1995. Macrofauna were collected by kick sampling using a 0.5 mm mesh rectangular net measuring 44 cm x 22 cm. The net was held on the bottom facing upstream in flowing water and the substrate was agitated for a distance of 1 m upstream over a period of one minute. Larger stones were also wiped clean when deemed necessary. Contents of the net were transferred to a sample jar and preserved with formalin for later picking, identification, and enumeration. Samples were collected from two locations at each station (a rapidly flowing riffle and a less rapid run) and composited.

In the lab the composite samples were rinsed with tap water and distributed evenly over a 35 cm x 40 cm pan marked with 5 cm squares. Organisms were removed from randomly selected squares until 200 organisms were obtained or the entire sample picked. The selected organisms were sorted into ethanol-glycerine, identified to family, and enumerated. Macroinvertebrate rating was calculated following the guidance of the EPA bioassessment manual (Plafkin et al. 1989).

Fish bioassessments were conducted using RBP V (Plafkin et al. 1989) at the same general times as macroinvertebrates. At each site a 200 m reach containing riffles, runs, and pools was measured from a reference point such as a bridge crossing or other easily distinguishable landmark. In some cases the entire stream reach of interest was less than 200 m and the entire distance was sampled. Sampling was accomplished using backpack-mounted battery-powered electroshocking gear. Boundary nets were set at either end of the stream reach when the reach boundaries coincided with deep pools or a wide channel. Once collected, the fish were identified and enumerated. The incidence of hybrids and diseased or anomalous individuals was also noted.

Index of Biotic Integrity was calculated using the procedure outlined in Plafkin et al. (1989). The only modifications were substitution of percent generalists for percent omnivores and percent specialists for insectivores.

Habitat assessments were conducted using the EPA protocol. Habitat index was calculated based on eight metrics which rate substrate and instream cover, channel morphology, and riparian and bank structure. The flow metric was not included. Habitat assesments were made at the same time as fish sampling.

Watershed land use was calculated using ARCView 2 and GIS files obtained from Prince William County. Data in the County's GIS files was augmented using a global positioning system.

Results

Reference sites exhibited two distinct groupings with regard to macroinvertebrate RBP index scores. Sites with watersheds greater than 1 km^2 yielded mean RBP index scores of 24-36 (out of a possible 36), while those with smaller watersheds averaged 9-17 (Figure 1). This difference was observed to some extent in all of the individual metrics with most showing little or no overlap in means between the two watershed size classes. Small watershed reference sites were numerically dominated by chironomids with substantial contributions from hydropsychid caddisflies, tipulid crane flies, elmid beetles, and dryopid beetles.

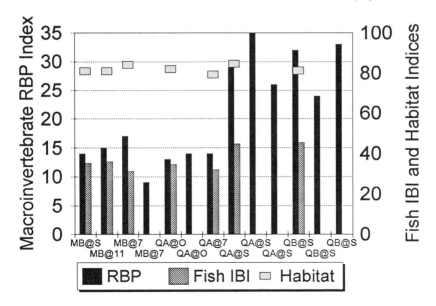

Figure 1. RBP macroinvertebrate index, fish IBI, and habitat index at reference stations.

Two to six stonefly families, 4 to 5 mayfly families, and 1 to 4 additional caddisfly families were observed. The larger watershed references were less dominated by chironomids and again had substantial numbers of hydropsychids, tipulids, elmids, and dryopids. Stoneflies were represented by 4-5 families, mayflies by 4-7 families, and caddisflies by 7-8 additional families.

Fish IBI scores in reference watersheds also fell into two groups, although differences were less pronounced (Figure 1). The smaller watersheds (<1 km^2) exhibited average values of 31-36 whereas the sites draining larger watersheds (>1 km^2) attained values of 44-46 (out of a possible 60). The smaller reference watersheds contained only eight species of fish. Blacknose dace, creek chub, and rosyside dace were found at all sites, often in considerable abundance. Cutlips minnow was found at all 5 smaller watershed sites, fallfish and creek chubsucker were found at 3 sites each, and American eel and common shiner at only one site each. The larger reference watersheds supported substantially more species with 12 found in QA@S and 16 at QB@S. Additional species commonly found at large reference sites included common shiner, redbreast sunfish, tesselated darter, and white sucker.

The habitat index at reference sites averaged 80-85 (out of a possible 115) and did not vary much between larger and smaller watersheds.

The most complete array of BMP-related sampling stations was obtained at the Parkway Pond. The four sites above the pond yielded mean RBP index scores of 14-16, well within the range of the small watershed reference sites in Quantico (Figure 2). FBI and EPT/Chironomids at the upstream sites were even more indicative of high quality than the reference sites whereas EPT index was substantially lower than at the small watershed reference sites. Chironomids and hydropsychid caddisflies were the most abundant families represented at PPUR and PPUL. Simulids and tipulids were also common and four families of stoneflies were present.

Below the Parkway Pond the benthic community exhibited signs of degradation. Fifty meters downstream of the pond outlet RBP index averaged 12 and declined slightly to 9 at 200 m and 400 m downstream. This decline was mainly due to an increase in FBI and a decrease in EPT/Chironomids, but other metrics were also generally lower below the pond. Chironomids and hydropsychids were still very abundant and simulids were obtained in much larger numbers than upstream. Caenid mayflies, not observed upstream, were frequently found at PPD1. Stoneflies were absent. It should be noted that sites below the pond drained more than 1 km^2 making comparisons with the large watershed reference sites more appropriate and making the degradation downstream relatively more severe.

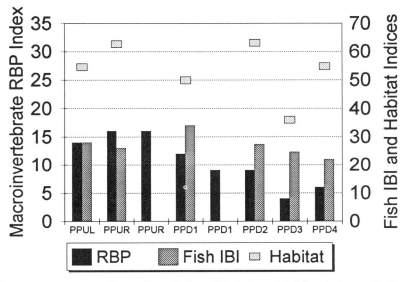

Figure 2. RBP macroinvertebrate index, fish IBI, and habitat index at Parkway Pond stations.

Further downstream, a large (84 inch) storm sewer draining a major shopping center discharged into the stream. Below this outfall RBP index dropped sharply, measuring 4-6 at two stations with substantial degradation in taxa richness and percent dominant taxon relative to the sites immediately upstream and in all metrics relative to sites above the pond. Chironomids were still fairly abundant. Hydropsychids and tipulids were present, but much reduced.

Above the Parkway Pond average IBI ranged from 25-28 (Figure 2). Immediately below the pond IBI peaked at 34, similar to the small watershed reference, but average IBI dropped steadily further downstream reaching a low of 22 at PPD4. Above the pond the fish community was represented by four species at PPUL and five species at PPUR. Blacknose dace was very abundant at both sites and bluegill and largemouth bass were also regularly observed at both stations. Green sunfish and warmouth were found sporadically at PPUL and PPUR, respectively. Least brook lamprey was consistently abundant at PPUR. At PPD1 and PPD2 blacknose dace, bluegill, and largemouth bass were again found in substantial numbers with blacknose dace substantially less abundant than above the pond. A consistent member of the fauna at PPD1 and PPD2 was American eel, whereas brown bullhead, black crappie, pumpkinseed, warmouth, least brook lamprey, and golden shiner were sporadically found. The fish fauna at PPD3 and PPD4, below the 84 inch storm sewer, was very depauperate with only blacknose dace, bluegill, and American eel represented in low numbers.

Above the Parkway Pond the habitat index averaged 55 at PPUL, while PPUR rated 63 due to better substrate and cover conditions. Below the pond habitat conditions appeared quite variable. Immediately downstream (PPD1) the habitat index was reduced to 50 with low values in almost all categories except vegetation. Further downstream at PPD2 habitat rebounded to an average of 63, although certain attributes such as embeddedness were quite variable between sampling rounds. At PPD3, immediately below the large stormwater outfall, habitat was very marginal remaining less than 40 on all three sampling dates. Habitat appeared to be substantially improved at PPD4, although this masked the fact that the reach is divided into three distinct regions, one of which is very exposed and that the stream is overflowing its banks and forming temporary channels.

GW, a wet pond site accepting drainage from relatively undeveloped, but open land, exhibited moderate degradation in RBP index relative to small watershed reference sites (Figure 3). Taxa richness, EPT index, and Sorensen's index were clearly reduced, but EPT/Chironomids was higher. Chironomids and hydropsychids were most abundant and simulids and tipulids were common. One mayfly family and one other caddisfly family were represented, but no stoneflies were found. GE, a wet pond site with over 50% commercial land use, was much more degraded. Only 4-6 taxa were normally present and percent dominant taxon averaged 76-78. Chironomids and hydropsychids were again the numerical dominants, but tipulids and simulids were less common than at GW. No mayflies or other caddisflies were present, but one stonefly was collected. WH, a site below two sequential wet ponds whose watershed was dominantly open space and medium and large lot residential, was somewhat degraded relative to small watershed references. EPT index was substantially reduced, but EPT/Chironomids was higher. Again, chironomids and hydropsychids were most abundant and tipulids and simulids were present. Physid and lymnaeid snails were often very common. Two mayfly families were found, but no stoneflies or additional caddisflies. The DM sites, located below a pair of dry ponds, showed substantial degradation. RBP index was only 5-8 with especially low taxa richness, EPT/Chironomid abundance, EPT index, and Sorensen's index and increased FBI. Chironomids were dominant with tipulids and hydropsychids being common. No stoneflies, mayflies, or additional caddisflies were observed.

Both GW and GE exhibited near reference IBI values (Figure 3). WH was slightly higher than the reference sites, while DM was substantially below reference level. GW and GE both yielded large numbers of blacknose dace; brown bullhead and pumpkinseed were also usually found. An assortment of other fishes were sometimes observed in small numbers including redbreast sunfish at

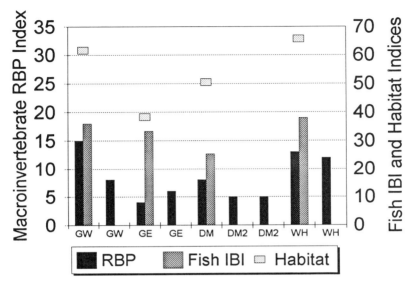

Figure 3. RBP macroinvertebrate index, fish IBI, and habitat index at conventional BMP stations.

GW and American eel, bluegill and margined madtom at GE. A lesser number of blacknose dace characterized DM with sporadic collections of redbreast sunfish, creek chub, and common shiner. WH was represented by a diverse assortment of 11 fish species of which 7 were considered tolerant and 4 intolerant. Highest numbers were found for creek chub, white sucker, green sunfish, largemouth bass, and blacknose dace.

At the conventional pond sites, habitat index showed only minor degradation at GW, whereas at GE habitat was strongly impacted. Substrate and instream cover as well as riparian and bank structure were very poor at GE. DM exhibited a wide range in habitat quality, averaging moderate degradation. Habitat index at WH exhibited only minor degradation.

The retrofitted culvert site, DB, was sampled only once upstream before beaver activity inundated the sample site (Figure 4). The RBP index and the individual metrics were also reduced relative to the reference sites indicating moderate degradation. Downstream sampling showed moderate to severe degradation. Both upstream and downstream chironomids were most abundant followed by hydropsychids and simulids. Sphaerid clams and sialid alderfly larvae were common below, but not above the culvert site. Sampling above and below

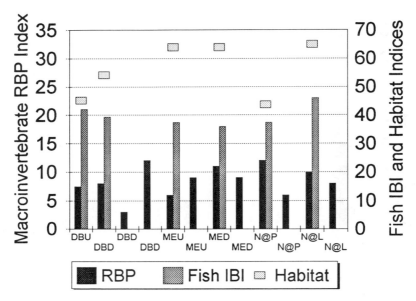

Figure 4. RBP macroinvertebrate index, fish IBI, and habitat index at other stations.

an unprotected storm sewer (MEU and MED, respectively) showed little evidence for enhanced degradation from the additional discharge. However, both sites were strongly degraded relative to the large watershed references. In addition to the usual chironomid, hydropsychid, and simulid dominants, two mayfly families were represented and isopods were common at both sites. Sites above and below a riparian park (N@P and N@L, respectively) exhibited clear degradation relative to the large watershed references. The riparian park appeared to have little positive impact on the benthic community as evidenced by little or no change in individual metrics or in the overall RBP index. Two mayfly families were represented at both sites, while one additional caddisfly family was found at N@L.

DBU and DBD both showed signs of slight to moderate degradation in IBI, with DBD somewhat lower (Figure 4). DBU, upstream of the retrofit culvert, had a limited fauna of golden shiner, creek chubsucker, green sunfish, and bluegill with small numbers of creek chub, white sucker, pumpkinseed and tesselated darter. At the downstream site, DBD, were found all of these taxa as well as substantial numbers of blacknose dace and rosyside dace with sporadic American eel, cutlips minnow, and redbreast sunfish. MEU and MED showed more substantial degradation relative to larger reference sites. MEU and MED had a similar fauna characterized by substantial numbers of blacknose dace, creek chub, and white sucker with a total of 12-14 species each. N@L, the site at the lower end of the riparian park, demonstrated a substantially higher average IBI than N@P, the site at the upper end. The IBI at N@L indicated reference conditions, while that at N@P suggested significant impacts. N@L, at the base of riparian parkland, possessed a much richer fauna than N@P, at its head, with 17 species observed in the former and 10 in the latter. Blacknose dace, creek chub, creek chubsucker, and tesselated darter were abundant at both, while cutlips minnow, common shiner, and margined madtom were intermediately tolerant additional species observed at N@L.

Sites around the retrofitted culvert exhibited moderate habitat degradation with somewhat higher habitat quality downstream, although the upstream site was sampled only once. Habitat quality at the unprotected storm sewer site showed minor degradation with little difference between upstream and downstream. The site at the uppermost end of the riparian parkland showed highly degraded habitat quality, while that at the lower end showed only minor habitat degradation.

Discussion

The Parkway Pond site had the most complete array of sampling stations and afforded the most comprehensive examination of BMP impacts. Relative to upstream stations and to the reference watersheds, downstream stations demonstrated a clear decline in macroinvertebrate RBP index. This was most clearly seen in the presence of 4 stonefly families above the pond and none below. The clear increase in simulids below the pond suggests an alteration in the stream food web from leaf-litter base to plankton base. The presence of caenid mayflies, a markedly tolerant species, below but not above the pond is further evidence for moderate degradation. The fish IBI was actually higher immediately downstream from the pond than above the pond. This is principally attributable to the addition of several species below the pond that were not present above. This may reflect the somewhat larger watershed, the sustained flows below the pond, the alteration in trophic structure by the pond, or the isolation of upstream habitat imposed by the dam and resulting pond.

Conditions observed further downstream demonstrate the clearly destructive effect of large pulses of stormwater. The 84 inch storm sewer entering below PPD2 is the outflow from an underground detention system. Nonetheless, severe bank erosion has occurred near the outfall and in areas immediately downstream. Chironomids remain, but hydropsychids and tipulids are reduced and other taxa are eliminated. Fish communities were very depauperate and habitat was marginal.

At this time, the Parkway Pond watershed is only partially developed and undisturbed forest borders the inflowing reaches immediately upstream. Relative to these somewhat impacted streams, the Parkway Pond is resulting in some degradation of stream quality from a biotic viewpoint. However, as land use intensifies in the upstream watersheds, these upstream reaches may come under increasing pressure. The presence of the BMP may help buffer downstream reaches against this increased impact. Continued studies will be necessary to allow determination of the pond's ability to mitigate the increased stormwater impacts.

Other wet ponds yielded varying results. GW and WH suffered only moderate degradation relative to small watershed controls. These sites drain relatively small catchments which have predominantly low to medium land use intensity. GE was more strongly impacted than the other wet pond sites, particularly in habitat and macroinvertebrate community. This site also drains a small catchment, but well over half of its land use is high density. This suggests that wet ponds are more effective in preserving downstream biotic quality when catchment land use is less highly developed.

The dry ponds did a rather poor job of mitigating stormwater impacts. Habitat, macroinvertebrate RBP, and fish IBI were all depressed relative to reference streams. The retrofit culvert had little impact on biotic quality. This was partially attributable to design and maintenance problems during the study period. The small storm sewer discharge at ME appeared to have minimal impacts on stream quality. The macroinvertebrate communities above and below the outfall were substantially degraded relative to the large watershed reference sites, but this appeared to be the result of general upstream watershed conditions rather than local immediate effects. The 2 km passage of Neabsco Creek through a riparian park resulted in a substantial enhancement of the fish community as well as an increase in habitat quality, while the macroinvertebrate community showed little change. The increase may be partially explained by the more than doubling in watershed size between the two stations which may have substantially increased the fish species pool. Potential improvements in the macroinvertebrate community attributable to the riparian park may have been offset by increased intensity of land use draining to the downstream site.

Alterations in the stream macroinvertebrate community in the suburban Neabsco watershed were clear at all stations even when RBP index values were near reference levels. Taxa richness was consistently lower in the suburban streams particularly in the key indicator groups of stoneflies, mayflies, and non-hydropsychid caddisflies. None of the BMP's assayed was able to restore families from these groups to the suburban streams. The elmid beetles constituted another group that was very consistent in reference streams and was found above the Parkway Pond, but was virtually absent from other suburban sites including those below BMP's. Taxa present in greater numbers at suburban sites, particularly below wet ponds, were the simulid blackflies and the physid and lymnaeid snails, reflecting the altered trophic conditions.

While hydropsychids were abundant at all stations, they were especially dominant in moderately impacted suburban streams. In fact, hydropsychid numbers were reduced in the reference streams relative to some suburban streams, resulting in lower EPT/Chironomid scores for the reference streams. This in turn tended to blur the distinction between the reference and suburban streams even when other metrics indicated a clear distinction. This indicates that EPT/Chironomid abundance may not be an appropriate metric when hydropsychids are very abundant. Alternatively, one could compute the metric but use EPT abundance exclusive of hydropsychids.

Fish communities were more difficult to interpret. Many of the difference in fish communities observed were related to stream size and degree of isolation. These factors clouded relationships to BMP's. Clearly, major inputs of poorly managed stormwater results in degradation of the fish community and its habitat as noted at PPU3 and PPU4. On the other hand, IBI values at many suburban stations differed little from the reference sites or from one another. Some of this is certainly due to the limited complement of species which utilize these small tributaries. In the watersheds of less than 1 km^2 even the reference sites housed a pool of only eight species, only four of which were found at all reference sites. This severely limited the resolution of the fish IBI.

In general, the data suggest that appropriately designed and properly sited BMP's can provide some mitigation of stormwater impacts on stream communities. However, no BMP's were able to restore the full complement of macroinvertebrate families found in the reference watersheds. The resulting communities reflect a fundamental alteration in stream biotic diversity, structure, and function.

Acknowledgements

The authors wish to thank Donald Kelso for his help in designing the fish sampling and confirming fish identifications. Art Spingarn and Fernando Pasquel provided leadership which initiated the Prince William Watersheds Project. Funding for this study was contributed by the U.S. Environmental Protection Agency and the Virginia Department of Conservation and Recreation.

Literature Cited

Benke, A.J., G.E. Willeke, F.K. Parrish, and D.L. Stites. 1981. Effects of urbanization on stream ecosystems. Completion Report Project number A-055-GA, Office of Water Research and Technology, U.S. Department of the Interior.

Crowther, R.A. and H.B.N. Hynes. 1977. The effect of road deicing salt on the drift of stream benthos. Environmental Pollution 14:113-126.

DiGiano, F.A., R.A. Coler, R.C. Dahiya, and B.B. Berger. 1975. A projection of pollutional effects of urban runoff in the Green River, Massachusetts. In: Urbanization and Water Quality Control, W. Whipple (Ed.) American Water Resources Association. Minneapolis, Minnesota. pp 28-37.

Duda, A.M., D.R. Lenat, and D.L. Penrose. 1982. Water quality in urban streams - what we can expect. Journal of Water Pollution Control Federation 54: 1139-1147.

Galli, F.J. 1988. A limnological study for an urban stormwater management pond and stream ecosystem. Masters Thesis. George Mason University. Fairfax, VA. 153 pp.

Galli, F.J. and R. Dubose. 1990. Thermal impacts associated with urbanization and stormwater management best management practices. Metropolitan Washington Council of Governments.

Gardner, M.B. 1981. Effects of turbidity on feeding rates and selectivity of bluegills. Transactions of the American Fisheries Society 110: 446-450.

Herlong, D.D. and M.A. Mallin. 1985. The benthos-plankton relationship upstream and downstream of a blackwater impoundment. Journal of Freshwater Ecology 3: 47-59.

Hilsenhoff, W.L. 1982. Using a biotic index to evaluate water quality in streams. Technical Bulletin No. 132. Wisconsin Department of Natural Resources. Madison, WI. 23 pp.

Jones, R.C. and C.C. Clark. 1987. Impact of watershed urbanization on stream insect communities. Water Resources Bulletin 23:1047-1055.

Jones, R.C. and D.P. Kelso. 1994. Bioassessment of nonpoint source impacts in three Northern Virginia watersheds. Final Report to U.S. EPA Region 3.

Lemly, A.D. 1982. Modification of benthic insect communities in polluted streams. Combined effects of sedimentation and nutrient enrichment. Hydrobiologia 87:229-245.

Mackay, R.J. and T.F. Waters. 1986. Effects of small impoundments on hydropsychid caddisfly production in Valley Creek, Minnesota. Ecology 67: 1680-1686.

Pitt, R. and M. Bozeman. 1982. Sources of urban runoff pollution and its effects on an urban creek. U.S. Environmental Protection Agency. EPA-600/s2-82-090.

Plafkin, J.L., M.T. Barbour, K.D. Porter, S.K. Gross, and R.M. Hughes. 1989. Rapid bioassessment protocols for use in streams and rivers: benthic macroinvertebrates and fish. U.S. Environmental Protection Agency. Office of Water. EPA/44/4-89-001

Robinson, C.T. and G.W. Minshall. 1990. Longitudinal development of macroinvertebrate communities below oligotrophic lake outlets. The Great Basin Naturalist 50: 303-311.

Vannote, R.L. and B.W. Sweeney. 1980. Geographical analysis of thermal equilibria: A conceptual model for evaluating the effect of natural and modified thermal regimes on aquatic insect communities. American Midland Naturalist 115: 667-695.

SESSION 2D: Impacts of Watershed Development on Aquatic Biota, Round II

DISCUSSION

The Impact of Urban Stormwater Runoff on Freshwater Wetlands and the Role of Aquatic Invertebrate Bioassessment
Anna Hicks

Only recently has research commenced on the application of an aquatic invertebrate bioassessment protocol suitable for wetland conditions. The preliminary results suggest that ecological integrity of wetlands is affected by the amount of impervious surface in the watershed, and that aquatic macroinvertebrate communities do serve as indicators of wetland condition.

Measuring and controlling impact from cumulative nonpoint pollution and hydrological alterations requires an integrated watershed approach - examining the wetland within a larger ecosystem, not just as a "black box".

In an integrated approach, three sets of factors must be considered:

1. The source of the impact - the drainage basin and immediate surrounding landscape
2. The nature of the sink - the characteristics of the wetland, and how effectively they can mitigate impact
3. The logical geographic scale for planning and management - the assessment unit, whether local, sub, or whole drainage basin.

In adapting Plafkin et al.'s Protocol II (1989) to wetland conditions a new design of Habitat Assessment with accompanying scoring criteria was developed to incorporate a watershed approach.

The integrated approach summarized with a simple direct visual provides a scientific, rapid assessment tool easily implemented by environmental engineers, and applicable to volunteer programs under supervision of a qualified coordinator.

Questions/Comments

Question:	How do you describe threshold?
Answer:	Rich Horner addresses this in his paper.
Question:	How can we use your approach to improve a degraded wetland.
Answer:	Read the reference literature, which better explains the techniques.

Can Environmental Impacts of Watershed Development be Measured
Bill Snodgrass

The objectives of this paper are to evaluate whether impacts upon aquatic ecosystems, particularly at a watershed scale can be measured, or whether the state-of-the-practice is one of being able to measure only a few effects using indicators, inferences and simulation models.

The state-of-the-practice of measurement appears to be at least half a decade to a decade from being able to actually provide the needed measurements.

The concept of a threshold effect between percent impervious and aquatic system response has evolved and is in widespread use in the field.

The influence of urban BMPs such as wet ponds upon stream systems is complex and requires several more years of research and field measurements to discern their explicit benefits.

The field is evolving toward using environmental indicators rather than mass measurements in many measurement efforts.

Predictions of changes due to watershed development are possible for hydrology, but that predictions, based upon a quantitative impact assessment methodology, are not presently possible for biological entities such as benthos and fish.

Questions/Comments

Question: How are we defining imperviousness, are we consistent in our definition?
Answer: I don't know, but we do need commonality of definitions.

Question: What is the meaning of "dynamic stability?"
Answer: Things are changing but the environmental setting remains unchanged, it follows the laws of nature; or, the geomorphological properties remain constant.

Question: Why do you recommend not using a reference site?
Answer: Because we don't know statistically if we have the same characteristics at each site.

Bioassessment of the BMP Effectiveness in Mitigating Stormwater Impacts on Aquatic Biota
          Christian Jones

Best management practices (BMPs) have been utilized to dampen erosive flow pulses and trap pollutants, but their ability to mitigate impacts on suburban stream ecosystems remains unproven. Macroinvertebrate and fish communities were used to assess the ability of urban BMPs to mitigate stormwater impacts in a suburban watershed. Results suggest that appropriately designed and properly sited BMPs can provide some mitigation of stormwater impacts on stream communities. However, the resulting communities differ greatly from those in undeveloped watersheds and reflect a fundamental alteration in stream biotic diversity, structure, and function.

Questions/Comments

Question: How old was the pond?
Answer: About two years old, and designed for control of the 2-yr storm.

Comment: Tailoring the matrix to this study, resulted in much better match, but
              it is not clear how to adapt them regionally without prejudicing the
              outcome to be favorable.

Question: Why did you use on-line ponds?
Answer: That was all that was available.

Urban Rivers in Arid Environments-Unique Ecosystems

Todd Harris[1], James F. Saunders, III[2], and William M. Lewis, Jr.[3]

Abstract

River ecosystems function by continuous processes, like a never ending card game rather than a single hand or event. The rules that govern many of the processes to which arid western rivers respond are unique to these ecosystems. Indeed in the West, the game maybe less stable, faster, and more violent. The watershed of the Rocky Mountains is much more impervious than that of the Cumberland Plateau. Water rights can trump and change both the water quality and quantity of water in arid streams more than precipitation events. Diversion and dilution to meet water supply and irrigation needs drastically alter western urban watersheds. Run-off from a storm on arid lands flows differently than run-off from a storm on mesic or humid lands. Minimally vegetated riparian corridors are unlike canopied stream banks. Intense radiation, drastic diurnal temperature and pH swings change biological productivity and biological productivity, in turn, modifies the effects of radiation, temperature, pH, and other water quality parameters. Diversity may not be natural. The saturation of gas in water differs with elevation. Turbidity, velocity, and salts all play differently in a xeroscape than in other landscapes. We are still in the process of discovering the unique rules by which western river ecosystems operate.

[1]Water Quality Officer, Metro Wastewater Reclamation District, 6450 York Street, Denver, CO 80229-7499

[2]Assoc. Dir., Center for Limnology, Univ. of Colorado, Boulder, CO 80309-0334

[3]Director, Center for Limnology, Univ. of Colorado, Boulder, CO 80309-0334

Introduction

The goal of the Clean Water Act (US Congress 1972) is to restore ecological integrity to our nation's rivers. Rivers have been described as having order and being the product of their tributaries (Strahler 1957). Many western rivers, especially in urbanized arid regions, only weakly reflect the characteristics of their tributaries. Rivers in developed areas are most often viewed as natural systems with problems that must be measured and controlled; urban streams in the arid West are, to a large extent, unnatural systems that represent opportunities for the design of new, stable environments.

To ensure the integrity and thwart degradation of our rivers, four major classes of environmental constituents are traditionally measured (Table 1): physical factors, chemical variables, biota and energy, and habitat. Measurement of these environmental constituents in urban rivers of the arid and semi-arid West illustrates that these systems are not natural. Urban arid watersheds present extremes of conditions that are far from natural for most biological communities. This paper reviews some of the unique characteristics of western rivers in urban environments using the South Platte River, an effluent-dominated stream, which flows north out of Denver, Colorado, as an example.

Physical Factors-Hydrology

Historically, most rivers of the arid and semi-arid West were intermittent or with highly irregular flows. Discharge ranged from a flooding torrent of snowmelt during springtime and during storm events to negligible or discontinuous during most of the year. These irregular flows, which carved the wide channels which are still present today, were directly related to climate. Pioneers following the Oregon trail would time their journey so as to be able to ford rivers such as the Platte at low summer discharge. No wagons could ford during spring-time flood flows (Parkham 1846). Today, discharge is sustained by water storage and wastewater or irrigation return flows and seasonal extremes are moderated by hydrologic control. The natural flow of the South Platte and other urban rivers in the West (Albuquerque-Rio Grande, Santa Fe-Santa Fe, Flagstaff-Rio de Flag, Phoenix-Salt, Boulder-Boulder Crk./St. Vrain, Colorado Springs-Fountain, Pueblo-Arkansas), have been affected by trans-mountain diversions, storage and flood control reservoirs, power developments, groundwater withdrawals, diversions for irrigation and municipal use, and return flow from irrigation. The majority of the surface water flowing from many of these urban watersheds is due to return flow from municipal use. For example, the South

PHYSICAL

- discharge/volume
- groundwater
- precipitation
- soils
- high/low extremes
- runoff
- velocity
- imperviousness/ground cover
- elevation
- temperature
- topography
- hyporheon
- diel variability

 HABITAT

- sinuosity
- current
- bank stability
- canopy
- channel width
- channel depth
- channel gradient
- riparian vegetation
- substrate
- instream cover
- rooted vegetation
- riffle/pool ratio
- flood plain

 CHEMICAL VARIABLES

- nutrients
- metals
- organics
- pesticides
- dissolved oxygen
- pH
- Redox
- hardness
- alkalinity
- chemical solubility
- suspended solids
- dissolved solids
- toxicity/bioavailability

 BIOTA/ENERGY SOURCE

- predation
- feeding
- parasitism
- disease
- life cycle length
- reproduction
- competition
- organic matter input
- $1°/2°$ production
- nutrient availability
- diversity
- sunlight
- population numbers

TABLE 1 - Environmental factors which affect biological integrity in an arid western river, or any river for that matter. Each of these are necessary to create the resultant integrity reflected in biological community performance...the ultimate goal of the Clean Water Act.

Platte now flows 365 days a year with an average discharge at Henderson, a gauging station 19.2 km downstream of Denver, of over 15.7 m^3/s (554 cfs) (USGS 1994a).

Average flows of an urban river in the arid/semi-arid West yield an unrealistic view of the hydrodynamics of these systems. Rivers in wet climates show discharge below the long term average 250-260 days a year. In a dry climate, with regulated discharges, average flows are achieved even less frequently. For example: flows upstream of Metro Wastewater Reclamation District's (Metro's) discharge. Average flows of 8 m^3/s (280 cfs) (based on over 18 years' data) were not achieved three hundred and fifty four days of the year (USGS 1994a).

A single thunderstorm over the upstream impervious surfaces of the watershed can cause the yearly average daily discharge to change dramatically. In 1992, a single precipitation event on the South Platte in Denver increased the daily discharge to 255 m^3/s(9000 cfs). Such an event can increase the daily average discharge 0.7 m^3/s/year (25 cfs/yr). A flow of 0.7 m^3/s/day (25 cfs/day) was not even achieved on the South Platte upstream of Metro during 250 days in water year 1994. Because of the influence that a single event can have on the average or mean discharge in arid and semi-arid urban rivers, it is more practical for some purposes to consider either the harmonic or geometric mean flow rather than the average flows.

In the arid West, river discharge is controlled by water rights for irrigation and industrial uses above, in and below developed urban areas. It is estimated that each drop of water in the South Platte is utilized for irrigation in Colorado three to seven times before it reaches the Nebraska border (Obmasic 1996). The effect of irrigation withdrawals, storage, usage, and subsequent return flows effect the hydrology of arid urban rivers and must be considered in any reviewing of the integrity of these systems. The waters of arid urban rivers are over-appropriated, there is greater legal right to use of these waters than there is water physically available. By 1900 there were over 17 m^3/s (600 cfs) of water appropriated from the South Platte in the 40 km segment below Denver (CDM 1991) for agricultural irrigation, yet in 1994, discharge averaged less than 8 m^3/s (280 cfs) (USGS 1994a). Fluctuation of flow may be extreme during periods of high irrigation demand.

Except during precipitation events, reservoir releases. and periods of snowmelt, other tributary flows contribute less than 10% of the total discharge of the South Platte below Denver. The only steady flow in this portion of the South Platte as in many rivers in arid western urban areas is due to point source discharges such as treated municipal effluent. During 1994, most of the surface water in the South Platte leaving Denver was due to discharge from Metro's central plant. The discharge from Metro directly to the South Platte averaged 5.7 m^3/s (200 cfs). Flows from Metro's outfalls represented over 90% of the daily total flow in the South Platte below Metro's discharge, 97% of the time during 1994. Municipal discharges follow daily pattern of use with a morning and evening peak and decreased usage at night. This daily effluent discharge pattern to the river results fluctuations in river flows below the District of over 5.7 m^3/s (200 cfs) during each twenty four hour period. These fluctuations in flow are visible more than 40 km downstream (Figure 1).

The South Platte is a gaining stream near Denver and receives water from the shallow South Platte alluvial aquifer. This aquifer has expanded in size from its historical boundaries due to controlled water storage and releases, irrigation, and trans-mountain diversions, and urban usage (Hurr & Schneider 1972). This aquifer contributes (based on mass balance equation and groundwater discharge measurements) (McMahon et al 1995a) a median incremental flow of 0.2 m^3/s/km (4.6 cfs/mi) to the river. This incremental groundwater discharge is very important to the overall water balance and water quality of the urban South Platte.

The rate of incremental groundwater flow to the South Platte fluctuates daily with stage height of the river. The stage height of the river changes due to the daily fluctuations of treated effluent discharge coming from the Metro District. Higher discharge results in increased surface water infiltration into the bed sediments underlying the river channel. A consequence of increased surface water infiltration into the bottom sediments is less infiltration and/or no infiltration of groundwater from the alluvium into the river. Lower river stages, caused by lower plant discharge, allow more groundwater to discharge from the alluvium into the river.

The localized movement of water in and out of the sediments under the stream bed of most alluvial arid rivers is correlated with a very active hyporheic zone. Surface water not only flows around and/or over meanders or obstructions, but also infiltrates the alluvium on the inside of meanders and beneath obstructions. The movement of groundwater to the river, the localized movement of surface waters through the hyporheon, and the mixing of the ground and surface waters are very important with respect to the water quality and dissolved oxygen dynamics of arid rivers.

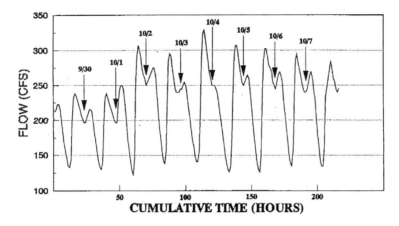

FIGURE 1 - Flow fluctuations in South Platte River due to effluent discharge measured 19.2 km downstream of Metro Wastewater Reclamation District's discharge. 9/30/92 - 10/7/92.

Physical Factors-Temperature

Temperature is one of the most important physical factors which regulates distribution of biota in running waters. The river temperature in the South Platte ranged from freezing in the winter months to above 30°C in the summer months. Daily fluctuations of temperature is more extreme due to low humidity and high elevation. Downstream of Denver, the river temperature varies on a daily cycle with flow and weather conditions. In most urban arid climates there are average daily differences in ambient air temperature of almost 17°C. The shallow waters of the South Platte 20 km downstream of Metro may demonstrate a 7°C range in daily temperature.

Controlled discharge of warm water can cause increased sedimentation. Warmer water is less viscous than cold water and fine suspended particles, such as silt and other fine solids will sink twice as fast at 23°C as at 0°C. As a result of this, often there is a greater deposition of finer silt particles and solids close to the discharges from warm effluents. These deposits often cause anoxic conditions in the sediments. Temperature also has a marked influence on both the chemical and biological processes of the South Platte. Increased temperature speeds up most chemical reactions. Below Metro's discharge, increased water temperatures prolong the growing seasons of benthic biota and aquatic plants. Fish species, which depend on temperature cues to trigger spawning, may spawn earlier in the year. The warmed water in the South Platte also provides suitable open-water habitat for many waterfowl species to overwinter. These extremely high (1,200 birds/km) (2,000 birds/mile) (Johnson et al 1993) waterfowl populations also have a pronounced effect on downstream water quality.

Chemical characteristics

The chemical content of most rivers is a reflection of the local geography and climate. The chemical content of the water of arid urban rivers, however, is affected by treated effluent discharged. During most of the year, chemical concentrations in the South Platte are a function of influent concentrations, treatment efficiency, and discharge volume from Metro.

1.   Dissolved Gases and pH

The two most important dissolved gases in all rivers are oxygen and carbon dioxide. Concentrations of these gases in solution are dependent on temperature, pressure, and reaeration due to production and flow rates. Concentration of these gases in water decrease with increase in temperature. In arid streams, which are typically shallow with limited tree canopy and subject to more intense solar irradiation due to high elevation and low humidity, large daily variations in dissolved gas concentrations often occur.

As temperature has a pronounced effect on the concentration of dissolved gases, so does elevation. Dissolved oxygen and carbon dioxide concentrations are lower (due to less atmospheric pressure) at higher elevation. Waters that make the Rio Grande in Albuquerque or the South Platte in Denver, "Mile High Cities," cannot contain as much oxygen as waters at a lower elevation. Dissolved oxygen and carbon dioxide maintain an inverse relationship in rivers. When the oxygen concentration is high, the carbon dioxide content is low, and when the dissolved oxygen concentration is low, the carbon dioxide concentration is high. This inverse relationship can be primarily attributed to the photosynthetic and respiratory activities of the biota. Probably the most notable effect of CO_2 concentrations in arid urban rivers is on pH. During photosynthesis CO_2 is removed from the water as O_2 is produced. This causes the river to become more basic during the daylight hours and more acidic at night. This increase in pH can raise the concentration of unionized ammonia in effluent dominated streams (Lewis & Saunders 1994).

The groundwater contribution, the numerous irrigation impoundments and pools, and numerous falls, weirs, and similar structures also have a significant effect on the concentration of oxygen and carbon dioxide in arid urban rivers. The South Platte north of Denver, unlike many arid urban rivers, is a gaining stream continually receiving a rather significant contribution to total flow from the groundwater alluvium. The groundwaters of the watershed, like most underground waters, are rich in carbon dioxide and for the most part almost devoid in oxygen. Groundwaters entering the river during periods of low dissolved oxygen in the surface waters enhance the oxygen deficit. The South Platte river has a very active hyporheic zone (McMahon et al 1995b). Water moving in and out of this zone through the sediments are stripped of oxygen.

The small man-made impoundments associated with the numerous irrigation diversions located on western urban rivers also reduce the oxygen content of the water and increase the carbon dioxide content of the water as it passes through them. When river water encounters a channel of decreased flow or velocity, there less reaeration and often less photosynthesis as well. Downstream of the irrigation structures on the South Platte there is generally an increase in the amount of dissolved oxygen in the water. The increase is caused by the drop of water over these diversion structures which causes more turbulence, more contact with the atmosphere, and the entrainment of bubbles. These drop structures enable restoration of equilibrium between the dissolved gases in water and the atmosphere and subsequently restoration of oxygen content. Dissolved oxygen concentrations at levels less than saturation are augmented by increased contact of the water with the atmosphere as water tumbles at an increased velocity over the structures.

The channel morphology and the biota can greatly enhance the daily fluctuation of CO_2, O_2, and pH in western urban rivers. Odum (1956) concludes that flowing, shallow rivers enriched by nutrients are possibly the areas of greatest aquatic production of O_2 on the planet. In quiet waters on arid rivers, there is an enhanced diurnal effect on gas concentrations due to the photosynthesis of benthic plants increasing dissolved oxygen concentrations in the daytime with corresponding higher levels of pH. Lower levels of oxygen and pH and increased levels of CO_2 occur at night not only because of plant respiration, but also due to sediment oxygen demand (SOD). The result is lower oxygen concentrations and higher carbon dioxide concentrations at night. In arid urban rivers with quiescent waters there is little or no physical reaeration and anoxic conditions can occur at night.

2.   Dissolved Salts and Nutrients

Under arid conditions, running waters generally contain much larger amounts of ionic salts due to high amounts of evaporation. Edwin James, the botanist in the Long expedition of 1820, described excess salts in the bottom lands of the South Platte Basin. "We passed extensive tracts nearly destitute of vegetation. The surface of these consisted entirely of coarse sand and gravel, with here [and] there an insulated mass of clay, highly impregnated with salt.... Some extensive portions of the immediate bottom land, along the river, were white with an effloresced salt" (James 1823). This was long before irrigation or urban development.

Urban rivers in arid environments are still typically high in dissolved salts and nutrients. Today most of the dissolved salts and nutrients present are not as much a function of local geography or climate as they were when first viewed by Edwin James. Under normal flow conditions, the dissolved salts and nutrients in developed urban arid river segments are, for the most part, a function of controlled discharges. In effluent dominated rivers the concentration of ions varies throughout the day with variations in discharge. Ionic concentrations of the discharge and the river are due to type and volume of influent received at the plant as well as treatment efficiency. Irrigation and evaporation of these waters concentrates the dissolved salts and nutrients of these rivers.

The fertility of all watersheds, including the arid urban rivers, is due primarily to the availability of the three plant nutrient salts: phosphorus (P), potassium (K), and nitrogen (N). Potassium is a common constituent of all hard water, and readily available throughout the West. Nitrogen and phosphorus are the two plant nutrients that are the most important. In most watersheds the two main sources of nitrate and phosphate to streams are rainfall and surface drainage. Except in extremely infertile conditions, the amounts of these nutrients that enter a stream are considerably in excess of the amounts normally applied as fertilizer and do not limit plant growth. In arid urban watersheds, this is true also, the amounts of these two nutrient ions are in excess of that needed for plant growth. However, the source of these nutrients is different than in most watersheds. The primary source in the South Platte north of Denver is not rainfall or surface drainage, rather the discharge of treated effluent from Metro.

In most undisturbed watersheds, the concentrations of nitrate and phosphate actually in solution are quite low, because the ions are rapidly taken up by plants. This is usually not true in arid urban watersheds. The waters down-stream of most developed urban areas in the arid West contain excess nutrients both from point and nonpoint sources. Biological transformation of the nutrient salts takes time and is dependent on plant uptake. Most arid urban rivers provide an inhospitable environment for plant growth. Most lack suitable substrate for plant growth. This lack of substrate is aggravated by daily changes in flow, velocity and constant shifting of bed material. The nutrient loads received by arid urban rivers can be significant. The nutrient load which is received by the South Platte in Denver comes from treated effluent of over one third of Colorado's population. Even if the Metro District could afford all of the best available technologies for nutrient removal, and they were available, the sheer volume of nutrients still received by the South Platte would take a significant amount of time to be processed by the biological community.

The South Platte receives a significant contribution of nutrients from the groundwater aquifer. Surface waters in the South Platte north of Denver average 7.7 mg/L nitrate N, the groundwater recharge at certain locations has a measured average concentration of 27 mg/L nitrate nitrogen (McMahon et al 1995a). High down-stream nitrate levels present in groundwater are apparently caused mainly by leachates from certain nitrate bearing shales and salt meadows, which occur naturally (CH2M HILL 1974) and from manure and excess fertilizers applied to crops (USGS 1994b).

In Denver, the first flush peak of pollutants which characterize storm waters is not as dramatic as in more humid climates. Pollutants are present in near equal amounts throughout the storm hydrograph. (Blakely et al 1983). In an arid watershed, such as surrounds the South Platte near Denver, the erosional load due to precipitation is primarily due to particles in suspension, rather than in the dissolved state as would be the case in a vegetated, more humid climate. Suspended particles not filtered by vegetation and soils, have a more direct route to the river. Past studies have shown that high metals concentrations in the Denver area (DRCOG 1983) are primarily the result of non-point source urban runoff. Storm water discharge in arid urban streams rise and subside quickly with each precipitation event resulting in very flashy runoff conditions.

Concentrations of metals are often high in arid urban rivers. (Flagstaff-copper, Pueblo-selenium, Albuquerque-aluminum) In some cases this is due to past and present mining practices, local industry, urban storm flows, or other activities of man. In many cases, however, background metal concentrations exceed limits imposed for intended use. Natural concentrations of metals can be high due to leaching from ores, shales, or soils that underlie the aquifers that drain into these rivers. In some cases, the dissolved metals may even be imported into the aquifer by diversion waters that come from a different watershed (Harris 1993). Clear Creek, which flows through the three richest ore bodies in the United States, contributes increased amounts of copper, lead, and zinc to the South Platte (Dennehy et al 1993). Historical mining operations have exposed many of these ores. Weathering of shallower arid soils and exposed Mancos shale deposits lead to natural selenium concentrations in rivers several times higher than national chronic standards (Mueller et al 1991).

Habitat Structure

The South Platte River in Denver was an aggrading river located in a transition zone between the mountains and the plains. Present channel substrate reflects this transition, with the substrate upstream primarily composed of fairly large cobble and gravel with the channel substrate in the lower section primarily of finer sands. Coarse material is associated with steeper slopes and finer materials with flatter ones. Below Denver, the natural gradient or channel slope is approximately 0.0017 and finer materials were deposited as the river became slower and more braided as it reached the plains. Finer materials have also been deposited in localized reaches that have been modified due to gravel operations or irrigation diversions. These irrigation diversions, bridges, rip-rap, and flood control structures have greatly affected the channel character of the South Platte. Today, although the flood plain has been encroached upon, the river still flows in its former channel downstream of Metro. This historic flood channel reach, created by extremes in discharge was a rather straight, extremely wide, highly braided (reticulate) stream channel that was constantly aggrading from sediments carried from the mountains. Maintenance of this channel does not allow for shading, sinuosity, or bank stability. Urbanization, farming practices, channelization, and impoundments have not only diminished the wide floodplain of the river, but have also controlled the narrowed the active channel. The present South Platte is much narrower (87 m vs. 335 m at Kersey, CO.) and shallow with an average depth of less than 45 cm (Dennehy et al 1995). Present flows no longer fit the historical channel.

River channels are a product of the water that formed them and are more or less adjusted to the normal discharge patterns of a watershed over long periods of time. Unlike urban rivers in more temperate, moist climates like the Charles,

Cayuga, and Chattahoochee, urban rivers in the arid West are new, changing systems that have not yet adjusted to the changes in discharge patterns that came with development. The South Platte is no longer a river with seasonal, highly fluctuating flows. Upstream of Denver it has been regulated by three new reservoirs in the past 20 years. But downstream, it still has a wide floodplain, reticulate-unconfined channel, and highly permeable alluvium. The present meandering river with a constantly shifting bed is never able to reach channel equilibrium or provide suitable habitat for many species. The reduced flood flows of the South Platte have also eliminated wetlands and ponded waters that were historically linked to the river at times of high discharge. Elimination of these linkages, due to reduced flows and urban and agricultural demands on floodplain lands, has not only eliminated a rich supply source of plankton to the river, but also eliminated valuable spawning and nursery grounds for many fish species.

Tree dominated riparian corridors of many arid rivers in urban environments are something new. Historically, many reaches of rivers like the South Platte had naked banks (McClure 1897). The energy of spring flood flows, fires, herds of buffalo, and the arid climate did not allow for development of shaded banks. The first explorers to Colorado called the South Platte valley the Great American Desert (Werner 1993). Although groves of trees existed in the floodplain of many of these rivers, they were quickly used for construction, fuel and fence posts by early settlers. A true riparian corridor did not develop along the South Platte until the advent of irrigation in the 19th century. Riparian corridors became established due to more constant flows, the invasion of many shrubs, forbes, and trees traditionally not found in the West, and subsurface irrigation return flows. New, diverse riparian zones are difficult and expensive to establish. Few urban rivers in the arid West have a diverse tree-lined riparian zone. Most treed riparian corridors along western urban rivers do not provide diverse habitat, rather they consist of monocultures of introduced species such as autumn olive, tamarisk, Siberian elm, and nursery-grown "cottonless" cotton-wood. The lack of shading, filtration, woody debris, snags, allochtonus detritus, and habitat contributions of a treed riparian corridor still exists along most Western rivers in arid urban environments.

Energy Flow/Biota

One primary measure of the health of an ecological system is a measurement of its energy flow and the resulting biological community performance (Karr 1991). As the ultimate energy source is the sun, and Western rivers in arid, urban environments get more than their share, compared to non-arid regions. Fewer cloudy days has been one of the reasons for these western cities phenomenal growth. In rivers, primary productivity is ultimately dependent on the amount of available light which penetrates to the channel bottom. Urban arid rivers of the West have more available light, the ultimate energy source, than their eastern cousins. Shallow streams enriched with nutrients have great production. (Odum 1956). The shallow, unshaded, effluent dominated arid urban rivers under clear western skies have the potential to be extremely productive.

Daily, seasonal, and stochastic extremes in flow limit the establishment of attached algae populations and the development of diverse macroinvertebrate and species. The establishment of diverse macroinvertebrate and fish populations is limited by the extreme physical environment of these rivers and by availability of suitable habitat for their establishment. Habitat surveys of the South Platte report a lack of suitable fish or macroinvertebrate habitat with only occasional snag providing permanent habitat (HabiTech 1992). Fine substrates and shifting sand bottoms coupled with both daily and seasonal velocity extremes severely limit many scraper and filter-feeding populations.

Biotic communities in western rivers in arid urban environments are also limited by the size and amount of food available. These systems drain relatively sterile watersheds. Most are limited by organic input. Communities in these controlled rivers are often carbon limited. Allochtonous detritus, coarse particulate matter and woody debris are well recognized (Gippel 1995) organic carbon resources in river systems. Western rivers in arid urban environments are, for the most part, limited in organic carbon. The surrounding landscapes, limited riparian corridors, substrates, velocities, dams and diversions and filtered point sources provide an environment lacking in coarse particulate organic matter (CPOM). CPOM in these streams is limited due to the sources of the water which feed them. Dams, treatment plants, and even irrigation diversions limit the availability of CPOM. Surface water is drawn away from these systems by irrigation diversions, only to return as filtered ground water. The wide, braided, almost sterile channels are nearly void of natural vegetation. Hypolimnetic discharges from storage reservoirs and treatment plant discharges provide only fine, dissolved materials to the rivers supporting only communities that are able to utilize these resources. The organic carbon provided from treatment systems is rapidly utilized and scoured. In the South Platte, which receives an urban organic load from over one third of Colorado's population, processes such denitrification in sediments are carbon limited less than 35 km downstream from Denver (Bradley et al 1995).

Fish populations are also limited to extremely tolerant species in these rivers. Historically, only eleven fish species were known to be adapted to the extreme environments provided by the South Platte (Jordan 1891). Indices of biotic integrity rate all these systems poor, especially when looking at fish populations (Bramblett & Fausch 1991). The Denver mayor's committee on looking at managing a native fishery in urban Denver decided the only native fish species that were adapted and could survive to provide an urban fishing experience in the South Platte were the brown bullhead and orange spotted sunfish. The fish that can tolerate the environmental extremes of an arid west urban river can neither migrate up or downstream due to dams, diversions, and grade control structures. Fish populations are further reduced by diversion eggs and larvae carried by current into irrigation canals and away from the rivers.

Summary

Arid western rivers have many characteristics that distinguish them from rivers in more temperate areas. Furthermore, many of these characteristics are not very compatible with the concept of ecological integrity that is embodied in the Clean Water Act. Extensive flow regulation has altered the natural hydrograph by decreasing peak discharge and augmenting base flows. The resulting ecosystems, though extremely valuable are no longer natural. Flow depletion may leave little in the river in urban areas except treated wastewater.

The South Platte River, which passes through the Denver metropolitan area, provides an example of a western arid river in which flow is dominated by municipal effluent. In this system, nutrients are always abundant and the channel is generally wide, shallow, and unshaded. Primary productivity is high and produces diel variations in pH that can have a strong influence on unionized ammonia concentrations. At the elevation of Denver, saturation concentrations of gases like dissolved oxygen are only 84% of saturation at sea level. Oxygen dynamics are strongly influenced by biological processes associated with the substrate. In particular, daily fluctuations in effluent discharge exacerbate the oxygen demand expressed by the sediment. In addition, the numerous diversion structures exaggerate the longitudinal pattern of DO concentration by creating excess demand in impounded pools followed by reaeration across the structure. The high degree of anthropogenic control of flow and channel geometry undermines the concept of ecological integrity that is applied so readily to the less regulated streams of other parts of the U.S.

Ecosystems change. The ecosystems of most arid urban rivers are no longer the historical, depauperate products of an extreme environment, nor are they well established diverse products of evolution over time. They are young, controlled, and developing. They are a product of an altered environment, with recent changes in hydrology and energy flow. They are a product of dynamic development. They will continue to gradually change, and how they change will depend on how they are perceived and protected. Rules that apply to less regulated systems, in watersheds which receive adequate precipitation, must be tempered or changed when utilized to measure the integrity of these unique ecosystems.

References

Blakely, S.R., M.H. Mustard, and J.T. Doerfer. 1983. Analysis of the August 14, 1980 Rainstorm and Storm Runoff to the South Platte River in the Southern Denver Metropolitan Area, Colorado, USGS Water Res. Invest. Report 83-1438.

Bradley, P.M., P.B. McMahon, and F.H. Chapelle. 1995. Effects of carbon and nitrate denitrification in bottom sediments of an effluent dominated river. Water Res. Research Vol. 31, No. 4, p 1063-1068.

Bramblett, R.G. and Kurt D. Fausch. 1991. Variable fish communities and the Index of Biological Integrity in a Western Great Plains River. Trans. Am. Fish. Soc. 120:752-769.

CDM. 1991. Nitrification Alternatives Study - Final Report. Camp Dresser & McKee, Denver, Co. 142 p.

CH2M HILL. 1974. Advanced Wastewater Treatment. Report to Metropolitan Sewage Disposal District No. 1. Denver, p. 1-60.

Dennehy, K.F., Litke, D.W., Tate, C.M., and Heiny, J.S., 1993, South Platte River Basin--Colorado, Nebraska, and Wyoming, Water Resources Bulletin, Vol. 29, No. 4, p. 647-683.

Dennehy, K.F., Litke, D.W., McMahon, P.B., Heiny, J.S., and Tate, C.M., 1995, Water-quality assessment of the South Platte River Basin, Colorado, Nebraska, and Wyoming--Analysis of available nutrient, suspended-sediment, and pesticide data for water years 1980-92; U.S. Geological Survey Water-Resources Investigation Report 94-4095, 145 p.

DRCOG. 1983. Urban Runoff Quality in the Denver Region. Denver Regional Council of Governments, Denver, CO. 1-156 p.

Gippel, Christopher, J. 1995. Environmental hydraulics of large woody debris in streams and rivers. Journal of Environmental Engineering Vol. 121 No. 5, p 388-395.

HabiTech. 1992. Physical habitat composition as it relates to the fisheries of Segment 15-South Platte River. Technical Memorandum for Metro Wastewater Reclamation District, Denver, CO.

Harris, Todd. 1993. Segment 15-A Unique Watershed. CWRRI Information Series No. 76. CSU, Fort Collins, CO.

Hurr, T.R. and D.A. Schneider, Jr. 1972. Hydrologic characteristics of the valley-fill aquifer in the Brighton Reach of the South Platte River Valley. USGS open file Report, 2 p.

James, Edwin. 1823. Account of Expedition from Pittsburg to the Rocky Mountains performed in 1819-20 under command of Stephan A. Long. University Microfilms Inc., Ann Arbor, 1966.

Johnson, G., D. Young, W. Erickson, D. Strickland, and L. McDonald. 1993. Abundance, Species Composition, Habitat Selection, Behavior, and Food Habits of Waterfowl Wintering Within Segment 15 of South Platte River. *in* CDM Preliminary Design Stream Reaeration Improvements Seg. 15 South Platte River Final Report. CDM, Denver, CO.

Jordan, D.S. 1891. Report of explorations in Colorado and Utah during summer of 1889. U.S. Fish Comm., Bull. 9:1-40.

Karr, J.R. 1991. Biological Integrity: A long neglected aspect of water resource management. Ecol. App. 1:66-84.

Lewis, William A, Jr., and J. Saunders III. 1994. Improvements in the Colorado Ammonia Model by Simultaneous Computation of Extremes in Flow and Water Chemistry. CWRRI Report. CSU, Fort Collins, CO.

McClure, Col. J. 1867. *in* The South Platte River, Then-Now. Exhibit by Amy Gilpen for U.S. Army Corps Engineers, Chatfield Reservoir. Feb. 1995. Denver, CO.

McMahon, Peter B., K.J. Lull, K.F. Dennehy, and J. Collins. 1995a. Quantity and Quality of Ground Water Discharge to the South Platte River, Denver to Fort Lupton, CO. August 1992-July 1993. USGS Water Res. Investigations. Report 95-4110. Denver, CO. 71 p.

McMahon, Peter B., J A. Tindall, J.A. Collins, K.J. Lull, and J. Nuttle. 1995b. Hydrological and geochemical effects on oxygen uptake in bottom sediments of an effluent dominated river. Water Research Vol. 31, No. 10, p. 2561-2569.

Mueller, D.K., L.R. DeWeese, A.J. Garner, and T.B. Spruill. 1991. Reconnaissance Investigations of Water Quality, Bottom Sediment and Biota Associates with Irrigation Drainage in Middle Arkansas River Basin Colorado and Kansas 1988-89. USGS Water Resource Investigation Report 92-4060. U.S. Geological Survey, Denver, CO.

Obmasic, Mark. 1996. Future of the South Platte River. Denver Post 6/23/96. Denver, CO.

Odum, Howard T. 1956. Primary production in flowing waters. Limnology - Oceanogr. 1:102-117.

Parkham, Francis. 1846. The Oregon Trail. Reprinted 1985 Penguin. New York.

Strahler, A.N. 1957. Quantitative analysis of watershed geomorphology. Amer. Geophy. Union Trans. 38: 913-920

U.S. Congress. 1972. PL 92-500 Federal Water Pollution Control Act. October 16, 1972. Washington, D.C.

USGS Water Data Report CO-1994a-94-4 Volume 1 p. 82-3 NTIS Springfield, VA.

USGS 1994b, USGS NAWQA program *in* Western Resources Wrap-up Series XXXII, No. 15, p 1-4, Washington, D.C.

Werner, Brian. 1993. The Great American Desert: Popular myth and water development along the South Platte. CWRRI Information Series No. 72. CSU, Fort Collins, CO.

SOUTH PLATTE IN METROPOLITAN DENVER -
A RIVER IN TRANSFORMATION

By Michael A. Stevens[1]

ABSTRACT

The pristine South Platte River of 1860 was 300 feet wide, shallow, unstable, fringed with a corridor of cottonwoods and willows, and flowed throughout the entire year. Its features can not be recovered, having been changed initially by logging of its trees for building materials, and next by the loss of river flows, diverted to irrigation on its floodplain. These developments were followed by the construction of upstream storage reservoirs, transmountain diversions, channelization and straightening of the river, stabilizing of its bank, dams for flood control, and a variety of lesser controls to mitigate the effects of the greater developments.

Now, the 41-mile (66-kilometer) reach of the river in the Denver metropolitan area from Baseline Road near Brighton to Chatfield Dam has many forms, conveniently catalogued morphologically by the author as Rural, Suburban, Engineered, and Urban. Each has a distinct form as a result of different combinations of development. It is now virtually impossible to recreate the pristine character and hydrology of this river as it makes it way through the metropolitan area.

The expected changes due to past, present, and future developments are towards a narrower, deeper, straighter river with stabilized banks, a degraded and somewhat stabilized bed, with trails and paths, and better water which can support more diverse aquatic life. In the Rural and Suburban sections, some of the treasured features of the more natural riverine environment can be retained.

INTRODUCTION

The South Platte River of "old" is of interest again because there are those who would have us restore that morphology and aquatic ecology as our ideal. Generally, it is

---

[1] Consultant, P.O. Box 3263, Boulder, Colorado, 80307.

thought that the pristine river shriveled to a mere thread after spring runoff. "Half-mile wide and six-inches deep" is a common cry. Some more historical research to back up these views is certainly in order.

No matter what it used to look like, it is abundantly clear that the South Platte River has changed because of all the developments that have been undertaken since the discovery of gold in it and some of its tributaries 140 years ago. It is still altering itself in places. But development has given way to management.

The South Platte River in the metropolitan area looks like four distinct rivers, herein named the Urban, Suburban, Engineered, and Rural reaches. The four types of reaches offer different opportunities to create environments, some more natural and others more urban. Within nature's and our own constraints, great improvements to the riverine environment have been made lately. This can be continued, but it must be accomplished without much water; that has been claimed and is owned for use outside the river channel. The issues of the pristine river morphology, its changes, and what holds for the future are addressed in this paper.

PRISTINE RIVER

The pristine river is of great interest because there are those who would have us endeavor to reconstruct that former morphology and aquatic ecosystem. This unsullied icon is presumably what we must desire and strive for. That there existed a "...healthy river riparian ecosystem" is predicated on its being undisturbed. That such a system could be partially or completely lacking in what people think a river should be would be a problem to many.

Under the supposition that the Indians who camped along this river before us did not use the water and land resources to any significant extent so as to change them, the pristine South Platte River would be that in existence when our written historic record began. The explorations of Stephen H. Long in 1820 and Fremont in 1842 recall for us how the pristine river looked to them. The accounts by Smiley (1901) are another prime source of information because he was able to talk to many people who lived with the founding of Denver. Unless otherwise noted, the information herein is from Smiley (1901).

On June 22, 1820, the Long party crossed the North Platte River near its confluence with the South Platte (James 1823) after having travelled up the north side of the Platte from Council Bluffs. They waded across and recorded that the North Fork was 800 yards (730 meters) wide, was "shoal" and rapid like the Platte below. The bed was sandy.

On the next day, they crossed the South Platte "...with less delay and difficulty..." than in wading the North Fork. The stream was 900 yards (820 meters) wide and very rapid. They described it as so "shoal" that it was unnecessary to dismount or unpack the mules. When the party "...met with wood...", they camped.

"Intermixed in the narrow fringe of timber, which marks the course of the river, are very numerous trees, killed by the action of the beaver or by the effects of old age, their decorticated and bleached trunks and limbs strongly contrasting with the surrounding objects..."

The Long expedition journeyed along the south side of the Platte, probably on the advice of the guide. They made notes on the vegetation, bison swimming and wading across the river, the current speeding up, the sand and gravel bed becoming narrower, and the banks more "precipitous." Comments and measurements were made on water temperature (68 to 75 degrees F. or 20 to 24 degrees C. at 11 am and getting warmer as they ascended). The entry for July 3 was:

"We had often heard our guide, in speaking of the country, two or three days journey from the mountains, mention the *Grand Forest*, and were a little surprised on arriving at it, to find no more than a narrow but uninterrupted strip of timber, extending along the immediate banks of the river, never occupying the space of half a mile in width."

On this day the expedition passed "three large creeks" entering from the northwest. These are taken to be the Cache la Poudre, Thompson, and St. Vrain. Their junctions with the Platte are well within the range that the party could traverse in a day.

On July 5, the expedition was at the mouth of Clear Creek. Some members intended to

"...ascend the Cannon-ball creek to the mountains which appeared to be about five miles distant. The creek is rapid and clear, flowing over a bed paved with rounded masses of granite and gneiss [hence the name]. ....The channel is sunk from fifty to one hundred feet below the common level of the plain..."

"The detached party extended their walk about eight miles without finding the apparent distance to the base of the mountain had very considerably diminished ... they re-crossed the Platte, which was here about three feet deep, clear, and rapid, and arrived at camp after sunset."

In 1832, Louis Vasquez founded a trading post at the junction of Clear Creek and the South Platte. His abode was of cottonwood logs and a sod roof. One must suppose that there were cottonwoods and water nearby, and that these were perennial. Otherwise Vasquez would have been a very industrious man. What he traded in was probably the commerce of the time - animal pelts for hunting and trapping supplies.

In 1842, John Charles Fremont (1845) journeyed up the south side of the Platte in Nebraska coming to the South Platte River at its mouth on July 2. Some of the party crossed to continue up the North Fork while Fremont went south. Here is what was recorded:

"The stream [South Platte] is here divided by an island into two channels. The southern is four hundred and fifty feet wide, having eighteen to twenty inches of water in the deepest places. With the exception of a few dry bars, the bed of the river is generally quicksands, in which the carts began to sink rapidly so soon as the mules halted, so that it was necessary to keep them constantly in motion."

"...The northern channel, two thousand two hundred and fifty feet wide, was somewhat deeper, having frequently three feet of water in the numerous small channels, with a bed of coarse gravel."

At the mouth, the South Platte was different than what the Long party had seen 22 years previously and was in fact two different rivers, depending on which channel one was viewing.

Fremont came down the South Platte as far as St. Vrain's enclave on the South Platte at the confluence with the tributary now with his name. It is tempting to apply the descriptions of Fremont for Denver just another 38 miles (61 kilometers) upstream. But as the Long expedition had recorded earlier, the entire pristine South Platte River on the plains did not look and was not the same at all places.

A tale of navigation quandaries told to the Fremont party by trappers bringing their furs down the North Platte from the Laramie River is bizarre, even for river stories.

"They [the trappers] proved to be a small party of fourteen.... with their baggage and provisions strapped to their backs... Sixty days since, they had left the mouth of the Laramie's fork, some three hundred miles above, in barges ladened with furs... They started with the annual flood and, drawing but nine inches of water, hoped to make a speedy and prosperous voyage to St. Louis.... They came down rapidly as far as Scott's bluffs, where their difficulties began. Sometimes they came upon places where the water was spread over a great extent, and here they toiled from morning until night, endeavoring to drag their boat through the sands, making only two or three miles in as many days. Sometimes they would enter an arm of the river where there appears a fine channel, and, after descending prosperously for eight or ten miles, would come suddenly upon dry sands, and be compelled to return, dragging their boat for days against the rapid current and at others, they came upon places where the water lay in holes, and getting out to float off their boat, would fall into water up to their necks, and the next moment tumble over against a sandbar."

Finally,

"...they sunk their barges, made a *cache* of their remaining furs and property, in the trees on the bank, and, packing on his back what each man could carry, had commenced, the day before we encountered them, their journey on foot to St. Louis."

In 1858, Denver was just coming into existence because of the discovery of gold in the South Platte and some of its tributaries two years earlier. Denver, then named St. Charles, was at the junction of Cherry Creek and the South Platte, to the east of both. Opposite, on the west side of Cherry Creek, the rival enclave of Auraria struggled with Denver to attract settlers. Both "cities" were but real estate endeavors, platted pieces of prairie. Yet because of these twin developments, we are left with rich descriptions and stories of serious misjudgments .

In 1858, Cherry Creek was

"... a shallow trough, dry, about one hundred yards in ranging width, and with banks from two to three feet high."

The South Platte was decidedly dissimilar:

"... the course of the Platte river by the pioneer towns was somewhat different from its present way [1901]. Between our Eleventh and Twenty-third streets it described the outline of a pair of oxbows [meanders], one of which received the water of Cherry creek .... On the eastward side the other bow projected in as far as the [future] site of the Union Railway Station.... The map of the old towns shows the river's former **devious** course" [emphasis added].

"...In the autumn of 1858, ... the South Platte and its tributaries were fringed with cottonwoods and willows, and altogether the local landscapes were most pleasing to the eye. "

" ...The settlements at Cherry creek and the subsequent mining operations were followed by great havoc in the timbered areas. Logs for cabin-building and wood for fuel called for the destruction of many trees, and when the saw-mills got to work among them, trying to supply the insatiable demands for timber and lumber for building and mining purposes, the forests were eaten away with amazing rapidity, the growth of centuries disappearing in a few years, leaving the country around Denver in its present [1901] comparatively deforested condition...."

"But the first is the lovely part of the picture. There, spread out like carpet, glowing in the rich splendor of the autumn sunlight, lies the brown, swelling plain, cut in various directions by wood-fringed streams - a magnificent sight. The American has come to this hitherto unknown region, and the sound of the rifle and the axe, the lowing of stock and the falling of the timber around the smoke columns, are sounding the **death-knell** of the wild beast and the wilder man of the soil ..." [emphasis added].

The death-knell was not long in coming. The Great Invasion in the spring of 1859 brought the first big influx, people excited to find richness in gold. With this came the need and demand for more infrastructure. A ferry across the South Platte from Auraria

came into existence.   General Larimer proposed to build a bridge across the South Platte
"... with a dam between the piers to hold water for running a grist mill."   Other bridges
were proposed, even for Cherry Creek between the two towns. "... Ordinarily there was
little trouble in crossing the often dry bed of that stream, but at other times a bridge was
needed."

The need for building supplies was so pressing that an act was put forth "... to
incorporate the South Platte river improvement and Lumbering Company 'to improve the
South Platte river from Auraria City to a point twenty miles above the mouth of Platte
canon, for floating timber, lumber, and so forth down the river.'"

The map of Auraria and Denver (Fosdick and Tappen, 1859) had the two"devious"
meanders at the confluence in December of that year. The "dry bed of Cherry Creek" was
attached to the upstream bend in the South Platte. Both rivers were drawn as 300 feet (90
meters) wide. It is assumed that the South Platte boundaries were from bank to bank, the
same as for Cherry Creek. On 6 April 1860, Auraria became a part of Denver.

In 1861, many townships in the Denver region were surveyed. These 1-mile = 2-
inches maps have the sinuous South Platte as approximately 140 feet wide, but Cherry
Creek, Clear Creek, and Bear Creek were shown as lines. This is in conflict with the
earlier and larger map of the towns. The floodplains of all the major creeks and the South
Platte were indicated on the township maps by hatching. The South Platte meandered a
distance of 18.8 miles (30.2 kilometers) between the Bear and Clear Creeks junctions. The
valley distance was 13.8 miles (22.2 kilometers), making the sinuosity 1.4. There were
some 40 meanders (and one oxbow lake) in this same reach, so on the average, the meander
wave length was 13.8(5280)/40 = 1,800 feet (550 meters). If one employs Yalin's (1992)
geomorphic relation that the meander wave length is about six times the river width, the
South Platte would have been 300 feet (90 meters) wide, as the town maps indicated. But
why would the township maps show the rivers not to scale?

The flood of 19 May 1864 in both Cherry Creek and the South Platte was the first
warning that the river beds and floodplains belong to the rivers and not to the settlers. The
lesson was not well-learned for on 20 July 1875 another flood caused even greater damage,
especially to all those who had encroached onto the beds of the streams. Debris from that
torrent was rediscovered years later by excavators constructing foundations for new works.

The 1865 town map by Denver City Surveyor F. J. EBfit indicated that the 1864
flood took a short-cut across the bend where Cherry Creek had joined the Platte. Cherry
Creek still retained the lower part of that bend for its own service. The former bend was
marked on the map as "Old Bed of the South Platte River." The map also suggested by
markings that the downstream bend was about to be abandoned by another short-cut. Both
streams were drawn as 300 feet wide.

In anticipation of the arrival of its tracks, the Denver Pacific Railroad (1865) drew
up a map indicating its line and terminus near the Cherry Creek junction. The map
revealed that indeed the downstream bend was cut off and that the confluence was then a

wide shoaling area with seven bars interlaced with channels. The Kellogg and Bonsall map of 1871 had the same number of bars but in a different arrangement.

DEVELOPMENTS

Developments on the South Platte were quick in coming. Even before the first railroad train rolled into Denver in June 1870, "... there were near two hundred thousand acres of land under cultivation along the streams flowing eastward from the mountains." Without question, this was irrigated agriculture, the water coming from the adjacent streams; their floodplains were the fields.

Change in the South Platte River at Denver has been seemingly incessant. Causes of change are the usual suspects starting with deforestation of the river banks. The others include: diversion for irrigation (a major influence) and for municipal and industrial use; dams for water supply; dams and levees for flood control; channelization and infilling for real-estate development (another major influence); bank protection to prevent loss of land; structures for grade control; bridges for railroads and other traffic; mining for gravel; and structures for habitat control and recreation. Now stream modification for water quality enhancement, not a usual suspect, is underway. All are the outgrowth of population swell and vice versa.

Measurements of the streamflow in the South Platte River at Denver began in 1895. The Denver gage was placed about one-half mile downstream from the Cherry Creek confluence at 19th Street, an ideal spot in the middle of Metropolitan Denver. Now we have 100 years of record but this does not reflect the earliest changes. In this last century, the annual amount of water passing the gage has varied more than thirteen-fold. The driest year was 1901 with only 81 cubic feet per second (2.3 cubic meters per second), less than the minimum output of the Metro Wastewater Treatment facility at the Sand Creek junction. The South Platte could have dried up on the plains in 1901.

At the turn of the century, the Denver Water Board began its dams and reservoirs phase of development for domestic and industrial water supply. Cheesman (1902) and Antero (1907) were followed by Eleven-Mile Canyon (1932), all on the main stem of the South Platte River in the eastern slope mountains. The flood-control dams (Fig. 1) came later - Cherry Creek (1953), Chatfield (1975) and Bear Creek (1979). As time went by, recreation became a demanded function of these structures.

Transmountain diversions were begun when the waters of the eastern slope rivers became depleted. The Moffat Tunnel (1936) was the first. It was followed by others as the front range expanded without satiating its thirst.

There is nothing startling in the 100-year record of annual streamflow in the South Platte River at Denver, in spite of all the developments. Prior to 1925, the mean annual discharge was 30 percent larger than after. It is likely that part of this change was due to an alteration in regional precipitation. Virgin flows (flows adjusted to include the amounts diverted) in the Colorado and Cache la Poudre decreased a similar amount about 1930.

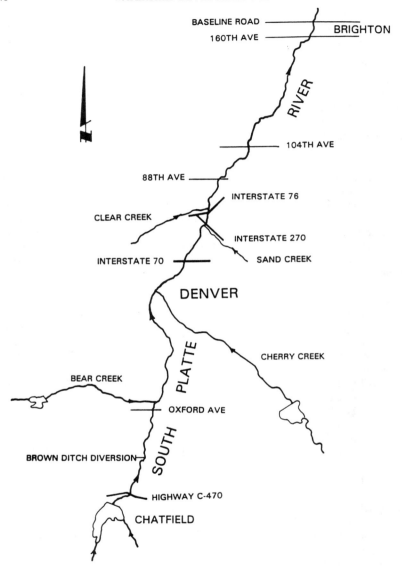

Fig. 1. Map of the South Platte River in the Denver metropolitan area.

The 100-year annual peak discharge record at Denver is far more interesting. There were three large floods clustered - in 1965, 1969, and 1973. All were the consequence of intense rainfall in the catchment. Prior to 1965, there was only one flood of significant discharge, that of 1933. Now the mean annual flood, approximately 5,000 cubic feet per second (150 cubic meters per second), can be accommodated easily within the South Platte channel.

In its 1964 assessment of the South Platte River drainage, the Missouri Basin Inter-Agency Committee proclaimed the catchment had attained almost maximum development. More than 90 percent of the river depletions is used for irrigation. Municipal and industrial users divert out less than five percent. At Denver, this milestone was reached sooner than for the catchment as a whole. Even so, Two Forks, Denver's most recent effort at damming the South Platte, was rejected.

Sediment records are much sparser than streamflow records. Historically, the sediment yield of the South Platte catchment at Denver has been low so there has been a lack of interest. The estimate made by the Corps of Engineers for planning the Chatfield reservoir was about 100 tons per square mile per year. The eastern slope is kind to us in this respect.

Lack of a plentiful sediment supply, other than its banks, has meant that the narrowing, straightening, and bank protection has caused a lowering of the river bed. Couple this degradation with infilling of the adjacent floodplain and new upstream reservoirs and the result is a river with much higher banks now. That is one reason why the average annual flood can be accommodated so easily and without much concern.

LAST DECADE

River morphology came to the forefront in preparation of the Master Plan for the South Platte River from Chatfield Dam to Baseline Road (Wright Water Engineers 1985). The Master Plan was titled Major Drainageway Planning and the founding premise was that

"The number one use of the South Platte River corridor is for flood control and drainage."

A long list of other uses was also presented. The goal of the plan was to optimize for: water supply; irrigation; aquatic life; recreation and aesthetics; and sewage dilution. Years of **development** had been completely replaced by a new direction, that of **management**.

The author's contribution in this effort was to describe the South Platte River in terms of its morphology and stability (Stevens 1983), why it looked as so, and what to expect in the future. Also troublesome degradation of the river bed had occurred in some areas, threatening bridges and other infrastructure under the river bed. This was also addressed (Stevens et al. 1990). The closures of the three flood control dams were expected to result in a lowering of the South Platte river bed. The dams decreased the supply of sediment, not large to start with, to the South Platte by an estimated 80 percent. In addition, miners had again moved onto the river bed in places, but this time to extract gravel, not gold, causing a rapid and large lowering of the bed immediately upstream.

Most of the degradation has been essentially arrested. Gravel mining of the river bed has been stopped and is done only on the floodplain. Timely and correct placing of grade-control structures now prevents the slower long-term degradation due to the reservoirs.

Four Rivers. In sizing up the South Platte, it becomes apparent that this river is distinct in different parts of the metropolitan area. That is, the river takes on appearances depending on the developments around it. One can classify and grade the river by where it is.

In 1983, the South Platte River though the Denver metropolitan area was classified as either Rural, Suburban, or Urban. In the last decade, a new member came into existence, the *Engineered* river. In summary, here is what they look like.

Rural River
- The river is not straight but meanders somewhat through its floodplain. Sinuosity is 1.26.
- The river is free to move laterally by eroding one bank and building the other.
- As a result of lateral movement, the river can have one high and one low bank.
- The river is wide and shallow, with sand, gravel, or cobble bars.
- Some bars are vegetated with pioneer weeds, grasses, and willows.
- On some banks, the natural succession of vegetation still exists.
- There is evidence of plentiful wildlife.
- Works of humans are few.

Suburban River
- Some river meander loops have been cut off, decreasing the sinuosity of the river to 1.20.
- Some eroding banks have been stabilized, in the past with rubble, but more frequently now with riprap.
- There are more bridges, weirs, grade-control structures, intakes, and sewer outfalls.
- There are some houses and other buildings on the historic floodplain.
- In places, the low bank has been filled with excavated materials.
- There are fewer bars and less native vegetation but still evidence of wildlife.
- The river is still wide and shallow in some places.
- There are many ponds and levees left by gravel mining operations.

Urban River
- The river has been straightened and realigned. Sinuosity is 1.09.
- The river is narrow and relatively deep compared to the Rural river
- The river bed is almost completely covered bank-to-bank. There are hardly any exposed bars except during very low-flow periods.
- There is much less riparian vegetation.

- Human works are everywhere.
- The river is bounded by very high-valued lands.

Engineered River
- The river has been realigned, straightened, and made prismatic.
- The entire river banks are protected with riprap.
- The bed is nearly flat with only a few low bars.
- The width is mostly uniform.
- Slowly, pioneer vegetation is appearing on the few bars and in the bank riprap.

Reach Classification. On the basis of river width, bar height, and other observed features, the South Platte River in the Denver metropolitan area is categorized as follows:

Type	Reach	Cross Section Numbers
Rural	168th Avenue (Baseline Road) to 104th Avenue	1 to 17
Suburban	104th Avenue to Interstate 270	18 to 28
Urban	Interstate 270 to Oxford Avenue	29 to 43
Engineered	Oxford Avenue to Brown's Ditch Diversion	44 to 48
Suburban	Brown's Ditch Diversion to Highway C-470	49 to 52
Rural	Highway C-470 to Chatfield Dam	53

The cross sections refer to transects established along the 41-mile (66-kilometer) reach of metropolitan river (Fig. 1), and surveyed annually by the Urban Drainage and Flood Control District. Monitoring began in 1986. Most of the cross sections are in the Rural (11 miles or 18 kilometers) and Suburban (8 miles or 13 kilometers) reaches where instability is anticipated to occur. The Urban reach (17 miles or 27 kilometers) is most stable.

Stability. In the Denver metropolitan area, the stability of South Platte River is of concern. Any river movement, whether up and down or laterally, affects many private and public properties and can threaten, damage, or destroy the existing infrastructure such as utility lines, hiker and biker trails, bridges, and adjacent roadways. The opinions on the stability of the river which follow are based on the first stability study and the annual monitoring of the 53 cross sections.

Width. The main geomorphic feature that illustrates the different modern South Platte River forms is river width. The width is defined as the distance across the river between top of banks. Shown in Fig. 2, the top widths vary greatly, being largest (up to 550 feet or 170 meters) near Brighton. The Urban and Engineered reaches are between 100 and 200 feet wide, with two exceptions. A short piece of river near the Chatfield Dam is Rural, and is approximately 350 feet wide . The Suburban reach, with "intermediate" widths, has become narrower since 1983. A long reach of Suburban river through Littleton was converted to Engineered river in the late 1980's by the U.S. Corps of Engineers. Recall that the pristine river was 300 feet (90 meters) wide.

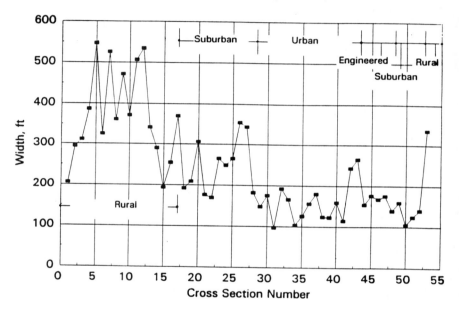

Fig. 2. The top bank width of the South Platte River at the monitored cross sections between Brighton (Cross Section 1) and Highway C-470 (Cross Section 53).

Bar Height. In the relatively sediment-free environment found in the study reaches of the South Platte River, it follows that high bars exist on the river bed only where the river is adequately wide. The *maximum bar height* in a cross section is taken as the difference between the level of highest point on the bar and the average bed level. The variation of the maximum bar height (Fig. 3) is great. In the wide river, bar heights are around six feet (two meters). In the narrow reaches, rarely are they greater than two feet (half a meter).

The initial perception was that the prolonged spring runoff in 1995 had built the bars to a new and higher level. An inspection of the change in maximum bars levels at the monitoring cross sections before and after the spring runoff does not support this

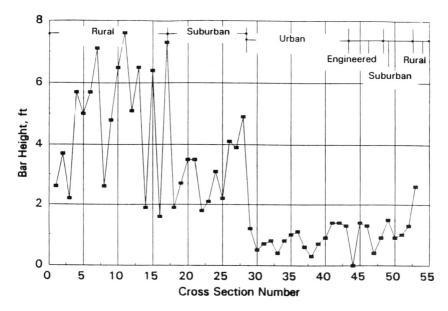

Fig. 3. The maximum bar height in all the monitored cross sections in 1996.

perception (Fig. 4). The freshness of the bar surfaces and the scouring out of bar vegetation was the most obvious result, and not bar raising.

Aggradation or Degradation. The *thalweg* is the lowest point in the river cross section. It is almost always, but not necessarily, in the low-flow channel. Where the river is wide, bars of river-transported and deposited sediment form on one or both sides of the low-flow channel. Commonly, a bar is composed of the upstream bank material if that bank is on the same side of the river and is eroding. If there are two low-flow channels in a cross section, the main and a side channel, the bar between them is a *middle bar*. Otherwise, the bars are *side bars*, or *alternate bars* if there is a sequence of bars on each side of a long straight reach of river. *Point bars* are deposits on the inside or convex side of bends . A middle bar that becomes vegetated and grows in height up to near floodplain level becomes an *island*.

The bars can be high or low, bare or vegetated. The upstream end of the bar is lower relative to the water surface than the downstream end. Bar sediment sizes grade from coarse to fine in the downstream direction and from water's edge towards the bank. Willows can colonize any part of the bar, but commonly grasses and weeds first appear at or near the water's edge. As a result of the scouring effect of the prolonged high spring runoff in 1995, the bar surfaces in 1996 are mostly bare and clean. New snags on bars are rare, but old snags often remain. These snags are usually old cottonwoods of stature.

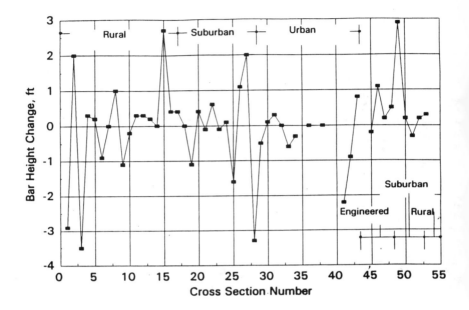

Fig. 4. Change in bar height in all the monitored cross sections in 1996.

The *riverbed level* is taken as the *average level* of the *active channel*. This includes the low bare bars but not the high or vegetated bars. In the Urban reach, the active channel width is the distance between the base of the two river banks. By definition, the river bed level is higher than the thalweg level.

The sediment on the bed of the South Platte River is small cobbles (two to five inches or five to thirteen centimeters in diameter) in the steeper upstream reaches to cobbles, gravel, or sand in the lower reaches. But deposits of sand, gravel, or cobbles can be found resting together in many places. In the Urban section, often there are chunks of concrete rubble or riprap stones on the channel bed. In a few spots, the weak rock of the Denver Formation outcrops on the river bed. The Denver Formation erodes very slowly so it controls riverbed degradation.

In the Rural segments, there is often a very small depression between a side bar and the river bank, almost always dry at low flow. This depression appears to be influential in maintaining a linear growth of woody vegetation along the base of the bank, mostly willows and cottonwoods.

Degradation is the general lowering of the river bed in a long reach of river over time. In contrast, *aggradation* is a general increase in the bed level. If the lowering is local, not over a long reach, it is *general scour*. General scour occurs, for example, in the river under a bridge with an opening less than that of the river upstream. The scour holes in the bed around bridge piers are *local scour*.

To help identify aggradation and degradation, the history of the thalweg level and average bed level at each monitored cross section was studied. If both the thalweg and the average bed level changed in one direction in the period of record, then degradation or aggradation is said to have occurred. The thalweg level can change but the average bed level does not. This is not aggradation or degradation. Sometimes, the average bed level cycles up and down from the mean without much net change during the period of record (Fig. 5). This cyclic behavior is judged stable, the same as if the bed did not change.

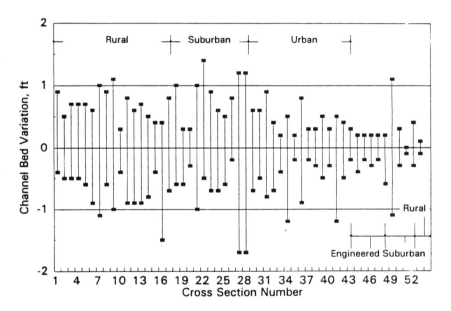

Fig. 5. Channel bed level variation from the mean bed elevation for the period of record.

In all, the river bed is judged to be stable. The severe degradation experienced in the early 1980's (Stevens et al. 1990) is no longer to be found. In 1996, fifteen cross sections had degradation, the most being 2.7 feet (82 centimeters) at Cross Section 38 (Fig. 6). Ten of these experiencing lowering are in the Urban reach between Cross Sections 29 and 43, and much of the lowering occurred in the last year, the apparent result of the long, high spring runoff in 1995.

Four cross sections experienced aggradation. Three of these are in the downstream Suburban reach between 104th Avenue and Interstate 270. The other one is in the Rural reach at the northern end of the metropolitan area.

of the metropolitan area. The most aggradation was 2.1 feet (64 centimeters) at Cross Sections 9 and 28.

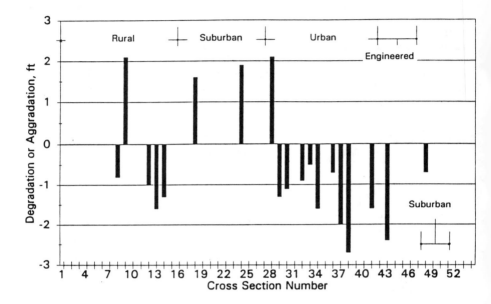

Fig. 6.   Values of aggradation (plus) and degradation (minus) at the monitored cross sections for the period of record.

In June 1986, the river was flowing through a 22-foot (6.7-meter) deep gravel pit at Cross Section 23. In November 1987, the pit had filled with river-transported sediments and has remained stable until construction on channel modifications for water quality improvement in this reach began in 1995. The modifications were undertaken by the Metro Wastewater Reclamation District at an estimated construction cost of $2,372,000. Aggradation immediately upstream at Cross Section 24 was the apparent result of the Metro project.

Construction of the Confluence Park river improvement has resulted, by design, in lowering the river bed 2.0 and 2.7 feet (60 and 82 centimeters) upstream at Cross Sections 37 and 38. In addition to the drops at Union Avenue, a number of *riprapped grade-control structures* with *boat chutes* have been constructed across the river to help control the riverbed level. Most of the drops are in the Rural and Suburban reaches.

In judging the sizes of sediments on the river bed in January 1996, it was found that the low-flow channel and its margins were often covered with algae, mostly brown in color, but green where the flow was faster and shallower. In addition to the algae, where the current was very low, black organic solids were deposited. In many places, the water was very murky because of suspended solids floating along. The bed could not be viewed through this water. The term *Urban Veneer* was adopted to describe this layer of algae growth and black sediment deposition on the river bed.

Flat Bed. Because of its rather straight alignment and the nature of it sediments, the bed of the South Platte River is relatively flat across. The thalweg is not much lower than the average bed level, meaning there are not many pools. This difference is the thalweg depth in Fig. 7. There are only three monitored cross sections where the difference is more than 2.5 feet (75 centimeters), all in the Rural reaches. This means that when the flow is low, it is very shallow.

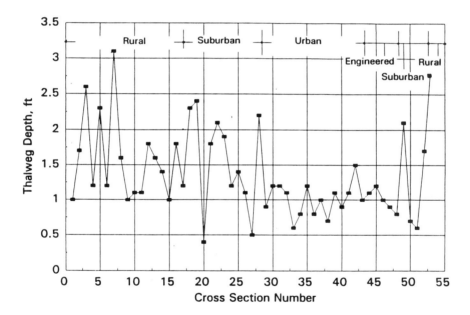

Fig. 7. The depth that the thalweg is below the average bed level. Values are averages for the period of record.

Bank Movement. Bankline movement is also a concern to those living along the river and to those who have facilities on or within the river banks. Most of the banklines of the South Platte River are erodible, being composed of gravels and sands previously deposited as bars and built up to floodplain level, or overburden from gravel mining operations. The bankline that does not erode is either away from the fast current or protected by riprap or concrete and asphalt rubble.

In the years of record, the 106 banks at the 53 monitoring cross sections have for the most part been stable. Of the nine eroding banks in 1996, six are in the Rural reach and three are in the Suburban reach. The most severe erosion of the banks is not at the monitored cross sections but in between, in the Rural reach.

Most of the Urban reach is completely protected by riprap or rubble. A lot of this aged protection is no longer visible, having been covered up by soil and vegetation. None are eroding at this time at the monitored cross sections, but there are several areas of bank erosion between sections.

The banks of the cross sections in the Engineered reach are riprapped and are undergoing the first stages of aging, with willows coming up at the base of the protection and grass higher up. They are stable.

Most banks have some type of vegetation growing on the them. Grasses do well except where the bank material is very dry - only weeds colonize these. Trees can become established in riprap to about mid-bank height. Older banks have trees up higher. Willows first appear at the base of the banks. Grass, willows, and weeds are found on the riprapped and rubble banks. The distinguishing factor for trees is their linearity. Trees grow in a row along the base of many banks.

THE FUTURE

New Master Plan. A revision of the Master Plan is appropriate. In almost all cases, it is impractical to have the major flood channel of the river in any place other than where it is now. There has been a large investment in infrastructure along the river as well as on the adjacent floodplain. These have to be accommodated as well as the needs of the river.

Any new planning should take into account the nature of the river, whether it is Urban, Suburban, Engineered, or Rural. In the Rural area, there is opportunity to preserve those features that are natural to some extent. The river is wide which allows more options. In contrast, the Urban river is narrow, offering very few options for a "natural" setting. This does not preclude changing the river in "urban" ways, as at Confluence Park. Improvement of fish habitat has been undertaken in the south Suburban reach by the Corps of Engineers. This work appears to have neither form nor function. The large loose stones in the river do not fit the image of the South Platte on the plains - they are out-of-place.

SUMMARY OF FINDINGS

Before 1860, the pristine South Platte River in the metropolitan area was a 300-foot wide unstable meandering stream with a riparian corridor of cottonwood and willows. Almost assuredly, it flowed year-round. The "half-mile wide and six-inches deep" Platte was at a different place, at a later time, or at a different place and later time. In 1860, the Great Invasion of people seeking wealth in placer gold brought an end to the pristine river. Change started immediately with the felling of the corridor of riparian forest for lumber and fuel. This was followed quickly by diversion of water for irrigation, since gold was not forthcoming in the hoped-for quantities. Even before 1870, when the first train rolled into Denver, there were near 200,000 acres (81,000 hectares) of land under cultivation and irrigation along the streams flowing eastward from the mountains. Thereafter, there were dams, more diversion weirs, narrowing, straightening, infilling, bank protection, sewer outfalls, grade-control structures, boat chutes and now structures to add oxygen. Years of development had been completely replaced by a new direction, that of management.

The pristine river can not be restored. The water was bought from nature and is now spent outside the river channel. Throughout much of the year the flow is effluent from wastewater treatment plants. Floods, a major geomorphic agent, are again smaller and farther between. In the Rural reaches, the river is now wider than the pristine was, but in most others, it is narrower. The old riparian lands are gone in the Urban area, converted to lots, blocks, and streets. The new South Platte River can accommodate the average annual flood now with ease. Even degradation has for the most part been arrested.

The Rural reach of the South Platte offers the most opportunity for preserving something of a natural setting. This reach is wide and accommodating, there are a number of features that are pleasing to nature lovers. Other features can be enhanced by suitable management. The Urban river offers a different challenge. There, the opportunity lies with creating a more favorable urban environment along and in the Platte. There is little room for more nature. The Engineered reach is growing from its sterile birth into a gentle juvenile. Age will still further improve its appearance. The Suburban reaches offer the most challenge. They can be made into the Urban type, but not Rural. Or they can be defined in their own manner. The choices are ours!

LIST OF REFERENCES

EBfit, F.J., (1865). "Map of Denver." No Publisher listed.

Fosdick, H.M. and Tappan, L.V., (1859). Plan of the Cities of Denver, Auraria, and Highland, Jefferson Territory." Publishers, Denver, 1 December.

Fremont, J.C., (1845). "Report of the exploring expedition to the Rocky Mountains [*in the year 1842, and to Oregon and North California in the years 1843-'44*]." Readix Microprint Corp, 1966, 327 pp.

James, E., (1823). "Account of an Expedition from Pittsburgh to the Rocky Mountains performed in the years 1819 and '20 under the command of Major Stephen H. Long." Compiled by James, vol. 1, 503 pp; vol 2, 442 pp plus appendices.

Kellogg, E.H. and Bonsall, J.H., (1871). "Map of Denver from Authentic Sources." Publishers, Denver.

Smiley, J.C., (1901). "History of Denver." Edited for the Denver Times, The Denver Times, The Times-Sun Publishing Co., Denver, 977 pp.

Stevens, M.A., (1983). "Stream Stability Investigation, South Platte River, Chatfield Dam to Baseline Road." Prepared for Urban Drainage and Flood Control District, Denver, Colorado.

Stevens, M.A., Urbonas, B., and Tucker, L. S., (1990). "Public-private cooperation protects river." APWA Reporter, September, 25-27.

Wright Water Engineers, (1985). "Major Drainageway Planning, South Platte River, Chatfield Dam to Baseline Road." Report prepared for Urban Drainage and Flood Control District, Denver, Colorado.

Yalin, M.S., (1992). "River Mechanics." Pergamon Press, New York, N.Y.

The Lower Truckee River, A System in Transition

J.J. Warwick[1], Member ASCE, A. McKay[2], J. Miller[3], P. Stacey[4],
M. Wright, Jr.[5], and C. Gourley[6]

Abstract

The lower Truckee River typifies riverine systems in the semi-arid west by draining inland to a terminal lake within the Great Basin. Lower Truckee River water quantity and quality are extremely important since the river is the sole freshwater source for Pyramid Lake which contains the endangered cui-ui sucker fish. The importance of cui-ui fish survival and overall recovery has been codified in Public Law 101-618, also know as the Negotiated Settlement Act. This Act attempts to accelerate the process of stream preservation and restoration. However, it would seem prudent to recognize the history of this system (where it has been, how it has been perturbed, and how it has responded to the imposed perturbations) prior to continuing in earnest with new imposed solutions. The paper will summarize the history of this important resource from both a physical and biological/chemical perspective. The importance of this resource will also be presented from the perspective of the Pyramid Lake Paiute Tribe. Past, present, and proposed stream restoration projects for the lower Truckee River will then be presented. Finally, a synthesis of observed system responses will be used to begin to evaluate the probable efficacy of restoration activities.

---

[1]Director, Grad. Prog. of Hydrologic Sciences, UNR.
[2]Research Scientist, Desert Research Institute.
[3]Research Professor, Desert Research Institute.
[4]Professor, Dept. of Env. and Resource Sci., UNR.
[5]Director, Pyramid Lake Water Resources.
[6]Truckee River Coordinator, The Nature Conservancy.

Introduction

Tributaries to the Truckee River head along the eastern flank of the Sierra Nevada, but the river itself "officially" begins at the outlet works to Lake Tahoe located near Tahoe City, California (COE, 1992). From the outlet, it flows 238 km in a north-northeasterly direction to its terminus at Pyramid Lake, the largest remnant of Pluvial Lake Lahontan (Fig. 1). Along its course it exhibits enormous contrasts in topography, geology, climate, and vegetation ranging from steep granitic environments of the Sierra Nevada, characterized by nearly 180 cm of annual precipitation, to low-relief valley floors of the Basin and Range Physiographic Province which often receive less than 25 cm of precipitation in any given year.

When John Fremont first visited the Paiute Tribe living along the river and around Pyramid Lake in 1844, he was impressed by the large Lahontan cutthroat trout (*Onchorhynchus clarki henshawi*), a native fish that populated both the river and lake. He also mentioned the cui-ui (*Chasmistes cujus*), another large native species, which is an omnivorous lake sucker endemic only to Pyramid Lake and the lower Truckee River. Both the cui-ui and the Lahontan cutthroat trout were economically essential to the existence of the Pyramid Lake Paiute Tribe. In fact, they called themselves the Kuyuidokado (cui-ui eaters).

The ensuing settlement of the Truckee Meadows upstream of Pyramid Lake led to extensive logging in the upper Truckee Watershed and large-scale water diversions for agricultural irrigation. These developments had severe consequences for both quality and quantity of water flowing into Pyramid Lake. The most significant adverse impacts were the result of the Reclamation Act of 1902 which authorized the construction of the Newlands Project in 1905, including Derby Dam which diverts Trucker River water to the lower Carson River Basin near Fallon, Nevada (Fig. 1). The Truckee River and Tributaries Project, initiated by the flood Control Act of 1954 and completed during the 1960s, involved widening and deepening the channel from Lake Tahoe to about 3,200 feet downstream, and through the Truckee Meadows, as well as removing the rock outcrop that controls the

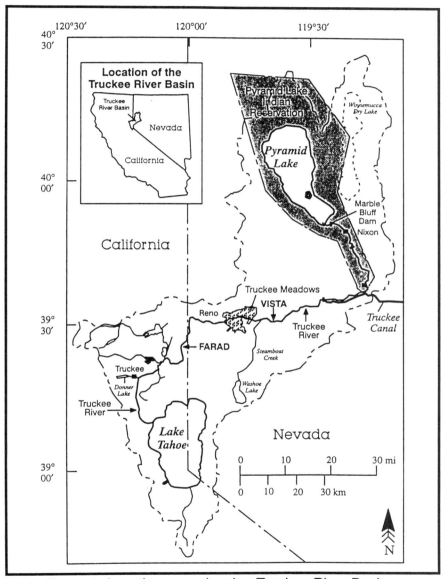

Figure 1: Location map showing Truckee River Basin.

river bed elevation at Vista. To compensate for the increased channel capacity, the U.S. Army Corps of Engineers (COE) implemented intermittent channel modifications (straightening and widening) from Vista to Pyramid Lake. After these modifications, the lower Truckee River channel was straighter, steeper, and more susceptible to erosive forces of the river.

Fluvial Geomorphology

<u>Physical Setting</u>

Much of the lower Truckee River is characterized by narrow alluviated valleys that have been cut in Tertiary-age volcanics. The primary exceptions include the broad valley floor of the Truckee Meadows in the vicinity of Reno and Sparks, and the laterally extensive plains located downstream of Derby Dam which is the diversion structure supplying water to the Truckee Canal (Fig. 1). Late Pleistocene glacial outwash deposits dominated by granitic lithologies form much of the valley fill from headwater areas downstream to at least Vista (Birkeland, 1967; 1968). Associated with the latest Pleistocene (Tahoe) outwash materials are local occurrences of granite boulders ranging from 12 m in maximum diameter at Farad to 1.2 m at Vista. Birkeland (1968) has argued that these boulders could only have been transported and deposited by one or more catastrophic flood flows (jökulhlaups) initiated by the near instantaneous release of waters from Lake Tahoe as glacial ice dams located in the Truckee River canyon below the outlet collapsed.

Downstream of Vista, the valley fill is primarily composed of colluvial and alluvial deposits dominated by volcanic lithologies as well as lacustrine sediments associated with Pluvial Lake Lahontan. The lacustrine deposits are most pronounced between Wadsworth and Pyramid Lake where floodplain and terraces of the lower Truckee River are inset into and rest upon a Pleistocene sequence of lacustrine sediment.

Anthropogenic Alterations

Since the 1850's, both the channel and watershed of the Truckee River have been extensively modified by human activities. In fact, it could be reasonably argued that it is one of the most altered watersheds in the western United States. Unfortunately, the timing and details of these modifications were rarely recorded by the U.S. Army Corps of Engineers (COE), the U.S. Bureau of Reclamation (USBR), or other responsible bodies. As a result, it is impossible to develop an exhaustive listing of anthropogenic modifications to the basin. Nevertheless, the most significant alterations to the channel and watershed are presented in Table 1 with the intent of illustrating the magnitude of human modifications to the Truckee River basin during the past 150 years.

Inspection of Table 1 suggests that watershed and channel modifications were initiated for two primary purposes: (1) to provide an amply supply of water for the region's urban and agricultural populations, and (2) to reduce the potential for flooding, particularly within the Reno-Sparks metropolitan area. The dominant methods invoked to accomplish these objectives include the construction of reservoirs, dams and diversion canals, the implementation of channelization programs which relied heavily on channel straightening, and the removal of local bedrock obstructions to flow (Table 1). In addition, many of these programs were accompanied by extensive bank stabilization efforts, dominated by the use of rip-rap.

Geomorphic Responses

One of the earliest and most significant modifications to the Truckee River was the completion of Derby Dam in 1905 as part of the Newlands project. Its intent was to divert Truckee River water, via the Truckee Canal, into the Carson River basin to irrigate large areas of the Fallon-Carson Desert. During approximately the past 80 years, about one half of the average annual flow downstream of Vista Reefs has been diverted from the Truckee River (COE, 1992). The net result has been a decline in water levels in Pyramid Lake of about 24 m between approximately 1910 and 1981. Although above normal runoff during the 1980s (particularly

Table 1. Summary of anthropomorphic changes to the Truckee River and Watershed

Date	Construction Location	Reference
1850s	Initial irrigation structures in Truckee Meadows	Townley, 1980
1863	Intense logging initiated Truckee, CA	Townley, 1980
1870	First dam at Lake Tahoe	Townley, 1980
1873	Marlette Lake Dam/Reservoir	CDWR, 1991
1875	Dam at Verdi, NV	Townley, 1980; CDWR, 1991
1876	Echo Lake Dam/Reservoir	CDWR, 1991
1877	First control structure at Donner Lake	CDWR, 1991
1883	Irrigation structures at reservation	Townley, 1980
1880	Diversion dam upstream of Verdi, NV	USDA, 1977
1899	Water power plant and diversion 2 miles west of Reno Town center	Townley, 1980
1900	Dam at Floriston, CA	USDA, 1977
1905	**Derby Dam**, NV	CDWR, 1991
1911	Verdi hydropower plant	CDWR, 1991
1913	Outflow structure at Lake Tahoe	CDWR, 1991
1937	Boca Dam/Reservoir	CDWR, 1991
1937	Revetment along Riverside Ave, Reno	Cooper, 1938
1938	River bed cleared and channelized through Reno	COE, 1946

Table 1. Continued.

Date	Construction Location	Reference
1930	Outflow structure at Donner Lake	CDWR, 1991
1938	NDOT partially removes rock reefs (Vista Reefs) downstream of Sparks, NV	CDWR, 1991
1939	Independence Lake Dan/Reservoir	CDWR, 1991
1945	Diversion dam near current location of Marble Bluff Dam	USBR, 1973
1954-1963	River channelized from Nixon bridge to present location of Marble Bluff Dam	Gregory, 1982
1954-1970	River straigthened from Fellnagle Dam to Dead Ox Wash	Gregory, 1982
1959	River straightened, dredged, and rip-rapped from E. 2nd St. to Vista	Gregory, 1982
1962	Prosser Creek Dam/Reservoir	CDWR, 1991
1963	Vista Reefs lowered	COE, 1992
1970	Stampede Dam/Reservoir	CDWR, 1991
1970s	S Bar S new diversion dam	Gregory, 1982
1971	Martis Creek Dam/Reservoir	CDWR, 1991
1976	Marble Bluff Dam/Fishway	CDWR, 1991
1988	Stampede Dam/hydropower plant	CDWR, 1991

1983 and 1986) and changes in water management programs has allowed lake levels to recover slightly during the past decade, Pyramid Lake levels are still more than 12 m below pre-diversion lows.

Some of the most pronounced effects of declining lake levels have been channel incision and the redeposition of the eroded sediments in a prograding delta, as the local base level of the Truckee River was rapidly lowered. Also associated with this episode of downcutting has been the conversion of the relatively narrow and deep, single channel system to a braided river flowing through a broad valley. Other effects have included the removal of riparian vegetation, consisting in large part of cottonwood trees, from the channel margins and the its replacement by a sequence of small, discontinuous, erosional terraces that are typically capped by a couple of meters of alluvium. The highest terraces, located immediately upstream of Pyramid lake, occur more than 11 m above the modern channel, indicating significant downcutting has occurred along the lower most reaches of the Truckee River during the past 80 years. Born and Ritter (1970) indicate that the development of these terraces is a product of varying flow conditions with incision occurring during years of "normal" flow and lateral bank erosion and tread formation occurring during unusually high flow events.

Data suggest that the impacts of base level lowering have been extensive, affecting approximately the lower 15 km of the channel bed. However, it is quite possible, if not probable, that channel incision and terrace development is related to factors other than base level lowering, including channelization, as will be described below. This raises the question as to the upstream distance to which the effects of base level lowering has impacted channel form and process. It is, perhaps, a question that cannot be definitively answered. Nevertheless, it is clear that the impacts of a 24 m drop in base level have not been as significant as might have first been guessed. This can be attributed, in part, to three primary factors. First, the Bureau of Indian Affairs (BIA) constructed a small earthen dam in 1941-42 approximately 460 m downstream of the modern Marble Bluff Dam site (WET, 1991). The structure diverted flows toward the east, away from the historic channel, but was eventually breached during a flood in 1950 (WET, 1991). In 1976, Marble Bluff Dam and fishway (Fig. 1) was constructed by the U.S. Bureau of Reclamation in an attempt to intercept channel incision resulting from base level lowering and to aid in the re-establishment of spawning access to the river by lake dwelling Cui-ui and Lahontan Cutthroat Trout.

These dam structures are likely to have partially limited the extent of channel incision along the lower most reaches of the Truckee. It should be recognized, however, that neither of these structures were completely successful in arresting channel bed incision, in part, because entrenchment had already passed upstream of the dam sites before their construction (as revealed by 1938 aerial photographs). Moreover, between the failure of the BIA structure and the construction of Marble Bluff Dam, incision is likely to have impacted reaches upstream of the dam sites. Continued downcutting during this period is supported by the maps constructed by Born and Ritter (1970) which show that two new terrace levels had formed along the lower reaches of the Truckee River between 1961 and 1965. It should be noted that channel incision upstream of the dam sites continues today, even 20 years after the construction of Marble Bluff Dam which was intended to arrest channel bed degradation.

Perhaps a more significant impact of base level lowering on channel incision is associated with the development of the prograding delta into Pyramid Lake. WET (1991) noted that delta progradation associated with the deposition of sediment eroded from upstream reaches during channel incision resulted in a total channel lengthening of approximately 4 km since the turn of the century.   This channel lengthening may have compensated for as much as one-half of the total base level lowering, thereby significantly reducing the magnitude of channel incision.   In addition, WET (1991) presents cogent arguments suggesting that channel incision was limited by fine-grained, erosionally resistant deposits that outcrop along the channel bed.

As eluded to earlier, not all of the recent changes in channel form and process can be attributed to base level lowering associated with declines in Pyramid Lake water levels.   Under authorization of the Flood Control Act of 1954, the U.S. Army Corps of Engineers implemented a number of "flood control" measures along the Truckee River including clearing and snagging, channel straightening, and enlargement (COE, 1992). The impacts of these procedures are, perhaps, most apparent between Wadsworth and Dead Ox Wash (Fig. 1) where selected reaches of the channel were straightened, resulting in an increase in channel gradient and bedload transport, and a reduction in sinuosity (Gregory, 1982). In addition, cottonwood trees were harvested for a 1.5-2 year period. Both the channel modifications and the timber harvests were completed before the flood of 1963, a flood that remains the largest on record for this reach of the river (COE, 1992). According to the COE (1992):

> "[t]he channel had not had time to adjust to these modifications before the 1963 flood...The damage to the freshly modified channel and riparian corridor was by all accounts catastrophic. There is no question that channel modifications had de-stabilized the river prior to the 1963 flood."

The lateral instability resulting from these modifications is clearly delineated on 1965 aerial photographs, riparian vegetation having been significantly removed along the channel margins and extensive point bars, suggestive of rapid channel migration, having formed. Channel modifications were also performed downstream of Numana Dam and, again, probably contributed to most of the changes in channel position observed in this area.

While channel reaches upstream of Wadsworth appear to be relatively stable, a few local areas of channel instability have been identified, the most significant of which is the reach extending roughly from the East McCarran bridge in Sparks downstream to the bedrock outcrops referred to as Vista Reefs (Fig. 1). This reach is characterized by steep vertical banks, toppled trees, and low-relief knickpoints, suggestive of channel incision. In addition, repeat cross-channel and longitudinal surveys compiled by WET (1990) shows a progressive decrease in channel bed elevations from 1959 to 1989.

The most likely cause of entrenchment is the lowering of local base level caused by the removal of bedrock, referred to as Vista Reefs, which was exposed in the channel floor immediately downstream of the Truckee Meadows (Fig. 1). The bedrock obstruction to flow was removed by the U.S. Army Corps of Engineers in 1963 to reduce flooding potential in the Reno-Sparks metropolitan area. Although it is easy to attribute the observed channel incision along this reach to base level lowering, channel modifications were also performed in this area and may have been partly responsible for the observed phase of channel degradation. Although the exact cause(s) of downcutting cannot be precisely determined, channel entrenchment is likely to continue.

In light of the above, it is clear that the geomorphic responses to channel and watershed modification have been locally significant. Nevertheless, it is surprising that the effects have not been more laterally extensive, as the most sever geomorphic adjustments are confined to either the modified reach or the stream section immediately up- and/or downstream of the reach. This can, perhaps, be best illustrated by examining the changes in channel position during the past several decades.

Miller et al. (1994) have quantitatively documented rates of lateral channel migration by comparing the position of the Truckee River along 38 reaches between Verdi and Pyramid Lake on aerial photographs. The data presented here were primarily obtained from 1965 and 1991 photos, although upstream of Reno, 1965 photographs were absent and, therefore, the analysis focused on 1977 photographs (see Miller et al., 1994, for a more detailed description of the methods used to determine channel change). These data show that the reaches characterized by relatively high rates of channel migration are located downstream of Wadsworth and immediately upstream of Vista Reefs, both of which represent areas directly impacted by base level lowering, channelization, and flow diversions (Fig. 2). Reaches located between the affected areas exhibit relatively slow rates of lateral channel change.

Undoubtedly, the changes in channel migration have been limited to some degree by bank stabilization efforts, particular the pervasive use of rip-rap. Nonetheless, relatively minor changes in channel position generally coincide with reaches characterized by narrow valley floors composed of coarse-grained colluvial, alluvial fan, and/or glacial outwash deposits. For example, the reach located between Vista Reefs and Wadsworth flows through a narrow valley cut in Tertiary volcanic rocks, and the river locally

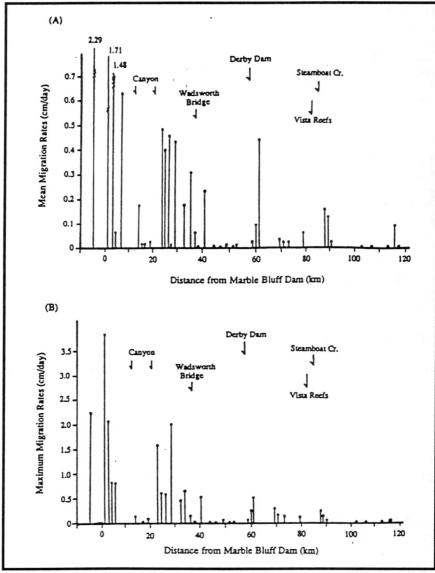

Figure 2: (A) mean rates of channel migration, and (B) maximum rates of channel migration. (Miller, et al., 1994)

abuts colluvial and alluvial fan materials derived from the surrounding uplands. In fact, at some sites, the influx of debris from side-valley alluvial fans has forced the river away from the fan complex and toward the far side of the valley floor. Bedrock also outcrops locally along the channel bed and banks. It seems probable that the narrow valley widths, and the coarse-grained valley fill deposits have restricted lateral changes in channel position (Fig. 2). The one notable exception along this reach is located immediately upstream of Derby Dam (Fig. 2); a reach both disturbed by human activities and characterized by a relatively wide valley floor.

Channel bed degradation may also have been limited, at least locally, by the influx of colluvial, alluvial fan, and outwash debris to the channel which cannot be transported by the current hydrologic regime. Miller et al. (1994) utilized the Shields' criterion to estimate the size of the clasts that could be transported by bankfull conditions along the river from Verdi to Pyramid Lake. These clast sizes were subsequently compared to: (1) the median diameter of the clasts in the channel bed (D50), and (2) the ten largest clasts in the channel bed (D90). Similarly, the tractive force required to transport D50 and D90 fractions of the bed material load were calculated on the basis of Komar's (1987) equation and compared with estimates of tractive force exerted on the channel bed under bankfull conditions. Both of these calculations suggest that, at least locally, the largest particles of the bed material load can not be transported by the modern hydrologic regime.

Biological

In semi-arid regions like the Great Basin, river systems and their associated riparian zones are centers of biological diversity and productivity. In addition to the aquatic zone itself, which is an extremely limited habitat type, the consistent source of underground moisture results in terrestrial plant communities with extraordinarily high primary productivity, consistent microclimates, and structural complexity. Animals in turn depend upon riparian areas not only for food and water, but also for places to raise their young, for cover to avoid predators, and for areas to spend the night or avoid the heat of day. Although riparian habitats typically comprise less than 2% of the total land area in the Western United States (Pase and Layser 1977) they are estimated to support up to 80% of the wildlife diversity in the region (Ohmart 1995). And because water is a limited and much sought after natural resource in the west, riparian zones have also been heavily impacted by human activities. It has been estimated that between 80-95% of the original riparian habitats in this region have been lost (Noss et al. 1995).

The biological diversity within a watershed is closely tied to the maintenance of normal hydrologic and geomorphologic processes within that system (e.g., Tellman et al. 1993). Many of the changes that have impacted the plant and animal diversity in western riparian zones can be illustrated by the history of the lower Truckee River. Much of the current restoration effort on the lower Truckee is driven by, and is directed at, the aquatic system, and in particular at endangered fish, the cui-ui (*Chasmistes cujus*). Although the cui-ui spends most of its life in Pyramid Lake, like many Great Basin lake fish, it moves into feeder rivers in the spring to spawn. Both the lower water level in Pyramid Lake and increased silt deposition from upstream erosion (which in turn was the result of channelization and the loss of riparian vegetation) created a delta at the south end of the lake that blocked both the Cui-ui and another important fish, the Lahontan cutthroat trout (*Oncorhynchus clarkii*) from access to their spawning grounds in the Truckee. Because reproduction was no longer possible, the cutthroat trout in the lake went extinct, and by 1967, only three year classes of cui-ui remained (1942, 1946 and 1950; Scoppettone et al. 1986). Only recently has successful spawning by cui-ui begun again (see below).

The loss of spawning runs by both cui-ui and cutthroat trout was not only a major natural and cultural problem for Pyramid Lake and the Paiute Indians, but it also removed a major component of the fish biomass from the Truckee River itself, at least during certain times of the year. This not only deprived many river residents of a rich source of food (from top animal predators like bald eagles, *Haliaeetus leucocephalus*, river otters, *Lutra canadensis*, and, historically, grizzly bears, *Ursus arctos*, to scavengers including many aquatic insects), but it also disrupted an important means of nutrient cycling between the river and lake.

There have been numerous other changes in the aquatic system of the Truckee, although these have been less well studied. The introduction of exotic fishes for sport fishing has probably resulted in the extinction of many native fish species that were once found in the river. Channelization has not only damaged spawning habitats for the Cui-ui and trout, but it also reduced the amount of stream-bed habitat for other aquatic organisms, thus reducing both productivity and diversity. The diversion of water for agricultural and urban uses has led to periods on nearly zero stream flows in certain reaches of the river during drought years; this in turn has probably caused a shift in the aquatic plant and invertebrate animal communities from organisms that depend upon constant submersion to ones that can tolerate drying out, although the extent of these changes is not well known.

Although much of the concern about the lower Truckee in the past has been based upon problems with the Cui-ui and Lahontan cutthroat trout, more recently attention has also been focused upon the entire riparian zone itself. This stems both from the recognition that the aquatic and terrestrial ecosystems are closely interrelated (for example, healthy stream side vegetation can provide shade that will lower in-stream water temperatures more effectively than can increased releases of cold water from upstream storage reservoirs), as well as the fact that the terrestrial communities are themselves important centers of biological diversity that are in need of extensive restoration.

Changes in land use practices and the extensive alteration of the hydrologic and geomorphologic processes in the lower Truckee watershed has had a major impact upon the biological diversity of the terrestrial riparian zone. Much of the flood plain has been cleared of vegetation and developed for agricultural and urban use, and extensive livestock grazing has altered much of the remaining areas (e.g., Kauffmann and Krueger 1984). The river channel has been straightened for flood control, a practice that largely eliminated the biologically important "oxbows" (bogs leftover from old meanders), which comprised a large proportion of the original wetland habitats. Water flow regulation, channelization and down cutting also have isolated stream banks and the flood plain from the periodic flooding that is necessary for maintenance and recruitment of many riparian plant species

(Debano and Schmidt 1990). With the reproductive failure of both cottonwoods (primarily *Poplus fremonti*) and willows (*Salix* spp.), much of the once continuous riparian forest habitat has disappeared, and the remaining patches are both small and highly fragmented (Stacey 1995). Increased erosion and down cutting has lowered the water table in much of the original flood plain, leading to the invasion of many upland plant species like sage brush, that have long taproots and are drought tolerant.

Collectively, all of these changes have led to the loss of many habitat types (particularly wetlands and sandbars or mudflats) as well as specific resources within the remaining habitats. In many river systems, the impact of these changes on biological diversity must remain a matter of speculation, although the direction of the change, toward lower diversity, is not in doubt. However, with the lower Truckee River, we have a unique opportunity to directly track the change in the biodiversity of a major group, the bird community, both before, during and after restoration of the river system. Birds are often used as indicators of the health of an ecosystem, because they are particularly sensitive to both the structure and resource base of the habitats in which they occur (Pianka 1967, Dobkin and Wilcox 1986, Finch 1991). Stacey (1995) reviewed the changes that have occurred on the number and kinds of birds found along the river over a hundred-year period, using data collected by R. Ridgway (1877) during a survey of the west for the U.S. Biological Survey in 1868, and by D. Klebenow and R. Oakleaf between 1972 and 1976 along the same part of the Truckee River (Klebenow and Oakleaf 1984). Of the 91 original species observed by Ridgway in 1868, 42 species were not recorded at all by Klebenow and Oakleaf during 1972-76, and 11 other species that were either common or abundant in 1868 had become rare by 1976 (only 1-3 individuals recorded over five summers). Although the survey methods used in the two studies were not identical, the magnitude of differences between the two reports represents an extraordinary change in the biological diversity along the river, with a 58% loss of the species that were originally present. And although 17 new species were recorded by Klebenow and Oakleaf in 1972-76, most of these were either introduced species (e.g., the English House Sparrow,

Passer domesticus) or birds that depend upon human activities (e.g., the Ring-billed Gull, *Larus delawerensis*).

Equally important, entire groups or guilds of birds had disappeared between 1868 and 1976, and these losses appear to reflect specific alterations of the riparian habitat (Stacey 1995). For example, in 1976 most species of shore birds recorded earlier were no longer present, because channelization of the river course has prevented the formation of the sandbars upon which many of these birds normally nest. Almost all of the marsh birds had disappeared, as a result of the loss of the oxbow wetland habitats. Many species that feed upon river fish, like the Osprey (*Pandion haliaetus*), were absent, probably because of the decline in native fish populations (including the Cui-ui). Similarly, most of the birds that depend upon a well developed willow shrub layer for either nesting or cover from predators had also either declined or had become locally extinct, including species like the Willow Flycatcher (*Empidonax traillii*) and many ground-nesting ducks (e.g., Gadwalls, *Anas strepera*, and American Widgeons, *A. americana*). These data not only document the widespread loss of biological diversity along the Truckee River between 1868 and 1976, but they also provide a good illustration of how the presence or absence of individual bird species can be used to understand how the riparian habitat itself has changed over time. While some of these species are now slowly returning to the Truckee as the riparian habitat is being restored, it is unknown whether the full range of biological diversity will ever return.

Water Quantity/Quality

Water Quantity

Figure 3 is a 10-year hydrograph for river discharge at the Vista Gage, 4 miles southeast (downstream) of Sparks, Nevada. Flow values are mean monthly discharges (ft^3/sec) compiled from U.S. Geological Survey Water Resource Annual Data Reports. The most striking aspect of Figure 3 is the 7-year period from 1985-92 where mean flows are well below the long-term (69 year) average of 819 ft^3/sec. While short-term records often reflect the impacts of upstream reservoir storage and control structures, river

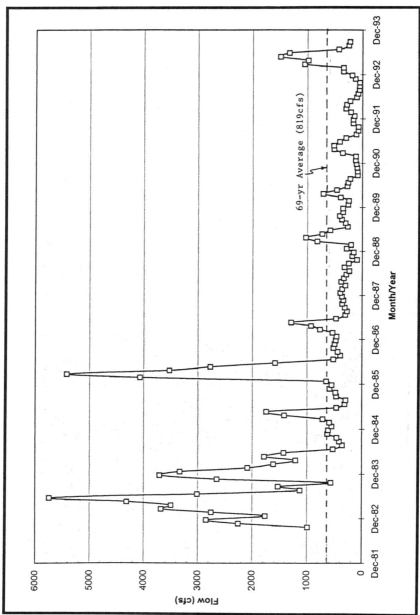

Figure 3. Mean monthly flows at Vista, Nevada.

discharge from 1985-92 is indicative of climatic conditions in the Truckee River Basin during the period. Precipitation was well below average for this period, and the snowpack in the higher elevations of the drainage basin reflected these "drought" conditions.

Water Quality

Figures 4 through 6 display the 10-year record for selected water quality parameters (total dissolved solids, nitrate, and phosphate) at two different locations on the Truckee River. The Farad (California) site is approximately 13 miles upstream from the cities of Reno and Sparks, and is expected to reflect water quality conditions prior to the impacts of agriculture, municipal and industrial (M&I) activities in the Truckee Meadows. By contrast, the Vista site downstream should reflect the full range of agricultural and M&I activities in the mid-river basin. Both the Farad and Vista sites are sample locations in the Desert Research Institute's (DRI) long-term Truckee River Monitoring Program which DRI conducts on behalf of the State of Nevada. The off-scale value of nitrate in Figure 5 is 7.5 mg-N/L, which occurred during a period of time when the Reno-Sparks Water Reclamation Facility was experiencing significant operational difficulties.

Prior to discussing the potential impacts of anthropogenic activities on river water quality, it is worthwhile to note the relationship between river discharge (Fig. 3) and all three quality parameters. For the period between 1987-93 in particular, there is an obvious increase in all three constituents which coincides with the below-average river flows for that same period. As expected, the relationship is most obvious when examining the plot of total dissolved solids (TDS, Fig. 4). Even at the upstream Farad site, the relationship between TDS and flow is quite well established.

More noteworthy, however, is the dramatic increase for each constituent between the upstream and downstream sites, independent of flow regime. The three figures are graphical proof that rivers are indeed, integrators of all activities which occur in a watershed. Phosphate and nitrate concentrations at the Vista site reflect the impacts of activities which include municipal wastewater discharge and drainage from agricultural lands in the Truckee

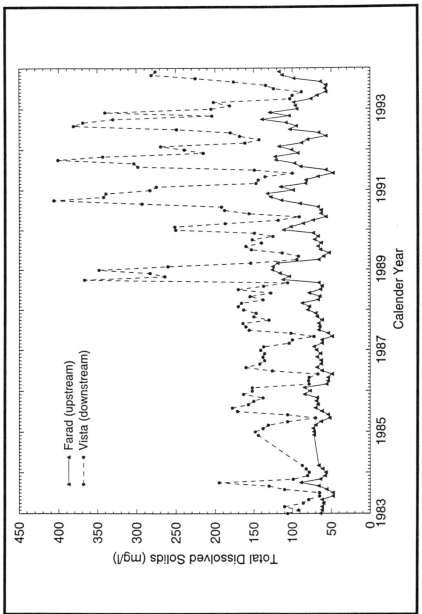

Figure 4: Total Dissolved Solids concentrations at Farad, California and Vista, Nevada.

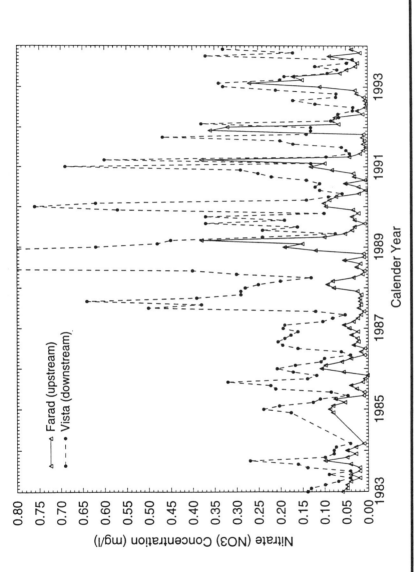

Figure 5: Nitrate Nitrogen concentrations at Farad, California and Vista, Nevada.

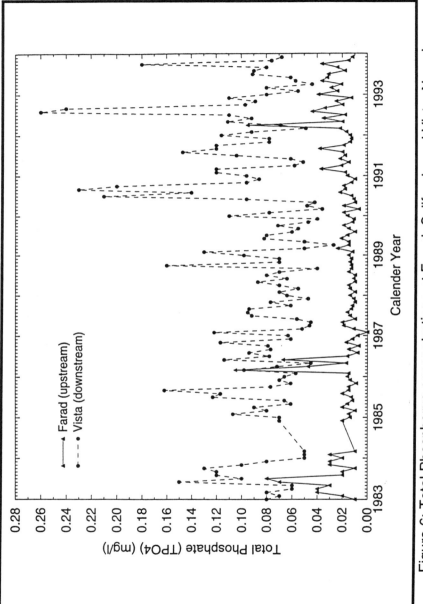

Figure 6: Total Phosphorus concentrations at Farad, California and Vista, Nevada.

Meadows. The plot of TDS (Fig. 4) shows an even greater increase (as a percentage of overall magnitude) between the two sites, particularly for the more recent period of record. This could be due to a variety of causes, including the impacts of increased urbanization and subsequent "storm" runoff, which would not be reflected as clearly in the nutrient data.

Native American Perspective

From the beginning of time, Indian People who lived in Northern Nevada cared for the natural resources with respect and honor. Over time the People may have changed and through that change an adaptation to the desert climate required teachings to utilize life sustaining resources for survival. The teaching of survival techniques derive from a higher power that we, as a People, accepted through belief with our creation. Our creation is based on natural law of our creator. Those of us here today represent those who lived here many years ago. We inherited a duty to be responsible for the resource that continues to belong to the land we call our home. We did not "own" rights for land or water use, but rather considered ourselves care-takers of everything encompassing our Mother Earth.

Today, there is a struggle between what we understand of the natural law and what we understand of man-made laws, the written laws of the United States of America. Public Law 101-618 is aimed to correct past wrongs of Congressional actions taken since the turn of the century. To understand the government to government relationship between the United States and Indian Tribes, one must understand the historical account of Western Civilization. The laws as they were developed, and as we feel the effects, have one focus. That focus is to develop and preserve, at any expense, Western Civilization based interests.

The taking of water resources from Pyramid Lake caused much hydrologic damage to the ecosystem. Water diversions to reclamation projects to hydropower generation and municipal water supplies have been developed at the expense of "trust assets" once owned by no one individual. The Truckee River has all the complexity of legal and technical issues that require much involvement to resolve historical problems. Water politics of the

West is another burden that is unavoidable. There is no solution to politics, even with the best technical information available.

Water quality, riparian management, recovering threatened and endangered species and river operation management are issues involving the Tribe's participation. In having full comprehension of the goals required to achieve objectives restoring the aquatic habitat, our energy and technical resources, along with cultural awareness are combined in hopes of improving the conditions of the lower Truckee River. By utilizing forecasts, anticipated snow melt, municipal and industrial demands, agricultural diversions and fishery demands we can anticipate the functions of a river basin both scientifically and politically. Our intent is to restore water back to Pyramid Lake through provisions of Public Law 101-618, primarily through water right acquisitions and river restoration efforts

The Pyramid Lake Paiute Tribe knows that the population and water demands in this region will only increase. The challenge is how we can best manage a limited resource without further damage and destruction to an already fragile environment. Becoming educated with modern society's curriculum, we have the capability to stand up to the forces that intend to further exploit the resources.

Restoration Activities

Because the Truckee River's degraded condition is related to changes in hydrology, channel morphology, and land management, river restoration efforts have specifically addressed these issues. The Lower Truckee River Restoration Steering Committee (LTRRSC) (Pyramid Lake Paiute Tribe, USFWS, COE, USBR, Nevada Department of Environmental Protection, U.S. Natural Resource Conservation Service, U.S. Environmental Protection Agency, and The Nature Conservancy) has been assigned to direct restoration efforts on the lower Truckee River. The Committee has initiated studies to determine how changes in the hydrologic regime influences river channel geometry and water quality. Also, the relationship between river hydrology and the recruitment and persistence of riparian plants is currently being examined at University of Nevada, Reno. Upon completion of these studies in 1997, the LTRRSC will make specific in-stream flow recommendations to restore riparian vegetation and to maintain fluvial processes. These recommendations will be based on the recognition that a variety of flow regimes, including droughts and floods, are required to sustain the river system. Key elements in the flow recommendations will be magnitude, timing, frequency, and duration of flow, and rate of flow change.

Riparian Vegetation

Stewart Rood of the University of Lethbridge and The Nature Conservancy of Nevada have adapted a common hydrologic model for successful cottonwood seedling establishment to the specific conditions of the Truckee River (Rood and Gourley, 1996). To develop the model, annual tree growth rings were counted in order to age the new stands of cottonwoods that exist in narrow bands along the banks of the lower river. According to preliminary data, the age of these trees correlated with the 1987 spring season when flows were released from Stampede Reservoir to facilitate cui-ui spawning.

The artificial flows implemented to restore fish populations also created the conditions most suitable for cottonwood seed germination and sapling survival. These findings are particularly notable, because they suggest that once natural flow regimes are reestablished, the entire system that may have co-evolved with these regimes responds positively. A review of several annual hydrographs revealed that the 1987 flow pattern was different in that the rate of flow decline of spring and summer flows, on average, did not exceed 1 inch per day, which appears to be the maximum root growth rate potential for cottonwoods (McBride, et al. 1988; Segelquist, et al., 1993). Earlier records suggested that, as a result of water diversion and storage, the rates of flow decline usually exceed those necessary for cottonwood recruitment. Consequently, significant cottonwood recruitment along the lower river was not possible for several decades prior to reservoir releases for cui-ui spawning (Lang, et al. 1990).

Record precipitation in 1994-1995 provided an opportunity to test the cottonwood recruitment model. The LTRRSC presented flow prescriptions to the USFWS, USBR, and the Office of the

Federal Water Master as these entities oversee Truckee River operations. In coordination with The Nature Conservancy and the LTRRSC, the Federal Water Master closely regulated declining flows in the Truckee River during the 1995 spring runoff. River operators supplemented unregulated spring and summer flows with flood water that was detained in upper basin reservoirs. The rate of river stage decline was, on average, less than 1 inch per day.

Initial germination and seedling recruitment were according to the hydrologic recruitment model, but during the late summer and fall other factors influenced seedling growth and survival. Peak flows and initiation of declining flows were closely coordinated with peak cottonwood seed production. As flows receded from the floodplain and exposed wet soils, dense stands of cottonwood seedlings were established in continuous bands along the banks where conditions were suitable for recruitment. However, the summer of 1995 had a record dry period (129 days with no measurable precipitation) with above average temperatures. These extreme conditions were unfortunately lethal for many seedlings. Although the new recruits experienced a high mortality, as determined by a field survey in October 1995, it was also established that a considerable number of seedlings survived in areas where moisture conditions were more favorable, i.e. areas that were lower, partially shaded, or consisted of fine grain soil. While the surviving new trees will not form a continuous canopy along the lower river, they should add a new age class of trees to the riparian zone.

Re-establishing a broad, dense zone of riparian vegetation will be the key criterion for restoration of the Truckee River. Most of the significant changes in both the river and adjacent terrestrial environments are related to the decline of riparian forest. To reverse these trends, restoration efforts must center around re-establishing and preserving riparian vegetation. Successful restoration will not only require changes in the annual flow pattern, but also changes in land management, such as reducing livestock grazing, preventing wood cutting, and decreasing conversions of riparian forest into agricultural land. Because the condition of the riparian vegetation appears to relate to overall bird species diversity and the presence of indicator species, results of bird surveys currently conducted on the river will be used for specific guidelines for habitat restoration and to identify essential sites for protection or enhancement.

Other Restoration Projects

Because changes in land management and hydrology alone cannot resolve all of the Truckee's problems, restoration will require active manipulation, such as constructing fish bypasses and reshaping the river channel. For example, while the population of cui-ui is recovering, their spawning grounds are still limited by restricted river access. Because of the cui-ui's limited swimming ability, traditional fish ladders that bypass Marble Bluff and Numana Dams have failed to allow unrestricted passage. Currently the LTRRSC has initiated the testing of a new fish ladder design and a constructed meandering fish channel (Fig. 7). If these systems are effective then these proven designs may be constructed to replace the existing facilities.

Other large scale construction projects include the building of several wetlands to replace habitat that once existed along the Truckee. In addition to providing wildlife habitat, the constructed wetlands create areas for flood attenuation and treatment systems for irrigation and drain water. Finally, in areas where existing banks are vertical and property ownership limits allowable river migration, the LTRRSC is proposing to create the physical conditions required to maintain a riparian buffer by sloping the river banks and planting riparian vegetation.

Synthesis & Conclusions

The water quantity and quality of the lower Truckee River has been perturbed by mans' activities for the better part of the past century. Upstream storage reservoirs and downstream diversions limit flows within the lower Truckee River and finally to Pyramid Lake. Channel modifications, executed with the goal of reducing the severity of flood damage to urbanized areas, have resulted in morphological changes which reduce the extend and vitality of both aquatic and terrestrial habitats. Water quality degradation occurs from municipal treatment plant discharges and agricultural non-point source pollution, while the impacts (both to

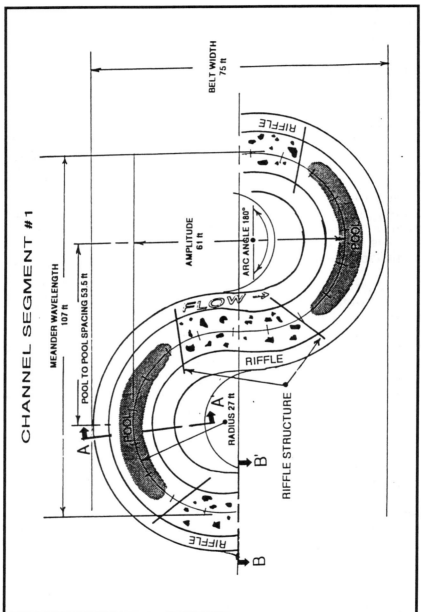

Figure 7: Nature Conservancy meandering fish channel design.

in-stream flows and associated water quality) of stormwater runoff are mostly undefined.

Local efforts are attempting to integrate the ideas from a variety of groups and currently focus upon the recovery of the cui-ui population within Pyramid Lake. Beyond this rather singular focus, we must begin asking and answering some fundamental and yet rather difficult questions. What do we really mean by the word "restoration"? What are we trying to restore the system state to, conditions prior to 1800's? Is such a restoration even possible or financially feasible? How can we begin to define the most cost effective options to affect at least partial restoration? How will we know that our efforts have been successful?

The questions given above are challenging enough from a purely technical perspective. However, our challenge goes beyond merely the technical, for restoration has a spiritual component as well. While some may look at the lower Truckee River based upon western water law others see the system based upon natural law. For restoration activities to be truly successful we must understand and appreciate all perspectives, with sound guidance from both the technical and spiritual domains.

References

Born, S.M. and Ritter, D.F., 1970. Modern terrace development near Pyramid Lake, Nevada, and its geologic implications. *Geological Society of America Bulletin*, 81: 1233-1242.

CDWR (California Department of Water Resources), 1991. Truckee River Atlas.

COE (U.S. Army Corps of Engineers), 1946. Interim survey report for flood control of Truckee River and tributaries in California and Nevada, Draft.

COE (U.S. Army Corps of Engineers), 1992. Lower Truckee River, Nevada. Reconnaissance Report. Sacramento District.

Cooper, J.A., 1938. Men at work. Fed-City of Reno Relief Program with Photographs.

Debano, L.F., and Schmidt, L.J., 1990. Potential for enhancing riparian habitats in the southwestern United States with watershed practices. *Forest Ecology and Management*, 33/34: 385-403.

Dobkin, D.S., and Wilcox, B.A., 1986. Analysis of natural forest fragments: Riparian birds in the Toyiabe Mountains, Nevada. *In:* J. Verner, M.L. Morrison, and C.J. Ralph (eds.), Wildlife 2000: Modeling habitat relationships of terrestrial vertebrates. Univ. of Wisconsin Press, Madison, Wisconsin. pp. 293-299.

Finch, D.M., 1991. Population ecology, habitat requirements and the conservation of neotropical migratory birds. GTR PM-205, USDA Forest Service, Fort Collins, Colorado.

Gregory, D., 1982. Geomorphic study of the lower Truckee River, Washoe County, Nevada. Unpublished M.S. thesis, Colorado State University, Fort Collins.

Kauffmann, J.B. and Krueger, W.C., 1984. Livestock impacts on riparian ecosystems and streamside management implications...a review. *J. Range Manage.* 37:430-438.

Klebenow, D.A., and Oakleaf, R.J., 1984. Historical avifaunal changes in the riparian zone of the Truckee River, Nevada. *In:* R.E. Warner and K.M. Hendrix (eds.) California Riparian Systems: Ecology, Conservation, and Productive Management. Univ. of California Press, Berkeley, Calif.

Komar, P.D., 1987. Selective gravel entrainment and the empirical evaluation of flow competence. *Sedimentology*, 34: 1165-1176.

Lang, J., Chainey, S., O'Leary, B., Shaul, W., and Rucker, A., 1990. Channel stabilization and restoration plan for the lower Truckee River, Nevada. Jones and Stokes Assoc., Inc., Sacramento, CA.

McBride, J.R.,Sugihara, N., and Morberg, E., 1988. Growth and survival of three riparian woodland species in relation to simulated water table dynamics. University of California, Berkeley, CA.

Miller, J.R., Orbock, S.M., and Zonge, L., 1994. Regional long-term assessment of channel stability along the Truckee River, Nevada, from Verdi to Pyramid Lake: Implications to the potential for catastrophic bridge failure. Final Report to the Nevada Department of Transportation, 74 p.

Noss, R.F., LaRoe, E.T., and Scott, J. M., 1995. Endangered ecosystems of the United States: A preliminary assessment of Loss and Degradation. USDI National Biological Service. Biological Reports 28. Washington, DC.

Ohmart, R.D., 1995. The effects of human-induced changes on the avifauna of western riparian habitats. *In:* J.R. Jehl, Jr., and N.K. Johnson (eds.), A century of avifaunal change in western North America. Studies in Avian Biology No. 15. Pp. 273-285.

Pace, C.P. and Layser, E.F., 1977. Classification of riparian habitat in the southwest. pp. 5-9 in R.R. Johnson and D.A. Jones, Technical Coordinators. Importance, preservation and management of riparian habitat: A symposium. USDA Forest Service GTR RM-43.

Pianka, E.R., 1967. Evolutionary Ecology. Prentice Hall, New York.

Ridgway, R., 1877. Ornithology. Pp. 303-669 in C. King, ed. Ornithology and paleontology. U.S. Geological Explorations 40th Parallel 4. Washington, D.C.

Rood, S.B. and Gourley,C., 1996. In-stream flows and the restoration of riparian cottonwoods along the lower Truckee River, Nevada. Report to USFWS, Reno, NV.

Scoppettone, G.G., Coleman, M., and Wedemeyer, G.A., 1986. Life history and status of the endangered cui-ui of Pyramid Lake, Nevada. *Fish and Wildlife Research* , 1: 1-23.

Segelquist, C.A., Scott, M.L., and Auble, G.T., 1993. Establishment of Populus deltoides under simulated alluvial groundwater declines, *American Midland Naturalist*, 130: 274-285

Stacey, P.B., 1995. Biodiversity of rangeland bird populations. In N. West, ed. Biodiversity of Rangelands. Utah State University Press, Logan, Utah.

Tellman, B., Cortner, H.J., Wallace, M.G., DeBano, L.F., and Hamre, R.H., 1993. Riparian management: common threads and shared interests. A western regional conference on river management strategies.1993 Feb 4-6; Albuquerque, NM. Gen. Tech. Rep. RM-226. Fort Collins, CO: U.S. Department of Agriculture, Forest Service, Rocky Mountain Forest and Range Experiment Station. 419 p.

Townley, J.M., 1980. The Truckee basin fishery 1844-1944. Nevada Historical Society, Publication 43008.

USBR (U.S. Bureau of Reclamation), 1973. Final Environmental Statement for Marble Bluff Dam and Pyramid Lake Fishway, Washoe Project, Nevada.

USDA (U.S. Department of Agriculture, Nevada River Basin Study Staff), 1977. Flood Chronology, Truckee River subbasin, 1861-1976.

WET (Water Engineering and Technology, Inc.), 1990. Geomorphic analysis of the Truckee River from RM 56 (Ambrose Park in Reno) to RM 46 (Vista). Report to the U.S. Army Corps of Engineers, Sacramento District, Contract No. DACW05-88-D-0044.

WET (Water Engineering and Technology, Inc.), 1991. Reconnaissance geomorphic investigation of Truckee River from Vista to Pyramid Lake, Nevada. Report to the U.S. Army Corps of Engineers, Sacramento District, Project No. 91-009-01.

FOCUS 3: WATERSHED DEVELOPMENT EFFECTS IN ARID AND SEMI-ARID REGIONS

Chair: Ben Urbonas

Urban Rivers in Arid Environments

Todd Harris

We are still in the process of discovering the unique rules by which western river ecosystems operate.

Urban arid watersheds present extremes of conditions that are far from natural for most biological communities.

Historically, most rivers of the arid and semi-arid West were intermittent or with highly irregular flows.

A single thunderstorm over the upstream impervious surfaces of the watershed can cause the yearly average daily discharge to change dramatically.

In the arid West, river discharge is controlled by water rights for irrigation and industrial uses above, in and below developed urban areas.

Biotic communities in western rivers in arid urban environments are also limited by the size and amount of food available. These systems drain relatively sterile watersheds. Fish populations are also limited to extremely tolerant species in these rivers.

The ecosystems of most arid urban rivers are no longer the historical, depauperate products of an extreme environment. They are a product of an altered environment, with recent changes in hydrology and energy flow. They will continue to gradually change, and how they change will depend on how they are perceived and protected.

Questions/Comments

| Question: | How will you know when you have re-established the natural system? |
| Answer: | There is no such thing as the reestablishment of nature, we can only make the present situation beautiful. |

| Question: | What do you mean that you are being held to Eastern biocriteria? |
| Answer: | Colorado being required to define reference reaches, but there are no reference reaches. It is a question of adapting the river to what serves the community best. We should measure the improvements, and compare it against goals, but not compare it against another site. |

We need to look at the water and determine which critters we want to grow, and make the environment suitable for them.

Comment: Do we need a reference site, or can we build the bioindicator matrix from measurements at another site, or can I do it from the fundamental knowledge that I currently have of biology?

South Platte in Metropolitan Denver - A River in Transformation
Mike Stevens

The pristine South Platte River of 1860 was 300 feet wide, shallow, unstable, fringed with a corridor of cottonwoods and willows, and flowed throughout the entire year.

No matter what it used to look like, it is abundantly clear that the South Platte River has changed because of all the developments that have been undertaken since the discovery of gold 140 years ago. It is still altering itself in places. But development has given way to management.

The South Platte River in the metropolitan area looks like four distinct rivers, herein named the Urban, Suburban, Engineered, and Rural reaches. The four types of reaches offer different opportunities to create environments, some more natural and others more urban. The pristine river can not be restored.

The Rural reach of the South Platte offers the most opportunity for preserving something of a natural setting. The Suburban reaches offer the most challenge. They can be made into the Urban type, but not Rural. Or they can be defined in their own manner. The choices are ours!

Questions/Comments

Comment: There is a lot of information available to create a new master plan for the South Platte, using Mike Steven's data and Todd Harris's data.

Question: Can you relate from your experience any other place where the rivers are what the South Platte used to be?

Answer: Yes, the rivers in Chile; however, they are probably not what EPA would want the Platte to look like. These rivers wander around, are unstable, ... We are far away from pristine, and we cannot get back because of water rights, stabilization of the river etc. And we probably don't want to.

Question: Did you try to relate aggregation and degradation to location of tributary inflows, storm drains, location of metro, etc.?

Answer: Yes and each has its own story, which is too long to tell here.

Question: Did the bars move downstream after the flood of 1995?

Answer: Yes, they seemed to move some, and appear to be cleaner now than before the flood.

Question: Which reach is closest to pristine setting?

Answer: Probably the suburban reach, but even it does not have enough sinuosity because of man-made changes to the riverbed.

The Lower Truckee River, A System in Transition
John Warwick

The water quantity and quality of the lower Truckee River has been perturbed by mans' activities for the better part of the past century. Upstream storage reservoirs and downstream diversions limit flows. Channel modifications have resulted in morphological changes which reduce the extent and vitality of both aquatic and terrestrial habitats.

Local efforts are attempting to integrate the ideas from a variety of groups and currently focus upon the recovery of the Cui-ui population within Pyramid Lake. Beyond this rather singular focus, we must begin asking what do we really mean by the word "restoration"?

Our challenge goes beyond merely the technical, for restoration has a spiritual component as well. For restoration activities to be truly successful we must understand and appreciate all perspectives, with sound guidance from both the technical and spiritual domains.

Questions/Comments

Question: You've suggested that the cut throat trout and Cui-ui are the critical biota, do we know what is required to protect them?

Answer: Yes, we understand what is required well enough to take action to improve the resource; the question is how to get the riparian vegetation to reestablish itself in this deeply incised channel.

Question: Is there some wiggle room within the water rights allocation to provide the necessary flows?

Answer: Yes, but the users want the flexibility to sell the water, not just give it away.

The Use of Retention Basins to Mitigate Stormwater Impacts on Aquatic Life

John Maxted[1]
Earl Shaver[2]

Abstract

Physical habitat and biological measurements were taken in nontidal streams below eight stormwater management pond facilities (BMPs) during the Spring of 1996. Two of the sites were predominantly in commercial land use while the remaining six sites were in residential land use. The results were compared to 38 sites with no BMPs sampled in the Fall of 1993. Three replicate macroinvertebrate samples were collected in riffle habitats at each BMP site using a 800 μm kick-net. Biological quality was determined from six metrics using 100-organism subsamples identified to the species level. Physical habitat quality was determined from twelve metrics that defined the condition of the channel, stream bank and riparian zone. These metrics were compared with the mean values derived from three reference sites to produce summary index scores for each site reported as "percent of reference". The overall macroinvertebrate community, as measured using the Community Index, was not significantly different between BMP and non-BMP sites. A similar result was found using the Sensitive Species Index. The BMPs did not prevent the almost complete loss of sensitive species (e.g., mayflies, stoneflies, and caddisflies) after development. The BMPs did not attenuate the impacts of urbanization once the watershed reached 20% impervious cover. More data are needed to determine whether these controls attenuate impacts at lower levels of development (0-15% impervious cover). Three of the eight BMP sites had

[1] Environmental Scientist, Division of Water Resources, Delaware Department of Natural Resources and Environmental Control (DNREC), 89 Kings Highway, P.O. Box 1401, Dover, Delaware, 19903
[2] Environmental Engineer, Division of Soil and Water Conservation, DNREC

Habitat Index scores comparable to the reference condition indicating improvement in physical habitat provided by the BMPs; five of the sites had impacted physical habitat similar to the non-BMP sites. <u>A data set of this size and complexity should not be used to derive definitive conclusions on the ability of stormwater controls to protect stream biota and habitat.</u> Stormwater management design criteria varied between different sites, and impacts may have been generated by construction site impacts where the stormwater facilities have not been in long enough to allow for restabilization and repopulation. The lack of comparable studies of other regions of the US and collection of data on other measures of stream health below stormwater controls suggests the need for additional research in this area.

Introduction

Over the last 90 years, the population of the United States (US) has increased 300% from 76 million in 1900 to 249 million in 1990 (United States census). This period has also seen a dramatic shift in the way people live and use the land. In 1900, the majority of the US population (60%) lived in rural areas, while today (1990) the majority (75%) live in urban areas. This trend continues today, although at a slower rate (Figure 1). But even as the rate levels off, roughly three-fourths of the estimated 25 million people that will be added to the population over the next decade will likely live in urban areas. Delaware's population has experienced a similar rate of population increase (185,000 to 666,000) and shift in land use (Figure 1).

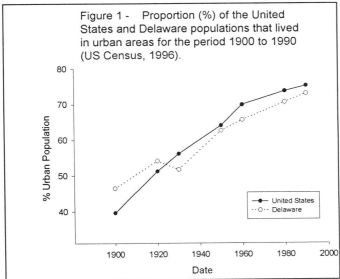

Figure 1 - Proportion (%) of the United States and Delaware populations that lived in urban areas for the period 1900 to 1990 (US Census, 1996).

This change in demographics and land use has brought about profound changes in the physical, chemical and biological integrity of nontidal streams in Delaware. The objective of this research was to determine the effectiveness of stormwater controls, principally retention basins, to protect stream resources after urbanization. This study focused on wadable, nontidal streams and the use of macroinvertebrates and physical habitat as indicators of overall stream health.

The placement of impervious surfaces in the watershed (roads, parking lots, rooftops, driveways, sidewalks, etc.) increases peak flows during storm events and reduces base flows during drought periods. Hollis (1975) found that urbanized watersheds with 20-30% impervious cover had 10-15 times the frequency of small flood events (1-year recurrence interval) compared to nonurban watersheds. Large flood events (100-year) doubled in size after urbanization. This change in stream hydrology affects the physical structure and stability of stream channels through accelerated erosion and sediment deposition. The removal of riparian vegetation (e.g., yards, golf courses) and the placement of structures (e.g., buildings, bridges) along stream channels in the floodplain further impacts the geomorphology of urban streams.

Stormwater contains water quality contaminants (metals, nutrients, organics) that further stress the aquatic community. During the summer months, exceedences of temperature criteria occur in channels that receive runoff from unshaded impervious surfaces. Exceedences of dissolved oxygen criteria occur in streams and ponds through nutrient enrichment and the removal of shade. Both physical and chemical factors associated with urbanization contribute to the overall biological condition of urban streams.

The effects of urbanization on the aquatic community (e.g., invertebrates and fish) has been quantified by several researchers (Jones and Clark 1987, Klein 1979, Limburg and Schmidt 1990, Pedersen and Perkins 1986, Booth and Jackson, Weaver and Garmen 1994, and Garie and McIntosh 1986). The use of aquatic organisms as indicators of stream health in Delaware was presented at the last Engineering Foundation Conference (Shaver and Maxted 1994) based upon data collected in 1993. Data collected at 19 sites established an association between selected biological measures and the degree of urbanization in the watershed. This data set was expanded to 38 sites and used as the basis for evaluating the effectiveness of stormwater controls monitoring in 1996.

Evaluation of Stormwater Facility Performance Using Aquatic Communities

The effects of stormwater runoff from a water quality perspective are fairly well documented, as is the ability of a variety of stormwater management facilities to provide for water quantity control and water quality treatment. There has, to date, been an inherent assumption that water quality treatment and pollutant capture would be translated into receiving system protection. That assumption has never been validated.

In addition, stormwater treatment facilities do remove pollutants, but the level of performance is highly variable, and needs to be expressed in ranges rather than in specific levels of treatment that can be expected. Chemical monitoring is expensive, and large amounts of data are needed in terms of number of events sampled and constituents monitored to make reasonable statements regarding function and performance (Urbonas, 1995).

The collection of large amounts of data also can present problems in terms of the accuracy of measurements. Stormwater management facilities often have multiple inflow points and may receive overland flow which makes data collection difficult. Monitoring each inflow point and the facility outfall, increases the potential for error in data collection or analysis in addition to dramatically increasing the cost of data collection. This difficulty in collecting data must also be considered in conjunction with the variability of the data from storm to storm.

What is needed is a simple, long term approach to system assessment which minimizes the cost of data collection, and provides a framework for system health evaluation. Presented in this paper is a framework for assessing health of aquatic ecosystems and performance of stormwater facilities by considering the habitat and organism diversity within the receiving system and downstream of stormwater facilities. Evaluation of ecological health using aquatic organisms has become fairly widespread, but evaluation of stormwater facility performance through the same approach is fairly new and will be explored in some detail here.

Baseline Condition - Stream Health Without Stormwater Controls

In 1993, the Delaware Department of Natural Resources and Environmental Control (DNREC) surveyed 38 wadable nontidal streams in the urbanized Piedmont Region of the State. This region encompasses approximately 180 square miles in Northern Delaware and includes portions of the Christina River, Naamans Creek, and Shellpot Creek watersheds. Based upon 1984 land use data, about half the region (48%) is in urban land use; 33% is undeveloped and 19% is in agriculture. Stormwater controls have only recently been included as part of new developments in the region. Therefore, these data represent conditions that existed before the implementation of regulatory programs for controlling stormwater runoff.

The macroinvertebrate community was used as the principle biological indicator. The Community Index, reported as a percent of reference, measured the overall condition of the macroinvertebrate community using six biological measures: taxonomic richness, EPT (Ephermeroptera, Plecoptera, Trichoptera) richness, % EPT abundance, % chironomidae, % dominant taxon, and the Hilsenhoff Biotic Index (HBI). The Habitat Index, also reported as a % of reference, was derived from 12 measures of the physical structure of the channel, bank and riparian zone. Scores for each site were calculated and placed into one of three classification categories ("Good", "Fair", or "Poor") using procedures developed by the U.S. Environmental Protection Agency (EPA) (Plafkin 1989).

The assessment concluded that three-fourths (74%) of the 171 miles of nontidal streams in the region had degraded ("Fair" or "Poor") biological quality (Figure 2). An even larger percentage (90%) was found to have degraded physical habitat. These results lead to further investigation into the role of urban land use in effecting stream ecological condition.

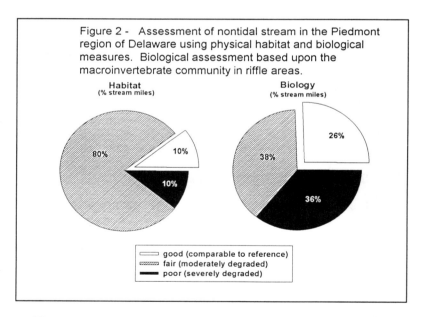

Figure 2 - Assessment of nontidal stream in the Piedmont region of Delaware using physical habitat and biological measures. Biological assessment based upon the macroinvertebrate community in riffle areas.

The association between biological quality and physical habitat quality provided the first indication that hydrology and its effects on the physical structure of the channel was an important stressor effecting the macroinvertebrate community (Figure 2). Streams in urban areas were observed to have a particular visual signature; eroded banks along both bends and runs, uniform and shallower depth, wider channel, and newly deposited sediment in the channel. The association between physical habitat quality and biological quality ($r^2 = 0.35$) provided objective evidence that the changes in physical structure were effecting the aquatic community (Figure 3).

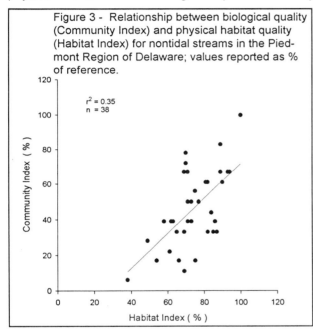

Figure 3 - Relationship between biological quality (Community Index) and physical habitat quality (Habitat Index) for nontidal streams in the Piedmont Region of Delaware; values reported as % of reference.

The association is further illustrated using classification information. A high proportion of sites classified as either "Good" or "Poor" for one measure were classified as either "Good" or "Poor" for the other measure (Table 1). Three fourths (75%) of the sites with "Poor" or "Good" habitat quality also had the same classification for biology. All of sites with "Poor" biology had degraded habitat while none of the sites with "Good" biology had "Poor" habitat. In short, the changes in physical habitat appeared to be effecting the biota.

Table 1 - Association between biological quality and physical habitat quality using "Good", "Fair", and "Poor" classifications; Northern Piedmont Region, Delaware.

"Poor" Habitat	(n = 4)	75%	had "Poor" Biology
"Good" Habitat	(n = 4)	75%	had "Good" Biology
"Poor" Biology	(n = 14)	100%	had Degraded Habitat
"Good" Biology	(n = 10)	0%	had "Poor" Habitat

To further evaluate the relationship between stream biological condition and urbanization, land use data were compiled above each sampling station.

Percent (%) impervious cover estimates were calculated for each subwatershed using published estimates for each of the major urban land use categories (USDA 1986). The relationship between biological condition and percent impervious cover ($r^2 = 0.43$, $n = 34$) indicated an association between the two measures (Figure 4). On average, biological quality decreased by about 50% once the watershed reached 10-15% impervious cover. The association was particularly strong ($r^2 = 0.71$, $n = 19$) when looking only at low density urban sites (0-30% impervious cover).

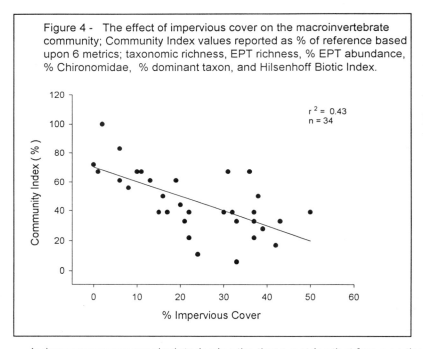

Figure 4 - The effect of impervious cover on the macroinvertebrate community; Community Index values reported as % of reference based upon 6 metrics; taxonomic richness, EPT richness, % EPT abundance, % Chironomidae, % dominant taxon, and Hilsenhoff Biotic Index.

Index scores were recalculated using the three metrics that focus on the most sensitive organisms or groups of organisms found in riffle areas. The resulting Sensitivity Index scores were derived for each site using the EPT richness, % EPT abundance and HBI metrics. The relationship between the Sensitivity Index and percent impervious cover was stronger ($r^2 = 0.56$, $n = 34$) than the relationship derived using the Community Index (Figure 5). On average, about 90% of the sensitive organisms were eliminated from the community after the watershed reached 10-15% impervious cover. The association was particularly strong ($r^2 = 0.78$, $n = 19$) when looking only at the low density urban sites (0-30% impervious cover).

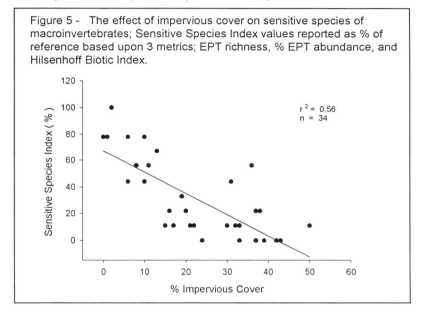

Figure 5 - The effect of impervious cover on sensitive species of macroinvertebrates; Sensitive Species Index values reported as % of reference based upon 3 metrics; EPT richness, % EPT abundance, and Hilsenhoff Biotic Index.

BMP Site Selection

BMP site selection consisted of a number of different elements including an existing stormwater facility inventory and discussions with individuals familiar with a number of facilities. Basic criteria on how sites were selected included:

Stormwater facility criteria

• Facilities being considered had to be retention or detention basins. The intent was to evaluate one specific type of BMP (ponds) as opposed to only have one data point per type of BMP. It was hoped that a trend for a specific BMP could be established.
• To the greatest extent possible, the facility had to meet current design criteria which included peak rate control (2, 10, 100) and water quality performance (24 hour detention for the first inch of runoff). If adequate facilities couldn't be found due to the recent nature of the State Stormwater Management Program (effective date July, 1991), older retention ponds serving development were used.

•   Facilities had to be at least 2 years old. The concern with newer facilities was that there was potential for construction related stream impacts. If a new facility had significant instream impacts, the cause of the impact would not be necessarily related to the performance of the BMP as opposed to excess runoff and sedimentation during site construction.

•   Impervious drainage area cover to the facility of at least 20% to attempt to see if receiving streams received a greater level of protection than would normally be seen if there were no stormwater management facility.

Receiving stream criteria

•   Discharge from the BMP represents the predominant flow in the stream

•   Riparian zone unimpacted (e.g., wooded/shaded). This criteria was extremely hard to achieve with a number of facilities eliminated due to downstream riparian area disturbance.

•   Perennial flow was necessary in the receiving stream. One difficulty with this criteria was the fact that these streams are first order streams and the facilities serve fairly small drainage areas.

•   Riffles were common so that stream sampling could be accomplished.

Table 2 provides information on the BMP facilities:

Table 2 - BMP Facility Data				
Site	Land Use	Impervious %	D.A. to Facility	Development
BMP 1[*]	Residential	25	88 acres	Corner Ketch
BMP 2[**]	Commercial	22	82.6 acres	Brandy. Comm.
BMP 3[**]	Commercial	65	36 acres	Core State
BMP 4[**]	Residential	30	31.8 acres	Hunt at Louv.
BMP 5[*]	Residential	28.3	383 acres	Veranda
BMP 6[*]	Residential	31	107 acres	Limestone Hills
BMP 7[*]	Residential	30	157 acres	Chestnut Hills
BMP 8[***]	Residential	28.3	330 acres	Jenny's Run

[*]   Project design based on peak control of the 10 year storm only
[**]   Project design based on peak control of the 2, 10, and 100 year storm in addition to 24 hour extended detention for the first 1" of runoff
[***]   Site in main stream of Jennys Run; considered stormwater flow from more than one development site

Study Methods

Samples were collected between May 2nd and June 6th, 1996. They were collected, processed and analyzed using the same procedures used to derive the baseline (without controls) data. Macroinvertebrate samples were collected using a 1 meter² kick net (800 μm mesh). Each sample was a composite of two 1 meter² riffle areas combined in a sieve bucket (600 μm mesh). Three replicate collections were made at each site while moving progressively upstream. A single 100-organism subsample was removed from each sample and identified to the species level using procedures developed by EPA (Plafkin 1989). QA/QC procedures followed those developed by Delaware DNREC. Six metrics were derived for each sample for quantitative analysis: taxonomic richness (TR); EPT richness (EPT); % EPT abundance (%EPT); %chironomidae (%C); % dominant taxon (%DT); and the Hilsenhoff Biotic Index (HBI) (Table 3).

Table 3 -   Biological metrics used to derive summary index scores for BMP and non-BMP sites.

DNREC Biological Metrics

Metric Name	Description	Type
taxa richness	total number of unique taxa	richness
EPT* richness	total number of EPT taxa	rich/tolerance
% EPT abundance	% of sample that are EPTs	tolerance/comp
% dominant taxon	largest % of a single taxon	composition
% chironomidae**	% of sample from this group	tolerance
HBI	composite tolerance by taxon	tolerance

* EPT - the orders ephemeroptera (mayflies), plecoptera (stoneflies), and trichoptera (caddisflies); large diversity or abundance indicates high quality.
** chironomidae - family of midges; large abundance indicates stress.

Three summary index scores were derived for each site following procedures developed by EPA (Plafkin 1989). Three reference sites were sampled during the same period and used as the basis for deriving index scores reported as "percent of reference". Habitat Index scores were determined by comparing the total habitat score for each BMP site with the mean total score for the three reference sites. Community Index (CI) scores were determined by comparing the six biological metric values for each site with the mean values from the reference sites. The CI was used to define the overall quality of the macroinvertebrate community. Sensitive Species Index (SSI) scores were determined using the three biological metrics that define the components of the community that are the most sensitive to pollution: EPT, %EPT and HBI. The SSI was used to assess this component of the macroinvertebrate community. Mean CI and SSI scores for each site were determined from the CI and SSI scores from each replicate sample.

Twelve measures of habitat quality were taken for each site and included measures of the channel, stream bank, and riparian zone. Those parameters, developed for piedmont streams and their "abbreviation" used for habitat assessment include the following:

CM - channel modification: stream is rated on whether or not it has been modified by man, e.g., channelized, and the extent to which it meanders.

BSC - bottom substrate/available cover: rates the amount and variety of available habitat throughout the stream segment.

E - embeddedness: the degree to which the substrate is surrounded or covered by fine sediment.

RQ - riffle quality: rates the dominant substrate in the riffle where the sample was collected, cobble being most desirable.

FR - frequency of riffles: rates the abundance of riffle areas in the stream segment.

SD - sediment deposition: the degree to which sediment is deposited in the stream channel. Rating is based on the presence of islands and point bars, or, sediment deposited along the substrate.

V/D - velocity/depth: rates the absence or presence of four categories of water; slow and deep, slow and shallow, fast and deep, fast and shallow.

BS - bank stability: measures erosion of the stream bank.

BV - bank vegetative type: rates the dominant vegetation on the stream bank, shrubs being most desirable, left and right sides are scored separately and then composited.

S - shading: rates the percent of the stream surface that is shaded throughout the day.

RZ - riparian zone width: measures the evidence of human activity along the edge of the stream, left and right sides are scored separately and then composited.

HCI - habitat comparison index: the assessment score, expressed as the percent comparison to the reference.

Other data collected at each site included temperature, pH, D.O., specific conductance, and flow.

Results

The biological and habitat data collected at the eight BMP sites are summarized in Tables 4 and 5 respectively.

Table 4 - Mean Values for Biological Metrics Determined in Streams below 8 BMP Sites

Site #	Replicate	TR	EPT	%EPT	%C	%DT	HBI	(%) CI	(%) SSI
BMP1	3	35	7	18.3	53.7	14.8	5.25	49	26
BMP2	3	21	2	25.8	52.2	29.5	6.10	35	15
BMP3	3	31	2	1.7	70.9	27.5	5.85	33	7
BMP4	3	26	6	16.9	60.0	27.1	5.33	39	18
BMP5	3	19	0	0.2	36.7	50.9	7.10	25	0
BMP6	3	22	7	14.5	38.0	46.6	6.36	31	7
BMP7	1	29	5	8.8	75.0	22.1	5.43	35	11
BMP8	3	23	8	26.7	60.3	19.9	4.71	51	33
Reference	10	24	10	56.3	14.5	23.8	2.94	--	--

Table 5 - Habitat Metrics for Streams below 8 BMP Sites

Site #	CM	BSC	E	RQ	FR	SD	VD	BS	BV	S	RZ	Total	(%) HCI
BMP1	12	19	17	16	18	16	10	13	10	20	20	171	100
BMP2	20	6	16	13	9	5	7	6	12	18	20	132	77
BMP3	15	13	11	13	12	7	9	11	10	15	12	128	75
BMP4	19	18	18	14	19	16	10	18	18	16	20	186	109
BMP5	19	19	9	11	19	6	10	20	10	16	17	156	91
BMP6	17	18	12	18	14	8	10	6	4	19	20	146	85
BMP7	11	5	19	10	1	6	11	20	10	17	13	123	72
BMP8	18	19	19	17	19	16	10	11	10	20	20	179	105
Reference	17	17	18	15	17	12	10	11	17	16	20	171	--
Possible	20	20	20	20	20	20	20	20	20	20	20	220	100

The 8 sites with BMPs did not appear to improve biological conditions using either the Community Index (Figure 6) or the Sensitive Species Index (Figure 7), as compared to sites without BMPs. Further, BMPs did not prevent the loss of sensitive species found at reference sites. In addition, the degree of urbanization did not appear to effect biological conditions at the BMP sites. The one BMP site with 65% impervious cover had a similar biological condition to the seven sites with 22-32% impervious cover.

Mean and standard deviation estimates for selected biological measures were calculated for the 8 BMP sites (22 replicate samples) and compared to the 21 non-BMP sites (29 replicate samples). Only non-BMP sites with land use similar to the BMP sites (20-65% impervious cover) were used in these calculations. Preliminary results showed no significant differences between the two groups of sites (Table 6). Further, both the BMP and non-BMP sites were significantly different from the reference condition for most measures (Table 6).

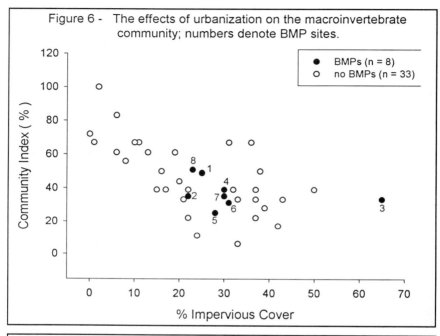

Figure 6 -   The effects of urbanization on the macroinvertebrate
community; numbers denote BMP sites.

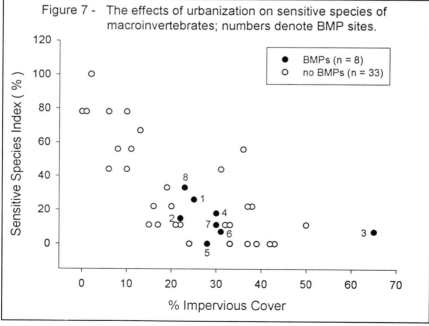

Figure 7 -   The effects of urbanization on sensitive species of
macroinvertebrates; numbers denote BMP sites.

Table 6 - Comparison of urban sites with and without stormwater controls (BMPs) using the mean ($\bar{x}$) and standard deviation (SD) for selected biological measures; Community Index (CI), Sensitive Species Index (SSI), taxonomic richness (TR), EPT richness (EPT), % EPT abundance (%EPT), and % chironomidae (%C). Data for reference sites provided for comparison with non-urban sites; n = number of samples.

		CI(%)		SSI(%)		TR		EPT		%EPT		%C	
	n	$\bar{x}$	SD	$\bar{x}$	SD	$\bar{x}$	SD	$\bar{x}$	SD	$\bar{x}$	SD	$\bar{x}$	SD
reference	10	--	--	--	--	23.6	4.5	10.3	1.8	56.3	11.7	14.5	12.7
BMPs	22	37.6	12.5	15.0	14.8	25.4	6.8	4.5	3.0	14.5	12.2	54.1	20.8
no BMPs*	29	36.2	14.1	14.1	14.5	19.9	4.7	4.7	2.7	27.5	18.2	27.7	23.4

* only sites with 20-65% impervious cover included.

To illustrate the effect of urbanization on stream biota, the data collected at all sites (BMP and non-BMP) were combined together and sorted into three groups; low density urbanization (0-14% impervious cover), medium density (15-29%), and high density (30-65%). Mean and standard deviation estimates for selected biological measures were calculated for each group. There was a significant difference between the low and medium density urbanization groups, while there was no difference between the medium and high density groups (Table 7). It appeared that the biological community changes significantly once the watershed reaches 10-15% impervious cover.

Table 7 - Biological quality of all sites surveyed (with and without stormwater controls) by density of urbanization; low (0-14% impervious cover), medium (15-29%), and high (30-65%). Mean ($\bar{x}$) and standard deviation (SD) provided for selected biological measures; Community Index (CI), Sensitive Species Index (SSI), taxonomic richness (TR), EPT richness (EPT), % EPT abundance (%EPT), and % chironomidae (%C); n = number of samples.

urban density	% imp	n	CI(%)		SSI(%)		TR		EPT		%EPT		%C	
			$\bar{x}$	SD	$\bar{x}$	SD	$\bar{x}$	SD	$\bar{x}$	SD	$\bar{x}$	SD	$\bar{x}$	SD
low	0-14	14	69.1	14.1	67.5	21.7	22.1	3.9	10.1	2.2	69.8	9.0	9.1	5.7
medium	15-29	29	40.6	12.7	17.1	13.1	22.2	5.8	4.9	2.7	25.1	15.1	31.7	25.0
high	30-65	32	36.4	13.3	13.1	14.7	22.0	5.9	4.9	2.7	22.0	17.8	37.4	26.4

The association between biological quality and physical habitat quality (Figure 3) was redone to include the BMP sites. The results showed that three of the BMP sites had better habitat quality than would have been expected using the biological information from the non-BMP sites (Figure 8). Five of the sites had biological and physical habitat conditions similar to the non-BMP sites. It appeared that some of the BMPs were providing improvements in physical habitat. Additional data and further analysis of the existing data are needed.

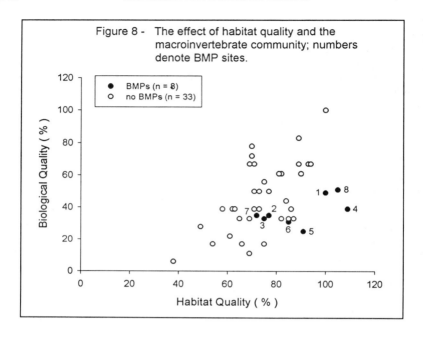

Figure 8 - The effect of habitat quality and the macroinvertebrate community; numbers denote BMP sites.

Discussion

The preliminary results of impacts to aquatic organisms and habitat downstream from a small number of BMP sites have been discussed in this paper. Similar data are needed at more sites in the Piedmont Region of the Mid-Atlantic US before making definitive conclusions on the effectiveness of stormwater basins to protect aquatic life and physical habitat. Data are also needed on other living resources (e.g., periphyton, fish, amphibians) as well as other measures of stream quality (e.g., geomorphology, hydrology, and water quality). Similarly, data are needed for other types of surface waters (e.g., wetlands) and physiographic regions of the US.

Site selection was more difficult than first anticipated. Many of the streams that received a discharge from a retention or detention pond were either too small to sample, had a degraded riparian zone (unshaded), did not have stream baseflow at all times of the year, or discharged directly to a stream with substantial upstream flow. Many of the BMP sites not selected for study had small drainage areas that would have intermittent flow even under predevelopment conditions. Further study is needed to determine whether intermittent streams below BMPs mimic natural systems.

Data did not show that retention and detention basins designed to control stormwater provided adequate protection of the macroinvertebrate community in wadable nontidal streams. This may be due to inadequate design or other factors not addressed in this study such as temperature, water quality (road salt, contaminants), inadequate controls during construction, age of the facility, and distance from the discharge. There was some indication that biological conditions improved with increased distance from the discharge. Half the BMP sites provided habitat conditions comparable to reference areas indicating their effectiveness in maintaining stable channel characteristics.

This method of analysis represents a different approach to assessing BMP effectiveness as compared to chemical monitoring. It represents a bottom-up resource based approach to gauge BMP performance which is a major shift from traditional approaches. Traditional approaches are top down in that performance is determined by how many pounds of pollutants are removed or whether water quality standards are attained. This approach looks at watersheds from the bottom up or from the perspective of the receiving system.

These results need to be placed in the proper context:

- There is only one data set
- The study followed a severe winter with record snowfall,
- Biota may still be impacted by construction related sedimentation, and
- Some chemical monitoring may be necessary (Temp., pH, etc.) to better understand results.

It is too early to panic. If further study indicates a continuing trend, BMP design criteria will have to be reconsidered to provide a greater level of protection to receiving systems.

It was difficult to find a representative sample of stormwater management BMP's that met modern design criteria due to the relative short time frame between initiation of the Statewide Stormwater Management Program and this monitoring effort. The only sites actually meeting the State stormwater management requirements in terms of when they were constructed and their design criteria were the two commercial sites and one residential site. The other residential sites were selected due to their being retention ponds, which is a preferred practice under the new State program. They were designed and constructed for peak control of the 10 year storm only, and did not have an extended detention component for water quality function.

Recommendations include the following:

- We need to monitor more facilities that meet current criteria for water quality treatment to obtain needed additional information that will provide a greater comfort in the widespread applicability of the results. This is important when considering program changes to improve stormwater facility performance in protecting downstream aquatic resources.
- Additional research is especially important for developments in pristine watersheds (< 10% impervious cover).
- We cannot just consider loadings, but must also focus on the resource.
- New measures must be used to assess resource protection efforts: biology and physical habitat are sensitive and cost-effective.

Questions Raised

How far downstream should the biological and habitat monitoring take place. Discharges from a stormwater management pond take water from one type of ecosystem (ponds) to another (free flowing streams). There may be a transition zone, where depressed populations can be expected. More work is needed to determine the linear extent of impacts in streams below detention facilities.

There is a period of time when a site has been developed where sediment from construction activities will impact on habitat, species abundance, and diversity. Construction related impacts must be expected and will occur as a result of increased stormwater discharges and elevated sediment loadings. There has to be a period of time after completion of construction before measurements can be taken to assess the response of a receiving system to site development and stormwater management facilities. We can make recommendations, such as a period of two years after construction completion, but that recommendation must be considered preliminary and subject to variation around the country due to climatic factors.

Final Comment

The importance of riparian zone protection cannot be stressed too much. More effective control over riparian zones must be provided if receiving systems are to maintain their function and value. The greatest difficulty in site selection resulted from downstream receiving systems from stormwater management facilities were in a typical suburban environment where the stream had grassed banks with little or no canopy. The majority of streams in Delaware had degraded physical habitat (Figure 2). Regardless of efforts to protect downstream receiving systems by proper design and construction of stormwater facilities, a failure to protect the downstream riparian zones from disturbance will cause a degraded resource condition.

Acknowledgments

The authors would like to thank Frank Piorko for providing assistance in selecting sites. Ellen Dickey provided assistance in sample collection, data management, and data analysis. Greg Mitchell and Terry Cole provided assistance in sample collection.

Literature Cited

Booth, D.B. and C.R. Jackson, "Urbanization of Aquatic Systems - Degradation Thresholds and the Limits of Degradation". American Water Resources Association Summer Symposium, Jackson, Wyoming. 1994.

Hollis, G.E.; 1975, "The Effect of Urbanization on Floods of Different Recurrence Intervals"; Water Resources Research, Vol. 11, No. 3.; pp. 431-435.

Weaver, L.A. and G.C. Garman; 1994, "Urbanization of a Watershed and Historical Changes in a Stream Fish Assemblage"; Transactions of the American Fisheries Society, Vol. 123; pp. 162-172.

Garie, H.L., and A. McIntosh; 1986, "Distribution of Benthic Macroinvertebrates in Stream Exposed to Urban Runoff"; Water Resources Bulletin, American Water Resources Association; Vol. 22, No. 3; pp. 447-455.

Jones, R.C. and C.C.Clark, "Impact of Watershed Urbanization on Stream Insect Communities", American Water Resources Association, Water Resources Bulletin 23 (6), December, 1987.

Klein, R.D., "Urbanization and Stream Quality Impairment", American Water Resources Association, Water Resources Bulletin 15 (4), August, 1979.

Limburg, K.E. and R.E. Schmidt, "Patterns of Fish Spawning in Hudson River Tributaries: Response to an Urban Gradient", Ecology 71 (4), pp.1238-1245, 1990.

Pedersen, E.R. and M.A. Perkins, "The Use of Benthic Invertebrate Data for

Evaluating Impacts of Urban Runoff", Hydrobiologia 139, pp.13-22, 1986.

Shaver, Earl and Maxted, John, "Watershed Protection Using an Integrated Approach", Stormwater NPDES Related Monitoring Needs, edited by Harry C. Torno, American Society of Civil Engineers, 1995, pp. 435-459.

United States Census; 1996; personal communications with Mike McGrath, Delaware Department of Agriculture, Dover, Delaware.

Urbonas, Ben R., "Parameters to Report with BMP Monitoring Data", Stormwater NPDES Related Monitoring Needs, edited by Harry C. Torno, American Society of Civil Engineers, 1995, pp. 306-328.

IMPERVIOUS COVER AS A URBAN STREAM INDICATOR AND A WATERSHED MANAGEMENT TOOL

Thomas Schueler and Richard Claytor, P.E.[1]

This paper reviews the utility of impervious cover as both an indicator of stream quality, and as a potential management tool to prevent cumulative impacts at the subwatershed level. A simple urban stream classification model is proposed to predict and manage urban stream quality, based upon the current or forecasted level of subwatershed impervious cover. Streams are classified as sensitive, impacted or non-supporting, based on a composite relationship between impervious cover and a series of individual indicators of stream quality that include: stream temperature, fish diversity, instream habitat, macroinvertebrate diversity, nutrient loading, channel stability, changes in stormflow peaks and frequency, spawning success and bacterial contamination. The next section explores a series of technical and policy implications that arise from the model, including scale effects, reference conditions, statistical variability, measurement of impervious cover, and geographic applicability. In addition, the influence of BMPs and riparian management in extending the impervious cover thresholds is discussed. The potential application of the model to other aquatic resources is also examined. Lastly, the paper explores the potential of the model to serve as a watershed management tool. The use of the model for forecasting cumulative stream impacts, devising more effective subwatershed management strategies, and conducting both sub-watershed-based zoning and impervious cover management are described.

Impervious Cover and Urban Stream Quality: a simple model

A growing body of research has demonstrated a close relationship between impervious cover and various indicators of stream quality (see Schueler, 1994 and this volume). Studies indicate that certain thresholds of stream quality exist, most notably at about

---

[1]Center for Watershed Protection, 8737 Colesville Rd. Suite 300, Silver Spring, MD 20910.

10% impervious cover, where sensitive stream elements are lost from the system. A second threshold appears to exist at around 25% impervious cover, where most indicators of stream quality consistently shift to a poor state or condition (i.e., aquatic diversity, water quality and habitat). The basic model is presented in Figure 1.

The thresholds inherent in the model represent a composite of multiple individual indicators (fish diversity, habitat, macroinvertebrates, pollutant loads, channel stability, stream temperature, changes in dry and wet weather flow, etc). This simple model classifies streams into one of three categories, based on impervious cover: sensitive, impacted, and non-supporting. The three stream categories can be expected to have the following characteristics:

Sensitive Streams. These streams typically have a watershed impervious cover of zero to 10 percent. Consequently, sensitive streams are of high quality, and are typified by stable channels, excellent habitat structure, good to excellent water quality, and diverse communities of both fish and aquatic insects. Since impervious cover is so low, these streams do not experience frequent flooding and other hydrological changes that accompany urbanization. It should be noted that some sensitive streams located in rural areas may have been impacted by prior poor grazing and cropping practices that may have severely altered the riparian zone, and consequently, may not have all the properties of a sensitive stream. Once riparian management improves, however, these streams are often expected to recover.

Impacted Streams. Streams in this category possess a watershed impervious cover ranging from 11 to 25%, and show clear signs of degradation due to watershed urbanization. Greater storm flows begins to alter the stream geometry. Both erosion and channel widening are clearly evident. Stream banks become unstable, and physical habitat in the stream declines noticeably. Stream water quality shifts into the fair/good category during both storms and dry weather periods. Stream biodiversity declines to fair levels, with the most sensitive fish and aquatic insects disappearing from the stream.

Non-Supporting. Once watershed impervious cover exceeds 25%, stream quality crosses a second threshold. Streams in this category essentially become a conduit for conveying stormwater flows, and can no longer support a diverse stream community. The stream channel becomes highly unstable, and many stream reaches experience severe widening, down-cutting and streambank erosion. Pool and riffle structure needed to sustain fish is diminished or eliminated, and the stream substrate can no longer provide habitat for aquatic insects, or spawning areas for fish. Water quality is consistently rated as fair to poor, and water contact recreation is no longer possible due to the presence of high bacterial levels. Subwatersheds in the non-supporting category will generally increase nutrient loads to downstream receiving waters, even if effective urban BMPs are installed and maintained. The biological quality of non-supporting streams is generally considered poor, and is dominated by pollution tolerant insects and fish.

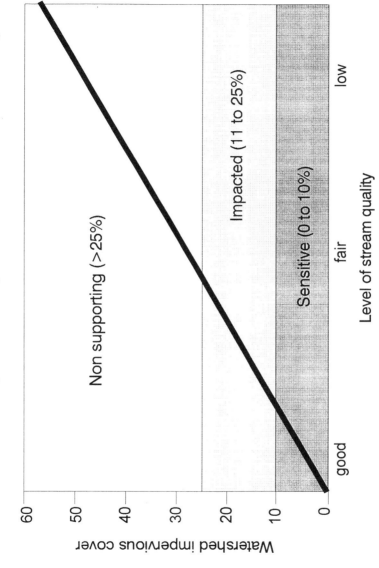

Figure 1. Schematic Showing Impervious Cover/Stream Quality Relationship Used in the Stream Classification Model

Technical Issues Associated with the Urban Stream Classification Model

Although the stream classification model is simple, it is important to consider some of the technical issues involved in its development and application. These are reviewed below:

1. Scale effect. The stream classification model should generally only be applied to smaller urban streams from first to third order. This limitation reflects the fact that most research has been conducted at the catchment or subwatershed level (0.2 to 10 square mile area), and that the influence of impervious cover is strongest at these spatial scales. In larger watersheds and basins, non-urban land uses, pollution sources and disturbances often dominate the quality and dynamics of streams and rivers.

2. Reference condition. The simple model predicts *potential* stream quality rather than *actual* stream quality. Thus, the reference condition for a sensitive streams is a high quality, non-impacted stream within a given ecoregion or sub-ecoregion. It can and should be expected that some individual streams, reaches or segments will depart from the predictions of the stream classification model. For example, physical and biological monitoring may find poor quality in a stream classified as sensitive, or good diversity in a non-supporting one. Rather than being a shortcoming, these "outliers" may help investigators better understand local watershed and stream dynamics. For example, an "outlier" stream may be a result of past human disturbance, such as grazing, channelization, acid mine drainage, agricultural drainage, poor forestry practices, or irrigation return flows.

3. Statistical variability. Individual impervious cover/stream quality indicator relationships tend to exhibit a considerable amount of scatter, although they do show a general trend downward as impervious cover increases. Thus, the stream classification model does not always accurately predict the precise score of an individual stream quality indicator, for a given level of impervious cover (particularly if the stream is near a threshold). Instead, the model attempts to predict the average behavior of a group stream indicators (hydrology, geometry, diversity, habitat, and water quality) within a range of impervious cover. In addition, the impervious cover thresholds defined by the model are not sharp breakpoints, but instead reflect the expected transition of a composite of individual stream indicators.

4. Measuring and projecting impervious cover. Given the central importance of impervious cover to the model, it is very important that it be accurately measured and projected. Yet comparatively relatively little attention has been paid to standardizing techniques for measuring existing impervious cover, or forecasting future impervious cover. In theory, these techniques should be relatively easy to develop. For example, the CWP utilizes a simple technique to measure current impervious cover from low altitude aerial photographs--anything that is not green or vegetated is considered impervious area. Others have defined an effective impervious area which may be lower than total impervious cover under certain suburban or exurban development patterns (Southerland, 1995).

If aerial photography cannot be obtained, then published land use/impervious cover relationships (NVPDC, 1978) or roadway density (Horner *et al.*, this volume) can be used as surrogate measures. Given the many GIS systems that have been developed, it may soon be possible to randomly sample blocks of land uses to derive accurate estimates of impervious cover. In particular, it is very likely that land use/impervious cover relationships vary, depending on whether the particular land use is situated in an urban, suburban or ex-urban area.

Projecting future impervious cover is a greater challenge. The most widely used technique has been to apply per-capita impervious cover creation rates to future population forecasts. However, actual per-capita impervious cover rates have seldom been calculated (but see Schueler, 1996b). The second technique has been to derive impervious cover assuming buildout of the prevailing land uses contained in the local zoning or comprehensive plan. Both techniques have their limitations and more research and sampling are needed to assess their validity.

5. Regional adaptability. To date, nearly all the research used to develop the model has been performed in the mid-Atlantic and Puget Sound eco-regions. In particular, very little research has been conducted in western or in midwestern streams, and only one study has examined the behavior in intermittent streams (Hansen, this volume). Further research is needed to determine if the stream classification model applies to these systems, and if so, what the thresholds may be. It is likely that defining the reference condition will be problematic, given the strong influence of agricultural drainage, channelization, erosion, water withdrawals, poor riparian management in shaping the ecology of these systems.

6. Defining thresholds for non-supporting streams. Most urban stream research has focused on the transition from sensitive streams to impacted ones. Much less is known about the exact transition from impacted streams to non-supporting ones. The stream classification model projects the transition occurs around 25% impervious cover for small urban streams, but more sampling is needed to firmly establish this threshold.

7. Influence of BMPs in Extending Thresholds. Urban BMPs could shift the impervious cover thresholds higher. The ability of the current generation of urban BMPs to shift them; however, appears to very modest according to several lines of evidence. First, a handful of the impervious cover/stream indicator research studies were conducted in localities that had some kind of requirements for urban best management practices; yet no significant improvement in stream quality was detected. Second, Maxted and Shaver and Jones (both in this volume) could not detect an improvement in bioassessment scores in streams served by stormwater ponds. Third, as Macrea notes in this volume, current BMP design criteria do not appear capable of mitigating downstream bank erosion. Consequently, instream habitat conditions often sharply decline which frequently leads to a drop in IBI or RPB indicator scores. Lastly, the current generation of BMPs appears to be subject to limitations in pollutant removal. A recent analysis of 42 published BMP performance studies indicates an irreducible concentration present in storm outflows from BMPs (See Table 1). Outflow nutrient concentrations, for example, are sufficiently high to pose a eutrophication risk to streams and downstream receiving waters.

8. Influence of Riparian Cover in Extending Thresholds. Conserving or restoring an intact and forested riparian zone along urban streams does appear to help extend the impervious cover threshold to a modest degree. For example, Steedman (1988) found that forested streams in Ontario had higher habitat and diversity scores for the same degree of urbanization than streams that lacked an intact riparian zone. Horner *et al.* (this volume) also found evidence of a similar relationship. This is not surprising, given the integral role the riparian zone plays in the ecology and morphology of headwater streams. Indeed, the value of conserving and restoring riparian forests to protect stream ecosystems is increasingly being recognized as a critical management tool in rural and agricultural landscapes (CBP, 1994). The clear implication is that urban watershed managers must implement both riparian and impervious cover management to protect urban streams.

9. Pervious Areas. An implicit assumption of the stream classification model is that pervious areas in the urban landscape do not matter much, and have little direct influence on stream quality. Yet urban pervious areas are highly disturbed, and possess few of the qualities associated with similar pervious cover types situated in non-urban areas. For example, it has recently been estimated that high input turf can comprise up to half the total pervious area in suburban areas (Schueler, 1995a). These lawns receive high inputs of fertilizers, pesticides and irrigation, and their surface soils are highly compacted. Although strong links between high input turf and stream quality have yet to be convincingly demonstrated, watershed planners should not neglect the management of pervious areas.

Pervious areas also provide opportunities to capture and store runoff generated from impervious areas. Examples include directing rooftop runoff over yards, use of swales and filter strips, and grading impervious areas to pockets of pervious area. When pervious and impervious areas are integrated closely together, it is possible to sharply reduce the "effective" impervious area in the landscape (Southerland, 1995). It is recommended, however, that this impervious reduction credit be granted only when such features are explicitly designed into a development site.

10. Application of Model to Other Urban Aquatic Systems. Some research points to the strong influence of impervious cover on other important aquatic systems such as shellfish beds and wetlands (Duda and Cromartie, 1982, Hicks, 1995 and Taylor, 1993). Interestingly, each study found degradation thresholds when impervious cover exceeded 10%. To date, no systematic research has been conducted to determine whether groundwater recharge or quality are predictably influenced by impervious

Table 1

Irreducible concentrations in storm outflows from four groups of urban BMPs
All values in mg/l plus or minus one standard deviation (adapted from Schueler, 1996b)

Parameter	Ponds	Wetlands	Filters	Open Channels
No. of Studies	11	11 -16	6-10	5
TSS	35.0 ± 19.0	32.0 ± 25.8	19.3 ± 10.1	43.4 ± 47.0
TP	0.22 ± 0.12	0.19 ± 0.13	0.14 ± 0.13	0.33 ± 0.15
OP	0.08 ± 0.04	0.08 ± 0.04	N.D.	0.16
TN	1.91 ± 0.56	1.63 ± 0.48	1.93 ± 1.02	1.74 ± 0.71
TKN	1.21 ± 0.36	1.29 ± 0.43	0.90 ± 0.52	1.19 ± 0.41
Nitrate	0.70 ± 0.36	0.35 ± 0.28	1.13 ± 0.55	0.55 ± 0.29

Table 3

Process for Watershed Based Zoning

Step 1. Comprehensive stream inventory
Step 2. Verify impervious cover/stream quality relationships
Step 3. Measure current levels of impervious cover
Step 4. Project future levels of impervious cover
Step 5. Designate subwatersheds, based on stream quality categories
Step 6. Modify master plan/zoning to meet subwatershed impervious cover targets
Step 7. Incorporate management priorities from larger watersheds/basins
Step 8. Adopt specific stream protection strategies for each subwatershed
Step 9. Long-term monitoring cycle to assess stream status.

cover. It is speculated that such relationships will be complex and hard to detect, since groundwater recharge and quality are also influenced by septic systems, wells, lawn irrigation, and sewer I/I.

It is also speculated that impervious cover may be strongly related to water quality in small urban lakes, where eutrophication is considered the primary measure of degradation (i.e, in nutrient-sensitive lakes). Some indication of the possible relationship is illustrated in Figure 2, which shows the urban phosphorus load as a function of both impervious cover and BMP treatment. If the lake's subwatershed has a low background phosphorus load, the Simple Method (Schueler, 1987) predicts that the post-development phosphorus load will exceed background loads once watershed imperviousness exceeds 20 to 25% impervious cover.

Urban phosphorus loads can be reduced when urban best management practices (BMPs), such as stormwater ponds, wetlands, filters or infiltration practices are installed. Performance monitoring data indicates that BMPs can reduce phosphorus loads by as much as 40 to 60%, depending on the practice selected. The impact of this pollutant reduction on the postdevelopment phosphorus loading rate from the site is also shown in Figure 2. The net effect is to raise the phosphorus threshold to about 35% to 60% imperviousness, depending on assumptions made about BMP effectiveness. Therefore, even when effective BMPs are widely applied across a subwatershed, a threshold of impervious cover is eventually crossed after which predevelopment phosphorus concentrations can no longer be maintained.

Impervious Cover as a Management Tool.

Impervious cover serves not only an indicator of urban stream quality but also is valuable management tool that local planners and engineers can use to prevent cumulative impacts within subwatersheds. Some of the ways that the stream classification model can be used in a planning context are described below.

1. Projecting Future Stream Quality

The stream classification model can be used to rapidly forecast the cumulative impact of future growth and development. This is done by projecting future stream quality conditions over all the stream miles in a watershed, based on estimates of the growth of impervious cover. The resulting stream quality projections can be of great value to planners at the local level.

The stream classification model was recently applied over the 571 square mile Occoquan watershed in Northern Virginia which has been experiencing rapid suburban growth (Schueler, 1996). An existing GIS system provided data on stream mileage and impervious cover within 15 major subwatersheds, based in 1989 aerial photography and regional land use/impervious cover relationships. Subwatershed impervious cover forecasts were made for the years 2005 and 2020, using population forecasts as a

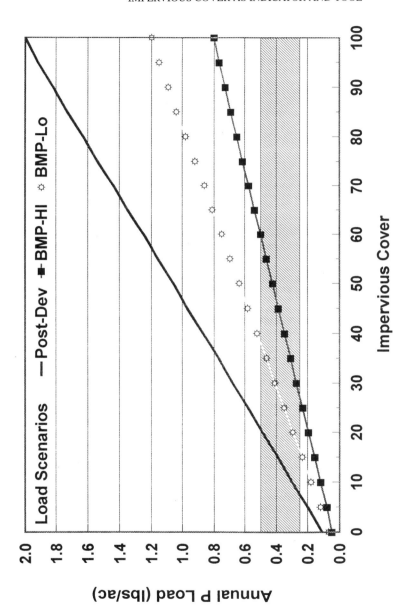

Figure 2. Relationship Between Impervious Cover and Phosphorous Load to Small Lakes; Under Different BMP Effectiveness Scenarios (Schueler, 1995b)

predictor of impervious cover. Creation of future impervious cover was estimated based on the historical per-capita impervious cover relationship measured in the watershed-- approximately one half acre of impervious cover is produced to serve each additional resident. The distribution of new impervious cover over individual subwatersheds was then modified to account for development patterns outlined local comprehensive land use plans. The stream mileage within each subwatershed was then assigned into one of the three stream categories, depending on the estimated impervious cover for each reference period.

The results of stream quality forecasts on the 1300 stream miles in the watershed are portrayed in Figure 3. During the base year (1989), 60% of all stream miles in the Occoquan watershed were classified as "sensitive" streams, 33% were considered "impacted", and only 7% of all stream miles were classified as "non-supporting". After 15 years of moderate growth within the watershed, stream quality conditions are expected to sharply deteriorate. Thus, by the year 2005, only 22% of the streams in the Occoquan are projected to remain in the sensitive category, while 64% shift to the impacted category and 14% become non-supporting. After an additional 15 years of urban growth (2020), over 80% of all stream miles in the Occoquan basin are projected to fall into the impacted or non-supporting category.

As this case study illustrates, the stream classification model can be a simple but powerful tool to convey to local decision-makers the likely consequences of long term growth and development on their inventory of streams.

2. Crafting More Effective Subwatershed Management Plans

The stream classification model can also be used as the foundation for more effective subwatershed plans. Since the water quality and resource potential of an urban stream is defined by its impervious cover, it makes sense to set realistic management goals using the stream classification model. For example, in sensitive streams, it is still possible to set a long-term goal to maintain or enhance predevelopment biodiversity and habitat structure. Achieving this goal requires the combined application of many stream protection tools to the subwatershed--including impervious cover limits, specific urban BMP designs, stream buffer requirements, biological monitoring and land acquisition. For non-supporting streams, however, it is only possible to achieve more restricted management goals--primarily to reduce downstream pollutant loads and flooding risk. Consequently, a different set of stream management tools should be applied to these subwatersheds--including BMPs designed to remove specific pollutants, greenways, pollution prevention and "hotspot" management. Some additional ideas for managing streams based on their impervious cover classification can be found in Table 2.

With intensive stream restoration and retrofitting, it may be possible to change a stream's designation from impacted to sensitive, or from non-supporting to impacted. Consequently, a restoration management strategy is followed in these subwatersheds

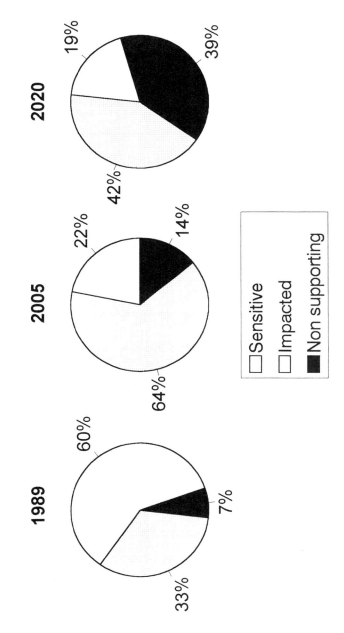

Figure 3. Projected Stream Classification in the Occoquan Basin as Impervious Increases from 1989 to 2020 (Schueler, 1996b)

Table 2

Subwatershed Development Criteria based on impervious cover.

Sensitive Streams (0 to 10 percent impervious cover)

Goal:	Maintain predevelopment biodiversity
Land Use:	Watershed and site impervious cover limits
BMPs	maintain predevelopment hydrology and recharge
	emphasis on ED and infiltration
	restrictions on wet ponds
	"country drainage"
Buffers:	widest stream buffers, protection sensitive areas
Monitoring:	biological, including single species (e.g., trout)
Other tools:	land acquisition, clearing limits, extra ESC control

Impacted Streams (11 to 25 percent impervious cover)

Goal:	Limit degradation of stream habitat and quality
Land Use:	Upper limit on sub-watershed impervious
BMPs	All: emphasize pollutant removal/channel protection
Buffers:	Standard three zone, variable width stream buffers
Monitoring:	biological and physical indicators
Other tools:	regional pond systems, low input lawn care, site planning techniques

Non-supporting Streams (26% percent or greater impervious cover)

Goal:	Minimize downstream pollutant loads/prevent floods
Land Use:	No watershed cap, redevelopment encouraged
BMPs	Maximize removal of phosphorus/metals/toxins
	No restrictions on ponds and wetlands
Buffers:	Greenway for recreation/flood protection
Monitoring:	water quality trends and loads
Other tools:	pollution prevention, illicit connections, "hotspot" management,

Example 4: Restorable Stream* (non-supporting or impacted stream)

Goal:	Restore stream biodiversity to impacted or sensitive levels
Land Use	Limited watershed redevelopment with full BMPs, some infill
BMPs:	Subwatershed restoration w/ stormwater retrofit ponds and wetland creation
Buffers:	Acquisition or easements on stream corridors, riparian reforestation
Monitoring:	biological monitoring, citizen monitoring.
Other tools:	pollution prevention, "hotspot" management, watershed awareness, fish barrier removal, floodplain wetland creation.

---

* potential candidate for restoration based on completion of subwatershed restoration inventory.

(see Table 2). The key question is whether or not the stream is a suitable candidate for restoration. At the present time, the feasibility of stream restoration can only be determined after a detailed screening process involving both desktop and field surveys of stormwater retrofit and stream restoration sites (Claytor, 1995). If a subwatershed emerges from this process as a strong candidate for watershed restoration, the stream classification model can be a useful tool to set expectations for restoration and define objective criteria to measure long-term trends. As such, it can provide a useful context in which to interpret the notoriously variable physical and biological data that are increasing being collected by watershed managers.

Several communities are now developing watershed management strategies based on variants of the basic stream classification model. One notable example is currently being implemented by Montgomery County, MD (1996). Some additional guidance on crafting subwatershed protection plans using the stream classification model can be found in Schueler (1996c).

3. Sub-watershed-based Zoning

The stream classification model can be used to test whether existing zoning can maintain or support aquatic resources in the future. The model provides a quantitative framework for evaluating future land use decisions. The entire process, known as subwatershed-based zoning, is outlined in Table 3. In short, a jurisdiction analyzes its inventory of subwatersheds, and classifies streams based on current and future impervious cover. If future growth is expected to downgrade a stream's classification, the current zoning or density in the subwatershed may need to be decreased to maintain stream quality. Thus, a sub-watershed wide impervious cover cap of 10% and 25% is set to maintain the quality of sensitive and impacted streams, respectively.

Non-supporting streams (that are not candidates for watershed restoration) have no watershed impervious cover caps, and new development or redevelopment is encouraged in their subwatersheds. Growth in impervious cover is only allowed in a subwatershed up to the impervious cover cap. If additional growth occurs, it must be shifted to other subwatersheds, that have additional room under their impervious cover cap, given their stream classification. In most cases, the growth should be directed to non-supporting watersheds that have the needed infrastructure to support it. Subwatershed zoning tends to complement growth management, as it provides an incentive to direct growth where it has occurred in the past, thus avoiding sprawl.

Several communities are now experimenting with subwatershed-based zoning as part of their comprehensive land use planning process, and an even greater number have imposed impervious cover caps to protect sensitive streams (see MNCCPC, 1995 for an urban trout stream). Still, the concept of sub-watershed based zoning is still in its infancy in most regions of the country, and it will be interesting to see what its final form may become.

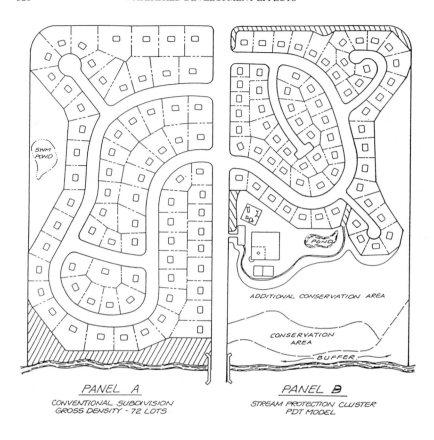

PANEL A
CONVENTIONAL SUBDIVISION
GROSS DENSITY - 72 LOTS

PANEL B
STREAM PROTECTION CLUSTER
PDT MODEL

Figure 4. Comparison of Impervious Cover Created by Conventional Large Lot Subdivision with a Stream Protection Cluster, with narrow roads (Schueler, 1995b)

4. Impervious Cover Management

In a general sense, the stream classification model drives planners and engineers to reduce the creation of impervious cover at the development site or a subwatershed, particularly if impervious cover in the watershed is close to cap.   Significant opportunities exist to reduce impervious cover without changing existing zoning (City of Olympia, 1996) through impervious cover management (ICM). ICM involves a thorough analysis of existing local subdivision, road, parking and drainage codes and criteria to reduce the amount of impervious cover generated by a unit of development. The four basic elements of ICM are:

> smaller residential street features,
> smaller and more integrated parking lots,
> more compact and clustered developments,
> and disconnection of impervious areas.

The potential reductions in impervious cover associated all four ICM elements can range from 10 to 50%, depending on initial lot size and subdivision layout. Performance criteria for ICM can be found in Arendt, 1994, Schueler 1995b. Some indication of the value of ICM can be seen in Figure 4 which shows a conventional residential subdivision compared to a cluster subdivision with narrow streets.

Summary:

It is not enough to identify and measure the cumulative impacts that impair the quality of urban streams; we must advocate measures that can prevent these impacts from being an inevitable consequence of growth. The low impervious cover thresholds evident in the stream classification model suggest that preventing cumulative stream impacts in the future will be a challenging task. Impervious cover, in its dual role as both an indicator of stream quality and as a management tool, provides planners, engineers and landscape architects and other involved in the development process a practical paradigm for managing urban subwatersheds.

References:

Arnold, C. and C. Gibbons. 1996. Impervious surface coverage: the emergence of a key environmental indicator. Journal of the American Planning Association. 62(2): 243-258.

Arendt, R. 1994. Designing Open Space Subdivisions--a practical step-by-step approach. Natural Lands Trust. W. Alton Jones Foundation. Media, PA. 148 pp.

Chesapeake Bay Program. 1995. Water Quality Functions of riparian forest buffer systems in the Chesapeake Bay watershed. U.S. EPA CBP. Annapolis, MD. 220 pp.

City of Olympia, 1996. Impervious Surface Reduction Study. Final Report. Olympia Public Works Department. Olympia, Washington.

Claytor, R. 1995. Assessing the potential for urban watershed restoration. Watershed Protection Techniques. 1(4): 166-172.

Claytor, R. and W. Brown. 1996. Environmental Indicators to Assess the Effectiveness of Municipal and Industrial Stormwater Control Programs. Final Report. Center for Watershed Protection. US EPA Office of Wastewater Treatment. Silver Spring, MD. 210 pp.

Duda, A and K. Cromartie. 1982. Coastal pollution from septic tank drainfields. Journal of the Environmental Engineering Division (ASCE) 108 EE6.

Galli, J and R. Dubose. 1991. Thermal impacts associated with urbanization and stormwater best management practices. MWCOG-DEP. Washington, DC. 188 pp.

Hicks, A. 1995. Impervious surface area and benthic macroinvertebrate response as an index of impact on freshwater wetland. M.S. Thesis. Dept. of Forestry and Wildlife Mgmt. University of Mass. Amherst, MA.

Horner. R, D. Booth, A. Azous and C. May. 1996. Watershed determinants of ecosystem functioning. (this volume)

Jones, R. C. 1996. Bioassessment of the BMP effectiveness in Mitigating Stormwater Impacts of Aquatic Biota. (this volume).

Macrae, C. 1996. Experience from morphological research in Canadian streams: is the two year frequency runoff event the best basis for stream channel protection? (this volume)

Maryland National Capital park and Planning Commission (MNCPPC). 1995. Upper Paint Branch watershed planning study--technical report. Montgomery County Planning Department. Silver Spring, MD. 68 pp.

Maxted, J and E. Shaver. 1996. The use of retention basins to mitigate stormwater impacts on aquatic life. (this volume)

Montgomery County, Maryland. 1996. Draft County-Wide Stream Protection Strategy. Department of Environmental Protection. Rockville, MD.

Northern Virginia Planning District Commission (NVPDC). 1978. Land use/runoff quality relationships in the Washington DC metropolitan area. Final Report. Occoquan/Four Mile Run NPS Correlation Study. MWCOG. 73 pp.

Schueler, T. 1987. Controlling Urban Runoff: A Practical Design Manual for Planning and Design of Urban BMPs.

Schueler, T. 1994. The Importance of Imperviousness. Watershed Protection Techniques 1(3):100-111.

Schueler, T. 1995a. The Peculiarities of Perviousness. Watershed Protection Techniques. 2(1): 233-238.

Schueler, T. 1995b. Site Planning for Urban Stream Protection. Center for Watershed Protection. Metropolitan Washington Council of Governments. Silver Spring, MD 222 pp.

Schueler, T. 1996a. The Limits to Growth: Sustainable development in the Occoquan Basin Center for Watershed Protection. Silver Spring, MD. 104 pp.

Schueler, T. 1996b. Irreducible Pollutant Concentrations Discharged from Urban BMPs. Watershed Protection Techniques. 2(2): 369-372.

Schueler, T. 1996c. Crafting Better Urban Watershed Protection Plans. Watershed Protection Techniques. 2(2): 329-337.

Southerland, R. 1995. Methodology for Estimating the Effective Impervious Area of Urban Watersheds. Technical Note 58. Watershed Protection Techniques. 2(1): 282-284.

Steedman, R. 1988. Modification and assessment of an index of biotic integrity to quantify stream quality in Southern Ontario. Canadian Journal of Fisheries and Aquatic Sciences. 45: 492-501

Taylor, B. L. 1993. The influence of wetland and watershed morphological characteristics to wetland vegetation communities. M.S. Thesis. Dept of Civil Engineering. University of Washington, Seattle, WA.

DISCUSSION

The Use of Retention Basins to Mitigate Stormwater Impacts on Aquatic Life
John Maxted/Earl Shaver

Physical habitat and biological measurements were taken in nontidal streams below eight stormwater management pond facilities (BMPs) during the Spring of 1996. The results were compared to 38 sites with no BMPs sampled in the Fall of 1993. Biological quality was determined from six metrics using 100-organism subsamples identified to the species level. Physical habitat quality was determined from twelve metrics that defined the condition of the channel, stream bank and riparian zone. The overall macroinvertebrate community, as measured using the Community Index, was not significantly different between BMP and non-BMP sites. A similar result was found using the Sensitive Species Index. The BMPs did not prevent the almost complete loss of sensitive species after development. The BMPs did not attenuate the impacts of urbanization once the watershed reached 20% impervious cover. More data are needed to determine whether these controls attenuate impacts at lower levels of development (0-15% impervious cover).

Questions/Comments

Question:	How did you compare temperature to the reference site?
Answer:	We didn't do it.

Comment: Suggest that you incorporate more physical information and statistics into performance evaluation.

Comment: Hydrologic criteria for capture and release should be based on land use. It appears that these BMPS may be designed to control the 2-yr storm. If so, the findings are not surprising since 90 percent (approximately) of the storms are smaller than the 2-yr storm.

Comment: John says that the water quality is still affecting the downstream biota, but the comment is that those BMPs are not functioning properly. The outlet hydraulics are not properly designed to handle small storms.

Shaver: Have we solved the hydraulics problem, yes; is the biota similar to the reference site, no. The biological indices do not show any difference between sites with BMPS and those without; but the visual examination shows much improvement for those site with BMPs.

Barbour: Have not been real clear on reference sites --- in beginning of week we were talking about smaller streams, there is not much we can do about the Mississippi. When we select reference sites, there is not much we can do about bringing back a water to its pristine condition, but we do need a place to start.

Yoder: We are not completely awash in this area, but we need the states to get on board and start developing the data. There needs to be some national leadership in this area so that the toolbox is there when we get there. I'm really impressed with Warwicks slides, they are right on. Attainable standards can change over time, how the potential changes, is strongly related to land use.

Maxted: Any arid region is going to have difficulty developing reference sites because the of the hydrologic modification of the streams due to irrigation withdrawals

Bannerman: In Wisconsin the best reference sites are found below lakes. This makes the point precisely; the residence time allows the pollutants to settle out.

Use Of Impervious Cover To Protect Streams In Urban Watersheds
Tom Schueler

Studies indicate that certain thresholds of stream quality exist, most notably at about 10% impervious cover, where sensitive stream elements are lost from the system. A second threshold appears to exist at around 25% impervious cover, where most indicators of stream quality consistently shift to a poor state or condition.

This simple model classifies streams into one of three categories, based on impervious cover: sensitive, impacted, and non-supporting.

Although the stream classification model is simple, it is important to consider some of the technical issues involved in its development and application.

Impervious cover, in its dual role as both an indicator of stream quality and as a management tool, provides planners, engineers and landscape architects and other involved in the development process a practical paradigm for managing urban subwatersheds.

Riparian Stewardship in the Post Regulatory Era
The Stick, The Carrot and the Light Bulb
Robert M. Searns, AICP[1]

Abstract

In the current political climate, there is a push to lessen the role of government in resource protection. There is a backlash against regulation and many put property rights first, regardless of the cumulative impact of individual land use decisions. Public funding of solutions is an increasingly daunting task with financially strained federal, state and local governments. Yet, more than ever, there is a need to preserve, protect, and enhance riparian resources as urbanization expands.

In the emerging "post regulatory era" new directions, solutions and currencies are being sought. The goal is to create durable institutions of stewardship that can endure both political and economic cycles. While regulation has its place, there is a need for positive incentives (as well as the elimination of the "perverse"ones). There is a need to build a broad-based ethic of watershed resource protection. Policy makers and professionals need inducements and reassurance to try new technologies and interdisciplinary approaches. This paper suggests some new directions in facing these challenges.

A Reminder of What Men Choose to Forget

> *I do not know much about gods: but I think that the river is a strong brown god--sullen untamed and intractable, patient to some degree... useful... as a conveyor of commerce... only a problem confronting the builder of bridges.*

---

[1] Principal, Urban Edges, Inc., a greenway planning, design and development firm, 1401 Blake Street, #301, Denver, CO 80202

Robert M. Searns

The problem once solved, the brown god is almost forgotten by the dwellers in the cities--ever, however, implacable, keeping his seasons and rages, destroyer, reminder of what men choose to forget.

--T.S. Elliot (quoted in *Taming the Flood* by Jeremy Purseglove)

Its a small world, but I wouldn't want to have to paint it.

--Steven Wright

In 1993, a "rogue" weather system moved through the American heartland triggering devastating floods in the Missouri and Mississippi River basins. With damages running into the billions, this may have been the epiphanal moment in the way policy makers think about storm water. For decades up until then, the paradigm focused primarily on the concept of conveyance--viewing storm water as a nuisance--something to carry away as quickly as possible. This view point was largely based on the economics of maximum land utilization and failure (or unwillingness) to consider the cumulative impacts and true costs of this approach.

That same year, 100 people died and thousands were sickened in Milwaukee, WI by an outbreak of cryptosporidium in that city's drinking water (Mitchell, 1996). Livestock waste in the watershed was the suspected culprit. Along with the floods and the illnesses, there was also the continued loss of vital riparian and wetland wildlife habitat. The 1993 floods and, perhaps, the cryptosporidium outbreak, brought home the message--a two-by-four over the head--that there is another basic component to storm water management--water storage! Many people began to take notice that something might be very wrong--and costly--in the way we manage watersheds and run-off.

At the same time, sadly and ironically, the *104th Congress*, in its zeal to cut back on regulation, stood poised to weaken the *Clean Water Act* and place "property rights" as paramount objective seemingly denying the cumulative impacts of bad land use decisions. A number of state legislatures seemed to be going in the same direction. Indeed, many in national and state level leadership seems transfixed by a new dogma of denial. In part, this attitude can be traced to special interests, and, some might say, opportunistic greed, but it is also the predictable result of past policies that did not always work, policies that were sometimes arbitrary and even abusive.

Deregulation, however will not make the flooding, wildlife habitat and water quality problems go away nor will poorly crafted regulation. Clearly new, enduring institutional structures are needed to build a lasting era of effective watershed management. Hopefully, by now, most of us have grasped the concept, at least in principal, that watersheds and the rivers and streams they generate are complex systems. They are perhaps, the single most important component of the

natural "infrastructure"--givers of life, conveyors of our wastes, and places of
solace. Yet, they are brutally unforgiving when we get in their way. For some of
us, to see the light is a good start, but how do we convey this concept and win
over irate property owners, elected officials and taxpayers?

The problem really boils down to two basic flaws in the way we address the
issues of run-off. First, we have not really understood, or chose not to understand,
how watersheds work. Second, because of relative wealth in both dollars and
natural resources, our society has been making imprudent choices in how to live
with drainage systems. We have viewed run-off primarily as a "nuisance" and we
have visualized rivers and streams monolithically, as "ditches"--utilitarian
conveyors of water. We are just now beginning to see them as the complex and
valuable geomorphologic and biological systems that they are. The general public,
thanks in a large part to the greenway movement, is also beginning to see them
as amenities.

Traditionally, "engineered" solutions have been considered a "good investment" because developers and policy makers believed the lost land use opportunities were too costly to justify staying out of the flood plain. Natural flood storage
in the watershed was rarely considered an option. If you can contain and carry off
the water quickly--the reasoning goes--then the adjacent land is too valuable to set
aside. In addition, because roads and utilities must cross flood plains, you must
channel and direct the flow to protect this infrastructure investment from flooding,
erosion and washout. In some ways, the latter argument does carry more weight
because communities must have the infrastructure connections even if the flood
plain is left undeveloped.

For decades, this paradigm seemed to work and has driven pubic policy.
Because devastating floods happened relatively infrequently relative to the human
life span (and, perhaps, the amortization period of a mortgage) and because local,
state and federal government had the disposable resources to bail people out, there
was little motivation to change. In this way, we have been able to institutionalize
irrational behaviors toward storm drainage.

Sadly, the problems caused by past mistakes tend to compound themselves
as good money follows bad. By favoring conveyance over natural storage, we
have created a massive self-perpetuating "flood control" infrastructure. This
system, by its nature, demands the creation of even more channels, culverts and
dams to contain and manage the increased flows, sediment and erosion that
conveyance oriented system generate. To compound this, many urban watersheds
are zoned for vast expanses of additional impervious surface and many flood
plains are already confined. Furthermore, there is a reluctance by design
professionals and decision-makers to take "risks" by trying new systems that
embrace storage and natural stream functions. Thus, we continue to transform

living streams into "dead" or near-dead culverts loosing wildlife habitat, clean water, places of beauty and flood storage in the process.

Making the institutional changes needed to remedy this may be like trying to change the course of a ship at full steam. It will not change easily or quickly and it will necessitate the introduction of new and powerful forces that demand change. It appears that in addition to the appropriate and necessary national and state level regulation, we need a new paradigm that begins at the local grass roots level and works its way up. The model should be based on four key elements: stewardship; positive incentives; accountability; and suitable technology. Hopefully education, economics, and the demand for outdoor recreational amenities will be the key engines of change.

Stewardship

Without a broad based understanding of how watersheds work and the consequences of their abuse, it will be difficult, if not impossible, to institute long term policies that protect them. The individual incentives for abuse, especially by those who have not or will not fall victim are just too strong. While most Americans have a high level of concern about environmental issues, sadly, less then 1 in 20 even knows what a watershed is, let alone understands its functions (Lavigne and Coyle 1995). Water storage and water cleansing benefits are not yet grasped by many citizens and elected officials. Amazingly, a recent national survey presented to the *American Rivers* organization showed that 30% of the population think that drinking water comes from the Ocean. Actually, more than 70% of our drinking water comes from rivers and streams (Lavigne and Coyle,1995).

It is clear that a public education process is needed because the notion of stewardship lies at the heart watershed management. There needs to be a broad understanding that healthy functioning natural systems such as flood plains and wetlands are a part of our "public infrastructure" and that they are as important economically as roads, water lines and power plants. This also implies the need for expanding the base of professional knowledge about watersheds and how they should function in urbanized areas.

An important place to begin is with instilling the notion that floods are not "acts of God". Rather, they are the consequences of humans altering natural drainage and absorption patterns and then placing themselves in the resultant path of destruction. To change this thinking implies changing the way we describe and the way the public grasps drainage systems. Indeed, as pointed out in a recent Lincoln Institute of Land Policy paper, authored by Scott Faber, even the language we use in describing watershed phenomenon has been misleading and destructive. Terms like "flood control" creates the dangerously false impression that we can actually control floods or that levees really guarantee protection. Many people also think that the "100-year flood" only comes every 100 years and that the extent of more frequent flooding is somehow immutable in the face of development in the watershed (Faber, 1996). "Flood disaster crap shoot zone" might be a better term because the flood can come anytime.

Until now, the message to the public has been conveyance in the channel as if storm water should be handled like traffic on a highway--the faster and straighter, the better. Ron Flanagan, a planner and pioneer of Tulsa's progressive storm water management program puts it simply yet eloquently understandable, when he says we must consider both *conveyance* and *storage* (Flanagan, 1996). We need to describe drainage systems as an interacting set of layers. At the core is the channel that conveys run-off water but there are also the various levels of flood storage, dynamic areas of geomorphologic equilibrium, and buffer zones that separate these areas from human settlement. There are also the intricately connected subsurface aquifer and water table systems. To see it as anything less that this multi-tiered system is to place ourselves ultimately in harm's way.

This message needs to be gotten out. Engineers, planners and public officials must get in the habit of consistently using holistic and systemic terms when presenting flood hazard reduction problems and programs. They also need to consistently identify the true costs of lost storage capacity and the civic benefits of setting land aside for storage in the form of parks, open spaces and greenways.

The term "greenways" may be the key "household word" in getting the concept of storage and stewardship across to a broad segment of the population. We have to talk about more than just flood control to reach people because flooding tends to be a concern primarily for people who are flooded--a relative minority and then only after they have been flooded. There is far less interest on the part of those whose properties generate the runoff that causes the flooding but don't live or work in the floodplain. The growing greenway movement has offered a very effective way to pull more people in. Greenways can help answer the "what's in it for me?" question many of those outside the floodplain might ask when called upon to support open space acquisition or other expenses associated with proper watershed management. Similarly developers might be more likely to accept streamside setbacks and other requirements when the "greenway card" is played.

Several years ago, a National Geographic article documented over 500 greenways projects underway nationwide (Grove, 1990). Today, that number may have increased several fold. While greenways got their start as primarily recreational amenities, the concept has expanded to embrace an array of objectives including: trails, habitat preservation, outdoor education, economic development, interpretation, and other benefits including flood control. By affording access, community involvement and interpretive opportunities, greenways teach people that streams are more than ditches.

Indeed, in many communities, the greenway movement has spawned the "multi-objective" approach to storm water management and drainage engineering. Now multi-disciplinary teams with urban designers, ecologists, and landscape architects do the planning--not just engineers, and engineers have gone beyond hydrology and hydraulics to embrace multiple objectives. Under the greenway banner, Denver, CO transformed its long abused South Platte River from a "storm sewer" into an amenity. Similarly, Arlington, TX said no to the extension of a hard-lined channel that devastated a reach of urban creek just downstream of the city limits. Arlington also raised over $8 million to create a riparian preserve and environmental education center that focuses exclusively on the function and value of sustainable floodplains and watersheds.

Community leaders in Toronto, Canada area have taken the greenway concept a step further in visualizing the "Greater Toronto Bioregion"--an integrated system of preserved drainageways and uplands that overlays the 4000 square mile metropolis encompassing of 15 river and stream system, the Lake Ontario Shore and the Niagara Escarpment. This "galaxy of greenways" includes trails, riparian preserves and wildlife habitat serving a project human population of six million. Sustainable watershed and subwatershed management that preserves ecological values and stormwater storage form the basis of the guiding principals that pull together nine metro Toronto jurisdictions. A look at a current Toronto street map demonstrates that they are well on their way to success because an earlier generation set aside the Don River Valley system of preserves, trails and parks (Taylor, Paine and Fitzgibbon, 1995).

The greenway concept can also inspire direct citizen participation. Streamside property owners, schools, service clubs and community volunteer groups can be important partners in watershed stewardship. Residents and business owners along Johnson Creek in Portland, OR received a *Stream Friendly Yards and Gardens* pamphlet that stresses the importance of healthy watersheds and provides tips, plant lists, and other advice so that each stream-adjacent individual can take on a personal role, voluntarily, by controlling erosion, not dumping contaminants and expanding habitat-supportive vegetation on to their own land (Johnson Creek Corridor Committee, 1994). Also, throughout Portland, the city is stenciling fish icons onto storm drain grates suggesting the link that every one in the watershed has with the quality of the streams that drains it. Near Denver, CO school kids, service groups and companies have teamed up to restore and revegetation stream banks and wetlands. Indeed, creeks and wetlands make great outdoor classrooms and schools can be great caretakers by adopting, monitoring and enhance their own nearby greenways.

Instilling the concept of stewardship also implies the need for new local management structures that reflect watershed structure and dynamics. Presently, most communities manage storm drainage on the basis of jurisdictional, not drainage basin boundaries. Each community concerns itself with its piece of the puzzle and not the big picture. The result is a patch work of stop-gap solutions often with the next community downstream bearing the brunt. Slowly the viewpoint is changing. The concept of watershed management rather than channel management is coming into its own. There are now hundreds of watershed councils, coalitions and programs across the nation and some communities have put institutional structures in place to better facilitate watershed thinking, effective planning and policy making on a drainage basin-wide basis.

One of the pioneering efforts is the metro Denver, CO six-county Urban Drainage and Flood Control District. Created under Colorado enabling legislation, the District has multi-jurisdictional property taxing authority to raise funds, plan and implement regional drainage solutions. Another example is the Lake County, IL. Stormwater Management Commission. Fed up with the $5.2 million per year in flood damage, Lake County has put in place a comprehensive watershed management plan. As part of the program the commission raises funds from property taxes and enforces a Watershed Development Ordinance in 36 communities. They are also pursuing non-structural, sustainable stormwater storage approaches such as wetlands and natural floodplain preservation (Miller, 1994). Springfield, MO has a coalition of public and private agencies that supports a watershed management agency with full time paid staff. This concept of watershed-based storm water planning, management, finance and enforcement needs to spread nationwide.

Finally, there is a need to instill the notion of stewardship amongst the professions who plan, design and manage stormwater infrastructure. This implies a major shift in the traditional approach to storm drainage planning and design and in the way we educate engineers, planners, geomorphologists, ecologists, landscape architects, and even lawyers. Each must understand that complex nature of drainage systems and their important social and economic values and the need to take an multi-objective, interdisciplinary approach.

Appropriate Incentives

There are two kinds of incentives--*sound* incentives and *perverse* incentives. Sound incentives promote outcomes that are beneficial to both the individual and society at large. Perverse incentives may benefit the individual in the short run but, in the long run, they are detrimental to both the individual and society (Davison and Ress-Moog, 1993). *The National Flood Insurance Program (NFIP)* and many *U.S. Army Corps of Engineers* funded channelization projects are examples of perverse incentives. Sadly, so is victim-relief funding. Each solves, or appears to solve, an immediate individual or community problem, but in the long run, some of these actions make the problem far worse. Established in 1968 and expanded in 1973, the NFIP program defined flood plains in communities across the nation and offered insurance to property owners in the floodway fringe.

Unfortunately, the NFIP has been flawed in several ways. Required mapping of the flood plain does not necessary take into account the effects of full development in the watershed. Emphasis is placed on protecting the flood channel and not the flood plain and government-subsidized insurance creates the false impression of security and incentive to reduce flood storage by filling in the floodway fringe. Perhaps, most glaring is the fact that the insurance pays for repairs of flood damaged structures but not for relocation or flood proofing. The National Flood Insurance Reform Act of 1994 attempted to address some of these shortcomings with stricter compliance requirements and a Community Rating System that lowers premium rates for jurisdictions that go beyond the minimum requirements but this does not go far enough (Faber, 1996).

Tulsa, OK set out to expand the flood plain regulatory zone beyond the NFIP 100 year flood map and is removing buildings from the path of floods (Schwab,1996). Other communities have embraced this idea. The city of Louisville and Jefferson County KY, recently published a *Multi-Objective Stream Corridor/ Greenway Plan* in which the concept of a regulatory flood plain is introduced. Based on calculated flows from a fully developed watershed, the Louisville concept will require land owners to prove that any proposed development in the flood plain or flood fringe will not cause flooding of up or downstream properties (Louisville and Jefferson County Greenways Advisory Committee, 1995).

Both federally funded channel projects through the Corps of Engineers and federally flood victim relief policies should be similarly modified. All federally funded projects and programs as well as state and local projects should require proof that the action will not ultimately result in increased flooding up or down stream. Littleton, CO set a precedent in this area in the late 1960's when they forced the Corps to consider a non-structural alternative to channelization of the South Platte River below a newly build flood control dam. The result is a 600-acre "flood plain" park the preserves a more natural setting for the river. The park now teams with wildlife, offers recreational trails and a nature center. Premium homes sprouting on the fringe of the park derive much of their value from the preserved flood storage amenity.

Harsh as it may sound, another important step is to stop subsiding flood hazard exposure. Flood victim relief payments should require and facilitate the relocation of flood prone properties out of the flood plain. Following Tulsa's example, a number of communities are instituting this policy. It needs to be a national policy.

On the private sector side, there are increasing economic incentives to stay out flood plains and wetlands. A recent survey of many of the nation's largest home builders indicates a strong market preference by home buyers for "natural" open space amenities, wildlife viewing and greenways (Harney, 1995). There are many examples of these kinds of projects. At *Ken Caryl Ranch*, a subdivision built near Denver, home buyers pay a substantial lot premium to have property facing preserved flood plain and open space areas that have been converted into amenities with trails and wildlife habitat. Near Portland, OR, the *Mentor Graphics* office park features views of a preserved wetland. Clustering of development to preserve natural infrastructure also saves substantial costs in building roads, utility connections and other human-built infrastructure. It is important that local planners and policy makers facilitate these kinds of positive incentives through supportive development ordinances and regulations.

Finally, there may be opportunities for the direct beneficiaries of sound watershed management policies to build effective partnerships and financially reward good stewardship. When the purity of New York City's drinking water was threatened by burgeoning development in the *Catskill/Deleware* watershed, the city faced a choice--spend $4 billion to $8 billion on a water filtration system or institute policies to protect the quality of the water at the source. The city chose to invest $1.5 billion in a sustainable watershed development and management program including land acquisition, wetland creation and run-off management (Strutin, 1996). The program encompasses multi-jurisdictional, multi-agency and multi-state partnerships where New York City rate-payers get clean water and watershed communities can better preserve and manage the quality of their communities with preserved open space and sustainable growth.

Accountability

For decades, private (and public) development has been able to occur relatively cheaply because of public ignorance about the cause of flooding and the fact that this cost could always be passed along to someone else. The right to a fair economic return on one's property had somehow become confused with taking license to damage someone downstream. In reality, realizing an economic return on property when storm run-off has not been contained has been at taxpayer and flood victim expense. This has been "deficit spending" in its worst form because we have severely compromised the natural infrastructure that we need for storm water storage and conveyance and safe drinking water, not to mention recreational and wildlife benefits.

Although, in some places, environmental regulation may be on the wane, the values of fairness and accountability, hopefully are not. The concept of holding individuals and entities directly responsible for the financial consequences of the environmental damage they do is not without precedent. There is a long tradition of water law that governs the stewardship and distribution of water largely through the courts rather than by regulation. This is especially true in the arid west where the rights of downstream users are carefully protected provided they have acquired "rights" to the water. The reason this system works is because water is an essential, but irreplaceable economic commodity.

Shouldn't downstream property owners have similar rights not to be damaged by flooding or erosion? Shouldn't downstream water consumers have rights to clean, safe drinking water not to have there water contaminated by upstream abuses? Shouldn't we all have the right to wildlife for viewing, fish to catch and natural stream corridors to enjoy? Unfortunately these values have not yet been able to muster the political and economic clout that water as a commodity has. This is the case, perhaps, because these resources are not yet scarce enough although they are becoming so. It is also difficult to build an economic constituency for resources that are hard to own or control. Nonetheless, those that damage these resources must be held accountable and we need to develop ways to measure the costs. If federal and state law makers are reluctant to regulate property owners, then maybe we need to further explore accountability through tort liability.

Economist Paul Hawkin in his book *The Ecology of Commerce*, writes about the imposition of fees on activities that adversely impact the environment. These "green fees", as he calls them, are a proportional assessment of environmental impact that are applied directly to mitigating the damage done (Hawkin, 1993). The benefit of such fees are two fold. First, they provide a strong financial incentive to minimize adverse impact and, second, they provide for accountability. In contrast to conventional regulation, impact fees also give the land owner or developer flexibility and the opportunity to be innovative--to come up with their own solutions provided the environmental standards are met.

The cost of creating impervious surfaces, a good example of an adverse environmental impact, that can be measured. Tulsa's Ron Flanagan calculates that lost flood storage can be priced at $100,000 per acre foot or more to replace (Flanagan, 1996). This lays a strong foundation for funding stormwater utilities through a "service charge" based on development impact on the system rather than paying for stormwater protection and flood damage mitigation with general tax dollars.

Effectively structured stormwater utilities a good start in this process. A 1992 survey indicated more that 200 stormwater utilities in 18 States. Fourteen of these were county-wide programs and, hopefully, this number is expanding (Miller, 1994). A number of these have storm water utility fees based on the amount of impervious surface. These fees pay for planning, enforcement and, in some communities, capital improvements. Why not expand the concept to cover the full cost of watershed management including using impervious surface fees to acquire and maintain open space, wetlands and riparian areas.

Ideally, each watershed should have a master plan that determines acceptable levels of storm water discharge, water quality and habitat preservation. The cost of creating and maintaining that condition is calculated and then assessed per square foot of existing or new impervious surface. Public entities, incidentally, should not be exempted. Roads, bridges, channelized stream and culverts, built by public entities, should be assessed and made equally accountable.

Again, the "greenway card" can be played here because an important side benefit of this kind of planning is the creation of parks and other outdoor public amenities. Ron Flanagan, interestingly points out that, in many parts of the country, floodplains, account for about 7-10 per cent of the land, this directly corresponds to the 10 per cent of land suggested for park, recreation and open space needs. Note also that surveys show 80% of the population prefer to recreate near water (Flanagan, 1993).

Appropriate Technologies

There are a number of technologies, some new and some not so new that can be pursued, improved upon and deployed. Ouray, CO, for example, installed a 2-acre wetland in lieu of a conventional secondary treatment plant. In the process, they created habitat, storm water absorption capacity and in the process they save $10,000 per year in chemical treatment expenses (Cox, 1996). Arcata, CA turned their wetland treatment facility into a 154-acre nature center a side benefit that attracts tourists each year. Designers of a Walmart store for Lawrence, KA pro-posed an at-grade stormwater collection and storage system that uses parking lot run off to irrigate landscape plantings and in the process reduces the need for underground stormwater piping by up to 60% and irrigation systems by up to 30% (Wenk, 1996). Use of pervious paving surfaces, on-site detention techniques, soil bio-engineering and cluster development are all examples of technologies to solve flooding problems.

We need incentives in both the public and private sector to continue to explore and implement these solutions and to build partnerships among different interests and disciplines. Part of the crux, however, of introducing new technolo-gies and approaches to watershed management is the issue of breaking traditions that have endured for decades. Changing a paradigm requires risk taking, sound and proven alternative solutions as well as strong evidence that the old approach does not work. With the passage of time and mounting flood costs the flaws in the old paradigm are slowly becoming evident but there also needs to be other incentives and encouragement for engineers, planners and policy makers to risk trying new or different ideas.

Watershed planning and engineering budgets should include funding to cover tinkering with more sustainable techniques that have not been field proven. This suggests that need for national, state and local funding sources, be they public or private, the provide incentives and perhaps a "risk pool" to mitigate help indemnify worthy demonstration projects.

Conclusion: It Best and Ultimately, Cheapest If We Can Stay Out of Nature's Way

Bruce K. Ferguson probably put it best when he wrote in a recent article: "if we can just stay our of nature's way it will work" (Ferguson, 1995). Now we are just learning how to stay out of nature's way. Unfortunately, past mistakes and current circumstance still, in many cases, make staying out of her way impractical or unfeasible. When that happens, we need to assess the true costs and insist that mitigation is addressed and paid for up front. Indeed, this should be our ethic: *no net degradation* of watershed systems. To do otherwise is, ultimately, immoral and should be illegal.

Obviously, regulation is important, but it cannot be solely relied on to assure that people do the right thing. Instilled values of stewardship, incentives, accountability and new technologies can go a long way toward solving problems before having to rely on regulation--although appropriate regulation is vital to steer public and private sector decision-making in the right direction. Over the long term, it is broadly held human values that will make the difference because laws come and go. The real and painful costs of flood damage, erosion and drinking water contamination are the true "stick". Greenways, open spaces, parks, and sound of songbirds are the true "carrot". The multi-objective approach is the true "light-bulb".

As planners, engineers, designers and policy makers we are the shapers of the landscape. When we transform a stream into ditch to expedite a short term political or economic objective, we are simply passing the along the problem (and the cost) to the folks downstream (and along to our children). Knowing what we now know, we can no longer plead ignorance. We must similarly advise and educate our clients and employers so that they too can no longer plead ignorance. It is imperative that we incorporate the stewardship and accountability ethic unyieldingly and uncompromisingly into all of our work. We can no longer be simply facilitators of development. Rather, we must be the intermediaries between human enterprise and nature--healers of the landscape.

References

Cox, Jack, Feb. 1, 1996. Wetlands Bring Bonus Benefits to Towns. The Denver Post. pp. 1E.

Davison, James Dale and William Rees-Mogg, 1993. The Great Reckoning. Simon & Schuster, New York.

Faber, Scott, 1996. On Borrowed Land: Public Polices for Floodplains. Lincoln Institute of Land Policy. Cambridge, MA..

Flanagan, Ron D., 1992. Water Quality By Plan: Stormwater Management, Community Open Space/Greenway Corridors and Trail Systems, R.D. Flanagan and Associates, Tulsa, OK.

Flanagan, Ron D., 1993. The Sight of Tears, The Sounds of Laughter: The Need For Multi-Objective Park, Recreation and Open Space Planning. R.D. Flanagan and Associates, Tulsa, OK.

Flanagan, Ron D., 1996. conversations between the author and Mr. Flanagan.

Ferguson, Bruce K. 1995. Preventing Problems of Urban Run-off, in Renewable Resources Journal. 13(4).

Harney, Kenneth, Jan 10, 1996. Homebuyer Preferences Changing. Denver Post. Denver, CO.

Johnson Creek Corridor Committee, 1994. Taking Care of Johnson Creek: A Stream-side Property Owner's Guide to Stream Stewardship. Bureau of Environmental Services, City of Portland, Portland, OR.

Grove, N., 1990. Greenways: Paths to the Future. Nat. Geogr., 177(6): pp. 77-99

Hawken, Paul, 1993.The Ecology of Commerce. Harperbusiness. New York.

Lavigne, Peter M. and Kevin J. Coyle, 1995. The Proceedings of the Watershed Innovations Workshop, River Network, Portland, OR.

Louisville and Jefferson County Greenways Advisory Committee, 1995. Louisville and Jefferson County Multi-Objective Stream Corridor/Greenway Plan. Metropolitan Sewer District. Louisville, KY.

Miller Ward S., October, 1994, Developing a Proactive Watershed Program. in Environment and Development, American Planning Association, Chicago, IL.

Mitchell, John G., 1996 Our Polluted Runoff. Nat. Geogr., 189(2): pp. 108-125

Purseglove, Jeremy, 1989. Taming The Flood. Oxford University Press, Oxford, England.

Schwab, Jim, Jan/Feb1996. Nature Bats Last: The Politics of Flood Plain Management. in Environment and Development. American Planning Association, Chicago.

Strutin, Michele, 1996. Springs of Compromise. in Landscape Architecture. 86 (5) pps. 66-72.

Taylor, James, Cecelia Paine, John Fitzgibbon, 1995. *From Greenbelt to Greenways: Four Canadian Case Studies, Greenways: The Beginning of an International Movement*, edited by Julius Gy. Fabos and Jack Ahern, Elsevier Press, Amstersdam, pp. 47-64.

Wenk, Bill., 1996. Walmart Alternative Storm Drainage Evaluation. a memo by Wenk and Associates, Denver, CO.

PROTECTING FULBRIGHT SPRING: CAN WE BEAT THE ODDS?

Timothy W. Smith, P.E.[1]

ABSTRACT: Springfield and surrounding Greene County (population 210,000) are located in the Missouri Ozarks, a region of limestone karst, characterized by thin stony soils overlying carbonate limestone. The area features many sinkholes, losing streams, caves, and springs.

Fulbright Spring was Springfield's original water supply. It still supplies about 20 percent of the city's water. Because of the area's karst geology, the spring is very vulnerable to changes in surface runoff water quality in the overlying South Dry Sac River watershed. Recent extension of a trunk sewer main up the river valley to serve a major transportation corridor along Interstate 44 and U.S. 65 is expected to accelerate urbanization in the recharge area.

If this resource is to be maintained, preventive measures must be aggressively pursued. The technical implications are relatively well understood. However, non-technical issues and constraints are frequently the true driving forces behind development of watershed management measures. This paper offers an examination of these factors and their implications with regard to implementing an effective watershed management program in time to preserve the spring.

INTRODUCTION

Fulbright Spring was Springfield, Missouri's original public drinking water source. It has been used continuously since 1883 and still supplies about 20% of the city's water.

---

[1] Greene County, Missouri, Stormwater Engineer

Springfield and surrounding Greene County are located in the Missouri Ozarks, a region of limestone karst. In this geologic setting, watershed management for protection of the spring's water quality presents a very intriguing technical issue.

As professionals working in the various disciplines relating to water quality management, our training and personal interest make us well suited to deal with such issues. Our technical inclinations, however, sometimes leave us poorly prepared to cope with, and easily caught off guard by, the non-technical issues which are unavoidable. Only after painful and frustrating experiences, have we learned to keep our technical instincts at bay and give appropriate attention to these other factors.

Watershed management and stormwater management are, in many cases, non-point pollution source issues. The solutions are often as diffuse as the pollution sources. Many of the aspects of "personal pollution" are nearly impossible to control by regulation and technology. The success of a watershed management project hinges on our ability to recognize both technical and non-technical factors and to formulate workable strategies to deal with them.

The purpose of this paper is to examine the non-technical issues which have an important influence on the management of the Fulbright Spring watershed. Though this watershed is unique, it is likely that our situation with Fulbright Spring is common in some fashion to watershed management efforts throughout the country. By sharing our experience we hope that we can give other managers and professionals additional insight into the issues which may influence their projects, while gaining additional insight into our project as well.

PHYSICAL SETTING

Fulbright Spring is located on the northern edge of Springfield. The spring recharge area is approximately 28 square miles in size. Spring discharges range from two to eight million gallons per day (MGD).

The karst terrain of the region creates a complex interaction between the surface and subsurface hydrologic regimes. The limestone weathers readily, forming sinkholes, caves, and other karst features. Losing streams, whose surface flow is "lost" into subterranean passages, are common. Nearly the entire Fulbright Spring recharge area is located in the surface watershed of the South Dry Sac River (a revealing name) and its principal tributary, Pea Ridge Creek.

An impermeable shale layer, the Northview, separates the lower, regional Ozark aquifer comprising Ordovician age sandstones and dolomites, from the Mississippian age calcium carbonate limestones of the upper, Springfield plateau aquifer. A number of faults trend in an east-west direction across the watershed. These provide a connection between the two aquifers, as well as a mechanism for formation of solution conduits within the limestone. Fulbright Spring is a component of the upper aquifer.

The Fulbright Spring recharge area is shown in Figure 1. The recharge area of the spring is not completely understood at present. However, it is likely that the majority of the flow in the spring enters the shallow groundwater in a losing reach of the South Dry Sac River known as the "swallow hole". During base flow conditions all of the flow in the river is pirated into solutional conduits underground, leaving only a dry stream bed downstream. Dye traces have confirmed the connection between the swallow hole and the spring.

Travel times for dye traces have ranged from 16 to 80 hours over the straight-line distance of 3.2 miles from the swallow hole to the spring. These short travel times are indicative of the well-developed solution channels which occur in mature karst. As would be expected, flows at the spring respond quickly to precipitation. With respect to water quality, the spring responds to surface hydrological events more like a surface water source than a groundwater source. It is questionable whether the shallow aquifer should even be classified as "groundwater". In fact, Fulbright Spring is now classified by the U. S. Environmental Protection Agency (EPA) as groundwater under the influence of surface water.

Greene County has a temperate climate with distinct seasonal variations in temperature and precipitation. The average temperature, precipitation, evaporation and relative humidity are 55.9° F, 43.2", 42", and 72.5%, respectively.

Topography of the watershed is rolling, with gently sloping uplands and steeper slopes (15 to 30 percent) bordering the valleys of the major streams and their tributaries. Elevations within the recharge area range from 1100 to 1460 feet above mean sea level (USGS datum). The South Dry Sac River has a gradient of 33 feet per mile and Pea Ridge Creek has a gradient of 44 feet per mile. These gradients are high enough to result in erosive stream velocities. Cut banks are common and the stream beds are generally composed of gravel and small cobbles.

OVERVIEW OF SPRINGFIELD'S WATER SUPPLIES

Springfield's water supply is operated by City Utilities (CU) Springfield, a municipally owned utility company. In 1995, CU's customers used an average of 28 million gallons of water per day (MGD), with a peak summer usage of 46 MGD. Average daily use is projected to increase to 42 MGD by 2020. City Utilities operates two water treatment plants, one at Fulbright Spring (capacity, 24 MGD), which is supplied with raw water from the spring and from McDaniel Lake, and Blackman Treatment Plant (30 MGD capacity) located in southeast Springfield. The plant at Fulbright Spring was completed

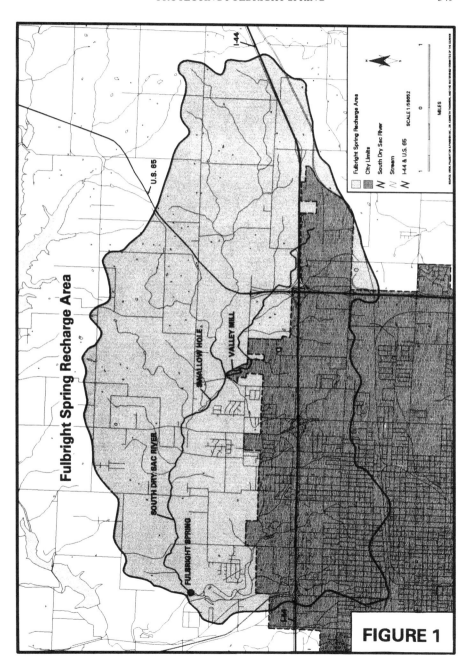

FIGURE 1

in 1941, and is a designated Water Landmark of the American Water Works Association (AWWA).

The 32-mile long, 36" diameter Nuccitelli pipeline, just completed this summer, can now deliver water from the Corps of Engineers' Stockton Lake to supplement the other supply sources during drought periods. It has increased the current drought yield from 27 to 42 MGD. The agreement with the Corps allows an increase of an additional 15 MGD in 2010, thereby serving projected needs through 2040.

Fulbright Spring History and Quality

Up to the present time, Fulbright Spring has been a reliable source in both quality and quantity of water. Interruptions of flow occurred during the 1880's, prior to the realization of the close connection between the Valley Water Mill/swallow hole area and flows at the spring. During that time, a mill still operated at Valley Water Mills. According to anecdotal accounts, flows at Fulbright would be interrupted if the tailrace at the mill was shut down. When the mill's tailrace was reopened, flow at Fulbright would recover, laden with husks and other debris from the mill. The Springfield City Water Company acquired Valley Water Mill in 1892, in order to gain more control over the flow at the spring.

Although extensive general water quality data have been collected at the spring, only limited data is available to relate specific sources or environmental conditions, such as wet weather flows, to water quality. During and shortly after runoff events, there are increased flow and increased concentrations of total and fecal coliform, turbidity, and total suspended solids at the spring. The current practice, although rarely used, is to temporarily take the spring off line after runoff events, when conditions warrant.

The existing treatment facility consists of a typical alum-coagulation/flocculation/filtration system supplemented by the addition of chlorine and carbon treatment to disinfect and remove tastes and odors. This system is used to reduce levels of conventional contaminants such as suspended solids and bacteria to acceptable federal and state levels.

Nitrates and many organic compounds which are commonly found in urban runoff are not effectively removed by this type of facility. Carbon adsorption, ion exchange, or other sophisticated processes would be required to significantly reduce concentrations of these contaminants. Such processes are expensive in terms of capital and operation and maintenance costs. Introduction of these contaminants into the raw water supply would force the upgrading of the present treatment process if Fulbright Spring is retained as a water supply source. It is, therefore, highly desirable to minimize the risk of contamination from surface runoff.

LAND USE AND DEVELOPMENT IN THE WATERSHED

Residential areas comprise 30% of the existing development in the watershed and are located primarily in the southern and western portions of the recharge area (Figure 2). Commercial and industrial development (20% of the present developed area) is mostly limited to the southern edge of the recharge area. The majority of the existing and historically developed area is located within the Pea Ridge Creek watershed, which is suspected to have only a weak connection to the spring. It is notable that relatively little development has occurred in the watershed above the swallow hole.

Transportation facilities comprise 40% of the existing land use. Two major highways cross the recharge area: Interstate Highway 44, running east-west along old Route 66; and U.S. 65, the main highway connection between Springfield and Branson. The emergence of Branson as a major tourist destination and of the southwest Missouri/northwest Arkansas region as a retirement area has contributed to rapid population growth in the Springfield area. Development pressures along U.S. 65 and I-44 have increased dramatically.

Development of the area near the I-44/U.S. 65 interchange has been hampered due to the lack of sanitary sewers (Greene County does not permit privately-owned small treatment plants). As early as the mid 1970's, a trunk sewer was planned for this area. Planning was slowed due to environmental concerns that sewer trenching could intercept the shallow groundwater conduits recharging Fulbright Spring. After extensive geologic and geophysical surveys, plans for the South Dry Sac Trunk Sewer were approved and construction was completed in 1994.

According to projections made for the Springfield-Greene County comprehensive plan, approximately 5 square miles of additional land in the watershed will undergo development during the next 25 years. Seventy percent of the expected development will be residential, the remainder commercial and industrial. The majority of the projected development area will be located in the southeastern portion of the recharge area, along the I-44/U.S. 65 corridors in the "Valley Mill" tributary watershed (Figure 2). This area is located upstream of the swallow hole.

With rapid development now expected to occur in the portion of the recharge area known to be directly connected to the spring, the risk of water quality degradation of the spring is greatly increased.

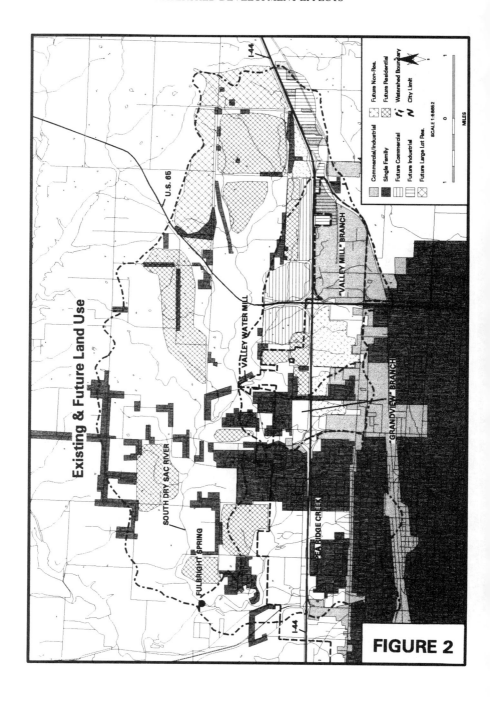

FIGURE 2

MANAGEMENT OF THE FULBRIGHT SPRING RECHARGE AREA

There are four primary local entities which share responsibility for management of the Fulbright Spring recharge area: the City of Springfield, City Utilities (CU), Greene County, and the Watershed Committee of the Ozarks. Approximately the southern one-third of the Fulbright Spring recharge area is located within the City of Springfield. The remainder is located within the unincorporated area of Greene County. Within the City, the Public Works Department is responsible for wastewater collection and treatment, stormwater facilities, and streets. Water, gas, and electricity are provided by CU. Land use planning, zoning, and subdivision administration are provided by the Department of Planning & Development.

Greene County is a non-charter first class county under Missouri law, giving it authority to administer building codes, zoning laws, and regulate subdivision of land in unincorporated areas of the County. County government is headed by a three-member commission. Land use planning, zoning, and subdivision administration are provided by the Resource Management Department. The County provides no stormwater management services per se. Review of stormwater plans associated with buildings and subdivisions is provided by the Resource Management Department. Storm drainage facilities within public road rights-of-way are owned and maintained by the County Highway Department. Ownership and maintenance of all stormwater facilities on private property are the responsibility of the property owner.

The County operates no wastewater or other utilities. These services are provided by the City of Springfield in the "Urban Services Area" which encompasses the suburban fringe around the City of Springfield. Other than electricity, which is provided through a number of rural electric cooperatives and utility companies, utility service is generally unavailable in the rural areas of the County.

Origin of the Present Watershed Management Program

In 1982, City Utilities' water customers experienced taste and odor problems due to an algae bloom in McDaniel Lake. Concerns over the quality of the water supply reservoirs led to the appointment of a Watershed Task Force. Among the recommendations made by the Task Force were adoption of an Urban Services Area and formation of a Watershed Management Coordinating Committee.

The Urban Services Area concept was adopted primarily to direct urban development into areas served by sanitary sewers and to discourage development on septic systems and small package treatment plants. Concurrent with the adoption of the Urban Services Area, the County amended its zoning regulations to require a minimum area of 5 acres for residential lots served by septic systems, to require all new commercial and industrial development to be served by sanitary sewer, and to require wastewater collection and treatment systems to be municipally owned and operated.

The Watershed Management Coordinating Committee originally consisted of the Presiding Commissioner of Greene County, the Springfield City Manager, the General Manager of City Utilities and two citizen members. The purpose of this group was to coordinate watershed management decisions in the municipal watersheds and to advocate good watershed management practice. (The name was later changed to the Watershed Committee of the Ozarks.)

The Watershed Committee has fulfilled and expanded its role in the ensuing years, and is now a major advocacy group for water quality. Though the Watershed Committee has no regulatory or enforcement authority, it plays an influential advisory role with the City, City Utilities, and the County.

THE PHASE 1 WATERSHED STUDY

In 1992, the Springfield City Council, hoping the increase in area tourism would bring long term stability to the sales tax base, proposed a large scale entertainment-based development. After considerable controversy, the City Council selected the northwest quadrant of the I-44/US 65 interchange for the location of the "entertainment park".

Among other concerns, opponents of the entertainment park cited the increased potential for water quality degradation. The entertainment park issue, in coincidence with the construction of the South Dry Sac Trunk sewer, pushed the issue of Fulbright Spring water quality into the spotlight.

In response to these concerns, the City Council and City Utilities appropriated $50,000 for a "Phase 1" study of the spring recharge area to be administered through the Watershed Committee. A consulting team led by Wright Water Engineers (WWE) of Denver, Colorado was hired in August 1994 to perform the study [2].The study was completed in early 1995. Among its findings and conclusions are the following:

- The effects of land use changes on water quality are well described in the literature and documented in studies nationwide (i.e., they are beyond question). New sources of pollutants and contaminants are produced by development, and the potential for transport of these pollutants to receiving waters is increased;

---

[2] In addition to Wright Water Engineers, the consulting team included Dr. Thomas Debo, Georgia Institute of Technology; Dr. Edwin Herricks, University of Illinois; Palmerton & Parrish Geotechnical Engineers, Springfield, Missouri; and Mr. Joe Wilson, University of Missouri-Rolla.

- Screening tests performed by the CU Central Laboratory showed volatile organic compounds (VOCs) in the more the fully-urbanized recharge areas of Jones Spring and lower Pierson Creek in east Springfield. These concentrations of volatile organic compounds (VOCs) are of the very type which would not be removed by present treatment methods at the Fulbright Plant;

- Though VOC's have not been a problem at Fulbright Spring, "pulses" of pollutants are not easily seen in the monitoring of groundwater with such high flow velocities. WWE suggested that additional wet weather "screening" data at Fulbright Spring (especially for target organic compounds) should be obtained;

- After reviewing the extent of existing and proposed development in the watershed, it is clear that the risk of degradation of Fulbright Spring water quality will increase as the recharge area develops, without carefully conceived runoff control measures;

- It is fortuitous that available spring water quality monitoring data generally show good water quality. It is likely that the remaining "safety cushion" in the recharge area is small;

- The optimum situation in public water system management is for those drainage basins which serve as raw water supply sources to be left undeveloped and undisturbed to the maximum extent practical. The Fulbright Spring recharge area is already partially urbanized, and will be to an even greater degree in the future as development increases. Consequently, if the present quality of the spring is to be maintained, a strong watershed management program is essential (One member of the consulting team observed that our concurrent goals of watershed protection and development could be described as "schizophrenic". Many communities find themselves in this position, however.);

- The State of Missouri already has strong numeric groundwater quality standards that appear to apply to the basin. An option is to rigorously enforce these standards within the recharge area, including the potential regulation of certain stormwater discharges via effluent limitations or on the basis of compliance with existing groundwater standards. Due to the expense, complexity, and difficulty in maintaining such a rigorous approach, WWE recommended a more moderate regulatory approach, at least initially;

- The use of relatively inexpensive non-structural and structural best management practices (BMPs) should be effective in maintaining spring water quality, but only if such BMPs are:

    (1) instituted before problems occur;

(2) implemented intensively, and without "cutting corners";

(3) designed to account for the variable risk levels that characterize any given development site; and

(4) thoroughly monitored and maintained.

This approach has been effective in other watersheds where implemented on a preventive rather than reactive basis;

- More expensive large-scale structural BMPs will be necessary if these initial measures are not taken in a timely manner. As Dr. Debo adroitly stated, "You can pay me now or pay me later".

STRATEGY FOR PROTECTING THE SPRING

Based upon their findings and conclusions, the consultants recommended a combination "non-structural/structural BMP" regulatory strategy in conjunction with careful monitoring of conditions at the spring and in the watershed as the appropriate and most effective starting point for protecting spring water quality. Key components of this strategy are described below.

Non-Structural BMPs

The consultants noted that since zoning, subdivision, and building codes have long been in use in both the City and the County, an effective framework already exists for application of non-structural BMP's. The following non-structural BMPs were recommended:

- Utilize existing ordinances, such as sinkhole protection ordinances, floodplain management ordinances, and sediment and erosion control regulations to their fullest extent;

- Utilize zoning techniques such as overlay zoning and redistribution of development density for protection of sensitive areas and impervious area limitations;

- Adopt watershed-based goals as a component of the comprehensive land use plan;

- Acquire or otherwise promote buffer zones (limited or no development) around sensitive environmental features, such as sinkholes, losing channel reaches, defined fractures, etc.;

- Implement "source controls" to limit the probability that pollutants will enter surface and groundwater;

- Develop an effective plan for responding to and managing unforeseen spills;

- Utilize the sensitive environmental features (SEFs) analysis procedure to help guide decisions on land development [3];

- Increase the level of awareness and involvement through public information and education;

- Provide for adequate maintenance, enforcement, staffing, and funding.

Structural BMPs

Structural BMPs were recommended to complement non-structural controls. The recommended initial approach is to require routine implementation of small, relatively inexpensive and easy-to-maintain BMPs, such as "slow flow" grass-lined channels and on-site dry detention ponds in conjunction with all new developments.

The basic intent of these practices is to "disconnect" impervious surfaces, slow down the rate of runoff, promote stormwater storage, and enhance the interaction of stormwater with vegetation. All of these actions reduce pollutant loading to receiving waters. Facilities of this kind can be readily incorporated into new industrial, commercial, and residential developments. Ample evidence from other communities demonstrates that this can be accomplished without imposing an undue financial burden on developers and in a way which is aesthetically pleasing. The key to the success of such measures is that they are designed to address only a small amount of storm runoff (in contrast to drainage/flood control BMPs) that contains the majority of the pollutant loads.

---

[3] As a part of the study, the consultants developed a procedure to assist in evaluating land use decisions in the recharge area. The SEF procedure quantitatively accounts for water quality influence potential by virtue of :

(1) potential pollution sources that will be present;
(2) the nature of the transport path for pollutants; and
(3) the geologic features that will permit entry of surface flows into the groundwater system that sustains the spring.

Baseline Data and Monitoring

Evaluating the success of the recommended BMPs depends upon monitoring water quality at the spring and in the recharge area, and adjusting the protection strategy if any decline in water quality becomes apparent. Reliable baseline data are needed in order to detect and track changes in water quality. The following key activities were recommended by the consultants:

- Provide additional dye traces and flow monitoring to better define the overall recharge area boundary, and refine the hydrologic mass balance for the spring;

- Monitor physical, chemical, and biologic parameters at key points in the watershed to develop a baseline database of water quality;

- Continue monitoring spring quantity/quality characteristics. Analyze more frequently for some subset of EPA's Priority Pollutants, especially under wet weather conditions, since data of this kind are currently limited. The single most effective way to determine the overall effectiveness of the watershed management plan is to monitor spring quality;

- Collect data during wet weather conditions to refine existing erosion and sediment control BMPs that have been implemented in Greene County;

- Monitor pilot stormwater quality BMPs such as grass-lined conveyances and dry detention ponds so that design criteria can be developed.

Long Range Needs

As can be seen from the foregoing, the recommended strategy appears to be an effective and fundamentally sound initial approach, given the present quality of the spring. Most important, much of the program can be implemented by utilizing existing resources. As the consultants noted, there remain important long term planning needs for the recharge area. These include:

- Developing a stormwater master plan for the watershed;

- Developing an applicable set of water quality standards;

- Reviewing and revising ordinances and regulations to provide for water quality protection in the recharge area;

- Developing technical design guidance for water quality BMPs;

- Improving operation and maintenance of stormwater facilities;

- Funding for implementation and management of the watershed plan.

Though need for these measures is obvious, they require a substantial amount of funding. This requires a high level of public input, involvement and support, as well as a high level of support from management and elected leaders. In the absence of a crisis, this degree of support can generally not be expected. The recommended strategy allows the program to be successfully started, and bases the need for an increased level of public funding and support on observed and documented changes of the water resource.

PROGRAM IMPLEMENTATION: OPPORTUNITIES & OBSTACLES

As we move forward with the Fulbright Spring protection program, it is important to recognize that many key factors for success of the watershed management program already exist. There is generally a high level of support and cooperation from managers and elected officials in both city and county government. The fact that only two local governments have jurisdiction over the entire recharge area is also a positive factor. The development community is generally cooperative, but does demand that additional costs to development be clearly justifiable. Further, as evidenced by response to public opinion surveys conducted by the Missouri Department of Conservation, there is a strong conservation ethic among citizens of the city and county, and a general concern for water quality.

A number of programs are currently in place or initiatives are already underway which contain components that are key to successful management of the Fulbright Spring watershed:

The City and County are currently in the process of adopting a new comprehensive plan - "Vision 2020". A joint City-County Water Resources Task Force has recommended that the strategies suggested by the Fulbright Spring study be incorporated into the plan. The Water Resources Task Force has also recommended improvements in regulations for sediment and erosion control, sinkhole protection and floodplain/riparian corridor management.

Earlier this year, the County adopted the "Northeast Development Plan" which encompasses the majority of the recharge area and which incorporates the recommendations from the Fulbright Spring study. The City and County are well along in development of a geographic information system (GIS).

Greene County has an active Local Emergency Planning Committee (LEPC). The LEPC conducted a successful emergency spill response exercise along I-44 earlier this year. The City's six-point solid waste management plan includes the Household Hazardous Waste Center, and the development of recycling drop-off points which are very popular. Partners in Pollution Prevention (P3), sponsored by the Environmental Collaborative of the Community Task Force works with local businesses and industries to reduce air and water pollution. The Watershed Committee was recently awarded an EPA grant to perform additional studies to better define the Fulbright Spring recharge area and to install and monitor the performance of pilot BMPs.

Missouri's highly successful Stream Team and Volunteer Water Quality Monitoring provide a well-established mechanism for citizen involvement in watershed issues. There are also several successful local programs that foster public education and involvement, such as Ozark Greenways, which provides assistance with development of linear parks, open space and greenways; and Greene County Groundwater Guardians, sponsored by the Groundwater Foundation.

Building upon these existing programs and focusing them toward the goal of protecting the spring provides our greatest opportunity for success. However, if we are to achieve our goal of protecting the spring, there remain some sizable obstacles to be overcome.

Perhaps the greatest of these is simply complacency. With the completion of the pipeline from Stockton Lake, Springfield no longer has its "back against the wall" with regard to providing adequate water supplies for the community's projected needs. This diminishes the impact of the potential loss of Fulbright Spring. The political will to implement far-reaching programs, such as those needed to protect the Fulbright Spring watershed, forms best in the face of a crisis. Ironically, good planning by City Utilities to avert a water supply crisis, may hurt efforts to protect the spring.

Our paradigm must be that no water source is expendable. This ethic has been largely absent in traditional engineering training. As a result, cost-benefit decision making is typically too narrow and short range in focus. Even though the community's water supply is now supplemented with water from Stockton Lake, Fulbright Spring will remain an important water source. From even a traditional cost-benefit point of view, a good argument can be made that water from Fulbright Spring provides a cheaper source, since it does not require the energy and transportation costs associated with pumping water from a source 32 miles distant and 450 feet lower in elevation.

Causes and effects of non-point pollution are not obvious. This leads to skepticism on the part of citizens, businessmen, and developers, as well as suspicion of over-regulation by environmental zealots. This attitude is not confined to the public at large. There remains widespread skepticism in the engineering profession that stormwater quality is really a serious issue. We must remember that many managers and consultants involved in key aspects of watershed management decision making are not water resource specialists, but come from diverse engineering and technical backgrounds. Their understanding and acceptance of this issue profoundly influence the success or failure of watershed management programs.

We must continually remind ourselves that non-point pollution issues deal as much, and in some cases perhaps more, with cultural change than technological change. Such things as grass swales, cluster development, and extended detention, are rarely viewed by the general public as a sign of enlightenment on the part of local planners. In this region, curbed streets are viewed by the public as a sign of quality. Developers willingly pay the additional cost of installing them. Grass swales are associated with the ugly, eroded "borrow ditches" which line roads in rural areas. Cluster developments are viewed as "more dense". This is anathema to the residents of suburbia where "sameness" in lot size and housing costs is considered paramount. Many of us have likely seen our "extended detention basins" often referred to by their neighbors as "snake pits" or "mosquito pits".

Complex issues such as these inevitably become politicized as they move through the public arena. Diverse interests and agendas become entwined. It was the politicization of the Entertainment Zone issue which resulted in the funding of the Phase 1 study in the first place. Was protecting Fulbright Spring really first and foremost in the minds of the individuals who protested that development of the Entertainment Zone would ruin the spring? Though this is a legitimate concern, the real objective was to defeat the development.

In Springfield, as in any other community, there is literally a minefield of political pitfalls which lay waiting to ambush any initiative to preserve the spring. Mirroring the mood of the country as a whole, the prevalent mood locally during the past couple of years has been anti-government, anti-tax. Criticism of long standing, even successful programs and institutions is widespread. This causes community leaders and managers to become defensive and sensitive to any issues which may draw controversy. As a result, there is reluctance to expend political capital on what may be perceived as a non-critical issue.

Experience in other communities is now confirming that the most common failure in stormwater water quality management programs is in maintaining BMP's. Both Springfield and Greene County have sufficiently strong regulatory programs to ensure that structural BMP's are designed and constructed properly. Beyond this, however, there is a high probability of failure. Greene County has no funding or legal authority for operation and maintenance of stormwater facilities, except in conjunction with public roads. The City of Springfield maintains the traditional storm drainage systems. However, small private drainage works are not City-maintained.

Though the recommended strategy envisions small, privately maintained BMP's, regulatory oversight and enforcement will be an absolute necessity. This eventually will require additional staff. Monitoring spring quality, water quality in the watershed, and performance of BMP's can be started with present staff and assistance from volunteers. However, additional staff will be required in order to sustain these efforts in the long term. Funding for these needs is essential if the recommended program is to succeed.

Funding presents a significant obstacle. Missouri has a tax limitation law, the "Hancock Amendment", which requires a public vote on any tax initiative. (User fees are exempt, however, and, thus far, legal challenges to this interpretation have not been successful). Any funding initiative will be forced to compete with a long list of community needs. Law enforcement, schools, roads, wastewater treatment and other needs place a seemingly constant stream of funding initiatives before the voters. Stormwater related tax issues have traditionally fared miserably in the community.

Though damage in this area was small in comparison to communities along the Missouri and Mississippi rivers, an outcry arose after the floods of 1993. In response, the City of Springfield offered both a 1/10 cent sales tax increase and a user fee to the voters. The user fee would have funded a stormwater utility. Similar to what is being done in other communities, the fee was based upon impervious area coverage. This encountered strong resistance from the business community, led by the Chamber of Commerce. It is interesting to note that most business persons felt that they were being "charged double" because stormwater detention was still required. There appears to be nearly a universal misunderstanding of the purpose and limitations of stormwater detention basins.

Timing of the vote also hurt the issue. Responding to the flood, the City formed a Citizens Stormwater Committee and moved quickly to offer the funding initiative. Concurrently, a "bias crimes" ordinance was being proposed. This issue mobilized religious conservatives claiming that it amounted to a veiled attempt to promote gay rights. The bias crimes ordinance appeared on the same ballot as the stormwater issues, ensuring their failure. (The sales tax failed with 48% of the vote. The "roof tax", doomed by resistance from the business community, failed by a wider margin.)

Opponents of the bias crimes ordinance struck a chord with voters when they claimed that this offered proof that the City Council was an "insider's club" no longer listening to "the people". This sentiment overflowed into the entertainment zone issue, assuring its failure, and spawning organized opposition directed at the City Council itself.

CONCLUSIONS

In reality, successful implementation of a protection plan for Fulbright Spring hinges on maintaining a strong sense of where this issue stands in a field of other issues, some directly related, some not. A strong intuition of how the issue will resonate with the public and a certain degree of agility in keeping the issue from being dragged off-course by unpredictable situations which inevitably arise, are necessary. Additionally, a strong commitment from, and strong leadership by both elected leadership and key staff is required to maintain focus on the objectives which must be attained.

Whether an elected leader, a key manager, or a consultant, success belongs to those who can "sense the balance" of all these factors. Our training and education do little to help us to recognize these issues. As a result, many of the best conceived programs and plans lay on the shelf, the money and effort largely wasted, because we failed to deal with non-technical side of the problem.

How can we guide a watershed protection plan through these pitfalls, and, as important, how can we maintain our confidence and morale as we deal with the social and political aspects of our projects? The following suggestions are offered.

1. Take a pragmatic approach.

Use the tools and resources that you already have while you lay the groundwork to obtain those which you need. The recommendations of the Fulbright Spring study are predicated upon this type of approach. Of course, it would be nice to begin by developing a watershed master plan, including a complete model for runoff quality and quantity, extensive chemical and biological sampling and the like. But, we might as well face reality. A study of this magnitude cannot be expected to be funded at this time, nor should it be funded until our needs are better focused.

We have the luxury of being "ahead of the curve" in the Fulbright Spring watershed. This allows us to the advantage of taking a proactive approach, and this is what we will do. Our greatest strength is in the application of land use controls, and we will base our approach on this strength.

2. Make sure the program goals are clear.

The goals which we propose to attain must be absolutely clear, not only to staff and consultants, but to managers, elected officials, and the public. Not only must the goals be clear, they must also fit the desires and aspirations of the stakeholders. Further the goals must be viewed by all as being *attainable*.

Our goal for Fulbright Spring is fairly clear: we wish to maintain its use as a municipal water supply source without having to provide more extensive and expensive methods of treatment. However, it is quite likely that not all of the players involved agree upon this goal, particularly those who might view Fulbright Spring as being expendable now that additional water supplies from Stockton Lake have been secured. We cannot ignore or overlook this, as we are frequently inclined to do.

3. Make sure your message is understandable.

We must be able to translate our understanding of the problem not only to elected officials and the public, but to our managers as well. While it is important that key staff understand and appreciate all of the technical details, page upon page of detailed technical information will lose the attention of the audiences we must reach. Most of us do a very poor job of summarizing our knowledge for the layman and underscoring what in means in terms of actions which must be taken. We must do this effectively if we expect community leaders and the public to respond.

4. Recognize the position that management and community leaders are in. Frame decision making within this context so that it is do-able.

There are very few bona fide "bad guys" among our managers and elected leadership. These folks generally want to do the right thing as much as any of us. We must recognize that their positions on our watershed management projects are influenced by many other factors, as illustrated above. Rather than becoming disenchanted or disgusted with leadership or management, we should concentrate on the things we can do to translate our needs into objectives upon which they can realistically be expected to take action.

A good technique to use is to separate actions and recommendations into "short term" and "long term" categories. Actions which cannot be taken with existing funding or authority should be categorized as "long term". This allows officials and managers the latitude to support initial steps which can be taken, while allowing time to develop support for new funding or authorization. To frequently we combine short term and long term actions in the way we word our recommendations. As a result, no action is taken simply because "the pill is too big to swallow".

5. Don't get too caught up in the "details".

We must remember that points of intense technical debate among us as scientists and engineers are of little interest or use to the public at large. There will, and should be, varied opinions among specialists in this or any field, and these should be subject to spirited debate. However, there is a point at which we must leave debate and uncertainty behind and focus upon making sure that we have informed management and the public about the important things we *do* know and agree upon. When we fail to do this, we end up merely "talking to ourselves".The poet Ogden Nash once wrote that, "Purity is obscurity". He was probably not thinking about the science of watershed management, but the shoe seems to fit.

6. Avoid professional or jurisdictional rivalries.

Successful watershed management requires close coordination and cooperation not only between different governmental agencies, but also different   professions. We must recognize that watershed management is not the sole province of any single discipline. We must rely on many types of expertise in dealing with the complex and intertwined array of technical and social issues which are involved Land planning is as important as engineering, biology is as important as land planning, and so on. We must appreciate and *respect* each other's expertise.

Because we have not traditionally managed water resources on a watershed basis, jurisdictional authorities over various aspects of watershed management are typically spread across several state and local agencies. It is naive to expect that each agency will have common goals or even a common perception of watershed management needs.

Rivalries and feuds between agencies and professions needlessly and senselessly distract our attention and energy. Even worse, they erode our credibility with the public.
We are too often prone to react to these situations by retaliating in some manner. Decisions or policies based upon hidden agendas or motives quickly become hard to support and provide easy ammunition for opponents and skeptics. The public is seldom fooled by these antics. Take the high road and stay on it!

These suggestions may seem trite or cliche, but we must remember that, when it comes to non-point source pollution issues, *simple is hard.* It is with the simple things, not the technical issues that we are failing. Given all this, can we really "beat the odds" with Fulbright Spring, or any watershed management program? We have no choice but to be optimistic, for to do otherwise concedes failure, and in the long term, failure isn't an option.

ACKNOWLEDGMENTS

I offer my sincere thanks to Dr. John Witherspoon, Director of Water Treatment & Supply for City Utilities of Springfield, and Mr. Loring Bullard, Director of the Watershed Committee of the Ozarks, two colleagues whose dedication gives me reason to be optimistic.

REFERENCES

McCartney, N. L., et al, "Report of the Watershed Task Force", Springfield, Missouri, 1983.

Moore, Beth A., Witherspoon, John T., Bullard, Loring L., Aley, Thomas J., and Rosenfeld, Jeffrey K., "Strategy for Delineation and Detection Monitoring of the Fulbright Springhead Protection Area, Springfield, Missouri", unpublished paper.

Thomson, Kenneth C., "Geology of Greene County, Missouri", Springfield, Missouri, 1986.

Waite, Loyd A., and Thomson, Kenneth C., "Development, Description, and Application of a Geographic Information System Data Base for Water Resource in Karst Terrane in Greene County, Missouri", U.S. Geological Survey, Water-Resources Investigations Report 93-4154, Rolla, Missouri, 1993.

Wright Water Engineers, "Fulbright Spring Protection Study", Denver, Colorado,1995.

FLORIDA'S EVOLVING STORMWATER/WATERSHED MANAGEMENT PROGRAM

Eric H. Livingston[1]

ABSTRACT

During the past 30 years Florida has experienced explosive growth and development resulting in more than a doubling of the state's population. This growth and the associated land use changes, along with the state's extensive agricultural and mining lands, makes protecting Florida's vulnerable water resources especially difficult. While a wide variety of resource management laws, regulations, and programs have been implemented during the past 25 years, they generally have promoted a piecemeal, site specific approach. While these programs certainly have helped to minimize the degradation of the state's natural resources, their environmental effectiveness has been less than optimal. This paper will discuss the state's continually evolving land and water resource management program as it moves towards implementation of an ecosystem management framework based on watershed units. A case study of the Tampa Bay watershed will be presented to demonstrate how the various pieces of the resource management puzzle are being fit together.

INTRODUCTION

Florida is blessed with a multitude of natural systems, from the longleaf pine-wiregrass hills of the panhandle, to the sinkhole and sand ridge lakes of the central ridge, to the Everglades "River of Grass", to the coral reefs of the Keys. Abundant surface water resources include over 20 major rivers and estuaries along with nearly 8000 lakes. Plentiful ground water aquifers provide over 90% of the state's residents with drinking water. Add the state's climate and its easy to see why many consider the Sunshine State a favored vacation destination and why the state has experienced phenomenal growth since the 1970's. Today, Florida is the fourth most populous state and is still

---

[1]Stormwater/NPS Management Section, Florida Dept. of Environmental Protection, Tallahassee, Florida 32399-2400

growing rapidly, although not at the rate of 900 people per day (300,000 per year) that occurred throughout the 1970s and 1980s.

However, Florida's natural systems, especially its surface and ground water resources, are extremely vulnerable and easily damaged. This is partially the result of the state's sandy porous soils, karst geology and abundant rainfall. The negative impacts of unplanned growth were seen as early as the 1930s, when southeast Florida's coastal water supply was threatened by saltwater intrusion into the fragile freshwater aquifer that supplied most of the potable water for the rapidly expanding population. By the 1970s, it was becoming all too clear that unplanned land use, development and water use decisions were altering the state in a manner that, if left unchecked, could lead to profound, irretrievable loss of the very natural beauty that brought residents and tourists to Florida. Extensive destruction of wetlands, bulldozing of beach and dune systems, continued saltwater intrusion into freshwater aquifers, and the extensive pollution of the state's rivers, lakes and estuaries were only some of the negative impacts of this rapid growth.

Fortunately, Florida's citizens and elected officials became educated about these problems and began developing programs to protect and manage the state's natural resources. Florida began serious and comprehensive efforts to manage its land and water resources and growth coincident with the increasing strength of the environmental movement in the nation and the state during the early 1970s. Florida's natural resources management programs have evolved over a twenty year period. Collectively, the individual laws and programs enacted during this period, can be considered "Florida's Watershed Management Program". In many cases, these laws have been integrated either statutorily with revisions to existing laws or through the adoption of regulations by various state, regional or local agencies.

The evolution of Florida's "watershed management program" typically involves the following sequence: concern about a specific "pollutant" or problem creates a resource/environmental management program which usually begins by focusing on "new sources" (site basis); over time, as new sources are controlled and the program administration and effectiveness increases, the focus shifts to cleaning up "older sources" (watershed or regional basis); focus shifts to integrating the program with similar ones to eliminate any duplication and improve efficiency and effectiveness.

FLORIDA'S WATERSHED MANAGEMENT PROGRAM - INITIAL EFFORTS

The initial efforts to develop and implement Florida's watershed management program, especially its stormwater management program, have been described previously (Livingston 1993, 1995). Key components of these efforts are briefly summarized below in chronological order:

1972 Package of land and water planning, regulation and acquisition programs:

•Chapter 373, F.S. The Florida Water Resources Act establishes the state's five regional water management districts, designates the Department of Pollution Control as the oversight agency for the WMDs, requires the development of a state water plan, and allows for the regulation of the water resource. WMDs financed by ad valorem property taxing authority of up to 1 mil ($1/$1000 value) which is set in the Florida Constitution. NWFWMD millage capped at .05 mil.

• Chapter 259, F.S. The Land Conservation Act establishes the Environmentally Endangered Lands Program and authorizes the state to purchase critical and sensitive lands. Envisioned as a 10 year long program investing $200 million funded by the sale of state bonds.

1973 • Chapter 403, F.S. The Florida Environmental Protection Act renames the Department of Pollution Control as the Department of Environmental Regulation and broadens its powers, duties and programs. This law is the state's general environmental protection act. It is amended almost annually as new environmental concerns and needs arise and as existing programs evolve.

1975 • Chapter 163, F.S. The Local Government Comprehensive Planning Act, the state's first growth management legislation, was recommended by the first Environmental Land Management Study Committee. The law requires all cities and counties to prepare a comprehensive plan to be submitted for review to the state's land planning agency, the Department of Community Affairs, which in turn sends the plans to other state agencies for review and comment. However, the LGCPA contains no "teeth". Local governments are under no statutory requirement to revise their plans to incorporate the comments and recommendations made by the state reviewing agencies. Furthermore, they are not required to pass land development regulations to implement their plans.

1979 • First components of the state's Section 208 Areawide Water Quality Management Plan, the Agriculture NPS Plan and the Silviculture NPS Plan, are submitted to and approved by EPA. These call for a non-regulatory approach with a regulatory backstop if BMPs required by farm conservation plans are not implemented or if the forestry BMPs required by the state's adopted Silviculture BMP Manual are not followed.

• Chapter 17-4.248, Florida Administrative Code (FAC), the state's first Stormwater Rule, is adopted by the state Environmental Regulation Commission (ERC) as a DER rule. This rule is intended as a temporary regulation until on-going research on BMP design and effectiveness is completed. The rule's adoption is controversial but data collected during from 208 program studies conclusively show that stormwater runoff, especially from urban land uses and highways, is a "pollutant" and therefore should be controlled pursuant to Chapter 403, F.S. Florida's continuing rapid growth makes it imperative that treatment of stormwater, using BMPs, be required for new stormwater discharges that would be "a significant source of pollution".

1981 • Through action taken by the Governor and Cabinet, the Save Our Coasts land acquisition program is established. The program proposes to spend $200 million over 10 years to purchase coastal lands such as beaches, shorelines and sensitive areas. Funding is provided by the sale of state bonds backed by documentary stamps as authorized in Chapter 375, F.S., which sets policy on how the Land Acquisition Trust Fund is to be administered.

   • Chapter 373, F.S., is amended with the creation of the Save Our Rivers land acquisition program. Administered by the WMDs, this program proposes to spend $320 million over 10 years to purchase wetlands, floodplains and other lands necessary for water management, water supply and the conservation and protection of water resources.

1982 • The state's Stormwater Treatment Rule, Chapter 17-25, FAC, is adopted by the ERC after two years of rule adoption workshops and 29 official rule drafts. The rule is technology-based rather than water-quality based, although the state's water quality standards remain as a backstop should a stormwater discharge cause violations. Compliance with the rule's BMP design criteria create a presumption that the desired performance standards will be met. This, in turn, provides a rebuttable presumption that water quality standards will be met. If an applicant wants to use BMPs other than those described in the rule, then a demonstration must be made that the BMP provides treatment that achieves the desired pollutant removal performance standard.

The basic stormwater management goal for new development is to assure that the post-development peak discharge rate, volume, timing and pollutant load does not exceed predevelopment levels. However, BMPs are not 100% effective in removing stormwater pollutants while site variations can also make this goal unachievable at times. Therefore, after review and analysis of various stormwater data, and after extensive public participation, the Department sets a stormwater treatment performance standard of removing at least 80% of the average annual pollutant load for stormwater discharges to Class III (fishable/swimmable) waters. A 95% removal level is set for stormwater discharges to sensitive waters such as potable supply waters (Class I), shellfish harvesting waters (Class II) and Outstanding Florida Waters. It is believed that these treatment levels will protect beneficial uses thereby establishing a relationship between the rule's BMP performance standards and water quality standards.

The rule creates design criteria for various types of BMPs including retention, detention with filtration, and wet detention. The actual design criteria are based on a number of factors including the state's overall stormwater quantity and quality goals, rainfall characteristics, runoff pollutant loadings, BMP efficiency and cost data. The rule also creates "general permits" if BMPs are built to the design criteria. Implementation of the rule is delegated immediately to the South Florida Water Management District allowing stormwater treatment requirements to be merged with stormwater quantity (flood control) requirements in one permit. During the next several years, implementation is delegated to three of the four other WMDs.

The current stormwater treatment volumes and some key design criteria for various BMPs are set forth in Table 1. Since adoption of the Stormwater Rule in 1982, the design criteria and treatment volumes have been revised several times as new information becomes available about the field effectiveness of the various types of BMPs.

TABLE 1. STORMWATER TREATMENT VOLUMES AND DESIGN CRITERIA FOR BMPs IN WATERSHEDS WITH CLASS III WATERS (FISHABLE/SWIMMABLE).

BMP	Treatment Volume	Key Design Criteria
Swales	Infiltrate 80% of a the runoff from a 3 yr-1 hr storm (about 2 inches or 5.1 cm)	Must be vegetated. Slopes 3:1 or flatter. Swale blocks or raised driveway culverts encouraged.
Off-line Retention (infiltration)	0.5 inch (1.25 cm) of runoff or the volume calculated by 1.25 times the percent imperviousness, whichever is greater.	Vegetated side slopes and bottom. At least two feet about water table. Percolate within 24-36 hours.
On-line Retention	Above volume plus an additional 0.5 inch (1.25 cm) of runoff.	
Off-line Detention with Filtration	1.0 inch (2.54 cm) of runoff or the volume calculated by 2.5 times the percent imperviousness, whichever is greater.	Filter material must be washed, have a uniformity coefficient of 1.5 to 4.0, and an effective grain size of 0.20 to 0.55 mm in diameter. Must have two feet of filter media.
On-line Detention with Filtration	Above volume plus an additional 0.5 inch (1.25 cm) of runoff.	Must recover treatment volume within 72 hours. Designed with a safety factor of at least two.
Wet Detention or wetlands	1.0 inch (2.54 cm) of runoff or the volume calculated by 2.5 times the percent imperviousness, whichever is greater.	14 day residence time permanent pool, 5 day drawdown, 30% vegetated littoral area.

1984 • Chapter 403, F.S., is revised to create Section IX which is known as the Henderson Wetlands Protection Act. This legislation expands the authority of the DER to protect wetlands; establishes administrative procedures to allow landowners to obtain legally binding "wetland lines"; allows the DER to consider fish and wildlife habitat, endangered species, historic and archaeological resource and other relevant concerns in wetland permitting; allows the use of certain wetlands for incorporation into domestic wastewater and stormwater management systems; transfers wetland regulation for agriculture and forestry activities to the WMDs; requires the WMDs to protect isolated wetlands and consider fish and wildlife habitat requirements.

• Chapter 186, F.S. The State and Regional Planning Act mandates the Governor's Office to prepare a State Comprehensive Plan and present it to the 1985 Legislature. It also requires the preparation of regional plans by the state's 11 Regional Planning Councils and provides $500,000 to them for plan preparation.1985

•Chapter 187, F.S. The State Comprehensive Plan contains important goals and policies in 25 different elements, including water resources, coastal & marine resources, natural systems & recreation, air quality, waste management, land use, mining, agriculture, public facilities and transportation.

1986 • Chapter 163, F.S., is amended with enactment of the Local Government Comprehensive Planning and Land Development Regulation Act of 1985. This law requires all local governments to prepare local comprehensive plans and implementing regulations, which must be consistent with the goals and policies of the state and regional plans. Numerous state and regional agencies review the local plans and submit their objections, recommendations and comments to the Department of Community Affairs for transmittal to the local government. The local plans must be revised to incorporate the objections, recommendations and comments. Furthermore, local governments face sanctions from the state which could result in the loss of state funding if adopted local plans are not consistent with the state and regional plans

Florida's revised growth management system is built around three key requirements: consistency, concurrency and compactness. Consistency establishes an "integrated policy framework", whereby the goals and policies of the State Plan frame a system of vertical consistency. State Agency Functional Plans and Regional Planning Council regional plans must be consistent with the goals and policies of the state plan while local plans are required to be consistent with the goals and policies of the state and appropriate regional plan. Local land development regulations (LDRs) must also be consistent with the local plans goals and policies. Horizontal consistency at the local level also is required to assure that the plans of neighboring local governments are compatible. Consistency is the strong cord that holds the growth management system together.

Concurrency is the most powerful policy requirement built into the growth management system. It requires state and local governments to abandon their long-standing policy of deficit financing growth by implementing a "pay as you grow system". Once local plans and LDRs are adopted, a local government may approve a development only if the public facilities and services (infrastructure) needed to accommodate the impact of the proposed development can be in place concurrent with the impacts of the development. Public facilities and services subject to the concurrency requirements are roads, stormwater management, solid waste, potable water, wastewater, parks and recreation and, if applicable, mass transit. Level of service standards acceptable to the community must be established for each of the types of public facilities.

• Chapter 403.0893, F.S., is created as the only surviving section of a stormwater management bill that was developed over a ten month process. The bill was an attempt to put into statute a cost-effective, timely process to retrofit existing drainage systems to reduce the pollutant loadings discharged to water bodies. Only the section creating explicit legislative authority for local governments to establish stormwater utilities or special stormwater management benefit areas is enacted.

FLORIDA'S WATERSHED MANAGEMENT PROGRAM - RECENT INITIATIVES

In the past ten years the focus of Florida's watershed management program has shifted to cleaning up "older sources" such as existing land uses, whether urban or agricultural and to integrating program components to eliminate duplication and improve efficiency and effectiveness. This has led to greater emphasis on more holistic approaches to address cumulative effects of land use activities within a watershed and to a greater emphasis on regional structural controls and the purchase or restoration of environmentally sensitive lands. Key institutional aspects of this changing focus include:

1987 • Chapter 373, F.S., revised to add a new section, the Surface Water Improvement and Management (SWIM) Act, which establishes six state priority water bodies. It directs the WMDs, under DER supervision, to prepare a priority water body list and develop and adopt comprehensive watershed management plans to preserve or restore the water bodies. Provides $15 million from general revenue sources and requires a match from the WMDs. Does not establish a dedicated funding source making the program dependent upon uncertain annual appropriations from the legislature.

1988 • The State Nonpoint Source Assessment and Management Plan, prepared pursuant to Section 319 of the Federal Clean Water Act, is submitted to EPA and approved. This qualifies the state for Section 319 NPS Implementation grants which are used for BMP demonstration projects and to refine existing NPS management programs. The delineation of the state's ecoregions, selection of riverine ecoregion reference sites, and modification of EPA's Rapid Bioassessment Protocols and metrics for use in Florida is initiated (See Livingston et. al, these proceedings).

1989 • Chapters 373 and 403, F.S., are revised as part of the 1989 Stormwater Legislation. The legislation clarifies the stormwater program's multiple goals and objectives; sets forth the program's institutional framework which involves a partnership among DER, the WMDs and local governments; defines the responsibilities of each entity; addresses the need for the treatment of agricultural runoff by amending Chapter 187, F.S., to add a policy in the Agriculture Element to "eliminate the discharge of inadequately treated agricultural wastewater and stormwater"; further promotes the watershed approach being used by the SWIM Program; attempts to integrate the stormwater program, SWIM program, and local comprehensive planning program (but doesn't succeed); establishes State Water Policy, an existing but little used DER rule, as the primary implementation guidance document for stormwater and all water resources management programs; and creates the State Stormwater Demonstration Grant Program with $2 million in funding as an incentive to local governments to implement stormwater utilities.

1990 • Chapter 17-40, FAC, State Water Policy undergoes a total revision and reorganization so that it can be used as guidance by all entities implementing water resource management programs and regulations. Section 17-40.420 (now 62-40.420) is created and includes the goals, policies and institutional framework for the state's stormwater management program. Key elements are:

• DER is designated as the lead agency with responsibility for setting goals for the program, for providing overall program guidance, for overseeing implementation of the program by the WMDs, and for coordinating with EPA, especially with the advent of the new NPDES stormwater permitting program.

• WMDs are the chief administrators of the stormwater regulatory program (quantity and quality). They are responsible for preparing SWIM Watershed Management Plans which include the establishment of stormwater pollutant load reduction goals (PLRGs) and for providing technical assistance to local governments, especially with respect to basin planning and the development of stormwater master plans.

• Local governments are the front lines in the stormwater/watershed management program since they determine land use and provide stormwater and other infrastructure. They are encouraged, but not required, to set up stormwater utilities to provide a dedicated funding source for their stormwater program. Their stormwater responsibilities include preparation of a stormwater master plan to address needs imposed by existing land uses and needs to be created by future growth; operation and maintenance activities; capital improvements of infrastructure; and public education. They are encouraged to set up an operating permit system wherein stormwater systems are inspected annually to assure needed maintenance is performed.

Important stormwater program goals include:
• preventing stormwater problems from new land use changes and restoring degraded water bodies by reducing the pollution contributions from older stormwater systems.
• retaining sediment on-site during construction.
• trying to assure that the stormwater peak discharge rate, volume and pollutant loading are no greater after a site is developed than before.
• 80% average annual load reduction for new stormwater discharges to most water bodies.
• 95% average annual load reduction for new stormwater discharges to Outstanding Florida Waters which are a special class of exceptionally high quality water bodies.
• reducing, on a watershed basis, the pollutant loading from older stormwater systems as needed to protect, maintain or restore the beneficial uses of the receiving water

body. The amount of needed pollutant load reduction is known as a "Pollutant Load Reduction Goal or PLRG".

●Chapter 375, F.S., is amended with the creation of Preservation 2000, a ten year land acquisition program with a goal of spending $300 million per year. The legislation divided available annual funding among seven programs: CARL, SOR, Florida Communities Trust, State Parks, State Forests, State Wildlife Areas and Rails to Trails. Funded the first year by state bonds backed by an increase in the documentary stamp fee. Between 1972 and 1991, the state's land acquisition programs invested over $1.5 billion to buy over 1.2 million acres. Equally important, as a result of the state land acquisition programs, fourteen Florida counties created local programs that currently commit up to $600 million for land conservation. Revenue sources for these local land acquisition programs include local option sales tax, impact fees, added property taxes, and local bonds.

1993 ● Chapters 373 and 403, F.S., are revised extensively to merge the DER and the DNR (Dept. of Natural Resources) to create the Department of Environmental Protection (DEP), and to streamline environmental permitting. The goals of the streamlining bill are to eliminate permitting duplication, increase administrative and environmental effectiveness by increasing delegation of programs from DEP to the WMDs, and assure greater program consistency and integration. The bill requires merging of the existing surface water/stormwater management permit with the wetland resource permit to create an "environmental resource permit", redefines wetlands based on their hydrology, vegetation and soils, and requires the development of a single wetland delineation method that will be used by the DEP, WMDs and local governments.

ECOSYSTEM MANAGEMENT - THE NEWEST INITIATIVE

As part of the 1993 legislation creating the Department of Environmental Protection, the agency was charged with developing a strategy to protect the functions of entire ecosystems. To develop this strategy, a consensus-building process was used which was open to anyone who wanted to participate. Twelve committees of interested parties were formed to develop the strategy. The were chaired by business interests, environmentalists, land owners, other agency representatives, and DEP employees. Over 300 people participated in these committees which met between June and October 1994. An Ecosystem Management Implementation Strategy Committee made up of the chairs of the committees, plus local government, water management district, state agency, Army Corps of Engineers, EPA, and Governor's Office representatives consolidated and prioritized the committee recommendations. The result was two documents. The first was a brief report to the Governor, Legislature, and general public explaining in general terms how ecosystem management will be implemented. The second was the Ecosystem Management Implementation Strategy (EMIS), an action plan for program managers (FDEP, 1995).

Stewardship is the central theme of the ecosystem management implementation strategy. Stewardship carries with it a strong sense of ownership in and responsibility for Florida's land, air, water, and other resources. This has been a central theme of the state's nonpoint/watershed management program which has long attempted to educate the public about "Pointless Personal Pollution" and the need for everyone to be part of the solution. A fundamental assumption is that government can not preserve Florida's ecosystems without the support and participation of all citizens.

Ecosystem management will rely upon four cornerstones: place-based management (watershed management), common sense regulation, cultural change, and ecosystem management "foundations" which include a statewide natural resource atlas, public infrastructure planning, science and technology, research, monitoring, education, training, and program audit and evaluation. For each of these cornerstones the EMIS includes a series of recommended activities and a schedule which will lead to implementation of this new strategy by 1998.

PUTTING THE PUZZLE PIECES TOGETHER: TAMPA BAY CASE STUDY

What Are the Puzzle Pieces?

Tampa Bay is Florida's largest open-water estuary, spanning almost 400 square miles. It's 2,200 square mile watershed, more than five times the bay's size, is home to nearly two million people whose everyday activities increasingly influence the bay's environmental health. From its headwaters of the Hillsborough River to the salty waters off Anna Marie Island, Tampa Bay encompasses a rich mosaic of underwater and coastal habitats that support thousands of species of plants and animals. Preserving and restoring these interdependent habitats, especially in the face of continued growth, presents a tremendous challenge that can only be met by a comprehensive watershed management approach.

The struggle to understand and protect Tampa Bay has evolved in less than 25 years from a grass-roots citizens effort to a complex, multi-layered regulatory network involving three counties, a dozen cities, a variety of regional, state, and federal agencies, and numerous special interest groups. Efforts began in the early 1970s with the formation of a citizens group called Save Our Bay which pushed for a halt to uncontrolled dredging and sewage disposal in the bay. Years later the Hillsborough Environmental Coalition was formed to coordinate and unify local environmental groups.

At the same time, the U.S. Environmental Protection Agency was formed and the Federal Clean Water Act of 1972 was enacted leading to new treatment requirements and funding to upgrade sewage treatment plants around the country. The City of Tampa received an EPA grant to upgrade its plant which had been discharging partially treated sewage into Hillsborough Bay for decades. These discharges were a major reason why Hillsborough Bay was clouded with noxious algae and in poor condition. The City installed an advanced wastewater treatment system, one of the first in the country, which greatly reduced the amount of nitrogen and other pollutants discharged to the bay. Subsequent state legislation (Wilson-Grizzle and Grizzle-Figg Acts) required all sewage discharges to the bay to meet advanced wastewater treatment standards. Additionally, the City of St. Petersburg began solving its sewage disposal problems by pioneering the first large-scale wastewater reuse program in the state, resulting in almost zero discharge to the bay.

The Environmental Protection Commission (EPC) of Hillsborough County, established in the late 1960s, has many responsibilities. An important one has been the implementation of a comprehensive long-term water quality monitoring program in the bay. Together with other monitoring efforts by state or regional agencies, the EPC data helps document changing water quality conditions allowing comparisons before and after implementation of management programs and providing documentation of Tampa Bay's progressive recovery.

In the early 1980s, in response to continuing pressure from residents to clean up the bay, the Florida legislature created a bay study commission to identify ways to improve bay protection. The study commission resulted in the formation in 1985 of an advisory group, the Agency on Bay Management. The 45 member group, which includes elected officials, regulators, and representatives of local governments and special interest groups, has been successful in focusing public attention on bay problems and in bringing together diverse and often competing bay users.

The state's Surface Water Improvement and Management (SWIM) program was established by the legislature in 1987. Tampa Bay was named in the SWIM Act as a priority waterbody within the Southwest Florida Water Management District. An essential early activity of the SWIM program was a watershed-wide assessment of pollution sources, especially stormwater, to identify "hot spots" - subbasins with high stormwater loadings - and to prioritize urban stormwater retrofitting projects (SWFWMD, 1990). The SWIM program has so far expended over $6 million in state and District funds, some being used as matching funds for Section 319 grants through the DEP nonpoint source management program, to treat urban stormwater and to restore bay habitats.

In 1990 Tampa Bay was adopted into the National Estuary Program by EPA to assist the region in developing a Comprehensive Conservation and Management Plan (CCMP) for the bay. Many of the agencies, citizen groups, and others long active in the restoration and management of Tampa Bay participated in the development of the CCMP. The recently developed CCMP builds on many of the region's ongoing environmental programs, from land acquisition to urban stormwater retrofitting to habitat restoration. It also identifies where unneccesary duplication exists in current environmental programs and provides recommendations which, when combined with the state's new ecosystem management initiative, can help to ensure that limited public funds are spent in the most environmentally effective manner. To no one's surprise, the primary area of duplication is in environmental permitting which often involves a multitude of federal, state, regional, and local agencies. Recognition of this problem led to the state's permit streamlining initiative which combined stormwater quantity, stormwater quality, and wetland resource permits into a single Environmental Resource Permit. The CCMP also recommends that increased monitoring and compliance become a higher priority with state, regional, and local agencies to increase the environmental effectiveness of regulatory programs.

The Hillsborough River and Bay ecosystem was designated in the EMIS as one of the state's Ecosystem Management Areas. The 1995 Legislature passed a bill establishing a demonstration project in this ecosystem to test the concepts of the EMIS. The demonstration project seeks to integrate the ongoing watershed management activities outlined above and to tests concepts such as team permitting, compliance audits, and linear infrastructure planning. Additionally, the project has developed a Coordinated Conservation Plan (HRICC, 1996) which identifies existing conservation areas, proposed conservation areas, and other core and buffer areas needed to complete protection of the Hillsborough River Greenway.

Shrinking public funds, combined with increasing demand for government services and increasing public scrutiny of expenditures, are providing new challenges for natural resource managers. Concurrently, attitudes about environmental management finally are shifting from an emphasis on piecemeal site management toward a more holistic approach that assesses the cumulative impacts of all human activities within a watershed. Whether called "watershed management" or "ecosystem management" this comprehensive approach requires redirection of efforts and funds towards more proactive projects and an emphasis on assessing overall impacts. A critical component of watershed management is using biological living resources, such as seagrasses, fish, and scallops, as a measure of a water body's health, with far less emphasis on traditional laboratory based water quality standards. This approach allows greater flexibility to achieve the desired ecological goals and provides taxpayers with a better benchmark to judge the return on their expenditures.

Ecological Assessment of Tampa Bay

As part of the development of the Tampa Bay SWIM Plan and the NEP CCMP, existing environmental information was assessed to determine the ecological health of the bay system. Major findings of these assessments are summarized below.

HABITATS

1. Since 1950, about half of the bay's natural shoreline and 40 percent of its seagrasses have been destroyed.
2. In 1950, the bay's shallow shelf supported about 40,000 acres of seagrasses. By 1982, only 21,600 acres remained and virtually all of Hillsborough Bay's 2,700 acres were gone. Seagrass decline is due to dredging and filling for residential development,

turbidity caused by dredging of the main shipping channel, and reduced light penetration caused by shading by algae fueled by excess nutrient discharges.

3. Since the early 1900s approximately 13,200 acres of bay bottom (3.6 percent of the bay's surface area) were filled, with over 90 percent of the activity occurring along the bay's shallow shelf where seagrasses once thrived. The surface area of Hillsborough Bay has been reduced by 14 percent.

4. Upgrading sewage plants in the 1980s to provide advanced wastewater treatment reduced nitrogen loadings, leading to a decline in phytoplankton, an increase in water clarity, and greater light penetration. Consequently, between 1982 and 1992 seagrass coverage increased by about 4,000 acres (18.5 percent) raising the bay's total acreage to over 25,600 acres.

5. About 43 percent (9,700 acres) of Tampa Bay's original saltwater wetlands were lost between 1950 and 1990, primarily because of dredging and filling for waterfront development. However, as many as 5,900 acres of new wetlands formed along causeways and other emergent land created by dredged spoil material during this period. Recent estimates of wetland habitat in Tampa Bay indicate that about 18,000 acres of mangroves and saltmarsh remain but many thousands of acres are damaged by invasion by exotic plants such as Brazilian pepper.

FISH AND WILDLIFE

1. Between 1966 and 1990, the harvest of 11 commercial species of fish declined by 24 percent, primarily because of smaller catches of mullet and sea trout. Each of these species is dependent on seagrass habitats.

2. Harvest of spotted sea trout declined by 86 percent between 1950 and 1990, from 487,000 pounds to 67,000 pounds. Similarly, red drum harvests plummeted from 80,000 pounds in 1950 to 15,000 pounds in 1986.

3. Tampa Bay's once-thriving shellfish industry has virtually collapsed, except for bait shrimping. Harvests of clams or oysters are restricted or prohibited throughout the bay because of high bacterial levels associated with stormwater discharges and septic tanks. The bay scallop, a highly pollution sensitive organism, all but disappeared from the bay in the 1960s.

WATER AND SEDIMENT QUALITY

1. While water quality has improved over the past ten years, primarily as a result of better wastewater treatment, water clarity, nutrients, and toxics continue to be a problem.

2. Because of natural circulation and flushing from the Gulf of Mexico water clarity is greatest in the lower part of Tampa Bay (2.5 m), and naturally decreases moving up the bay, dropping to an average of 2 meters (6.6 feet) in Middle Tampa Bay and Old Tampa Bay. The lowest average water clarity is in Hillsborough Bay (1.5 m) which has poor circulation and receives a larger share of nutrients and sediments from major rivers.

3. Excessive amounts of nitrogen continue to accelerate algal growth which subsequently reduces light penetration to seagrasses and contributes to oxygen depletion. The bay's total annual nitrogen load was estimated to be 2.5 times greater in 1976 than the load computed for 1985 to 1991 (Figure 1).

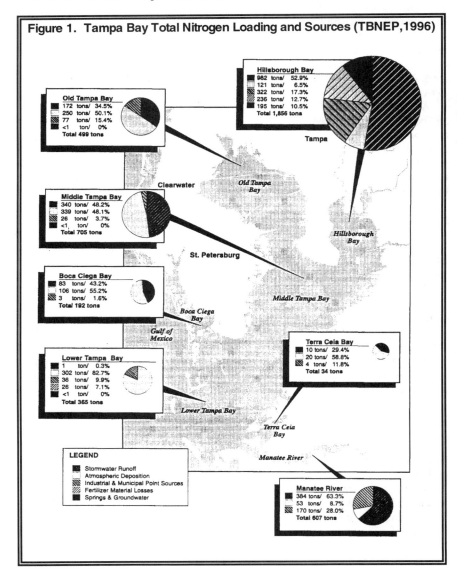

Figure 1. Tampa Bay Total Nitrogen Loading and Sources (TBNEP,1996)

4. Recent studies by NOAA, in cooperation with FDEP, provide an excellent overview of the levels and distribution of toxics in bay sediments (Long, et. al. 1991; FDEP, 1994). Compared to other urban estuaries, Tampa Bay has low to moderate levels of most toxic parameters. Contamination is centered around large urban centers, ports and marinas, and concentrations generally decrease from the top of the bay toward the Gulf of Mexico.

5. Generally, the highest levels of sediment toxic contamination occur in Hillsborough Bay, the bay's most industrialized area and home to the state's busiest port. Upper Hillsborough Bay has the highest levels of cadmium, copper, mercury, zinc, and lead, as well as the pesticide DDT. Concentrations in sediments at a site in northern Hillsborough Bay were the highest of any toxic pollutant measured in Tampa Bay. Two other bays with heavily urbanized watersheds, Boca Ciega Bay and Bayboro Harbor, also can be considered as hot spots of toxic contamination.

6. Figure 2 shows sites in Tampa Bay where concentrations of toxic contaminants in sediments exceeded Florida's Probable Effects Level (PEL) and No Observable Effects Level (NOEL). Sites above the PEL indicate a high probability for biological impact to marine organisms while those above the NOEL are considered "at risk" to biological impact (MacDonald, 1994).

7. Pesticides in sediments and oysters in Tampa Bay are in the mid-range of concentrations for sites sampled around the nation by NOAA. However, total chlordane and mirex concentrations in oysters from Tampa Bay are relatively high on a national scale. Sediments in Cockroach Bay in rural Hillsborough County contain the pesticides chlordane, DDT, and endosulfan. The latter two pesticides along with endrin also have been found in these surface waters.

POLLUTION SOURCES

1. Stormwater runoff from the Tampa Bay watershed contributes about 47 percent of the bay's total annual nitrogen load with urban runoff accounting for about 16 percent, or 680 tons. Residential areas, the watershed's largest land use, is responsible for over half of the nitrogen loading while commercial/industrial sites account for about 20 percent. Figure 3 summarizes sources of nonpoint nitrogen loadings to Tampa Bay.

2. About 28 percent of the bay's total nitrogen loadings, or 1,200 tons, are from atmospheric pollutants falling directly on the water. An additional 7,500 tons fall in the watershed, although no one can determine how much enters the bay via stormwater. EPA estimates that as much as 67 percent of the bay's total nitrogen load may be from the atmosphere.

3. Stationary sources, primarily power plants, contribute an estimated 50 percent of the anthropogenic NO_x emissions as compared to 35 percent from motor vehicles.

4. Domestic wastewater discharges still discharge about 8 percent (340 tons) of the bay's total annual nitrogen loadings, even though all plants provide AWT. Hillsborough Bay receives about two-thirds of the cumulative nitrogen load from the 36 billion gallons of effluent discharged to Tampa Bay each day.

Figure 2. Sediment Toxic Hot Spots in Tampa Bay (TBNEP, 1996)

Figure 2 shows sites in Tampa Bay where concentrations of toxic contaminants in sediments have exceeded Florida's Probable Effects Level (PEL) and No Observable Effects Level (NOEL) for biological impact. Sites registering above the PEL indicate that some biological impact to marine organisms is likely. Sites registering above the NOEL are "at risk" to biological impact.

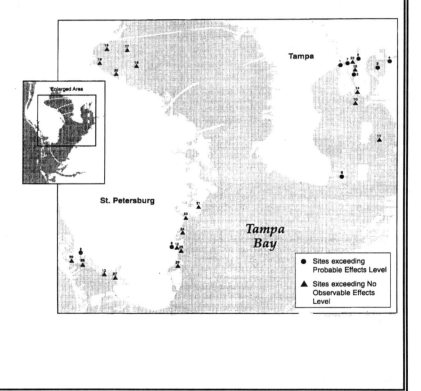

Figure 3. Total Nitrogen Loadings (TBNEP, 1996)

Per-Acre Nitrogen Loadings from Non-Point Sources

	% Loading	% Watershed	Yield lbs/ac/yr
Residential	11	15.5	4.52
Commercial Industrial/Institutional	5	6.4	5.26
Mining	4	3.2	4.97
Range and Pasture	13	28.4	2.81
Intensive Agriculture	6	6.5	5.63
Undeveloped Land	8	39.93	1.15

Total Nitrogen Loadings to Tampa Bay (1992-1994 average)

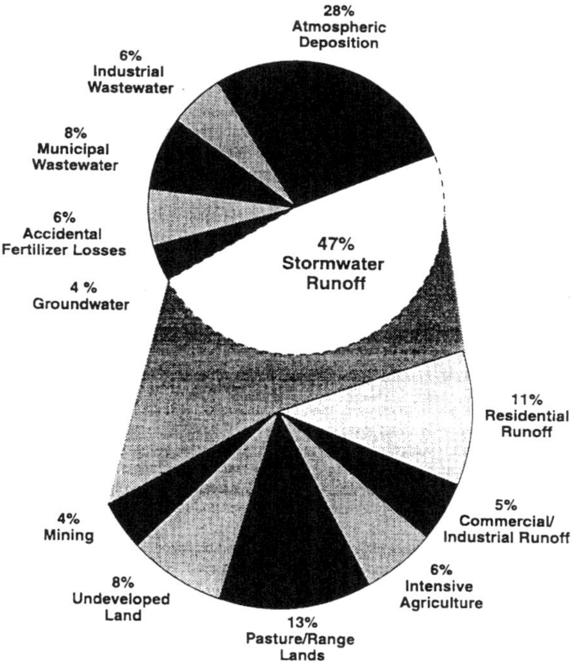

- 28% Atmospheric Deposition
- 6% Industrial Wastewater
- 8% Municipal Wastewater
- 6% Accidental Fertilizer Losses
- 4% Groundwater
- 47% Stormwater Runoff
- 11% Residential Runoff
- 5% Commercial/Industrial Runoff
- 6% Intensive Agriculture
- 13% Pasture/Range Lands
- 8% Undeveloped Land
- 4% Mining

5. Industrial wastewater discharges, primarily fertilizer manufacturing and shipping facilities, are responsible for about 6 percent of the bay's total annual nitrogen loadings.

6. Septic tanks, which serve about 20 percent of the watershed's population, are another important source of nitrogen and pathogen loadings, especially in some areas such as Allen's Creek and tributaries to McKay Bay.

7. Another 7 percent of the bay's total nitrogen loadings had been attributed to losses of fertilizer during shiploading and en route to port. However, this figure has declined substantially since 1991 as source control BMPs were implemented at the port.

8. More than 60 percent of the bay's annual loadings of chromium, zinc, mercury and lead, as well as significant amounts of petroleum hydrocarbons and pesticides are conveyed by stormwater.

9. Atmospheric deposition also is a major source of toxic substances accounting for 44 percent of the bay's total cadmium loading and about 17 percent of the copper and lead loadings. PAHs also enter the bay from the atmosphere.

10. Industrial and domestic point sources also contribute about 30 percent of the bay's total loadings of arsenic, cadmium, chromium, and copper.

GOALS FOR TAMPA BAY

Through the SWIM Plan and the Tampa Bay NEP CCMP, the primary overall goals have been established for the restoration and protection of Tampa Bay:

1. To reverse the environmental degradation of the Tampa Bay estuarine system.
2. To optimize water quality and other habitat values, thereby promoting the sustained existence or re-establishment of thriving, integrated, biological communities.
3. To ensure the maintenance of a productive, balanced ecosystem complimentary with human needs and uses of the resources.

To achieve these overall goals the following specific goals have been established:

1. The overall goal is to restore seagrasses to 1950s levels. This will lead to restoration of commercially important species such as the bay scallop, mullet, sea trout, and red drum.
2. To restore seagrasses to 14,000 acres of the bay. The ability of seagrasses to recolonize the bay depends on the amount of sunlight the grass species require, as well as shading factors such as the amount of drift macro-algae and attached algal growth on grass blades. For most seagrasses in the bay, an estimated 20 to 25 percent of the light striking the bay's surface must penetrate to target depths to allow seagrass regrowth. Reducing nitrogen loadings will reduce chlorophyll a concentrations thereby increasing the depth of sunlight penetration.
3. As many as 12,000 acres of seagrass can be recovered by maintaining recent water quality conditions. This will require local communities to reduce their nitrogen loadings to the bay by about 10 percent by the year 2010 to compensate for increases in nitrogen loadings associated with the watershed's population growth.
4. A coastal habitat master plan is being developed for the watershed which will help to coordinate and prioritize existing state, regional, and local restoration programs. The long term goal is to recover 1,800 acres of low salinity tidal marshes while

maintaining and enhancing salt marshes and mangroves at existing levels. A minimum goal is to restore 100 acres of tidal marsh habitat every five years.
5. Reduce sediment toxicity to minimize risks to marine life and humans. Using three tests - evaluation of sediment chemistry, sediment toxicity, and benthic community health - bay sediments will be characterized and prioritized for management.
6. Reduce bacterial contamination to levels safe for swimming and shellfish harvesting.

IMPLEMENTING MANAGEMENT PROGRAMS

The Tampa Bay SWIM Plan, the NEP CCMP, and the Hillsborough River and Bay Ecosystem Demonstration Project all have established a series of action steps to decrease pollution inputs to levels that will allow meeting the resource management goals above. Broadly, these actions are aimed at reducing stormwater loadings and atmospheric deposition, preserving critical habitats and environmentally sensitive lands through purchase, restoring habitats, and educating all citizens of the watershed about how they can reduce "Pointless Personal Pollution." Many of these actions are being implemented through cooperative partnerships involving state, regional, and local governments together with businesses and citizens. Some, such as the allocation of stormwater pollutant load reduction goals developed by the Tampa Bay NEP, will need formal agreements and perhaps even be tied to the NPDES municipal stormwater permits which nearly all local governments in the watershed are receiving.

Costs associated with the individual actions presented in the Tampa Bay SWIM Plan and the NEP CCMP are considerable. However, these should not automatically be construed as requirements for new sources of revenues, since some of these initiatives can and are being accomplished with existing resources or by redirecting current funding allocations to better address the bay's needs. A number of actions seek to improve coordination, cooperation, and planning among state and local governments, and the private sector. These may actually result in cost savings for currently funded activities.

A 1994 survey by the Tampa Bay NEP attempted to quantify how much money is spent to manage and monitor bay quality and administer environmental programs. Based on FY94-95 budgets, the study indicates that over $260 million is spent annually by federal, state, and local agencies on the restoration and management of Tampa Bay. As seen in Figure 4, the largest part of the funds (65% or $170 million) are spent on wastewater collection, treatment, and reuse. Approximately $35 million (13%) is spent by local governments and the SWFWMD on stormwater management, although much of this is still for traditional drainage purposes. Habitat restoration and land acquisition, two long favored and implemented environmental programs in the Tampa Bay region, account for over $27 million in expenditures.

Implementation of a variety of structural and nonstructural BMPs is underway including:

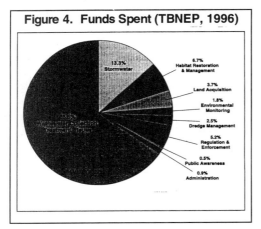

Figure 4. Funds Spent (TBNEP, 1996)

13.3% Stormwater
6.7% Habitat Restoration & Management
3.7% Land Acquisition
1.8% Environmental Monitoring
2.5% Dredge Management
5.2% Regulation & Enforcement
0.5% Public Awareness
0.9% Administration

Urban Stormwater BMP Projects

Using the basin prioritization conducted in 1991, along with information from the sediment and biological sampling efforts, several urban stormwater management projects have been built or are planned in priority subbasins (Figure 5). These projects are being undertaken with cooperative funding from DEP (using EPA Section 319 grants), SWFWMD, and local governments. Since vacant land in the highly urbanized area is scarce or extremely expensive, many of these projects are being conducted on existing public lands providing multiple benefits including regional stormwater management, open space, and recreation. Public education is a frequent component of these projects with the placement of signs depicting the effects of urbanization and the need for stormwater management.

In addition to building new structural controls efforts are ongoing to improve the effectiveness of BMPs required by Florida's stormwater management program. Since surveys have shown that up to 70% of the stormwater BMPs are not being properly maintained, assuring their long term operation and maintenance can greatly reduce stormwater pollution. Maintenance and operation of BMPs typically is the responsibility of private land owners and property owner associations. Unfortunately, DEP, SWFWMD, and local governments do not have enough staff to conduct regular inspections. To improve this deficiency, DEP is implementing, in cooperation with local governments and the WMDs, a training and certification program for individuals who wish to become erosion, sediment, and stormwater inspectors. Local governments also are encouraged to implement Stormwater Operating Permit systems which require an annual inspection and certification that the stormwater system has been maintained and is properly operating. As an economic incentive, some local stormwater utilities provide credits for individuals served by a properly maintained and operating system. Additionally, Hillsborough County as implemented the "Adopt a Pond" program to help educate stormwater system owners on how to maintain their systems.

Active Habitat Restoration Projects ●

1. Lowry Park
2. NE McKay Bay
3. Delaney Creek
4. Simmons Park
5. Hendry Fill
6. Peanut Lake
7. Bayshore Blvd.
8. Gandy Park
9. Cabbage Head Bayou
10. Boca Ciega
11. Cargil S. parcel
12. Mangrove Bay
13. Cockroach Bay
14. Little Bayou
15. MacDill AFB
16. Picnic Island

Stormwater Retrofitting Projects ■

1. Lowry Park
2. Horizon Park
3. Old Coachman
4. S. Pasadena
5. Jungle Lake
6. Pinellas Park
7. N. Redington
8. EMS Site
9. St. Pete/Clearwater Airport
10. Brushy Creek
11. Safety Harbor
12. 102nd Avenue
13. 94th Avenue
14. Lake Carroll
15. Delaney Creek
16. Haynsworth
17. 141st Avenue

Agricultural Stormwater Management

The SWFWMD, in cooperation with the USDA NRCS, is implementing the Agricultural Ground and Surface Water Management Program (AGSWM) to encourage farmers to develop and implement agricultural water management plans, improve compliance with agricultural water management plans, and improve BMP implementation. The program provides a streamlined, less cumbersome approach for farmers to comply with the intent of SWFWMD's wetlands and water quality protection rules. Additionally, in cooperation with the state's NPS Management Program, the Cooperative Extension System has developed the Florida Agricultural Information Retrieval System (FAIRS) to make it easier to obtain educational materials on the design and use of agricultural BMPs. The NPS Program in cooperation with the Hillsborough River Ecosystem Management Project, SWFWMD, NRCS, and local farmers also are installing and testing various innovative BMPs, especially on intensive agricultural operations such as dairies. Many of these focus on improved animal waste management BMPs, including composting, and improved fertigation and pesticide practices.

Habitat Restoration Projects

The restoration and protection of critical habitats is critical to achieving the ecological goals for Tampa Bay. A large number of projects have been or will be implemented through the state's land acquisition programs, local land acquisition programs such as Hillsborough County's Environmental Lands Aquisition and Protection Program, and the SWIM Program (Figure 5). These progressive, interdisciplinary, cooperative projects typically combine habitat restoration and enhancement with improved water quality through enhanced tidal flushing, water circulation, stormwater treatment, and public education and recreation.

Nonstructural BMPs

A variety of projects are being undertaken to help prevent or minimize "Pointless Personal Pollution" through the implementation of nonstructural controls. These included continued implementation of the Florida Yards and Neighborhoods Program,

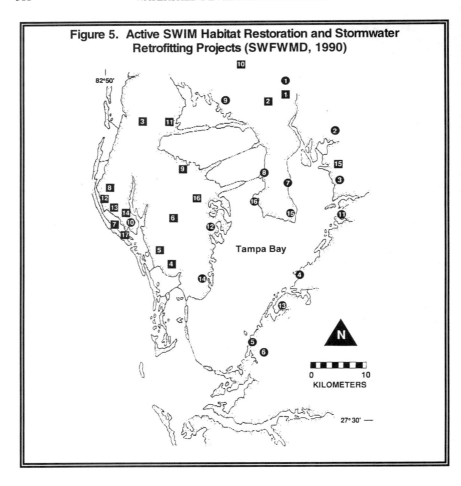

Figure 5. Active SWIM Habitat Restoration and Stormwater Retrofitting Projects (SWFWMD, 1990)

assisting businesses in developing and implementing pollution prevention plans with a focus on source controls, developing model landscaping guidelines for commercial landscapes, and incorporating model landscaping guidelines into local government site review processes for new development. The region's continuing rapid growth provides opportunities through local government comprehensive plans and land development regulations to promote compact development and to reduce impervious surfaces, especially parking lots at commercial developments.

DISCUSSION AND RECOMMENDATIONS

Florida has established a wide variety of laws, regulations and programs at the state, regional and local level to protect, manage and restore the state's incredibly valuable yet vulnerable natural resources, especially its water resources. There is no doubt that these programs have been effective in helping to reduce adverse impacts on natural resources resulting from the state's rapid and continuing growth over the past twenty years. However, even with the implementation of these programs, many of Florida's natural resources have been severely strained or degraded. Some of these adverse effects can be attributed to activities that occurred before the implementation of modern watershed management programs such as the channelization of the Kissimmee River and the creation of the vast drainage canal network south of Lake Okeechobee both of which are contributing to the decline of Lake Okeechobee, the Everglades and Florida Bay. Other adverse impacts, though, are directly related to the state's rapid growth and development during the last twenty years. These include water supply problems, water quality problems, declining habitat and impacts on endangered species such as the manatee and the Florida panther.

Why are these adverse impacts still occurring given the wide range of watershed management programs that have been implemented in Florida? What could be done to reduce these effects and possibly restore already degraded areas? The continuing evolution of Florida's land and water management programs into a more holistic approach which seek to manage cumulative effects can help to overcome many of the current program deficiencies. Cooperative efforts and partnerships, together with citizen education and involvement to improve the stewardship ethic of all Floridians is essential. With increased support and participation by all Floridians, the effectiveness of the state's programs can be improved helping to assure that our natural resources will be able to be enjoyed by future generations.

REFERENCES

Florida Department of Environmental Protection. 1994. Florida Coastal Sediment Contamination Atlas. Office of Water Policy, Tallahassee, Florida.

Florida Department of Environmental Protection. 1995. Ecosystem Management Implementation Strategy: An Action Plan. Tallahassee, Florida.

Hillsborough River Integration and Coordination Committee. 1996. Hillsborough River and Bay Ecosystem Demonstration Project Interim Status Report.

Livingston, E.H. 1993. Local Government Model Stormwater Management Program. Stormwater/NPS Management Section, FDEP. Tallahassee, Fl.

Livingston, E.H. 1995. The Evolution of Florida's Stormwater/Watershed Management Program. In Proceedings of the National Conference on Urban Runoff Management: Enhancing Urban Watershed Management at the Local, County, and State Levels. EPA 625/R-95/003. Cincinnati, Ohio.

Livingston, E.H., E. McCarron, T. Seal, and G. Sloane. 1995. Use of Sediment and Biological Monitoring. In Stormwater NPDES Related Monitoring Needs, Proceedings of an Engineering Foundation Conference. ASCE, New York.

Long, E. R., D. MacDonald, and C. Cairncross. 1991. Status and Trends of Toxicants and the Potential for Their Biological Effects in Tampa Bay, Florida. NOAA Technical Memorandum NOS OMA 58.

Long, E. R., D.A. Wolfe, R.S. Carr, K.J. Scott, G.B. Thursby, H.L. Windom, R. Lee, F.D. Calder, G.M. Sloane, and T. Seal. 1994. Magnitude and Extent of Sediment Toxicity in Tampa Bay, Florida. NOAA Technical Memorandum NOS ORCA 78.

MacDonald, D. 1994. Approach to the Assessment of Sediment Quality in Florida Coastal Waters. Final report submitted to the Florida Department of Environmental Protection, Tallahassee, Florida.

Southwest Florida Water Management District. 1990. Urban Stormwater Analysis and Improvements for the Tampa Bay Watershed. Brooksville, Florida.

Southwest Florida Water Management District. 1992. Tampa Bay Surface Water Improvement and Management Plan. Brooksville, Florida.

Tampa Bay National Estuary Program. 1996. Charting the Course for Tampa Bay: Draft Comprehensive Conservation and Management Plan. St. Petersburg, Florida.

SESSION 4B: Institutional Arrangements for Watershed Management

DISCUSSION

Riparian Stewardship in the Post-Regulatory Era
Robert Stearns

By now most of us in the professions have grasped the concept that watersheds and the rivers and streams they generate are complex systems. For us to see the light is a good start, but how do we convey the concept to property owners, elected officials and taxpayers?

The problem really boils down to two basic flaws in our thinking. First, we have not really understood, or choose not to understand, how watersheds work. Second, because of our relative wealth in both dollars and natural resources, we have been making imprudent choices in how we live with drainage systems.

While most Americans have a high level of concern about environmental issues, sadly, less than 1 in 20 knows what a watershed is let alone understand its functions. It is clear that a public education process is needed.

It is imperative that we incorporate the stewardship and accountability ethic unyieldingly and uncompromisingly into all of our work. We can no longer be simply facilitators of development. Rather, we must be the intermediaries between enterprise and nature and, increasingly, healers of the landscape.

Questions/Comments

Question:	Have you personally been involved in these riparian corridor developments?
Answer:	Yes, we need to include greenway development in the total development package, and create our developments with no net impact on the downstream resource.
Question:	How do you decide where the engineering ends and the soft engineering begins when you retrofit into the environment.?
Answer:	This is difficult in an established setting, but in a developing area, the planning can precede the design.
Comment:	It would be good if we could move the a regulatory setting that is incentive based rather than punitive based.

Protecting Fulbright Spring: Can We Beat The Odds
Timothy Smith

Up to the present time, Fulbright Spring has been a reliable source in both quality and quantity of water.

During and shortly after runoff events, there is increased flow and increased concentrations of total and fecal coliform, turbidity, and total suspended solids at the spring.

In 1992 the City Council and City Utilities approved a "Phase 1" study of the spring recharge area.

Based upon their findings and conclusions, the consultants recommended a combination "non-structural/structural BMP" regulatory strategy in conjunction with careful monitoring of conditions at the spring and in the watershed as the appropriate and most effective starting point for protecting spring water quality.

Florida's Evolving Stormwater/Watershed Management Program
Eric Livingston

During the past 30 years Florida has experienced explosive growth and development resulting in more than a doubling of the state's population. This growth and the associated land use changes, along with the state's extensive agricultural and mining lands, makes protecting Florida's vulnerable water resources especially difficult. While a wide variety of resource management laws, regulations, and programs have been implemented during the past 25 years, they generally have promoted a piecemeal, site specific approach. While these programs certainly have helped to minimize the degradation of the state's natural resources, their environmental effectiveness has been less than optimal. This paper discusses the state's continually evolving land and water resource management program as it moves towards implementation of an ecosystem management framework based on watershed units. A case study of the Tampa Bay watershed was presented to demonstrate how the various pieces of the resource management puzzle are being fit together.

AUTHOR INDEX

Page number refers to the first page of paper

SUBJECT INDEX

Page number refers to the first page of paper